# THE CONTINENTAL DRIFT CONTROVERSY
## Volume III: Introduction of Seafloor Spreading

Resolution of the sixty-year debate over continental drift, culminating in the triumph of plate tectonics, changed the very fabric of Earth science. Plate tectonics can be considered alongside the theories of evolution in the life sciences and of quantum mechanics in physics in terms of its fundamental importance to our scientific understanding of the world. This four-volume treatise on *The Continental Drift Controversy* is the first complete history of the origin, debate, and gradual acceptance of this revolutionary explanation of the structure and motion of the Earth's outer surface. Based on extensive interviews, archival papers, and original works, Frankel weaves together the lives and work of the scientists involved, producing an accessible narrative for scientists and non-scientists alike.

This third volume describes the expansion of the land-based paleomagnetic case for drifting continents and recounts the golden age of marine geology and geophysics. Fuelled by the Cold War, US and British workers led the way in making discoveries and forming new hypotheses, especially about the origin of oceanic ridges. When first proposed, seafloor spreading was just one of several competing hypotheses about the evolution of ocean basins, and every hypothesis left unexplained the newly discovered and wholly unexpected magnetic anomalies associated with mid-ocean ridges and in the Pacific Basin off the western coast of the United States.

Other volumes in *The Continental Drift Controversy*:

Volume I – Wegener and the Early Debate

Volume II – Paleomagnetism and Confirmation of Drift

Volume IV – Evolution into Plate Tectonics

HENRY R. FRANKEL was awarded a Ph.D. from Ohio State University in 1974 and then took a position at the University of Missouri–Kansas City where he became Professor of Philosophy and Chair of the Philosophy Department (1999–2004). His interest in the continental drift controversy and the plate tectonics revolution began while teaching a course on conceptual issues in science during the late 1970s. The controversy provided him with an example of a recent and major scientific revolution to test philosophical accounts of scientific growth and change. Over the next thirty years, and with the support of the United States National Science Foundation, the National Endowment for the Humanities, the American Philosophical Society, and his home institution, Professor Frankel's research went on to yield new and fascinating insights into the evolution of the most important theory in the Earth sciences.

To Nora

# THE CONTINENTAL DRIFT CONTROVERSY

## Volume III: Introduction of Seafloor Spreading

HENRY R. FRANKEL

*University of Missouri–Kansas City*

CAMBRIDGE
UNIVERSITY PRESS

CAMBRIDGE UNIVERSITY PRESS
Cambridge, New York, Melbourne, Madrid, Cape Town,
Singapore, São Paulo, Delhi, Mexico City

Cambridge University Press
The Edinburgh Building, Cambridge CB2 8RU, UK

Published in the United States of America by Cambridge University Press, New York

www.cambridge.org
Information on this title: www.cambridge.org/9780521875066

First published 2012

Printed in the United Kingdom at the University Press, Cambridge

*A catalogue record for this publication is available from the British Library*

ISBN 978-0-521-87506-6 Hardback

# Contents

# Foreword

Henry Frankel has a fine eye, and ear, for the interlocking aspects of the emergence, recognized evolution, and acceptance of that flowering of a worldwide phenomenon, continental displacement. I see the enlightening process as a scaled-up relay in which – by mid-1950s to early 1960s – the baton was being passed from terrestrial paleomagnetists to sea-smart marine geologist–seismologists and their shoreside arbiters. In retrospect, the number of participants is not great; Frankel has identified them plainly and rather fairly, in my view.

In the early 1950s still-young Earth science-trained graduates of several US, British, and European universities joined wartime agency-fostered academic installations, acquired institutional use of platforms capable of open-sea operations, taught themselves to carry out (to publication) fruitful exploratory reconnaissances and also magnified experiments at sea, viz., topographic–geological–geophysical collaborations even in the most remote and deepest parts of the oceans. Given the opportunities for actual *discovery*, and perhaps collegial renown, competition was vigorous but arguably constructive. We knew of, and could admire and apply, each other's accumulating refinements to the submarine macro-jigsaw puzzle.

Stemming from undergraduate exposure to ideas of Gutenberg, Vening Meinesz, and Benioff, I undertook to clarify and determine reliably the characteristics – bathymetry, crustal structure, geological processes and petrology – of several Pacific trenches. By late 1954–early 1958 we at Scripps Institution of Oceanography (Scripps) had demonstrated off Central America and Peru/Chile the "creeping crustal layers," igneous oceanic crust above the Moho passing diagonally beneath the trenches' inshore slope and continental shelf, presumably even farther (from Benioff–Wadati seismicity analysis).

But at just the same time an ultimately much more significant program – identification of seafloor magnetics – was being initiated by London University's Ronald Mason and his engineer Arthur Raff at Scripps. (Early incidents in a revolution can be, and should be, remembered.) People at Scripps had learned that the US Coast and Geodetic Survey's vessel *Pioneer* would undertake a multi-year electronically navigated, closely spaced, E–W track bathymetric survey for several hundred kilometers off the West Coast, from the Mexican to the Canadian border. Sensing a windfall opportunity, Mason and Raff got approval to provide and monitor a flux-gate magnetometer towed throughout those mid-1950s surveys. They, with their contoured magnetic anomaly plots, were the *discoverers* of the parallel-lineated seafloor magnetic field pattern associated with the world-girdling active igneous ridge systems, "stripes" found and dated in all the oceans, the key element via the Vine–Matthews hypothesis of Hess's "seafloor spreading," which he first proposed in a December 1960 preprint sent to many of us in marine geology and geophysics.

Until the mid-1960s other towering figures, well brought to us in this volume and by their own published recollections, built upon the cascading observations and implications, interpreted very often, and most convincingly, by Mycrofts ashore. The result: the mobilistically compelling "plate tectonics."

But on a poignant note: one morning in the mid-1960s, while climbing the stairs to a Scripps seminar, visitor Ron Mason turned to me and said, "Bob, if only we'd carried it one step farther. . ." Ron's enduring misfortune: only in their surveys' latest traverses, the Juan de Fuca Ridge sector off Washington State, is the East Pacific Rise's magnetic anomaly plot bi-laterally symmetrical, age-wise. In nearly all other areas, and most historically the Carlsberg Ridge, such patterns are mirrored, as demonstrated so fruitfully by Cambridge's Drum Matthews (aboard HMS *Owen* in the Somali Basin, 1961–2) with Fred Vine in their attic nook at Madingley Rise.

*Robert L. Fisher*
*Emeritus Professor, Scripps Institution of Oceanography*

# Acknowledgments

I could not have undertaken and completed this book without enormous help from paleomagnetists whose work led to the measurement of drifting continents. Ken Creer, Edward Irving, and S. K. Runcorn (deceased) answered many questions over many years about their work and that of others. I also could not have completed this book without the great help I received from Bob Fisher, Bill Menard (deceased), and Marie Tharp (deceased). They answered many questions about their own work, and that of others. I have greatly benefited from studying Menard's own retrospective work, *Ocean of Truth*. Robert Dietz (deceased) also answered many questions about his own work. Many others kindly and patiently answered questions about their work and that of others. I should like to thank Charles Bentley, Bill Bonini, James Briden, David Brown, Colin Bull, W. G. Chaloner, Ernie Deutsch (deceased), Richard Doell (deceased), Robert Dott Jr., P. F. Friend, Ian Gough (deceased), Ron Green, Tony Hallam, Warren Hamilton, Brian Harland (deceased), Lester King (deceased), Ian McDougall, Michael McElhinny, Eldridge Moores, A. E. M. Nairn (deceased), Charles Officer, Neil Opdyke, Edna Plumstead (deceased), Martin Rudwick, John Sclater, Alan Smith, Don Tarling, and Ian Tolstoy for discussing their own work and that of others. It is a pleasure to acknowledge their considerable help.

I should like to thank Ted Irving for critically reviewing the entire manuscript and Bob Fisher for critically reviewing the chapters on marine geology. Both made significant improvements, and Irving provided flash forward updates about the current status of various problems. I also want to thank Ken Creer, Ursula Marvin, Dan McKenzie, Keith Runcorn, and Fred Vine for critically reading selected chapters.

I thank Alan Allwardt for extended conversations about Hess and the development of his ideas. I also learned much from Annette Hess (deceased) about her husband's relationship with several fellow marine geologists at Lamont Geological Observatory, and the importance he put on friendship. I thank George Hess for allowing me to reproduce documents from his father's papers at Princeton University.

I thank my colleagues Bill Ashworth, Bruce Bubacz, George Gale, Clancy Martin, and Dana Tulodziecki for various discussions relevant to this work.

I thank Nancy V. Green and her digital imaging staff at Linda Hall Library, Kansas City, Missouri, for providing the vast majority of the images; Richard Franklin for color image of the *Time Magazine* representation of Creer's 1954 Oxford version of his APW path for Britain. I thank the reference librarians at Linda Hall Library, and the interlibrary staff at the Miller Nichols Library, UMKC. I would also like to thank Deborah Day, former archivist at Scripps Institution of Oceanography, for her aid and encouragement over many years. Her generous help to me and other historians of science over many years is unsurpassed.

I owe much to Nanette Biersmith for serving as my longtime editor and proofreader.

I am indebted to the United States National Science Foundation, the National Endowment of the Humanities, and the American Philosophical Society for financial support. I also thank the University of Missouri Research Board and my own institution for timely grants to continue this project.

I wish to thank Susan Francis and her staff at Cambridge University Press for believing in this project and for their great assistance throughout its production.

# Abbreviations

| | |
|---|---|
| AAPG | American Association of Petroleum Geologists |
| AGU | American Geophysical Union |
| AMSOC | American Miscellaneous Society |
| ANU | Australian National University |
| APW | Apparent polar wander |
| Caltech | California Institute of Technology |
| CRM | Chemical remanent magnetization |
| GAD | Geocentric axial dipole |
| GSA | Geological Society of America |
| IGY | International Geophysical Year |
| IUGG | International Union of Geodesy and Geophysics |
| *JGR* | *Journal of Geophysical Research* |
| Lamont | Lamont Geophysical Observatory |
| Ma | Million years |
| NAS | National Academy of Sciences (USA) |
| NEL | US Navy Electronics Laboratory |
| NRM | Natural remanent magnetization |
| NSF | National Science Foundation (USA) |
| ONR | Office of Naval Research (USA) |
| RAS | Royal Astronomical Society (UK) |
| RS1 | Research Strategy 1 |
| RS2 | Research Strategy 2 |
| RS3 | Research Strategy 3 |
| Scripps | Scripps Institution of Oceanography |
| UCLA | University of California, Los Angeles |
| UCRN | University College of Rhodesia and Nyasaland |
| USCGS | United States Coast and Geodetic Survey |
| USGS | United States Geological Survey |
| VRM | Viscous remanent magnetization |
| WHOI | Woods Hole Oceanographic Institute |

# Introduction

Volume II described the development of the paleomagnetic case for mobilism based on observations from continents. It explained how, in the second half of the 1950s, a few researchers took a small field of enquiry and transformed it into an important field, and used it to show that continental drift had occurred. Volume III describes how, during the first half of the 1960s, paleomagnetists further expanded the land-based paleomagnetic case for mobilism. My treatment of their case ends by describing how six notable participants in the controversy responded to it.

Leaving behind land-based paleomagnetism, the remainder of this volume describes the first half – late 1940s through the early 1960s – of the golden age of marine geology and geophysics, when a small group of Earth scientists, supported by governmental funding based on defense concerns during the Cold War, crisscrossed oceans, better familiarizing themselves with known features of the ocean floor and discovering entirely new ones. Prompted by these new discoveries, especially new information about mid-ocean ridges, four hypotheses were proposed. There was seafloor spreading by Hess and Dietz, followed by Menard's hypothesis of seafloor thinning; both were motivated by mantle convection and both had much in common with Holmes' 1928 theory of mantle convection. Then there was rapid Earth expansion by Heezen. Finally, Maurice Ewing proposed that the ocean ridges were produced by mantle convection but without continental drift.

During this decade, most Earth scientists did not acknowledge that continental drift had been shown to occur by paleomagnetists; their descriptions of their land-based evidence of drift fell mainly on deaf ears. Drift did, however, find support among a few marine geologists. Hess became a mobilist because of it, and took drifting continents as a constraint on his speculations about seafloor evolution. Dietz too accepted drift's paleomagnetic support. Menard appealed to it when proposing and defending seafloor thinning but, as we shall see in Volume IV (Chapter 3), ceased to do so when he rejected seafloor thinning and instead proposed a fixist hypothesis of seafloor evolution. Heezen eagerly endorsed mobilism's land-based paleomagnetic support until it became inconsistent with rapid Earth expansion. For Heezen and Menard, appeal to drift's paleomagnetic support was a luxury, jettisoned once it became an embarrassment. Ewing, an inflexible fixist, at this time paid no attention to paleomagnetism.

As in Volumes I and II, I shall describe how researchers acted in accordance with what I have identified as three standard research strategies (I, §1.13). Workers did not recognize or say that they acted in this way; these three research strategies are my retrospective description of how they went about their tasks, how they addressed their problems. Research Strategy 1 (hereafter, RS1) was used by researchers to expand the problem-solving effectiveness of solutions and theories. Research Strategy 2 (hereafter, RS2) was used by them to diminish the effectiveness of competing solutions and theories; RS2 was an attacking strategy used to raise difficulties against opposing solutions, and to place all possible obstacles in their way. Workers used Research Strategy 3 (hereafter, RS3) to compare the effectiveness of competing solutions and theories, and to emphasize those aspects of a solution or theory which gave it a decided advantage over its competitors.

Like the small group of researchers who turned paleomagnetism from a backwater discipline into one of central importance in the Earth sciences, this small group of marine geologists and geophysicists took another little-known field of enquiry and turned it into one of central importance in Earth history. It is a story about how these surveyors of the deep used new technologies to uncover the basic structure of seventy percent of the Earth's surface, about how these new discoveries fostered speculation about seafloor evolution, and about how this speculation later led to the further confirmation of continental drift.

# 1

# Extension and reception of paleomagnetic/paleoclimatic support for mobilism: 1960–1966

## 1.1 Introduction

In the early 1960s, paleomagnetists continued to buttress their case for mobilism by extending surveys, by conducting further field stability tests, by bringing magnetic cleaning into general use, by utilizing new radiometric ages, and by further enlisting evidence of paleoclimates. This was the acme of the contribution of continental paleomagnetism to the mobilism debate, and by the mid-1960s the global paleomagnetic test of continental drift was essentially completed and Wegener's general notion of continental drift confirmed. By then, work on continents was being overtaken by an avalanche of new data from the oceans, especially topographic surveys and surveys of geomagnetic anomalies reflective of reversals of remanent magnetization of the oceanic crust. There was also a renewal of interest in drift mechanisms.

Participants from North America continued to criticize continental drift harshly, but a few began to welcome paleomagnetism's support for it. There was also a renewal of interest in drift mechanisms. I shall trace the reception at this time of the paleomagnetists' case by certain notable figures in Earth sciences.

The Newcastle and Canberra groups continued to argue for continental drift and polar wandering; the London, Imperial College, group discussed continental drift in a manner that could include polar wandering without requiring it, but notably Deutsch questioned, as he always had done, the need for polar wandering at all, and argued for continental drift as the dominant process. A new paleomagnetic group in Salisbury (Harare), Southern Rhodesia (Zimbabwe), contributed importantly. Paleomagnetists organized symposia, wrote and edited books. There was the 1962 anthology, *Continental Drift*, to commemorate the fiftieth anniversary of Wegener's theory, and in 1963 the huge NATO symposium on paleoclimatology with proceedings edited by Nairn. Irving (1964) wrote *Paleomagnetism and Its Application to Geological and Geophysical Problems*, the first general text in English.

## 1.2 Dott reexamines the Squantum Tillite

I begin by returning to the controversy concerning the Squantum Tillite, an old obstacle to Wegener's continental drift (I, §3.10–§3.12). The Squantum Tillite of the

Boston area, purportedly of Permian age, was troublesome during the classical stage of the controversy, being seen as a glacial deposit in a region otherwise characterized by Late Paleozoic deposits laid down under a hot climate (I, §3.11). It became troublesome to the paleomagnetic case for mobilism for a similar reason, because it indicated cold climate at a time when paleomagnetically determined latitudes were low (I, §3.12). In timely fashion, in 1961, R. H. Dott Jr. revisited this longstanding problem.

   Born in 1929, Dott received his B.S. and M.S. degrees in 1950 and 1951 from the University of Michigan, majoring in geology.[1] He entered Columbia University in 1951 and received a Ph.D. in geology in 1956. He took a job in the oil industry in 1954 before finishing his Ph.D. thesis, and from 1954 until 1956 worked in the Pacific Northwest on turbidites. He spent 1956–7 as a first lieutenant in the US Air Force, stationed at the Air Force Cambridge Research Center, Bedford, Massachusetts, where, in his spare time, he studied the Squantum. Dott recalled:

First let me paint a little background. The Squantum interested me because I had been working in Oregon and California (oil industry) for two years prior to going to Air Force active duty in Boston (1956–57). While on the West Coast, I had been studying a lot of ancient submarine "mudflow" and turbidite deposits, and had become enthralled with their sedimentology. Of course these kinds of deposits had only recently been recognized through the seminal work of Ph. H. Kuenen, etc. and the marine folks like Heezen. As a graduate student at Columbia (and classmate of Heezen), these new wonders were very hot stuff. Against that background, the experience of California and Oregon was exhilarating to say the least. When I found I was to be in the Boston area for two years and not doing geology, I decided to look at the already controversial Squantum to compare it with what I had been seeing in the west. This began simply as something to do to keep in touch with rocks, but – at the urging of an older colleague when I got to Wisconsin – it evolved into my paper.

*(Dott, August 21, 2000 email to author)*

At the time, turbidity currents were a hot topic among oceanographers. Recognized by none other than Daly (1936), who called them density currents; they are underwater currents made denser by a substantial sediment load, which propels them rapidly forward under gravity. As they cross the continental shelf they erode underwater canyons and travel hundreds of miles out into the ocean floor. Ph. H. Kuenen (1937) produced them in the laboratory. After World War II, Kuenen and C. I. Migliorini (1950) revived interest in them. The next year, D. B. Ericson, M. Ewing, and B. C. Heezen (1951) of Columbia University mapped several huge submarine canyons that crossed the edge of the continental shelf off the eastern United States and Canada, and argued that they had been eroded by turbidity currents. After reviewing reports of the breakage of trans-Atlantic cables coincident with the 1929 Grand Banks (Newfoundland) earthquake, Heezen and Ewing (1952) proposed that unconsolidated sediments on the continental shelf had been dislodged by an earthquake and slid down the continental slope, forming a gigantic and rapidly moving current, turbid with suspended sediment that eroded canyons and then moved out across the ocean floor, breaking cables as it went (§6.5). Establishing that cables located closer to shore broke before those further out, they estimated the current's velocity.[2]

Because Dott had been at Columbia, he knew about the work of Heezen and colleagues. As a sedimentologist, he was interested in turbidites, the deposits laid down as turbidity currents slow down and their sediment falls out of suspension. He had already studied turbidites and submarine mass flow deposits in California and Oregon. Slumps, turbidites, and glacial tills were all poorly sorted and could mimic one another.[3] Indeed, J. C. Crowell, who had begun to express doubts about the glacial origin of some "tillites," had questioned the characterization of the Squantum as glacial.

The writer felt, for example, on visiting the Squantum "tillite" near Boston on a Geological Society of America excursion in 1952 that these rocks perhaps also formed by slumping and accordingly required re-investigation before their glacial origin could be accepted.

*(Crowell, 1957: 1005)*

Dott decided that the Squantum strata resembled turbidites more closely than tillites, and had more likely been deposited from local gravity-driven turbidity currents than from glacial activity. Dott did more, arguing that several key "glacial" indicators, i.e., scratched and polished underlying rock, and erratics, embedded in laminated mudstones, and derived from distant sources, were missing from Squantum strata.

The most convincing evidence, a widespread gouged and polished pavement beneath till-like deposits in clearly nonmarine strata, is lacking at Boston and in many other places ... The Squantum possess neither significant numbers of erratics in laminated mudstones nor any rock fragments that could not have been derived locally. Unlike the Gowganda [a Precambrian glacial deposit in western Ontario] and some of the better-established tillites of the southern hemisphere, all the debris in the Squantum could have originated merely by redeposition of local gravels with addition of torn-up and bent fragments of contemporaneous soft muds.

*(Dott, 1961: 1301–1302; my bracketed addition)*

Dott (1961: 1296) also questioned the dating of the Squantum, arguing that its assignment to the Permian, based primarily on two "poorly preserved casts of presumed fossil trees which lacked any bark impressions or internal structure," was likely mistaken. He pushed its age back to the Mississippian or Devonian, basing his estimate primarily on recent radiometric dating of associated rocks. This changed, but did not remove, the difficulty the Squantum beds presented to the mobilism argument because the paleomagnetically determined latitudes of eastern North America for the Devonian through Permian were all low.

Reminiscent of Köppen and Wegener (I, §3.15), Dott argued that warm climate findings from North America for the Late Paleozoic were inconsistent with the glacial interpretation, as were the paleomagnetic data which indicated that North America had been "rather close to the equator" throughout the Paleozoic.

The Squantum was thought to be [glacial] but it is unusual in its isolated geographic position ... [it is] particularly troublesome to advocates of continental drift, because plotting of [its] predrift [position] brings [it] rather close to the equator. [The] Past [position] of North America ... [is] ... more in accord with other Paleozoic evidence, such as paleomagnetic data, extensive coals, and somewhat younger evaporite deposits, than [is its] present [position].

*(Dott, 1961: 1290; my bracketed additions)*

But he did not take a position on mobilism, instead arguing, based on new knowledge of turbidites, that many ancient supposedly glacial deposits should be reexamined before they could be used to draw conclusions about continental drift or polar wandering.

Promiscuous postulation of ancient glacial periods still exercises imaginations of scientists; however, most supposed examples of ancient glacial deposits must be critically re-examined. Many others may have been formed by one of the several possible alternative mechanisms. To be valid, postulated glaciations must be compatible with paleogeographic and paleotectonic evidence. Theories of geotectonics, paleoclimatology, and paleobiogeography based wholly or in part upon supposed ancient glaciations, including continental drift and polar wandering, cannot be evaluated honestly until evidence for all such periods has been critically reanalyzed. If the present paper does nothing else, it underscores the great difficulty in interpreting glacial-like deposits.

*(Dott, 1961: 1303)*

Did this mean that Dott had come to doubt extensive Permo-Carboniferous glaciations in the Southern Hemisphere? No, although he thought that some of them were questionable, the evidence that most were glacial was excellent. In fact, they served as the touchstone for what counted as good evidence for ancient glaciation.

A preserved, extensive, grooved and polished pavement overlain by poorly sorted, till-like material – particularly if nonmarine – is the most compelling glacial evidence. Very large erratic boulders are suggestive of ice movement, as are abundant rafted erratic fragments in fine muds ... Independent biologic or isotopic cold-temperature indicators are sorely needed to strengthen glacial interpretations. General stratigraphic relationships and tectonic setting serve as important factors in judging probability of alternative interpretations, particularly in geosynclinal sequences. From these criteria, the writer judges that only the Permo-Carboniferous glaciation in certain parts of the southern hemisphere is firmly established. The Gowganda and some other Precambrian deposits are very likely glacial, but most postulated examples must be re-evaluated.

*(Dott, 1961: 1289)*

Dott (1961: 1303) had no problem with the evidence for Permo-Carboniferous glaciations in South Africa, India, cratonic Australia, and southeastern Brazil; they were geographically extensive, chiefly found in non-marine sequences on stable cratons as for the great Pleistocene continental glaciation. He did question the

Permo-Carboniferous glaciations in southeastern Australia,[4] Argentina, Chile, and Bolivia, because they occurred in what were then active orogenic belts, and therefore were suspect, but he (1961: 1303) thought that even they would likely be confirmed after reevaluation.

Dott had carefully discussed criteria for identification of glacial deposits, ranked them with regard to their relative importance, and considered other paleoclimatological findings about North America during the Late Paleozoic, including the paleomagnetic positioning of North America near to the equator at that time. Dott argued that Squantum strata probably had been deposited from turbidity currents that could have been in low latitudes and hence their presence was not inconsistent with mobilism. But he was not yet a mobilist; the paleomagnetic evidence and the widespread Permo-Carboniferous glaciation were not sufficient for him. As Dott (January 2001 email to author) recalled, "First my paper was not about drift but about sedimentology, and the research was done from 1957 to 1960 – before much of the paleomagnetic data had reached me. Second, being a North American geologist, I was cautious about drift . . ."

He first learned about continental drift as an undergraduate at the University of Michigan majoring in geology.

I had first been exposed to continental drift either in 1950 or 1951 at Michigan by Digby McLaren (Cambridge graduate, then Ph.D. candidate at the University of Michigan [later Director of the Geological Survey of Canada]). Digby McLaren is a guy who likes to provoke, and he reckoned we isolationists needed to hear this wild theory. I do not remember *ever* hearing of drift or Wegener until that day! He gave a fascinating noontime "brownbag" talk, which aroused my interest in the idea.

(Dott, January 2001 email to author; my bracketed addition)

When Dott arrived at Columbia University in the fall of 1951, he quickly learned of the general opposition to continental drift, but he maintained his interest in it.

Then I arrived at Columbia in the fall, 1951. Now I heard a bit about the opposition to Drift from various sources, and bought Du Toit's fascinating book [*Our Wandering Continents*] in a used bookstore in downtown Manhattan. I think I read it in one night.

(Dott, January 2001 email to author; my bracketed addition)

At Columbia, the influential Walter Bucher was strongly opposed to continental drift (see Volumes I and II), and Dott remarked on the reaction of graduate students.

Did I have courses from Bucher? Yes, indeed, at least 2 or 3, and he was on my dissertation committee; he also gave me a French reading exam. He was a charming man whose great enthusiasm was very contagious. But his opposition to Drift was well known and taken by students with amused deference.

(Dott, January 2001 email to author)

Dott recalled a debate that the graduate students arranged between Bucher and Lester King (I, §6.10 for other recollections of this debate).

Lester King came to town either in the fall of 1952 or spring of 1953. We graduate students arranged a kind of debate between King and Bucher on Drift. We students felt afterward that King clearly had "won" in terms of the beauty of his presentation. What a magnificent orator he was with a beautifully modulated voice and a commanding presence. I thought he should have been on the stage. I was also mesmerized by him again in 1963 during an Antarctic symposium I attended in Capetown. So I was by [1953] 1963 very sympathetic to the idea and open to pro-arguments.

*(Dott, January 2001 email to author; my bracketed addition)*

Dott got his next exposure to mobilism from Runcorn. Runcorn, on one of his North American jaunts, gave his "standard" talk on paleomagnetism, telling his audience what he and his cohorts had done at Cambridge and were now doing at Newcastle.

I was in Los Angeles back in the oil industry in the spring of 1958 (5 or 6 months before I began at the University of Wisconsin in September 1958), and Runcorn came to town. He gave a noontime talk to a large audience of petroleum geologists. I was fascinated by his talk! Most of the audience probably was either put off or asleep, but it really jolted me. As soon as I moved to Madison and started teaching, I began to watch the paleomagnetic data for pole positions as new papers appeared. In 1961 I began teaching Historical Geology, and immediately decided to compare "geologic" evidence of climate and latitude with the emerging paleomagnetic data. I made a series of paleogeographic maps for North America, which showed the comparisons, and they looked mighty good for Drift. I copyrighted my maps and began giving them to my class as a supplement to the textbook.

*(Dott, January 2001 email to author)*

Dott submitted his Squantum paper in April 1960, arguing that it was turbidite, not tillite. Perhaps if he had begun teaching historical geology a year earlier, he would have argued that his interpretation increased support for mobilism. Had he done so most readers would doubtless have dismissed it, as indicative of his bias toward mobilism. Dott's work was motivated by the topical question of whether ancient glacial tillites and subaqueous gravity flow deposits could be distinguished, not by paleomagnetism or continental drift, and it was understandable that he did not then and there take a position on mobilism.

Even if Dott's paper was an example of the "new" paleoclimatology that Nairn would soon call for (§1.3), Edward Bullard, the British geophysicist, who during the 1960s was to play a major role in advancing the fortunes of mobilism (§2.13–§2.15), was not convinced that Dott was correct. Bullard referred to his work in an address given to the Geological Society of London. Although Bullard had at the time begun to view mobilism with some favor, he did not think that paleoclimatology could provide a definite answer to the drift question.

... how good is our knowledge of past climates? Can we recognize an ice age with certainty? Even supposing these [Permo-Carboniferous] ice ages to be genuine are we sure, for example, that the Squantum tillite does not represent a contemporaneous ice age in New England? Dott (1961), who has studied the rocks recently, says that it does not, but can we be sure – perhaps he has been influenced by the improbability of ice occurring 90° of latitude from that in South

Africa, and therefore presumably somewhere near the Equator ... Such arguments and doubts are endless and it is not profitable to work through them again ... Clearly it is necessary to break away from this well-trodden circle of ideas.

*(Bullard, 1964: 2–3; my bracketed addition)*

Bullard's point was not that Dott's work was careless, but that some paleoclimatological studies, no matter how competent the practitioner and how good the exposure, were necessarily based on incomplete and, to a degree, ambiguous data. Needless to say, paleoclimatologists probably put little stock in Bullard's view of the limitations of their science, because he was, after all, a geophysicist with little knowledge of paleoecology or sedimentology.

Bullard's negative assessment, coming from someone who had only recently become disposed toward mobilism, illustrates, I believe, how many outside of paleoclimatology had come to feel that it alone was incapable of providing definitive solutions to problems of past climate. Perhaps Nairn might have agreed with Bullard, but he would still have thought the practice of paleoclimatology capable of improvement. I doubt if Dott would have found Bullard's skepticism acceptable, because he accepted as firmly established much of the Gondwana Permo-Carboniferous glaciation (the cornerstone of the classical argument for drift). Dott (1961: 1301) also wrote of his hypothesis about the origin of the Squantum as "seem[ing] more plausible than glaciation." Evidently he believed that there were useful statements that careful paleoclimatologists or paleoecologists could make about their data when viewed alongside the independent paleomagnetic data; Nairn would have approved.

## 1.3 Comparisons of paleomagnetic and paleoclimatic evidence: the 1959 Newcastle symposium and its 1961 publication *Descriptive Palaeoclimatology*

Creer, Irving, and Nairn (1959) obtained a Lower Permian pole from their paleomagnetic survey of the Great Whin Sill in northeastern England and found latitudes derived from it compatible with the Permian climate of Europe. Irving (1964), working mainly with his Ph.D. student James Briden, extended the comparison of paleomagnetic and paleoclimatological data in a number of ways. He also worked with David Brown on the distribution of Late Paleozoic amphibians, and argued that the consilience between the paleontologic/paleoclimatic and paleomagnetic data strengthened support for mobilism (Irving and Brown, 1964). Opdyke (1961, 1962) and Runcorn (1961, 1964a, 1965) wrote further about paleowind directions, and broadened their involvement in paleomagnetic and paleoclimatic comparisons. Blackett (1961), more or less following the plan of attack first utilized by Irving, appealed to paleoclimatology to support specifically the geocentric axial dipole (GAD) hypothesis and generally the reliability of the paleomagnetic data. Nairn, besides making paleoclimatological comparisons in his own work (Schove, Nairn, and Opdyke, 1958), began to work with paleoclimatologists and paleontologists.

Encouraged by Runcorn, Nairn edited two volumes in paleoclimatology, *Descriptive Palaeoclimatology* (Nairn, 1961) and *Problems in Palaeoclimatology* (Nairn, 1964a). Both grew out of Newcastle symposia attended by paleomagnetists and paleoclimatologists.

I begin with *Descriptive Palaeoclimatology*, then turn to Blackett's 1961 foray into paleoclimatology, and close with an examination of *Problems in Palaeoclimatology*. I shall consider the attitude of the symposiasts toward paleomagnetism and mobilism to ascertain the reception of the paleomagnetic case for mobilism in the early 1960s. Opdyke, Runcorn, Irving, and Creer participated in one or other of these Newcastle symposia, and I can follow their efforts.

In 1959 Nairn assembled a symposium in Newcastle from which the book he edited, *Descriptive Palaeoclimatology*, arose; sixteen authors wrote fifteen essays, divided into fourteen chapters. In his introduction, he provided a brief history of paleoclimatology: "The early history of palaeoclimatology can be divided into two phases, the turning point being the appearance of Wegener's theory." During the first phase only Northern Hemisphere geology was known, and old glaciations were unrecognized; it was generally believed that Earth's climate had been mild until the Late Tertiary when extensive glaciation began.[5] This phase ended with the discovery of large-scale Permo-Carboniferous glaciation in India and across the Southern Hemisphere, ushering in the second and present phase with its ever-present backdrop, the mobilism controversy. Nairn argued that paleoclimatologists needed now to advance to a third stage where their interpretation of paleoclimatological data would not be dictated by their attitude toward fixism or mobilism. By this he did not mean that paleoclimatologists should not make claims about mobilism or fixism, but that such claims should be conclusions, not initial premises. Nairn wanted to see climatology based on sound meteorology. Before deciding in favor of mobilism or fixism, he wanted paleoclimatologists to develop better criteria for determining the reliability of their assessments of past climates. In this way Nairn believed that paleoclimatology could then mature into the third phase, and it was the aim of the symposium and the book to further that development. He aptly entitled it *Descriptive Palaeoclimatology*.

The third phase of paleoclimatology must therefore be free from prior assumptions about land–sea distribution. A meteorologically acceptable framework is needed to which references about past climates can be referred. This information in turn must be acquired by methods whose reliability and whose limitations have been established by the critical examination of criteria involved. When world cross-sections of the climatic sequences of different areas can be assembled it must prove possible to piece together climatic zones and in so doing palaeoclimatologists will have made a significant contribution to the history of the earth's crust ... It is with the descriptive part of this scheme, the meteorological framework, the climatic criteria and the regional climatic histories, that this volume is concerned.

*(Nairn, 1961: 2)*

The essays fell into three groups: the meteorological framework, paleoclimatic indicators, and regional climatic histories. Authors' attitudes toward mobilism differed. The first group comprised Nairn's introduction and H. H. Lamb's presentation of the present energy budget of Earth's surface. There were essays on paleoclimatic indicators: by Opdyke on desert sandstones and paleowinds, Robert Green on evaporites, F. B. Van Houten on red beds, R. F. Flint on evidence for cold climates, N. Thorley on the use of $O^{16}$-$O^{18}$ ratios to determine paleotemperatures, Nairn on paleomagnetically determined paleolatitudes, and by A. S. Romer, G. Y. Craig, and R. Krausel, respectively, on the use of vertebrates, invertebrates, and plants as past climate indicators. Regional paleoclimatological histories were presented by M. Schwarzbach for Europe and North America, T. Kobayashi and T. Shikama for the Far East, Lester King for Gondwana during the Paleozoic and Mesozoic, and E. D. Gill for Gondwana during the Cenozoic. Flint, Romer, Gill, Krausel, and Kobayashi and Shikama were silent on mobilism and its paleomagnetic support. That Romer failed to comment on drift is at first sight surprising because he was one of the few vertebrate paleontologists from North America who showed sympathy toward mobilism (I, §3.8). However, he wandered little from his basic concerns – the assessment of the information about past climates obtainable from vertebrate fossils and vertebrate evolution. He concluded that they provided some useful information, but not nearly as much as sedimentology and paleobotany, so his silence is perhaps not surprising.[6]

As expected, Opdyke and Nairn argued in favor of mobilism. Opdyke summarized the results of his own and others' paleowind studies, and drew comparisons with paleomagnetism. Nairn devoted most of his section of his joint paper to a summary of paleomagnetism (Nairn and Thorley, 1961). He emphasized the consilience between paleomagnetic and paleoclimatic results, singling out Irving's early comparisons and citing Opdyke and Runcorn's work on paleowinds. He also (Nairn and Thorley, 1961: 180) advocated that maps of individual landmasses based on paleomagnetic data (his "drift diagrams" (II, §5.4)), even if incomplete, "may be a useful basis for further research and form a preliminary framework to which other climatic data may be added"; he wanted paleoclimatologists to use paleomagnetically determined paleolatitudes as a tool for paleoclimatic studies.

Again as expected, Lester King argued in favor of mobilism, writing a persuasive essay on the success of mobilism to account for the Permo-Carboniferous glaciation and the distribution of *Glossopteris* flora. He welcomed the paleomagnetic support for mobilism.[7] Noting its independence from classical arguments, he argued that the mobilist interpretation of the paleomagnetic data offered further support for paleogeographies of Gondwana based on paleoclimatic and paleontological data. He hoped this new evidence would persuade Earth scientists who had been reticent to accept mobilism because of its lack of an adequate mechanism.

Some critics have been reluctant to accept Drift because they could find no mechanical explanation for it, recently developed palaeomagnetic techniques independently establish relative movement between the southern continents, and reinforce the concept of Gondwanaland during the Palaeozoic era ...

*(King, 1961: 311)*

King believed the best way to establish that something is possible is to demonstrate its occurrence. He thought that paleomagnetism established the existence of Gondwana during the Paleozoic, and hence the need to posit continental drift.

Martin Schwarzbach, then at the University of Cologne, had written an important monograph on paleoclimatology, *Das Klima der Vorzeit*, first published in 1950, expanded in 1961 and translated into English two years later (Schwarzbach, 1963). Schwarzbach discussed mobilism and paleomagnetism in his contribution to *Descriptive Palaeoclimatology* as well as in his book. He described the consilience of the paleoclimatic and paleomagnetic data.

The way in which the climatic conditions of Europe and North America fit into the whole picture of the earth at the time can be particularly well illustrated by the example of the Devonian period: both these northern continents belong to a warm climatic zone, in contrast to South America and Central and South Africa. The position of the geomagnetic pole, which is determined palaeomagnetically, falls in the south Atlantic region; if one assumes that the geographical pole is also in the same position at that time the peculiar climate conditions can be explained, for the equator would then pass through North America and Europe.

*(Schwarzbach, 1961b: 261–262)*

Turning to paleowinds, he referred to papers by several German workers. He noted that Poole (1957) favored broadening the belts of trade winds extending them further to the north, while Shotton (1956), Opdyke and Runcorn (1959), and Schove, Nairn, and Opdyke (1958) preferred to move continents. He (1961b: 262) concluded, "many more observations of wind directions are required." He (1961b: 262) supported Wegener's view about the proposed Late Paleozoic glaciations in North America (I, §3.10, §3.12, §3.15) and that they and the Squantum Tillite (§1.2) "do not fit well into the climatic picture of early times. They could at most have originated from mountain glaciers, but certainly not from continental sheets." He criticized Stehli's fixist argument (RS2).

Stehli made use of the various species in the Permian fauna to infer a cool climate in what are now polar regions, but in doing so he has grossly underrated the element of chance involved in finding fossils.

*(Schwarzbach, 1961b: 262)*

Schwarzbach's views remained unchanged in *Das Klima der Vorzeit*.

Modern investigations of paleomagnetism have furnished unexpected arguments in favor of shift of the poles ... Another important result is that pole positions, as estimated in different continents, are somewhat different. This can only be interpreted as showing that the position of the continents relative to one another has changed. Paleomagnetism therefore affords evidence not only of an overall polar wandering, but also of continental drift.

*(Schwarzbach, 1963: 246–247)*

In his discussion of paleomagnetism, he repeated the familiar difficulties (RS2), stepping very close to Billings' misapprehension about the time-averaged form of the field, the GAD hypothesis (II, §7.12).

The following difficulties stand in the way of the geological application of this method: (a) the measured values are always extremely small, and only relatively few rocks, mostly those containing magnetite, are suitable. (b) The remanent magnetism may not have remained absolutely unaltered ... (c) In many cases, the magnetic field has reversed, i.e. the north and south poles have changed places. The cause of this phenomenon is still uncertain ... (d) It is not certain that the magnetic pole always lay in the vicinity of the geographic pole as it does at present ... There are theoretical grounds for believing that the pole of rotation and the magnetic pole are causally related; the terrestrial magnetic field probably arises from currents in the interior of the earth, which will be symmetrically disposed with regard to the axis of rotation. Thus the earth represents a sort of gigantic dynamo. Conditions over the last few centuries, however, show that we must reckon with a deviation of at least 20°.

*(Schwarzbach, 1963: 246–247)*

Schwarzbach, although cautiously favoring mobilism and impressed with the new paleomagnetic evidence, was wary, taking something of a wait-and-see attitude. Had he read Cox and Doell's Geological Society of America (*GSA*) review (II, §8.4, §8.7)? Yes, but when? It is neither named nor referenced in the 1961 edition, but it is in the English translation (Schwarzbach, 1963).

Van Houten, a red beds expert from Princeton University, discussed Permo-Triassic red beds, paleomagnetism, and continental drift. Although he expressed reservation about the common view that red beds formed in the tropics and subtropics from the erosion of lateritic soils, he thought it was the best available. This led him to look with favor on paleomagnetic work on red beds because for the most part it placed red beds in low latitudes, but he was cautious.

With few exceptions the magnetic inclination of ancient red beds that have been analyzed is small and indicates that they lay within 25° of their equator when they were magnetized. Although these are very limited data they do agree with the lateritic origin of detrital red beds, and thus appear to support the idea of polar wandering in the past. Nevertheless, so many poorly understood factors are involved in producing the inclination that this observation is no more than suggestive at the present time.

*(Van Houten, 1961: 124)*

Van Houten also recognized several apparent conflicts between paleomagnetic and paleoclimatic results. There was the troublesome matter of northern China with its red beds apparently far from the equator during the Permian as determined from Europe; he may not have known that Nairn already had disposed of the problem (II, §5.8), and there was Stehli's study to which Runcorn had replied (II, §7.10), but Van Houten mentioned neither.

Gordon Craig, an invertebrate paleontologist from the University of Edinburgh, wrote cautiously and thoroughly. He expressed no opinion about mobilism or paleomagnetism, but did raise a difficulty that plagued many paleoclimatological interpretations, namely that their comparisons were made over entire geological periods, and may be too broad-brush.

The main difficulty in interpreting the climate of an entire system is that an enormous length of time is generally involved. The Permian spreads over a period of some 45,000,000 years, about equal to that of the Oligocene-Present, during which time exceedingly severe climatic changes took place. It is therefore preferable to work within as narrow a time interval as geological dating will allow.

*(Craig, 1961: 217)*

This was a familiar difficulty, also raised against comparisons of "equivalent" paleo-poles from different regions. Finding sufficiently well-dated samples for comparison was everybody's problem.

Discussing evaporites, Robert Green from the Research Council of Alberta, Edmonton, Canada, remembered:

It is considered significant that the distribution of evaporites through geological time is explicable on the basis of a gross trend in climatic change, an explanation which necessitates large-scale change neither in the relative positions of continental masses nor in positions of the rotational poles of the earth.

*(Green, 1961: 86–87)*

Although Green admitted that placing Europe and North America near the equator as their apparent polar wander (APW) paths required agrees with the distribution of evaporites in both places, he noted that the postulation of a generally warmer Earth as an explanation of the paths brings about an anomalous distribution of evaporites in South America and Southeast Asia.

If the distribution of evaporites is considered in relation to lines of latitude constructed using palaeomagnetic data [here he cites Runcorn, 1956a], a general northward movement through geological time is indicated for the mean position of arid climatic belt in the northern hemisphere. This infers an over-all warming trend in the earth's climate through time, an interpretation which is not in agreement with interpretations based on specific evaporite deposits. Certain deposits, such as the Carboniferous evaporites of Brazil, the Permian and Triassic evaporites of south-east Asia, the Cretaceous evaporites of Bolivia and Jurassic and Cretaceous evaporites of Argentina occupy anomalous positions relative to the paleomagnetic poles of their particular periods, anomalies which are explicable only by means of continental drift.

*(Green, 1961: 1962; my bracketed addition)*

Green was either confused about or unaware of recent work in paleomagnetism, especially from the Southern Hemisphere; he drew his paleomagnetic information mainly from one of Runcorn's 1956 pre-drift papers, but he also cited Runcorn (1959c), who made it clear that continental drift was needed to explain diverging APW paths. His comments are interesting because they illustrate the improbable consequences of assuming polar wandering only.

Lester King, Opdyke, and Nairn strongly supported the paleomagnetic argument for mobilism; Schwarzbach raised specific difficulties with it but he and Van Houten were cautiously optimistic about its potential. Green trusted evaporites rather than paleomagnetism as indicators of past climates. Other symposiasts voiced no opinion on mobilism.

## 1.4 Reviews of *Descriptive Palaeoclimatology*

Reviews of *Descriptive Palaeoclimatology* by Bucher, A. G. Fischer, and Irving soon appeared. Examining the first two helps to assess whether the two fixists thought that the many comparisons of paleoclimatology and paleomagnetism had enhanced the case for mobilism. Irving's mobilist review was different. As I have already noted, Bucher, a structural geologist and staunch fixist, was not only interested in paleo-magnetism, but had helped Runcorn make known British work in the United States. Without mentioning continental drift by name, he questioned the ability of current paleomagnetic research to settle the controversy.

This is a trim book with an adroit title succeeding admirably in describing the essence and present state of paleoclimatology including the perplexing ambiguities inherent in the climatic interpretation of most records. It also exemplifies when synthesis in time and space are attempted, the wide gap between the groping, tentative suggestions of some and the triumphant faith of others in one hypothesis.

*(Bucher, 1962: 296A)*

Bucher had two main criticisms, both reflecting his antipathy to mobilism. First was the absence of a chapter on paleoclimatology's methodological framework that would show the many instances in which Earth's current climate does not conform to zonal climatic belts. According to the old fixist:

The omission of this crucial chapter reflects no doubt the paleomagnetist's overvaluation of latitude as a climatic factor, which is also evident from the wording of the introductory chapter [written by Nairn].

*(Bucher, 1962: 296A; my bracketed addition)*

He provided several examples, including the occurrence of salt beds in the Gulf of Kara Bugaz (Caspian Sea) at only 33° N latitude in a region with more severe winters than the Gulf of Finland 27° further north. He objected to the reliance that paleo-magnetists placed on climatic zonation. The second omission was the role of turbidity currents in forming glacial-like deposits in a possible fixist solution to the origin of the Permo-Carboniferous glaciation (RS1).

In Part II, the discussion of "geological evidence of cold climate" does not even specify in its title the pressing problem of recognizing "glacial" conditions in the geological record and does not face realistically the hard stratigraphic facts which must be presented impartially and analyzed critically before a claim of glacial origin can be considered valid. *This applies especially to many of the far-flung glacial deposits in the southern hemisphere* which lie intercalated in marine formations, often with rich faunas. In the last quarter century, the recognition of the existence, magnitude and ubiquity of turbidity currents has thrown severe and justified doubt on most such occurrences. There exist a rich literature on the results of the modern exploration of the sea floor and the systematic study of pertinent fossil occurrences. Yet the author index does not contain the names of those who have led in this development. The index does not contain the word "turbidity current" nor does the concept appear in the text or the table of criteria of "till and till-like sediments." ... *Since "glacial deposits" in unlikely places have furnished the chief leverage with which to force continents to move in men's minds, this omission is serious.* In view of the well-established tendency of turbidity currents to travel long distances parallel to a submarine trench or trough, the presence of minerals and rock fragments foreign to a given locality does not prove that land must have been where there is now the ocean floor.

(Bucher, 1962: 296A, 300A; emphasis added)

Perhaps he was stung by the lack of reference to the turbidity-current works of his colleagues at Columbia University's Lamont Geological Observatory (hereafter, Lamont) (§6.5). But he had more in mind: here was a new possible solution that would render questionable the very existence of the large-scale, far-flung Permo-Carboniferous glaciation of the southern continents and continental drift as a figment of the imagination. He might have had Dott's study in mind, but if he did, he so grossly exaggerated what Dott had written that he ended up contradicting him. After all, Dott was confident that most of the Permo-Carboniferous glacial evidence in the Southern Hemisphere in places remote from contemporary orogenic belts had been correctly interpreted; his work certainly did not weaken, but strengthened, mobilism. Finally, Bucher was unwilling even to admit that developments in paleomagnetism improved the case for mobilism.

In contrast, Fischer, of the Department of Geology, Princeton University, who worked in paleoclimatology, paleontology, paleoecology, and evolution, believed that the fortunes of mobilism had improved, and he attributed this to paleomagnetism.

Nothing since Wegener has hit paleogeography with such impact as paleomagnetism. It is building up global patterns with considerable coherence, patterns which must mean something.

(Fischer, 1963: 268)

But he urged caution and found unrestrained zeal unhelpful. He noted that data used to support mobilism were often unreliable and could be explained in terms of fixed continents. Furthermore, he claimed that more specific drift reconstructions were needed if mobilism were to be conclusively proved, and he was not yet ready to support mobilism.

In retrospect no sweeping conclusions have been reached in this volume. The present state of paleoclimatology has been well illustrated. We are making progress in the understanding of past climates, but as yet the fragments have not been fitted into a single coherent picture of climatic evolution, which would contain within itself the keys to the major problems of global paleogeography. The case for permanence appears to be on the defensive, but drift remains a loose working hypothesis. Due to its spectacular nature it has always attracted, among others, geologists whose imagination has consistently outrun the factual basis and who have grasped at factual straws to support this hypothesis. The survival of the theory has been in spite of, not because of, these overzealous apostles. Many supporting data used by Wegener and others appear naïve in the light of more knowledge, and most of the geological observations – climatological or otherwise – which are cited in support of drift are not thoroughly documented by modern standards or are amenable to alternative explanations. The counterargument is that whereas perhaps no lone line of evidence is wholly convincing, the weight of many lines of evidence, pointing to drift, is conclusive. But it will be wholly conclusive only if these independent lines of evidence support not only drift in general but a specific temporal and spatial pattern of drift.

*(Fischer, 1963: 268)*

Fischer made clear that this call for more caution and less zeal applied to Opdyke's discussion of paleowinds, Lester King's Gondwana essay, and Nairn's view of paleomagnetism. Opdyke's account (RS2), he claimed, faced a reliability difficulty; it was not easy to correctly identify sandstones as aeolian (of which Opdyke was well aware (II, §5.10–§5.14)); it faced the same theoretical difficulty that Bucher had raised about applying climatic zonation too rigidly; and, according to Fischer, Opdyke's assumption of the underlying global nature of current wind patterns was dubious.

Not all of this [Opdyke's] story is as straightforward as one might be led to believe. There is, for example, the question of recognizing aeolian sandstones. Extensive cross-bedding is found in water-laid sands as well, and thus need not be related to prevailing wind directions. The criteria for recognition of aeolian sands need to be clearly stated, and do not appear to be diagnostic in all cases ... The other source of difficulty lies in the generalizations concerning present-day wind directions. How reliable are they? Concerning the Peruvian coastal desert, where I happen to have made observations, Opdyke's map and generalizations are misleading. He states that "Dune movements lower in latitude than 22° should be from the ... south, southeast or east to north, northwest and west, respectively, in the southern hemisphere," and for the Peruvian desert he shows a due south to north movement. This is the movement in the immediate vicinity of the beach ... but as the winds move inland from the ocean, they and the dunes which they construct are deflected in a clockwise wheel-about, so that along the Andean front the winds and sands move from west to east. Most of the dunes thus show a southwest to northeast movement. If found in the fossil state, Opdyke might well assign to this area (located between 5° and 10° S Lat.) a latitude of more than 22°.

*(Fischer, 1963: 284; my bracketed addition)*

Fischer was even harder on King.

To King the evidence for classical Wegener–Du Toit break-up of Gondwana and drifting apart of its fragments is so complete that alternatives need not be considered. He tells his story skillfully, but the critical reader would like to know much more of the details such as the firmness of the stratigraphical dating with which King establishes a migration of glaciation (presumably the result of Gondwana's drift relative to the pole) and the reliability of the glacial interpretation of the large number of boulder clays. One gets the feeling that King has stated many possibilities and probabilities as facts ... Although King's chapter is most persuasively written, it is not likely to make converts in the hard core of anti-drifters. It is a pity that the geological anomalies of the "Gondwana" continents have not been dealt with in a more objective and soul-searching way, by someone who is not so obviously committed to a theory and who will point out the weaknesses as well as the strengths in the case for drift. Not all drifters will agree with the patterns of drift proposed by King, which differ somewhat from those suggested by paleomagnetic work.

*(Fischer, 1963: 291)*

Turning to paleomagnetism, Fischer thought that Nairn had overstated the strength of its support for mobilism; it was not so cut and dried.

Nairn's flat statement that "the latitude and orientation of land masses in the past can be derived from the study of permanent magnetism of rocks" is unfortunately dogmatic in a chapter meant to be a plea for objectivity. Paleomagnetism as a paleogeographical tool remains a theory – a fascinating one, with implications of continental drift. Neither the theory nor its proponents will derive credit from parading it at present as established fact.

*(Fischer, 1963: 283)*

He noted the longitude ambiguity in paleomagnetic paleogeographic reconstructions. He objected to Nairn's maps.

A point that Nairn might have brought out more explicitly is that if drift be permitted, paleomagnetism may yield latitudes and orientations for a given continent, but not longitude, and therefore cannot offer unique paleogeographic reconstructions of the distribution of continents. In that connection, one may wonder how he obtained the longitude values of his paleomagnetic reconstructions of Europe.

*(Fischer, 1963: 288)*

Despite these criticisms, Fischer thought paleomagnetic and other approaches showed promise.

The future of paleoclimatology rests with *criteria* and their *application*. Nairn's volume summarizes the present status of criteria and points to the necessity for refining existing ones, and for adding new ones. The reliability of some of the most promising ones – such as paleotemperature work, the significance of wind-direction studies, the meaning in invertebrate growth patterns – are, like paleomagnetism, still subject to certain reservations, and will have to be tested chiefly in their application.

*(Fischer, 1963: 292)*

Fischer used Dott's work to throw doubt on mobilism's interpretation of Permo-Carboniferous glaciation.

A number of ancient "tillites" have been shown to be definitely or probably nonglacial; since the appearance of Nairn's volume additional doubt has been thrown on the Squantum "tillite" (Dott, 1961). Thus many of the presumed tillites of the southern hemisphere are in need of critical study.

*(Fischer, 1963: 287)*

But Fischer, like Bucher, did not mention that Dott had not questioned that most of the purported Permo-Carboniferous glacial beds in the Southern Hemisphere were truly glacial. It is as if Fischer focused on Dott's (1961: 1303) concluding comment that "most supposed examples of ancient glacial deposits must be critically re-examined" to the exclusion of his main remark (1961: 1289) that "the Permo-Carboniferous glaciation in certain parts of the southern hemisphere is firmly established."

Irving (1962a) thought quite differently about King's contribution. Without mentioning names, he clearly had King's contribution in mind when writing about the paleoclimates of Gondwana.

The book is concluded by four useful and informative chapters in which the variations of the past climate in different regions of the Earth are described. The forthright statement of Gondwanaland palaeoclimates should serve as a timely reminder that it is in the southern continents that palaeoclimatology presents the most formidable challenge to the view that the continental positions have not changed.

*(Irving, 1962a: 268)*

As if to underscore the idea that "out of sight, out of mind" applies to those who venture to express opinions about Gondwana palaeoclimates without having extensive firsthand field experience of the relevant strata, Irving singled out Van Houten's study of red beds as a very careful review. Nevertheless, he, from firsthand experience, corrected Van Houten's account of the very limited Lower Triassic Australian red beds describing it as misleading. Irving, however, was most critical of Nairn's essay on paleomagnetism and its application to paleoclimatology.

The central theme of this topic is that in the same general region the palaeoclimatic evidence is consistent with the latitude obtained from palaeomagnetic studies, which therefore serve as an independent check on conclusions based on palaeoclimatic evidence. But this basic feature receives no proper description – it is allowed only two lines in the penultimate paragraph.

*(Irving, 1962a: 268)*

Irving was disappointed that Nairn had not presented a much more compelling account of the numerous comparisons within regions that paleomagnetists already had made and which showed consilience between contemporaneous paleomagnetic and paleoclimatic data. Perhaps he felt Nairn had missed a marvelous opportunity to spread the word to paleoclimatologists.[8]

## 1.5 Speculations on mechanism in the early 1960s

No, I don't think [I ever discussed my ideas about convection and long term stresses] with Jeffreys ... Undoubtedly Vening Meinesz influenced my views on convection and the likelihood of it occurring. But, I think that the reason I talked about long term behavior was that ... I had been very interested in solid state physics so it was very natural for me, as it were, to reconcile the ideas of flow in the mantle with convection, which had been, of course, talked about by Holmes and Vening Meinesz, and even by geologists. It was natural for me to tie them in with solid state creep ... You'll find that Jeffreys argued very much against continental drift from the fact that the crust of the Earth could hold mountains up and otherwise behave like a rigid body over long periods of geological time. In a way I was obviously answering him. I'm not sure if I specifically mentioned Jeffreys in this article, but certainly I was answering him.

Certainly, you can see from that article that I thought that I was saying something new. Not that plastic behavior of the Earth had not been talked about before. But I was saying that we have a very sound foundation for creep, and creep occurring at very low rates of stress and strain. We have evidence from the most fundamental aspects of solid state physics that this is not merely possible but that is to be expected on the long time scale we're talking about.

*(Runcorn, 1984 interview with author; my bracketed addition)*

With the maturation of the paleomagnetic support for mobilism, the search for a mechanism for continental drift was reopened by Runcorn and others during the early 1960s. I shall focus on Runcorn's ruminations in which he sought to synthesize a wide range of topics related to mobilism, attempting to stitch them together. Earlier he had discussed the cause of polar wandering, even touching on drift's mechanism problem (II, §3.9). Now he began thinking about them again, announcing that he had done so by appending brief notes on mantle convection and the ideas of Vening Meinesz and Harold Urey at the end of one of his reviews of the paleomagnetic and paleoclimatic evidence for mobilism (Runcorn, 1959c). He gave a second version of his speculations on mantle convection as an add-on to his lengthy review of paleoclimatology and paleomagnetism (1961), and another more complete version a year later (1962a). He continued to develop his ideas throughout the 1960s, taking into account emerging information about ocean ridges, the evidence of strongly magnetized material centered in the Mid-Atlantic Ridge (1962b), radiometric dates and mountain building episodes (1962c), and Earth's gravitational field as measured by artificial satellites (1963). In his eclectic lectures and reviews there was something for everyone.

Initially when Runcorn (1955a) had advocated polar wandering only without continental drift, he thought Vening Meinesz's mantle convection might be its cause. Upon adopting drift the following year, he (1956c) thought mantle convection might be its cause too; he thought of it as a rejoinder to Jeffreys' adamantine opposition to mobilism.

Before presenting his convection hypothesis, Runcorn, like du Toit (1924), van der Gracht (1928), Wegener (1929), and Rastall (1929) (I, §5.13), attempted to remove mechanism difficulties by appealing to historical precedent (RS1). Jeffreys had said that there had been "no tenable dynamical theory of continental drift since the hypothesis was first put forward" that would allow for the occurrence of continental drift so late in geological time. Runcorn saw no sense in rejecting phenomena just because they cannot be explained.

These points are certainly not arguments against studying the evidence for continental drift seriously; no one in the past few centuries has disbelieved in the existence of the Earth's magnetic field in spite of there being no theory of it.

*(Runcorn, 1962a: 311)*

He then raised a general methodological difficulty with Jeffreys' dismissal of continental drift (RS2). Reminiscent of Carey, prompted as he had been by Paterson (II, §6.10), Runcorn argued that mechanism should be approached from the viewpoint of solid state physics.

Sounding like a veteran of the classical stage of the mobilism controversy, Runcorn noted that Wegener had understood the significance of isostasy as a process that acts over the long term, and was distinct from short-term phenomena such as earthquakes and nutations in which the effects of stresses applied over millions of years need not be considered. He hinted that Jeffreys was behind the times, having failed to appreciate advances in solid state physics that allowed for the possibility that Earth's mantle, being a solid near its melting point, could flow in response to long-term stresses. Runcorn's attack is worth quoting because it echoes Jeffreys lecturing Joly (I, §4.11), except that Jeffreys was now the pupil.

Wegener took the first step towards understanding the mechanism of continental drift by his emphasis on the significance of the isostatic equilibrium of the continents. The re-adjustment of the heights of continental areas consequent on changes of loading involves behaviour of a fluid nature in the upper mantle below the Moho, and of course, by the continuity equation horizontal as well as vertical flow must occur ... The geophysicist studying the response of the Earth to short-period stresses (seismic waves of seconds and minutes period and the nutations of 12 and 14 month periods) has not needed to assume other than an elastic behaviour of the Earth (for example, Jeffreys). Jeffreys experiments with a delayed elastic equation to account for the damping of the nutation and the rotation of the satellites of planets. He assumes a behaviour similar to that found experimentally in many metals and other materials in which, on applying a stress $p$, small compared with the yield stress, an instantaneous strain $p/E$ (as in classical theory) occurs, followed by an additional strain increasing with time, as shown in Fig. 1 [reproduced in Figure 1.1], by line A, to a constant value. On removing the stress the strain decreases with time, as shown by line B, leaving no permanent set.

Jeffreys shows that, as would be expected, a material with such delayed elastic behaviour is not subject to convection and on these grounds excludes the movements in the mantle which will be seen to be required to cause continental drift. However, Jeffreys' description of the behaviour of solids is incomplete from the point of view of modern physics. In the long run, if the stress is maintained, we expect to find that the solid creeps, as shown in Fig. 1 by line C. Thus, the behaviour of the Earth in response to the stresses of seismicity and nutation is totally irrelevant

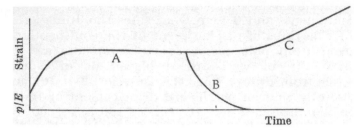

Figure 1.1 Runcorn's Figure 1 (1962a: 311). Time response to a constant stress. Line C represents the creep or flow of a material to applied stress over a long period of time. Line B shows what happens if the stress is removed. Runcorn claimed that Jeffreys' description of the behavior of solids was incomplete, allowing only for a response in accordance with Line B. Solid state physics suggests flow or creep in accordance with Line C if the stress is applied constantly for long periods of time.

to stresses applied constantly for millions of years: geological field observations of folds and contortions in hard rocks are sufficient evidence of this.

In fact, the behaviour described by classical elasticity would appear to be quite exceptional and illustrates the difficulty geophysicists frequently have in adjusting their thinking from ordinary experience in the laboratory to the large scale (both in space and time) of planetary problems. If a solid be put under a shearing stress, the resulting distortional strain deranges the order in it, but in the long run various physical processes occur to re-establish order ... The constantly applied stress will, therefore, cause the re-establishment of fresh strain, and creep or flow will occur. Thus it seems incontestable that in the long run any material will flow accordingly to curve C. That most of the mantle is likely to be at temperatures not much below the melting point is further cause for suspecting that such behaviour will occur.

*(Runcorn, 1962a: 311; my bracketed addition)*

According to Jeffreys, Joly had failed in the late 1920s to understand the physics of heat. According to Runcorn, Jeffreys failed thirty years later to appreciate modern solid state physics.

Runcorn continued clearing the way, theorizing that mantle convection would occur, given sufficient temperature differences; he argued a difference of only $0.2\,°C$ between up-going and down-going currents would produce one convective overturn in approximately $10^8$ years, enough, he contended, to cause the motions of continents required by paleomagnetic results. Next, he turned to the difficulty presented by seismic discontinuities in the mantle. If they reflect chemical changes, as Birch (1952) believed, then mantle convection would be of insufficient scale to drive continental drift; if the discontinuities represent phase changes, a possibility that Birch (1952) was unwilling to rule out, then convection currents could be of the appropriate scale to drive continental drift.

Runcorn then drew on ideas from G. F. S. Hills and, especially, Vening Meinesz.[9] Both had proposed large-scale convection cells. Hills (1947) postulated that early in Earth's history a single cell had swept sialic continental material together into one hemisphere, but Jeffreys, originally attracted to Hills' hypothesis, had objected that single-cell convection could not occur in the presence of a dense core of present radius.

During the early 1950s, Vening Meinesz had blended his ideas about convection with the study of Earth's surface elevations by Adalbert Prey. Prey was a physicist who succeeded Egon von Oppolzer at the astronomical observatory at the University of Innsbruck during the early nineteenth century and who moved later to Charles University in Prague. He expressed surface elevations in terms of spherical harmonics that mapped elevated and depressed areas, and that Vening Meinesz (1951, 1952a, b) had identified as places where mantle currents rise and sink, respectively. This led him to conclude that convection cells were much larger than he had earlier thought, and that Earth has undergone periods of intense convection separated by long quiet periods. At an early stage a huge single convective cell (degree $n = 1$) had brought about the separation of the core and mantle and the formation of a single sialic proto-continent above converging and descending convection currents. This convective cell was not subject to the objection Jeffreys had made to Hills' model, in which convection did not begin until core and mantle had separated. A second stage of convection tore the proto-continent apart, scattering fragments into their current positions (1952a: 379). Vening Meinesz (1952b: 530) further proposed, "similarities of continental coasts, as e.g., those of South America and Africa can thus be explained as the result of the rents in the proto-continent and dragging apart of its pieces by the second stage current-system." He was not proposing classical continental drift; all his drift happened much much earlier than that of Wegener; he firmly distinguished his view from Wegener's.

We may perhaps ask whether it would be possible that the system of currents has taken place in a much more recent period and in such a way that the resulting movement and redistribution of the continental shields would fit in Wegener's theory. It seems to the writer that such a hypothesis would be difficult to accept, mainly because of its being hard to believe that in a recent period the oceanic parts of the crust could have become sufficiently plastic to allow the transportation by the currents of the sialic shields as assumed in this paper without these shields being fundamentally deformed themselves.

*(Vening Meinesz, 1951: 226)*

Vening Meinesz proposed a third convective system that brought about polar wandering, without continental drift. As Earth's rigidity increased, subsequent convection caused mountain belts, island arcs, and oceanic trenches. He (1952b: 551) remarked of his hypothesis, "It is hardly necessary to stress its speculative character."

Runcorn recalled the work of Vening Meinesz (who at the time was a fixist), and its further development by Chandrasekhar (1952).

Vening Meinesz ... became interested in Prey's demonstration that there is a certain regularity about the arrangements of the continents. Prey took values of the height of the land above and the depths of the ocean below sea-level, counting the latter negative, and expressed them as a series of spherical harmonics. The predominant term is, of course, of degree $n = 1$, expressing the fact that the continents are concentrated in a single hemisphere. The terms $n = 2$ and those greater than $n = 5$ are relatively weak, the terms $n = 3$, 4 and 5 being strong. Terms of odd degree have opposite signs at antipodal points; therefore Prey's analysis simply gives mathematical expression to the fact, which seemed of significance to an older generation of geologists, that the continents are antipodal to oceans ... This geophysical fact that only 3 per cent of the area of continent is antipodal to continent demands explanation. Vening Meinesz reasoned that this element of regularity in the positions of the continents to-day could result from a large-scale, regular pattern of convection motions in the mantle ... Vening Meinesz and Chandrasekhar ... show, as is intuitively obvious, that as the ratio of the radius of the inner spherical boundary to that of the outer increases, the convection which is excited at marginal stability is characterized by harmonics of higher degree. Chandrasekhar shows that for a core of the present radius, 0.55 of Earth's radius, harmonics $n = 3$, 4, and 5 are almost equally likely to be excited at marginal stability. Vening Meinesz argued, therefore, that a continental distribution such as Prey found arises naturally from the mathematical theory of convection, and as this depends simply on the geometry of the problem, Vening Meinesz, in his earlier discussions, felt that this showed continental drift to be unlikely.

*(Runcorn, 1962a: 312–313)*

Following a suggestion of Urey's, Runcorn invoked an expanding core, which would progressively increase the number and decrease the size of convection cells.

Urey has advanced reasons why the Earth must have had a cold origin: for example, the volatile elements would not otherwise be present in the Earth's crust in their observed abundances. He recognized that a theory of the Earth's evolution which envisages it starting from a mixture of planetesimals, presumably similar to the iron and stony meteorites, has difficulty in explaining the separation of the core, and Urey supposed that this occurred slowly: perhaps by the iron creeping downwards along the boundaries of the silicate crystals. Urey speculated that the growth of the core might still not be complete.

*(Runcorn, 1962a: 313; Runcorn cited Urey, 1952)*

In fact, Urey had already made a similar, albeit incomplete, proposal in his 1952 book, *The Planets*, which Runcorn cited, and in which Urey had everything but continental drift. In 1953, Urey reviewed earlier works, especially Vening Meinesz's incorporation of Prey's work, and that of Chandrasekhar (1952), who had studied the effect of growing core convection patterns. Urey offered this brief history of events.

Convection occurred during a first epoch of the earth's existence in a single cell with more dense parts moving to the interior. The less dense surface areas floated to one hemisphere and formed one continent. This convection proceeded for about one cycle because the higher density material remained at the center of the earth. The large gravitational energy was converted to heat and raised the temperature throughout the earth to values approaching those of the present time. The iron-nickel and iron sulfide melted and began sinking through the silicates to form a consolidated core. Such separation produced high temperatures and chemical inhomogeneities which initiated new convection, which now, since a core had formed, took place in two cells and partially broke up the single continent to form the present continental masses. Following this, other convections added their effects to the spherical harmonics of the earth's surface ... These convections are continuing at the present time and constitute the principal machines for mountain building and continent maintenance.

*(Urey, 1953: 943)*

Although Urey accepted the formation of a primordial continent during the first convective cycle, and its later breakup during the second cycle, he, like Vening Meinesz, was not advocating Wegener-style continental drift; his proposed breakup also occurred much earlier in Earth's history. Subsequent cycles caused only mountain building and "continent maintenance," which, as Griggs also argued, were needed to rebuild continents; rebuilding was achieved by sweeping back onto continental margins the seafloor sediment that had been eroded from continents (I, §5.10).

Just as Urey adopted the idea of changing convective patterns brought about by a growing core to explain early Earth history, Runcorn now adapted them to explain late drift as well. He did this by proposing a model for the core's rate of growth that would yield appropriately sized convection cells. From the Early Precambrian to the Late Paleozoic, just before the breakup of the single supercontinent Pangea, the ratio of the radii of the core to the whole Earth would best match the harmonic $n = 3$. Continued core expansion excited higher harmonics, $n = 4$ and $n = 5$, and the ensuing change in the convective pattern broke up Pangea into the present continents and moved them to their current positions (Runcorn, 1962a: 313).

Irving privately criticized Runcorn, recalling that he raised the following difficulty with his former supervisor.

Runcorn did not take into account certain paleomagnetic data and arguments based on them. For him all major motions of continents occurred late in Earth history, all of them post-Paleozoic. This was in flat contradiction to the arguments of Irving and Green (1958) that relative motions of continental blocks also likely occurred much earlier in the Paleozoic and Precambrian because the paleomagnetically determined APW paths for Europe, North American and Australia for these times have very different forms. Runcorn referenced the Irving and Green paper and had himself been involved in obtaining data from these Paleozoic and Precambrian rocks.

*(Irving, October 2008 note to author)*

Irving added that "Keith may have been so pleased with what he had just done, that he just forgot!"

Pursuing his hypothesis in the anthology *Continental Drift* (§1.6), Runcorn attempted to account for mid-ocean ridges, offering an explanation of the large magnetic anomaly above the central rift valley of the Mid-Atlantic Ridge. He found that the present ridge pattern matched $n = 5$, and suggested that they mark upwelling convection that caused continents to break up. He had to modify his original idea that the continents had been brought to their former position (Pangea) by a convective system of degree $n = 3$ to $n = 4$.

The fact of the oceanic ridges suggests that there are lines in the ocean floor where there are concentrations of lighter material, which may represent sialic material collecting near the upwelling convection currents. Oceanographers have recently found evidence that the Mid-Atlantic ridge has a central valley over which there is a strong magnetic anomaly. This suggests that the ocean floor is parting and new basaltic material is rising into the rift and on cooling acquires a strong thermo-remanent magnetization giving rise to the observed magnetic anomaly. The pattern of these ridges ... suggests convection of degree $n = 5$ ... Such a continental distribution would be characterized by the $n = 3$ or 4 harmonics being much more important relative to the $n = 5$ harmonic than at the present day. One may therefore conclude that before the continental drift deduced in the geological record took place the continents were positioned essentially by a convection pattern of the $n = 4$ type, and this gave way in the last 200 million years to convection of $n = 5$ type.

*(Runcorn, 1962b: 34–35)*

Although it might appear that Runcorn was adopting or grafting seafloor spreading onto his convection hypothesis, he maintained (1984 interview with author) that he did not at the time know about Hess's idea of seafloor spreading. A letter from Hess to Runcorn dated February 1, 1962, also suggests that Hess had not sent him a preprint of his paper.

I did have a review of your current ideas secondhand from Girdler. I was fascinated by them particularly the concept of increasing number of convection cells with a growing core. Starting from another set of premises or prejudices, I would arrive at partly different conclusions, but this is what makes problems so interesting. We could have had a splendid time debating the pros and cons of conflicting systems.

But Runcorn surely knew of Dietz's version of seafloor spreading because he had already asked Dietz to contribute a chapter to the forthcoming *Continental Drift*. Nevertheless, he did not seem to have Dietz in mind. Twenty years later, when I asked Runcorn if he had been influenced specifically by the idea of seafloor spreading, he denied it.

I don't think it [seafloor spreading] did. Because, you know, it takes time to digest new ideas and I obviously wrote this over a period of time. I'm almost certain that I had the idea and wrote it in the summer of 1961. Presumably after finishing some field work in Arizona, I went to La Jolla and sat on the beach for a few days, and I remember this idea of changing convection cells came to me there, and I remember meeting Menard and telling him about it. By then Menard was very much interested in convection coming up under the ridges. I would have submitted it after the summer of '61.

*(Runcorn, 1984 interview with author; my bracketed addition)*

Runcorn's discussion of what transpires at ridges lacks sufficient detail to categorize it as the same as seafloor spreading or some other geological hypothesis. He had upwelling basalt, a parting of the ocean floor along the rift valley, and powering of continental drift by convection. But this isn't enough for seafloor spreading. There is no mention of the new basalt moving outward along the back of convection currents.

Runcorn's would have much in common with others' accounts of the origin of mid-ocean ridges that did not involve seafloor spreading; for example, Menard and Girdler invoked convection in their oceanic ridge models. Both were in contact with Runcorn, because he (1963: 630) thanked them "for many discussions on the significance of the ocean ridges."

Nor was Runcorn foreshadowing Vine and Matthews. Even though he discussed the central anomaly over the Mid-Atlantic Ridge, attributing its origin to thermo-remanent magnetization, he made no appeal to geomagnetic field reversals. Runcorn did not concern himself with the pattern of high and low magnetic anomalies in the northeast Pacific offshore from the United States and Canada, and, at the time, there were no published zebra-pattern surveys over areas known to contain ridges. In answer to my question about this extension of his mantle convection hypothesis to marine geology, he refreshingly suggested that he felt somewhat ill at ease with it because he was working outside his field.

... although I knew that evidence for the ocean floor was relevant I was, you know, not as at home with the ocean floor evidence as other people. You know if you're writing outside of your field you don't necessarily write with the confidence that a person within the field would. I think in discussing the magnetization, of course, I was concentrating attention on the fact that these highs over the ridge were known and had to have some explanation in terms of the thermo-remanent process.

*(Runcorn, 1984 interview with author)*

But, even though Runcorn was not proposing the hypothesis that Vine and Matthews were to suggest a year later, he was one of the first of two to suppose that the central anomaly of oceanic ridges was caused by remanent rather than induced magnetization – Ron Girdler was the other (IV, §2.4).

## 1.6 The 1962 anthology *Continental Drift* and MacDonald's review of it

This anthology appeared fifty years after Wegener announced this theory. I shall interweave accounts of the essays in it with MacDonald's review of them, and in order to capture at the outset something of the mood of the times, I shall begin with MacDonald's opening statement: as far as general nastiness goes, it rivals anything found in the proceedings of the 1928 American Association of Petroleum Geologists (AAPG) symposium. Here are his opening salvos.

The hypothesis of continental drift requires that relative positions of permanent or semipermanent continents have changed during geologic time. There are two parts of the story. The relative positions in the past must first be inferred by using the methods of paleomagnetism and paleoclimatology. Then a mechanism by which the continents pull apart and sail over the globe must be imagined.

Continental drift has many appealing features: it is, for example, a favorite topic of pundits condescending to the lay public; it is a grandiose theory involving vast changes in the familiar face of the Earth, eminently suitable for a "Wonders of Nature" series. At the same time, it has so many degrees of freedom in its complexity that it can allow for changes in data as well as drifts in styles of thinking. Any evidence that can be adduced in support of the hypothesis must be gathered from every branch of geology and geophysics and the assumed changes in the crust influence virtually all of the properties of the Earth – its rotation, gravity field, internal thermal state, climate. Those that deal with the subject, therefore, have an appearance of universal genius which mankind has not seen since Elizabethan times.

Wegener, a meteorologist by trade, did much to popularize the story of wandering continents. His first book, published in 1915, attracted the attention of the more speculative geologists. Physicists were repelled by the morass of paleontological and paleoclimatological detail as well as the lack of a plausible physical mechanism. The last ten years have seen a change in attitude. The development of paleomagnetic methods by which the position of the magnetic pole in ancient times may be determined yields tables of numbers presentable to physicists. Today, as the editor of the volume under review readily admits, there is no plausible mechanism, yet many physicists have adopted continental drift as their own. The authors of the present volume reflect the change in attitudes; one finds physicists, geophysicists, mathematicians, oceanographers – but where are the professional geologists?

*(MacDonald, 1963a: 602–603)*

MacDonald was especially disturbed by the interest in mobilism awakened by paleomagnetists and their enthusiasm for it. Before the rise of paleomagnetism, Jeffreys had succeeded in deflating the enthusiasm of many geologists for mobilism. With the rise of paleomagnetism, MacDonald aimed to stop these "universal geniuses" from fooling geologists into believing that paleomagnetism had made mobilism respectable and preferable to fixism.

   MacDonald categorized the authors and, except for Opdyke, was more or less correct. Dietz and Heezen were oceanographers, but geologists by training. Benioff and Hodgson were seismologists. Runcorn, Vening Meinesz, Vacquier, and Gaskell were geophysicists. Chadwick, Chamalaun, and Roberts were applied mathematicians. However, although Opdyke already had retrained as a paleomagnetist, he was and remains a geologist, and his work in paleoclimatology was as a geologist using paleomagnetic data not, as MacDonald said, a geophysicist turning to geology. MacDonald, however, was right about the source of the new support for mobilism, originated as it was by paleomagnetists and by oceanographers who had by then begun to make contributions. But, if he also was implying that no key workers in either field were geologists or had been trained in the subject, he was wrong. Irving and Nairn had been trained as geologists. Dietz, Hess, Heezen, and Menard were more at home in geology than in geophysics. Girdler, Vine, and Matthews were all geology graduates. What they did so successfully was to gather and use geophysical data to solve geological problems.

MacDonald made more specific comments, distinguishing the contributions both in terms of their scientific content and enthusiasm for continental drift, which he perceived as inversely related.

The individual contributions in the collection are most uneven both in scientific content and dedication to the cause of continental drift. Vacquier presents a fascinating account of the evidence for large horizontal displacements in the floor of the Pacific. A detailed tracing of magnetic anomalies establishes without a doubt that major movements have taken place in the oceanic crust. The tracing of these faults into continental regions remains a problem. Benioff discusses the faults bounding the Pacific and combines detailed description with speculative thoughts. Hodgson covers the question of the movement along fault planes, a problem in which great progress has been stimulated by the demands of high politics.

*(MacDonald, 1963a: 602)*

Vacquier, whose work I shall examine in more detail later (§5.9; IV, §3.3, §3.8), matched the pattern of magnetic anomalies across several fracture zones in the Pacific floor off California, and argued that they demonstrated a total relative horizontal displacement of 1500 km. But, unsure how these movements related to the San Andreas Fault in California onshore, he refused to relate them to continental drift.

The significance of a new fact of observation in geology is usually discussed from the author's personal bias. This reporter is not a geologist and has at this time no all-embracing system for which lateral displacements of 1500 km in the ocean floor provide the missing keystone. It is likely that enough factual information of this kind will be added in the next ten years for starting meaningful speculations. In the meantime we can review some logical implications without attempting a final synthesis. Two possible points of view can be adopted. One postulates that the lateral displacements in the ocean floor propagate into the continent at the time of their occurrence; the other assumes that although there is some alignment of the oceanic fractures with continental features of the present day, the displacements do not propagate into the continent.

*(Vacquier, 1962: 139)*

Vacquier's restraint must have pleased MacDonald. Vacquier was right about one thing, namely, the ten years he thought it would probably take before there would be enough factual information to make meaningful speculations about the "all-embracing system for the lateral displacements." Wilson needed only three years; Morgan, McKenzie, and Parker took five. However, a proper and fairly complete explanation linking the evolution of the Pacific Ocean and onshore Californian tectonics was not made until the work of Atwater (1970), and in this respect Vacquier's prediction was about right.

Despite Vacquier's reluctance to speculate about the origin of ocean floor displacements, he did suggest that they might be explained in terms of mobilism. Although he said nothing definite about how the displacements occurred, he argued that because displacement had happened, the mechanical obstacle to them happening was removed; their discovery had freed mobilists of any prior need to demonstrate a mechanism (RS1) – how very refreshing.

The hypothesis of convection currents in the mantle has been advanced without proof to permit continents to move through the more rigid material of the ocean floor. Now the measurement of displacements in the floor of the ocean has demonstrated the existence of a kind of mobility of the crust that disposes of the objection that the strength of the oceanic crust prevents the continents from moving.

*(Vacquier, 1962: 144)*

His move recalls Galileo's use of his discovery of Jupiter's moons to counter the theoretical difficulty raised at the time against the possibility of Earth revolving around the Sun while retaining its moon. Galileo argued that even though he did not know how it happened, the moons of Jupiter showed that bodies could revolve around another moving body. Vacquier was silent on paleomagnetic support for mobilism.

Neither of the two seismologists, Hodgson and Benioff, leaned toward mobilism. Neither mentioned its paleomagnetic support. Hodgson was one of the pioneers of deriving fault plane solutions using first motion studies, which had received a boost, as MacDonald pointed out, "by the demands of high politics." This arose from the need to distinguish between natural earthquakes and underground nuclear tests in order to monitor observances of nuclear test ban treaties. The initial shock wave produced by underground detonation of bombs is a push directed radially away from the explosion; in a natural earthquake both pushes and pulls are produced. In concluding his essay, Hodgson (1962: 99) claimed, "It will be apparent from the foregoing that seismologists are not in a position to supply any final, hard-and-fast rules by which to judge a theory of continental drift." Little did he suspect that within five years first motion studies and the seismic slip vectors derived from them would become a cornerstone of plate tectonics (IV, §7.14).

In the 1962 anthology, Benioff discussed movements along circum-Pacific marginal faults that caused shallow earthquakes. He argued that transcurrent faults are pre-dominately dextral, and that with the exception of the San Andreas Fault they intersect the surface under the ocean roughly 50 km from the coasts. This led him to speculate that the Pacific Ocean basin rotates as a whole relative "to the surrounding continental mass with the marginal faults representing the contact between the two moving systems." But he (1962: 130) characterized the idea as a "simplification" and "not without difficulties." Moreover, he had nothing in mind remotely like continental drift or seafloor spreading, for in the latter case the required symmetry was lacking.

All of these departures from circular symmetry make the concept of a simple rotation of the continental mass relative to the oceanic mass difficult, if not perhaps untenable … Thus we are left with a series of separate transcurrent fault segments without a satisfactory explanation as to how they may be integrated into a single simple system.

*(Benioff, 1962: 131)*

Benioff also discussed the origin of the circum-Pacific deep and intermediate earthquakes, which are associated with island arcs and oceanic trenches. Their epicenters together with those of shallow earthquakes form seismic zones that slope downward beneath continents.

On the basis of the geometry of the marginal profiles, the writer assumes that orogenic characteristics were generated by a relative encroachment of the continental margins on the ocean basins. This may have been a result of a mass movement of the continents toward the Pacific Ocean described as continental drift. However, the writer prefers the hypothesis in which the spreading or encroachment is a result of accelerated continental growth at the margins by accretion from below either as a physical change or chemical differentiation. As the continental margin grows and so increases in volume, the upper portion is forced to override the adjacent denser oceanic mass by gravitational forces. Consequently, the contact between the two becomes a fault dipping under the continent. The oceanic trench and the parallel mountain range are expressions of this marginal faulting activity.

*(Benioff, 1962: 132–133)*

Continental accretion, through evolution into mountain belts of island arcs marginal to continents, was an old and still respectable idea, and he preferred it to continental drift. Apparently, the paleomagnetic support for mobilism was insufficient for him to think otherwise, but he at least did mention continental drift.

Turning to those in favor of mobilism, MacDonald singled out Opdyke and Runcorn for their biased selection of evidence.

The enthusiasm for continental drift has led several of the contributors to a careful selection of data. A few examples may suffice. Opdyke discusses the paleoclimatological evidence for continental drift. There is no mention of the detailed studies of Stehli [1957] of Permian brachiopods and fusulinids which show a latitudinal distribution consistent with the position of the present pole and the present distribution of continents. Similarly, the work of Durham and Axelrod is passed over. Runcorn reviews the paleomagnetic evidence, but in his discussion there is no reference to the scholarly and unbiased investigation of Cox and Doell [1960].

*(MacDonald, 1963a: 602)*

This was an old tactic used by both Wegener and his critics; they selected studies favorable to their own position but omitted those opposed (RS2). However, MacDonald, like Munk and MacDonald in their 1960 *The Rotation of the Earth*, neglected to note that Runcorn (1959b) already had replied to Stehli, and that Blackett (1961) had given reasons for preferring Runcorn's analysis to Stehli's. The readiness with which fixists embraced the 1960 *GSA* review is again evident.

MacDonald also discussed some of the contributors' essays on mantle convection, warning readers that all was not as it may seem.

Several of the contributors appeal to thermal convection in the mantle as the mechanism for continental drift. Vening Meinesz reviews his earlier work on thermal convection. The problem of the onset of thermal convection in a viscous, impressible fluid possessed of zero strength provides a neat stability problem, and the mathematical details are reproduced by

Chamalaun and Roberts. There is a danger that the unwary will take this kind of analysis as a firm basis for a theory of thermal convection. Nothing could be further from the truth. Even if the unrealistic physical model is accepted, the treatment considers only the onset of instability. The numerical parameters used by Chamalaun and Roberts guarantee *a priori* a turbulent convection, so that the destruction by nonlinearities of the symmetric flow pattern eliminates the very purpose of the convective hypothesis, that of providing an organized dragging force on the base of the crust.

*(MacDonald, 1963a: 602–603)*

T. Chamalaun and P. H. Roberts (1962: 177) were sympathetic toward mobilism. They thought that paleomagnetism had "lent impressive support to Wegener's hypothesis of continental drift," and cited Collinson and Runcorn (1960), Irving and Green (1957b), and several papers by Runcorn before beginning their mathematical treatment of convection. MacDonald's comments had a familiar ring, for Jeffreys had raised similar difficulties with Holmes' convection model, admitting it was possible but "more like the nature of a fluke." Chamalaun and Roberts' model, which could provide currents of sufficient size and stability to move the continents on *an* Earth of some sort, did not, MacDonald argued, apply to *our* Earth.

MacDonald said nothing about the contributions by Dietz, Heezen, Chadwick, or Georgi. Georgi spoke of Wegener the man. Dietz wrote about seafloor spreading, Heezen argued in favor of Earth expansion, and Chadwick compared competing theories of mountain building. Here I consider only their stances on the paleomagnetic case for mobilism, and delay examination of the remainder of their contributions (§4.9, §6.9).

Chadwick (1962: 212) was impressed with the paleomagnetic support for polar wandering but ignored its support for drift. He cited the *GSA* review (1960) and Runcorn (1956a), the latter before converting to continental drift. Heezen said nothing about the paleomagnetic support for mobilism, only commenting disbelievingly on Cox and Doell's paleomagnetic study critical of Earth expansion.

Cox and Doell (1961), comparing paleomagnetic results from Europe and Siberia concluded that the Permian radius of the earth was nearly identical with the present radius. But similar calculations based on data from other parts of Eurasia would clearly give radically different values.

*(Heezen, 1962: 285; Cox and Doell (1961) is my Cox and Doell (1961a))*

Dietz, the mobilist, welcomed paleomagnetic support, claiming (1962a: 294–295), "Former scepticism about continental drift is rapidly vanishing, especially as a result of palaeomagnetic findings and new tectonic analyses," presumably having in mind his own version of seafloor spreading (§4.9).

MacDonald closed his attack on the Wegener fiftieth anniversary anthology by repeating a difficulty that he and Munk (1960b) had already raised, namely that very new gravity measurements using artificial satellites reinforced Jeffreys' earlier argument that the mantle was of sufficient strength to prohibit convection; none of the authors who favored mantle convection mentioned them and he wanted to let readers know:

A further and surprising omission is the lack of any discussion of the gravitational potential of the Earth as determined by Earth satellites. These data make it clear that the Earth's mantle is supporting large stress differences, confirming Jeffreys' earlier inferences from limited terrestrial data. The strength of the mantle led Jeffreys to dismiss continental drift. The editor and several of the contributors prefer to ignore the data.

*(MacDonald, 1963a: 603)*

As it happened, Runcorn was in the process of working through the satellite data, a story that relates more to the mid-1960s than to the early 1960s, and I shall return to it in the next chapter (§2.7).

## 1.7 Blackett turns to paleoclimatology

Blackett (1961) undertook a comparison of paleoclimatic and paleomagnetic data.

A systematic comparison has been made of the ancient magnetic latitudes of Europe, North America, India, Australia and South Africa with the evidence of ancient climates as deduced from geological data, in particular from the distribution of salt, glaciations and fossil corals. In spite of some discrepancies, the general agreement is close enough to lend support to the assumption that the ancient magnetic latitudes, calculated on the hypothesis of an axial dipole field, do represent also the ancient geographical latitudes. This support for the reliability of the magnetic data as a whole gives support for the hypothesis of continental drift and is opposed to the hypothesis that the earth's ancient field differed greatly from that of a dipole.

*(Blackett, 1961: 1)*

Such comparisons had been initiated by Irving and pursued by others (II, §3.12, §4.2, §5.10–§5.14). He thought his account more systematic than that of others, and in one particular he was right; it was certainly different and it authoritatively and very effectively spread the word.

Blackett made two approaches. First, he compiled the indicators used to identify past climate – fossil corals, evaporites, red beds, Late Paleozoic glaciation, and Stehli's Permian marine invertebrates. He then compared them with paleomagnetically determined latitudes for North and South America, Europe, India, Africa, Antarctica, and Australia, the landmasses from which the paleomagnetic data had come; as others had done, he compared the paleoclimatic evidence itself directly against paleomagnetically determined latitudes. Although he admitted that his results were tentative, and recognized the incompleteness of fossil records and other difficulties, he argued that for each landmass there was significant agreement between them. Second, he developed a model of climate zones based on the present Earth and, placing these same climatic indicators in their appropriate zone, estimated the past latitudes and the variations through time of the above continental blocks. He

then compared the climatically and paleomagnetically determined latitudinal variations, one set of latitude determinations against the other: this was something others
had not done. He found that both approaches gave agreement between findings from
paleoclimatology and paleomagnetism.

Blackett did not campaign for outright acceptance of mobilism. Instead, he
characterized his conclusions as "tentative."

So far as their limited precision allows, there is on the whole rather good agreement
between the measurements of ancient latitudes, as deduced from the rock magnetic data
on the assumption of an axial dipole field, and the evidence for the ancient climatic zones,
though some possible discrepancies have been noted. However, more precise and extensive
magnetic measurements are very desirable as are also more detailed studies of the ancient
climates of the main land masses: until these are available, all the general conclusions
reached here must be considered as tentative. The possible gross distortion of the distribution of ancient corals and salt, due to the distribution of rock types must be remembered.
Though legitimate doubt may be entertained about any one piece of climatic evidence, the
rough agreement of so many with the magnetic data does make it rather unlikely that the
agreement is purely fortuitous.

(Blackett, 1961: 24–25)

Adding a new twist, he claimed, categorically and somewhat surprisingly, that
although the consilience demonstrated the reliability of the paleomagnetically measured inclination, it did not for declination.

Unless one supposes, rather improbably, that the agreement between the climatic and
magnetic latitudes is accidental, then one can use the climatic data to give support to the
assumption that the ancient magnetic latitudes, as calculated from the magnetic inclination $I$,
give us the ancient geographical latitudes. Though no similar confirmation by geological
means of the reliability of the magnetic rotation $\psi$ as indicating a real rotation of a land
mass, is yet certainly available, it can rather safely be assumed that it does: for $\psi$ is less likely
to have been subject to errors due to physical processes than the inclination $I$. So one
concludes that the climatic data to a considerable extent demonstrate the reliability of the
magnetic data as a whole, and so makes highly probable that the continents have in fact
drifted relative to each other as well as relative to the geographical axis. For the alternative
possibility of polar wandering (that is, that the crust as a whole has moved relative to the
axis) is ruled out by the magnetic data, since the pole positions calculated for contemporary
ancient rocks in different land masses do not agree.

(Blackett, 1961: 25)

It is surprising because paleowind studies (II, §5.10–§5.14) had, in some instances,
confirmed that the declination of the time-averaged field was a good estimate of the
paleogeographic meridian in the sampled region (II, Figure 5.14). Blackett's not
mentioning it may reflect the uneasiness with paleowind results that he had expressed
at the 1957 meeting of the Royal Astronomical Society (RAS) (II, §5.13). Finally, he
argued that the agreement between the paleomagnetic and paleoclimatic evidence
increased the reliability of the paleoclimatic results.

Conversely, the agreement between the two sets of data can be held to give support to the rough reliability of the climatic data. In particular, the fact that the latitude widths of both the ancient coral belt and the ancient northern salt belt appear to have always been about the same as today suggests that the mean temperature and the climatic zoning of the earth's surface up to about 60° of latitude have also been the same.

*(Blackett, 1961: 25)*

He believed that the consilience of the two independent sets of latitude indicators not only enhanced the reliability of the paleomagnetic data and helped substantiate the assumption that the geomagnetic field has always been axial and dipolar, it also enhanced the reliability of the paleoclimatic evidence and endorsed the assumption that planetary climatic zones have during the Phanerozoic remained primarily controlled by latitude. But he went too far when he said that this consilience indicated that climate zoning "up to about 60° of latitude" had "been also the same" as at present; this was not supported by the paleoclimatic evidence (Brooks, 1949) and the paleomagnetic evidence at the time was insufficient to have said so. However, within a few years, Briden and Irving (1964) and Irving and Brown (1964) showed that climate zones, although primarily latitudinally controlled, had, in the past, varied substantially in width above about 40° latitude (§1.18). Nonetheless, Blackett made a strong case that moving continents were much to be preferred to the gross distortion of climatic zones that fixists required.

## 1.8 Deutsch proposes continental drift without polar wandering

Creer, Irving, and Runcorn recognized that the paleomagnetic results could be interpreted solely in terms of continental drift, but they chose to attribute the apparent similarity in shapes of APW paths to polar wandering, not that it was necessarily so, but that it was possibly so. Blackett, Clegg, Almond, Stubbs, and others associated with the group at Imperial College wrote in terms of continental drift but remained uncommitted as to whether strict polar wandering had also occurred. Deutsch moved away from them, interpreting paleomagnetic data in terms of continental drift only. In fact, Deutsch seems to have been skeptical of the evidence for polar wandering as opposed to continental drift as far back as 1954, soon after he had joined the Imperial College group. After hearing Runcorn's presentation at the Rome International Union of Geodesy and Geophysics (IUGG) meeting in 1954 in favor of polar wandering based on the findings of Creer, Hospers, and Irving, he commented about a forthcoming paper (Clegg *et al.*, 1954a) from his group, and voiced his skepticism of polar wandering.

I wonder whether the occurrence of obliquely directed magnetizations vectors, observed only over a small area of Great Britain, is alone sufficient to support the theory of polar wandering. In this connection, I would like to mention some of the results obtained by Prof. Blackett's group at Imperial College . . . It is suggested in this paper that the whole land mass which now constitutes England has rotated clockwise through 34° relative to the earth's geographic axis. Finally, if pole wandering has occurred, as Dr. Runcorn suggests, does this not imply that at the times concerned the rocks outside the British Isles should also have been magnetized in directions differing considerably from that of the present geomagnetic field? Graham, on the other hand, has suggested that the direction of the magnetic azimuths of rocks from certain parts of the United States has remained roughly northward for 200 million years. This seems to me to contradict the polar wandering theory.

(Anonymous, 1957: 22)

Graham was wrong, the northward trend of directions was due to viscous remanent magnetization (VRM) overprinting (II, §1.9), but that is beside the point. Deutsch said nothing in print about his skepticism until early 1963, long after he had left Imperial College. However, he must have continued thinking about it all along because before March 17, 1959, he wrote to Irving (a letter which Irving did not keep), who replied as follows:

You may be right that continental drift is the whole story, and my bias towards there being a "common" effect due to polar wandering may be a built in prejudice installed at the time we produced the first estimate of the European curve. However I am always amazed that the paths back to the Carboniferous all have a similar form – that is the ones for Europe, N. America, Australia and India, like the spokes of a wheel. The rate of polar movement in the first three is the same, but in India seems to be about twice as great. Nevertheless the form is the same. I prefer to attribute this similarity to polar wandering which is the simpler hypothesis. Of course you are perfectly free to say all is due to continental drift, but in this case a certain component of the drift of each continent will, unless I am very much mistaken, be the same relative to all continents, the remaining proportion will be special effects. I may be wrong here and I confess that I have not tried all possibilities (there are too many). However I would very much like to see alternative reconstructions. It is important to separate these two effects if they exist because we shall need to know them when a mechanism is sought for.[10]

(Irving, March 17, 1959 letter to Deutsch)

Deutsch's response came in two papers, one (1963b) with the evocative title, "Discussion: polar wandering – a phantom event?" He agreed with Irving's closing comment about the need to separate the two effects. He began with the familiar, namely, that paleomagnetic data could be explained solely in terms of continental drift, or by drift and polar wandering together, and that they cannot be explained without invoking continental drift, unless one hypothesizes exceedingly rapid polar wandering occurring not just as a special event or as a series of special events, but throughout much of the Phanerozoic. He himself found rapid polar wandering unpalatable, remarking that most paleomagnetists invoked continental drift, but singled out Cox and Doell as exceptions who favored rapid polar wandering. Deutsch did not mention Graham.

There still exists no conclusive evidence which would enable one to reject with finality one or the other mechanism, for the Palaeozoic and Mesozoic eras at least, but it appears that ... rapid polar wandering, would demand coincidence to such a high degree as to nearly forfeit its claim as a serious alternative. Therefore, the second choice, involving some kind of continental drift, will be adopted here, though admittedly such acceptance still rests upon a judgment of probability. Most workers in palaeomagnetism have reached the same conclusion, though Cox and Doell (1960) after a lengthy analysis of that subject, arrive at an interpretation partly at variance with this view.

*(Deutsch, 1963b: 11)*

Deutsch next distinguished between polar wandering as a shift in Earth's axis of rotation with respect to the whole Earth, or a shift of the entire Earth's crust with respect to its interior. He then raised an internal theoretical difficulty with polar wandering in the latter sense (RS2).

Concurrency of such a movement [i.e., a shift of the Earth's axis with respect to the entire crust] with continental drift is clearly a contradiction, in terms, since the phrase "entire crust" in the definition becomes meaningless whilst the crust suffers disruption.

*(Deutsch, 1963a: 196; my bracketed addition)*

He then argued that polar wandering was superfluous, the paleomagnetic data could be solely and better explained in terms of continental drift only. Deutsch, in an attempt to downplay Irving's point about the similarity in shape of APW curves, claimed correctly that APW paths are probably not similar with regard to their rates.

Further, the resemblance of the polar curves, whilst undoubtedly a striking feature, is perhaps not as significant as it may seem, for the similarity applies to *shape*, but probably not to the *rates of motion*, which the curves represent.

*(Deutsch, 1963b: 14–15)*

Irving thought that rates of motion (relative to the geographic pole) were about the same for Europe, North America, and Australia, but admitted that the rate of motion for India was about twice as fast, and invoked latitudinal change of India relative to the other landmasses. Deutsch correctly countered by appealing to simplicity.

This apart, I find it difficult to justify preference for complex motions when a simpler reconstruction on the basis of continental drift alone might do as well. It would be quite straightforward, for example, to compare the European and Indian curves in terms of predominately northward drift on the part of both land masses, with rotation also; the alternative quoted [polar wandering and continental drift] requires continental drift of great magnitude as the primary east-west component, together with secondary motions, plus polar wandering.[11]

*(Deutsch, 1963b: 15; my bracketed addition)*

Deutsch then presented an analysis of the paleomagnetic data either solely in terms of continental drift or in terms of drift plus very minor polar wandering. Here he was guided by Lester King's reconstruction of the continents as presented at the 1956 Hobart symposium (II, §6.16).

Deutsch also discussed mechanisms. He acknowledged the feasibility of Gold's (1955) idea that polar movement could result from changes in mass distribution in Earth's crust, and admitted (1963b: 8), "In theory, polar wandering appears more likely to come about than does continental drift." In an attempt to diminish the apparently greater theoretical likelihood of polar wandering, Deutsch suggested that polar wandering might be restricted by the variability in the redistribution of land-masses during continental drift. He appealed to mantle convection as the cause of drift, mentioning not only Holmes' and Runcorn's work, but also Hess's and Dietz's seafloor spreading.

> Perhaps the most promising approach comes from the hypothesis of subcrustal convection currents, as advanced by Holmes ... and since developed by several authors. Runcorn ... suggests that continental drift may have been initiated by a change in the number (and hence pattern) of convection cells in the mantle, this resulting from expansion of the Earth's iron core ... New discoveries in oceanography have led to the formation of a "spreading sea floor" concept (H. H. Hess ... Dietz ...), according to which the sea bottom, being essentially "outcropping mantle" rather than "oceanic crust," is strongly coupled with overturn of the convection currents; these would rise under the mid-ocean ridges and descend below continents. Major obstacles against continental drift would thus be removed, for the sialic blocks no longer need execute the impossible task of floating through the sima; rather they would move "along with" it, or else remain balanced while the sima, transported by opposing convection currents, sheared underneath them. So far results of thermal, seismic and magnetic studies of the ocean floors ... appear to be largely consistent with the concepts of Dietz and Hess.
>
> *(Deutsch, 1963b: 6–7)*

Deutsch had been reading papers on marine geology, prompted by his evident desire to understand how drift happened. The importance of Deutsch's prescient thoughts was not sufficiently recognized at the time or since.

### 1.9  The 1963 Newcastle NATO conference

With Runcorn's encouragement, Nairn had already edited *Descriptive Palaeoclimatology*. Neither, however, were finished. Runcorn wanted Earth scientists generally and paleoclimatologists and paleontologists in particular to get together and to reexamine the question of mobilism. With this objective in mind he became the driving force in bringing about a meeting in Newcastle upon Tyne, which was much larger than that in 1959. As early as 1960, Runcorn had talked with Bucher about organizing such a meeting. His later choice of Bucher as a lead speaker for the conference was a good one; no one could argue that the speakers list was biased toward mobilism. Despite his qualms about paleomagnetism, and his long-standing anti-mobilism, Bucher recognized the importance of the new field (II, §3.8, §4.2). He and Runcorn were friends, and he had already helped

Runcorn promote the new paleomagnetic work (II, §2.13). Runcorn recalled what happened and his early failed attempts to raise funds for it.

I had a lot to do with Walter Bucher, and he was again very keen on paleomagnetism without being convinced about continental drift. You see, he was a person who thought that polar wandering was entirely acceptable and understandable, and he thought that the data were obviously very good on the basis of the European and North American data. Now, Bucher and I had discussed, in particular, the comparisons with paleoclimates. He, of course, very quickly agreed that it was very important. But he cautioned me and said the data on paleoclimates, in particular the glaciations in the Southern Hemisphere, were obtained a long time ago. We know quite a lot more about sedimentary processes. And, he always said you may well find that the glaciations in Australia, and Africa, and India were produced by turbidity currents, which had just been [further investigated] at Lamont [Geological Observatory]. And, on those grounds Bucher and I agreed to run a joint conference, which turned out eventually to be the 1963 NATO meeting in Newcastle. Of course, he came over to it. In fact, we had an honorary degree ceremony and we gave Ewing and Bucher honorary degrees – and the German paleontologist whose name I've just forgotten. Walter Bucher was very much behind the holding of that meeting. Initially, after we discussed it, he said, "Let's get the NSF to finance it." It wasn't then going to be a NATO meeting. He got a letter back from the NSF [National Science Foundation] – Walter Bucher was, of course, a very distinguished geologist – but he got a letter back saying that they were not in favor of discussing this matter. So he then had the idea of writing to Tuzo Wilson, who was President of the International Union of Geophysics, and ask him whether they would sponsor it. Tuzo Wilson was very interested in a whole variety of geophysical questions but he did balk at this one, and he wrote back a letter to Walter Bucher – maybe it's in Walter Bucher's archives. Tuzo Wilson wrote back saying that he didn't see any advantage in having a meeting to discuss paleoclimates in relation to continental drift as continental drift was an entirely discredited theory. It must have been around 1960 because we held our NATO meeting in January of 1963. It was very soon after the negative reply from Tuzo Wilson that I heard about the NATO scheme, and I said to Walter Bucher that we would organize it. The time scale for doing this is that you put in an application, and if it is accepted you have to do it within about two years. Because they give you a grant to be used in the following year, we must have had the NATO committee agree to it in '62, and I must have [applied] in '61.[12]

*(Runcorn, 1984 interview with author; my bracketed additions)*

Runcorn certainly knew how to get key people involved – dangle an honorary degree in front of them as an inducement to support the conference.[13] Runcorn (1964c: v) talked with Bucher about the arrangements at the Helsinki Assembly of the IUGG (July 1960). Bucher and he also talked to W. E. Benson of the NSF, which declined to fund the conference, and to J. R. Balsley, an influential and ardent fixist then at the United States Geological Survey (USGS) (II, §7.4, §8.3).

The NATO conference was finally held in January 1963 at Newcastle. Seventy Earth scientists presented papers or participated in the many reported discussions. Nairn edited (1964a) and contributed to the proceedings (1964b).

With anti-mobilism forces in the ascendant in most places, it is a tribute to Runcorn's persistence and persuasiveness that it got off the ground at all. The proceedings include fifty-four papers. Many who spoke had already expressed their opinion about mobilism or would later play significant roles within the controversy. Six of the participants in the "Descriptive Palaeoclimatology" conference attended – Craig, Kraüsel, Lamb, Nairn, Schwarzbach, and Van Houten – all gave papers. Bucher, Fischer, and Irving, three reviewers from the earlier conference, were authors. Creer, Girdler, Nairn, and Runcorn presented paleomagnetic/paleoclimatic papers; James (Jim) Briden and Irving, from Australia, authored a paper but could not attend. Three geologists from South America, Maack, Bigarella, and R. Salamuni, gave papers. Maack, who had sent Irving paleomagnetic samples for analysis (II, §5.6), spoke. So did Bigarella, who had helped Creer in South America (II, §5.6). McKee, Poole, and Runcorn talked on paleowinds, and Shotton was a discussant. Ewing and Heezen (with C. Hollister) from Lamont joined Bucher and R. W. Fairbridge from the Geology Department at Columbia University. Westoll served as a host and discussant. Stehli gave two papers. Colbert and Chaney, who had previously supported fixism, spoke. Harland spoke, and arranged for the young geologist Martin J. S. Rudwick, later an historian of science, to contribute to the proceedings even though he did not attend.

Opinions expressed about mobilism or its paleomagnetic support were divided. Of the fifty-three participants who authored or co-authored papers in the proceedings, twenty-four expressed no opinion about mobilism or fixism. Of the twenty-nine who did, sixteen favored mobilism, eleven were against, and two were neutral. Twenty-nine participants did not mention paleomagnetism, twenty-four did: fifteen presented or reported paleomagnetic results, noted their agreement with paleontologic or paleoclimatic findings, or spoke favorably about them; seven were critical of paleo-magnetic results; two mentioned paleomagnetism, but expressed no or mixed opinion; and two mentioned the mobilist-fixist controversy, but expressed no preference.

Of the sixteen pro-drift papers, nine were by non-paleomagnetists. Harland and Rudwick favored mobilism (§1.11). Here is what the others had to say. The South American geologists, Bigarella and Salamuni (1964) found agreement between their paleowind study of Botucatú Sandstone in Brazil and Uruguay and Creer's South American paleomagnetic results. Maack (1964), a longstanding mobilist (I, §6.16), documented the similarities between Devonian strata of Brazil and South Africa. J. C. L. Hulley (1964), a geophysicist from Cambridge, drew a correlation among gravity anomalies, transcurrent faults, and paleomagnetically determined pole positions. Fischer (1964) reevaluated Ting Ying Ma's work on Silurian corals,[14] and concluded (1964: 615) that their distribution was incompatible with permanency, but consistent with some degree of polar wandering and continental drift. He used paleomagnetism to determine paleolatitudes, and because of the lack of coincidence among poles from different continents, had a slight preference for drift; but it was only slight, the results being of varying reliability. McKee (1964), the geologist from the USGS who had helped Opdyke and Runcorn get started on their own investigation of paleowinds, wrote very cautiously, reviewing various evidence of past arid and hot climates, and mildly supported drift. Although he stressed the need for more data, he noted that paleomagnetically determined latitudes are consistent with the distribution of ancient salt and limestone deposits. Fairbridge (1964) imagined several "paleo-geochemical" revolutions, some the result of evolution in organic metabolism. He discussed their effect on calcium carbonate deposition and paleoclimatic implications and drew some vague, highly speculative paleogeographic consequences (1964: 473, 477); when considering changes to the current positions of the continents, he appealed to paleomagnetism. E. A. Bernard (1964), a geophysicist from the Catholic University of Louvain, Belgium, explained the principles of physical climatology, emphasizing that paleogeographic reconstructions should not violate them; he made no reference to paleomagnetism. He believed in the permanence of climatic zonation, advocating moving continents instead of varying climatic zones, offered no paleogeographic reconstructions, and was the only mobilist who did not appeal to its paleomagnetic support.

The eleven participants who rejected mobilism were predominately North American, a legacy of regionalism: Bucher (USA), Erling Dorf (USA), E. S. Barghoorn (USA), Chaney (USA), Heinz A. Lowenstam (USA), J. Shirley (UK), M. Ewing (USA), Colbert (USA), Franz Lotze (West Germany), F. G. Stehli and his co-author C. E. Helsley (both USA). Four did not comment on paleomagnetism; seven did, and I consider them first. Bucher (more on him later) raised theoretical difficulties with the paleomagnetic case for mobilism (RS2). Lowenstam (1964), from the California Institute of Technology, favored polar wandering, appealing to paleomagnetism only when it supported his reconstruction, otherwise arguing against it. Employing Urey's oxygen isotope method of determining paleotemperatures, he argued that the relative positions of Australia, India, Madagascar, and North Africa relative to Europe and North America remained unchanged since the Cretaceous. He favored some degree of polar wandering. He noted that his positioning of Europe and North America was in agreement with their paleomagnetic data, but raised difficulties with the paleomagnetically determined positioning of other continents, and cited, in support, Hibberd (1962) and Cox and Doell's *GSA* review (RS2) (II, §7.9, §8.4). Shirley analyzed two Devonian marine invertebrate faunas in a fixist framework, and found them likely inconsistent with paleomagnetic results; he (1964: 257) considered the Triassic pole, determined from South African rocks by K. T. W. Graham and Hales (II, §5.4), "hardly possible." Colbert (1964: 626) presented a fixist interpretation of the distribution of fossil land-living vertebrates. Regarding paleomagnetism, he asked, "How does the distribution of fossil terrestrial vertebrates fit the data of paleomagnetism?" Without referring to any paleomagnetic sources, he argued that the present position of continents fitted the distribution of Triassic terrestrial tetrapods and their latitudinal spread was the same as or less than that determined paleomagnetically. He concluded (1964: 626), "palaeomagnetism solves no problems with relation to tetrapod faunas and temperature belts," and that his solution had a slight advantage over that based on paleomagnetism (RS3). (See §1.14 for more on Colbert.) Stehli was true to form (II, §7.10); he and C. Helsley, his co-author (Helsley and Stehli, 1964: 562) renewed the attack on paleomagnetism. According to their analysis of the distribution of Permian marine fossils, continents had not moved and Permian paleomagnetic poles were grossly inconsistent with their fixist positioning of the continents (RS2). They doubted that Earth's magnetic field had always been a geocentric axial dipole (RS2).

M. Ewing reported mainly on some of the recent studies at Lamont of deep-sea sediments and what they indicated about Pleistocene climate and glaciation, placing it outside the scope of any discussion of the paleomagnetic support for mobilism. According to Runcorn, the main reason he came to the meeting was because:

He had one person [Ericson] looking at the paleontology of sediment cores, which made him interested in paleoclimatology, and that was why he came to our 1963 meeting ...

*(Runcorn, 1984 interview with author; my bracketed addition)*

Although at the time Ewing was a staunch fixist, he did not directly attack the paleomagnetic case for mobilism, but did raise a difficulty against it (RS2). Admitting that, as presently known sediments in the Pacific and Indian oceans were only 50–100 m thick, he thought it possible that thick sedimentary deposits might yet remain to be discovered there because Lamont ships with other interests in mind had not yet crossed them in the most likely places. He also argued that the presence of several very thick sedimentary deposits in the Atlantic is incompatible with its recent opening.

Ewing (1964: 353) predicted that the distribution of deep-sea sediments "will contribute even more definitely than palaeomagnetic data to the debates about continental drift and polar wandering." Although deep-sea sediment distribution was important, especially when it became known that the sediments thickened away from ridges (§3.15) and were later found to contain important records of reversals of the field (IV, §6.4), it was the zebra pattern of magnetic anomalies over the seafloor caused by the natural remanent magnetization (NRM) (the paleomagnetism) of ocean floor basalts that recorded reversals of the geomagnetic field which turned out to be much more critical (IV, §6.2, §6.5–§6.7). Ewing's mistaken prediction soon became a casualty in a very fast-evolving debate.

How did Ewing come to object to the paleomagnetic case for mobilism? I asked Runcorn if Ewing discussed such issues at the meeting, and he responded:

Yes. He was interested, and ... the sorts of arguments that he would have produced were that paleomagnetism was a most interesting method that yielded very interesting results, but he wasn't himself willing to interpret them the way we were ... I think he would have probably said that there are a number of steps in the paleomagnetic method. You first have to assume that what you're measuring is due to the field and not to local magnetization processes. Then he would have said, "Well, even if the field were lying along the axis of rotation in the Tertiary, can you be sure of it in earlier times?" In other words, does polar wandering have something to do with the mechanics of the Earth or the peculiar behavior of the field? Undoubtedly, Ewing thought very strongly that if continental drift had occurred then one ought to see some record on the ocean floor. I remember him saying to me that the northern movement of India surely would have left a record.

*(Runcorn, 1984 interview with author)*

Runcorn's speculation is interesting and perhaps correct.

And then there were the four fixists who said nothing about the paleomagnetic support for mobilism. Three of them, Dorf, Barghoorn, and Chaney, concentrated on the distribution of plants during the Cenozoic in Europe and North America; they believed in neither polar wandering nor continental drift (see I, §3.8 for Chaney's long-held fixist beliefs). Runcorn, never one to let things go, reminded the audience that mobilists posited only small changes in the latitude of these two continents since the beginning of the Tertiary. Lotze (1964), the fourth fixist, favored polar wandering but not drift, based on his analysis of the distribution of Northern Hemisphere evaporites from the Precambrian to the present. He said nothing about paleomagnetism. Perhaps he was unfamiliar with the literature. Nevertheless, he should not be dismissed as a counter-example to our hypothesis about the widespread use of the three research strategies. Commenting, Shotton argued that Lotze's analysis would work regardless of whether the Atlantic were present or not, and Runcorn added that if the continents were positioned in accordance with the paleomagnetic data, evaporite distribution still showed a convincing zonation. So, eliminating those fixists who concentrated on the Pleistocene (Ewing) or Tertiary (Dorf, Barghoorn, and Chaney), six out of seven of the fixists who presented paleogeographic arguments attacked the paleomagnetic case for mobilism – they surely saw it as a threat.

### 1.10 Bucher continues to criticize mobilism at the NATO conference

I am singling out Bucher's contribution because he raised specific and original difficulties against the paleomagnetic support for mobilism, and because of his strong influence on Ewing and Earth scientists generally, especially in the United States. Earlier, Bucher had argued that the Permo-Carboniferous strata of Gondwana that had been described as glacial were in fact turbidites, and that paleomagnetists had overstated the case for latitudinal zonation of past climates (§1.4). At the NATO conference he broadened his attack. At bottom, he believed that paleogeographic reconstructions based on paleomagnetism were not as well founded as those based on paleoclimatic and paleontological evidence and the conflicts between them were the fault of paleomagnetism.

Describing himself (1964: 5) as "one of those who still doubts the reality of continental drift" and putting on his historical hat, he categorized the new challenge to geologists from paleomagnetism as the third confrontation between conflicting geological and geophysical evidence. The first, between the physicist Kelvin and geologists over Earth's age, was, following the discovery of radioactivity, resolved in favor of geology. The second, won by geophysicists, arose over Suess' idea of large submerged continents; Bucher claimed that this controversy was not settled until Vening Meinesz made accurate gravity measurements at sea showing that isostatically this was not possible (I, §5.6, §8.14), and Ewing and co-workers, from seismic studies at sea, had demonstrated that oceanic and continental crust are fundamentally different (§6.3; I, §5.11). Bucher gave Wegener credit for recognizing the problem, but none for leading geologists astray by introducing continental drift. The current confrontation, between paleomagnetists and geologists, Bucher compared to the former confrontation, in which geologists had been victorious and correct, and he urged that in this third confrontation they would again be the final arbiters. With breathtaking hubris, Bucher then claimed that paleomagnetism could not even determine whether the continents had moved; proof of mobilism must come from geology, and only then would paleomagnetism be shown to be an acceptable method for determining the past continental positions!

Ultimately the proof must come from the geologic and palaeontologic record. In some ways the present situation is reminiscent of Kelvin's day. Many geologists simply accept the physicist's conclusion and re-orient their thinking, some enthusiastically. Others are not so easily swayed. As historians of the geological past, they call for a systematic search in the record for observations on floras, faunas and sediments that provide unambiguous, decisive proof concerning specific palaeoclimatic factors. Each of these must then be tested in as many regions as possible. When plotted on maps, their distribution will show whether it is consistent with the present relative positions of the continents or not; whether the face of the Earth, or the Earth's magnetic field, has changed in the last 2000 My.

*(Bucher, 1964: 4)*

But presciently, Bucher went on to explain that if his historical analogy is truly correct, then physicists working in some other area would eventually settle the present confrontation between geology and paleomagnetism, just as the discovery of radioactivity by physicists and the development of the radiometric timescale had resolved the first controversy.

Then came the dramatic decade at the turn of the century, when the wholly undreamed world of atomic energy opened up in the physical laboratories. By 1905 the study of spontaneous disintegration of radioactive materials had led Rutherford ... and Boltwood ... independently to the conclusion that that lead was the stable end-product of the disintegration of radium and thorium. Following Rutherford's suggestion, Boltwood determined the uranium–lead ratio of 27 specimens of ore minerals from many countries and found them ranging in age from 92 to 500 My. None of the results were contradicted by the geological data on the relative ages of the different deposits ... Seldom in the history of science have the results of two disciplines, using wholly different methods, been thus tested against each other.

*(Bucher, 1964: 3)*

And so, ironically in view of Bucher's predilection for fixism, it turned out to be in the mobilism debate; it was the geophysical study of marine magnetic anomalies, of earthquake slip vectors, and of the application of Euler's point theorem that eventually led to Bucher's "unambiguous decisive proof" – plate tectonics, the "gift" of geophysics to geology.

Bucher claimed that no contradictions arose between geological and radiometric estimates of Earth's age, despite the fact that both changed considerably as the discussion evolved, and he gave equal importance to both the geological and radiometric timescales. But, there is more. Bucher wrote as if geologists and geophysicists spoke with unanimous voices, that geologists were united in opposition to mobilism, whereas, in fact, some paleoclimatologists, paleontologists, stratigraphers, structural geologists, and geomorphologists favored it. His sanguine belief that the practitioners of historical geology would provide an "unambiguous, decisive proof concerning specific palaeoclimatic factors" was unrealistic. Bucher wrote his piece after the conference, and although he should not be criticized for not foreseeing what was about to happen (the hypothesized world of Vine and Matthews had just appeared in *Nature* and was three years from confirmation), it still would, at the time, have been more reasonable of him, in view of what he had heard at the conference, to moderate his confident prediction that the controversy would be settled by paleontologists and paleoclimatologists. If the conference was at all typical of the state of play in the early 1960s, then he should have realized that paleoclimatology had a long way to go and would not be able to go there alone. Consider, for example, his own strictures about the need for careful work in distinguishing glacial tillites from non-glacial turbidites. He (1964: 6) noted the strong disagreement voiced at the meeting about the nature of Eocambrian (Infra-Cambrian) glaciation, whether it was global or even glacial at all. There were difficulties in distinguishing tillites from turbidites, yet with careful work it could be done. Here is what Heezen had to say on this vexing topic.

There has been considerable recent discussion of the possible confusion of tillites and turbidites. It is hard to understand how any confusion could exist, for the two types of deposits are as different as deposits could possibly be . . .

(*Heezen and Hollister, 1964: 101*)

Crowell (1964), who had earlier expressed doubt about the glacial origin of the Squantum (§1.2), repeated it in his attack on Infra-Cambrian glaciation; however, even he did not question Gondwana Permo-Carboniferous glaciation.[15] Moreover, Dott believed that evidence for Permo-Carboniferous glaciation in South Africa, India, Australia, and southeastern Brazil was good (§1.2). Coleman, who had seen the outcrops and was a staunch fixist, also acknowledged that the Gondwana Permo-Carboniferous glaciation was real and had been widespread (I, §3.11). Although there may at times be somewhat baffling cases such as the Squantum strata, experts can, under most circumstances where outcrops are extensive, distinguish turbidites from tillites.[16]

Heezen and Hollister (1964: 107–108) also noted that there had been ten times fewer turbidity currents since the Pleistocene glaciation ended, indicating a strong correlation between them. Turbidity currents require an abundant source of easily eroded sediment, and consequently are commonly associated with mountains and/or unconsolidated freshly exposed glacial deposits.

Nevertheless, there is a certain imperfect relationship between turbidity-current deposition in the deep sea and glaciation, just as there must be a similarly imperfect relationship between turbidity-current deposition in a geosyncline and the stage of uplift of the adjacent land. Thus, just as the occurrence of thick fluvial deposits is not positive evidence of glaciation in an adjacent area, the occurrence of thick turbidites in deep-sea sediments is not proof of glaciation in the adjacent lands. Both extensive fluvial deposits and thick turbidites can be expected from any glaciation, and although not positive indicators such as till or glacial-moraine, their presence may suggest the possibility of glaciation in adjacent areas.

*(Heezen and Hollister, 1964: 108)*

It should therefore come as no surprise that turbidites occur among the marine glacial deposits associated with the Permo-Carboniferous glaciation in the Southern Hemisphere, and certainly should not cast doubt on the existence of that glaciation.[17]

Returning to Bucher's attack against paleomagnetism; doubtless feeling the need to justify his claim that the paleomagnetic method was less trustworthy than paleo-climatology, he raised a theoretical difficulty with the GAD hypothesis. Like Runcorn, he appealed to Urey's suggestion of an expanding core, but, unlike him, he did so to throw doubt on the GAD hypothesis.

If the Earth's core had thus been increasing in size through geologic time, with its outer part gradually melting, presumably by radioactive heating, then the energy available for convection currents, its distribution around the circumference of the core and the resulting pattern of such currents should have differed the more from the present pattern, the farther one goes back in geologic time. Since this is what the remanent magnetism suggests, the geologist feels justified in doubting the basic conviction which dominates the present thinking of geomagnetists ... This assumes that the present condition of the Earth's field has remained essentially unchanged through recorded geologic time, viz. a very weak non-dipole field superimposed on the dominant dipole field, on which it causes only minor secular variations. But this is not to be expected if formerly the Earth's core was substantially smaller and cooler.

*(Bucher, 1964: 7–8)*

Suprisingly, Bucher buttressed his attack on the dipole hypothesis by appealing to Ari Brynjolfsson's study of the behavior of the geomagnetic field during a reversal. The change in polarity appeared to have taken 1000 to 3000 years, and Brynjolfsson tracked the rapidly shifting direction of the geomagnetic field in twelve successive lava flows.

Even in its present condition, the whole dipole field is capable of swinging across the face of the Earth at times when its polarity is reversed. Brynjolfsson (1957: 247–250) measured the pole directions in a succession of weakly magnetized Tertiary lavas on Iceland ... The clockwise path [recorded by Brynjolfsson] recalls the clockwise traces of the secular variation and suggests that the time involved in the reversal may have been of similar duration, namely 1000–3000 years ... So, for a short time, the irregularly looping path in which the magnetic North Pole circles about the geographical pole [as recorded by Brynjolfsson] must have been disrupted while the reversal took place. Moreover, while the magnetization in the lavas from which these records were taken was weak, the reversing field retained enough strength to impress on all the lavas a definite measurable remanent magnetism, even when the North Pole had come to lie on the Equator. To the geologist it seems reasonable to consider the present strong dipole field as conditioned by the large volume and high temperature of the outer core. Smaller volumes and lower temperatures may have produced a more complex, dominant non-dipole pattern.[18]

*(Bucher, 1964: 8; my bracketed addition)*

Brynjolfsson would not have agreed with Bucher's very loose interpretation. He had raised no difficulties with the GAD hypothesis, in fact he defended it elsewhere in the same paper. Here, with his emphasis added, is what Brynjolfsson said:

As a result of special studies of the tertiary basalts, I will here mention the result of an investigation of an area between normally and reversely magnetized lava-groups. This result shows that *the direction of magnetization changes gradually, from reversed to normal* ... It is clearly seen that the change is gradual from the southern hemisphere, over the equator, and to the geographic north pole. The clockwise traces of the variation remind one of the present clockwise traces of the secular variation. Perhaps this similarity indicates that the reversal of the field took place during a period of 1000–3000 years. (Experimental result XII.) It is difficult to understand how such changes could be caused by some self-reversal during the cooling process or during the time passed, as we would then expect random variations in the direction of magnetism. These lavas were all weakly magnetized ... The intensity is about five times smaller than is otherwise the case in tertiary basalts.

*(Brynjolfsson, 1957: 164)*

Paleomagnetists who interpreted their results in terms of mobilism did not claim that the field never changed. They did not deny secular variation or changes in polarity. Instead, they maintained that regardless of polarity the *time-averaged* geomagnetic field was dipolar, and symmetrical about the axis of rotation: reversals were swift and the amount of time taken to reverse was small.

Bucher closed his discussion of the GAD hypothesis with an appeal to Blackett, Clegg, and Stubbs (1960).

Smaller volumes [referring to Earth's core] and lower temperatures may have produced a more complex, dominant non-dipole pattern. In a recent review of rock magnetic data, Blackett and co-workers point out that the possibility of such a transformation cannot be excluded at present on (magnetic) "phenomenological grounds alone."

*(Bucher, 1964: 8; my bracketed addition)*

Actually, Blackett and company said no such thing, they simply acknowledged the possibility of a non-dipole field and that such a possibility cannot be excluded on phenomenological grounds alone.

In general then, one can conclude that it is not at present possible to exclude, on phenomenological grounds alone, the possibility that the rock magnetic data now available might be explained by the two assumptions of a time-dependent non-dipole field together with the fixity of the continents.

*(Blackett, Clegg, and Stubbs, 1960: 308)*

Bucher ended his paper by lamely admitting that paleomagnetism was outside his area of expertise (it had not stopped him from raising many alleged difficulties) and recommended that judgment be withheld until resolution is reached.

These speculations concerning matters outside the writer's competence are here mentioned only to make clear that the present confrontation of opposing views challenges both sides. Both should withhold judgment until the riddle is solved.

*(Bucher, 1964: 8)*

Of course, to claim that the mean geomagnetic field before the Neogene was dipolar as well as axial *is* an assumption. No paleomagnetists ever denied it; they simply worked hard travelling across the world to provide empirical evidence for it. As noted above, they showed in studies from six continents that the time-averaged geomagnetic field during the later Cenozoic (Late Miocene or younger) was axial and dipolar.

It is natural to assume that the mean geomagnetic field is dipolar as well as axial, as this is the simplest type of field. The theory of the geomagnetic field does not provide a simple argument for this assumption which would be acceptable to the skeptic. There is strong evidence that since middle Tertiary times the mean geomagnetic field has been a dipole along the present geographical axis ...

*(Runcorn, 1962b: 9–10)*

Runcorn was also suggesting that fixists could agree with the paleomagnetic results back to middle Tertiary because they indicated (with the then available technology) that there had been no appreciable continental drift since then, but could only dismiss the pre-Tertiary paleomagnetic support in favor of mobilism by denying that the time-averaged geomagnetic field had then been dipolar and axial.

Fixists could always dismiss the paleomagnetic support for mobilism by supposing a non-dipole field. They could posit a highly irregular and changing pre-Tertiary geomagnetic field, and still maintain fixism. Irving discussed just such a move a year later, and raised a methodological difficulty (RS2). He categorized the non-dipole hypothesis as a "hypothesis of desperation," which is really "no hypothesis at all."

Finally, it should be mentioned that the [paleomagnetic] directions could be explained by assuming that the regions of observation have remained unchanged in position and that the ancient field was highly irregular and similar in form to the nondipole field nowadays, but varying perhaps a million times more slowly. It would be easy to choose arbitrarily a time-variable distribution of magnetization within the Earth that would give the observed directions. This *nondipole hypothesis*, as it is sometimes called, is a hypothesis of desperation, useful at this stage only to those anxious to avoid the implications of paleomagnetism. No conceivable observation can refute it. In fact, it really is no hypothesis at all, since it does not predict any observable geological or geophysical effects, and so cannot be checked by independent observations ... Nothing is gained by considering the nondipole hypothesis until the simple refutable hypotheses have been tested.

*(Irving, 1964: 132–133; my bracketed addition)*

In an accompanying footnote, Irving discussed Popper's criterion of demarcation, and his distinction between scientific or refutable hypotheses and pseudoscientific or irrefutable hypotheses. Irving classified the non-dipole hypothesis as non-scientific.[19]

Irving's point was not that fixists should not use RS2 to question the hypothesis that the mean geomagnetic field has been dipolar and axial; it was that in doing so they had to provide an argument to explain why the hypothesis was dubious in light of the theoretical and empirical support that he and others had secured over the previous decade; if they offered an alternative hypothesis, they needed to support it by arguments and tests. Irving wanted more than the question-begging of fixists who were so confident that continents had not moved that they felt able, by fiat, to will the pre-mid-Tertiary geomagnetic to be non-dipolar; that was not science, it was wishful thinking.

Runcorn suggested that Earth scientists simply were not ready to regard the mean geomagnetic field as a stable entity, let alone as dipolar and axial throughout its history. The prominent geologist Marland P. Billings of Harvard University was not alone (II, §7.12). In Bucher, Runcorn had found another "Billings" who thought Earth's magnetic field was as variable as the weather.

Perhaps a more general point was that as I explained when we were talking about the mine experiments: you know the Earth's magnetism was very much a closed book to a lot of people including initially, of course, ourselves. The only kind of known great work on geomagnetism was Chapman's in which all the emphasis was on what we now call the magnetosphere as the origin of all the interesting phenomena that were being investigated. So people, you know, were not ready for our presentation on paleomagnetism because that had to do with the main field. I'm sure psychologically, you see, people thought of the Earth's magnetic field as something which slipped around all over the place, which, of course, it does a bit from interactions of the solar wind. People had got a feeling that the earth's magnetic field was very variable, almost like the weather.

*(Runcorn, 1984 interview with author)*

Irving, in his recent account of Hospers' contribution, said it somewhat differently:

When Hospers began research in 1949, the geomagnetic field was depicted as highly variable over centuries. However, by adopting a stratigraphic approach and with brilliant assistance from statistician Ronald A. Fisher, professor of genetics at the University of Cambridge, Hospers was able to show how solid, long-term information could be derived from such a variable phenomenon.

*(Irving, 2008: 457)*

The notion that the geomagnetic field was so variable over the short term, yet with a remarkably stable dipolar configuration over the much longer term, was not easily grasped. And in the mid-1960s many skeptics remained.

### 1.11 Harland and Rudwick link mobilism, the Great Infra-Cambrian Ice Age and the burgeoning of Cambrian fauna

Harland graduated from the University of Cambridge in 1938 with a first in natural science and geology, and soon after took part in a small expedition to Spitsbergen.

A Quaker, he was a conscientious objector during World War II, and spent 1942 to 1946 at a Quaker mission in West China Union University, Chengdu. He returned to Cambridge in 1946, accepting a faculty position in the Department of Geology. During the 1950s, Harland was one of the few geologists at Cambridge who spoke favorably of continental drift. P. F. Friend, who began his undergraduate work at the University of Cambridge in 1954, and had Harland as his supervisor, noted that Harland "was already an enthusiast for continental drift."

He always argued in general terms, particularly using southern hemisphere, continental shape, and glaciation arguments. His main concern was that we, as students, kept an open mind. He read very widely, and I suspect that his interest in drift was generated by reading and thinking in China. The north Atlantic-Arctic disjunct was certainly not clear enough to have generated the idea in his mind.

*(Friend, January 3, 2007 email to author)*

Irving also recalled that during the late 1940s Harland told students to keep an open mind about continental drift (II, §1.16). Harland became an expert on the geology of Spitsbergen. Before retiring in 1984, he helped to organize over forty yearly expeditions to Spitsbergen, and personally led twenty-nine of them (Friend, 2004: 39). It was his work in Spitsbergen that turned him into an active mobilist. Like Bailey, Holtedahl, and Wegmann (I, §8.10–§8.13), Harland (1958) thought continental drift offered the best explanation for Caledonide disjuncts in Spitsbergen, Greenland, and Norway.

   Harland also supported another controversial idea, worldwide Infra-Cambrian (Eocambrian or latest Precambrian) glaciation, which he defended in 1956.

If it be assumed, as seems reasonable, that most of the Eo-Cambrian tillites throughout the World represent one major ice age then we have the basis of a very satisfactory datum for correlation. The observations recorded in this paper add to the cases where a marine (or at least an aqueous) tillite is followed apparently conformably by marine Lower Cambrian beds and preceded by a conformable series of limestones barren, except for algal remains, yet with lithology suitable for the preservation of fossils. We must envisage widespread refrigeration probably accompanied by fall of sea-level and worldwide peneplanation. This would be followed by a general climatic amelioration accompanied by a general transgression which together may partly account for the arrival of Lower Cambrian faunas both in transgressive deposits and in continuous marine successions.

*(Harland and Wilson, 1956)*

At the 1963 NATO meeting, Harland again defended worldwide Infra-Cambrian glaciation. If he was correct, the Paleozoic was bookended by huge glacial episodes. Infra-Cambrian strata of possible glacial origin had been discovered around the turn of the century, and their explanation has had a checkered history; he identified them from all continents except Antarctica, and argued that some were deposited near the equator, appealing for support to paleomagnetic work that he and his Ph.D. student D. E. T. Bidgood had done on strata from Norway

(Harland and Bidgood, 1959, 1961b). A little later, he reviewed the paleomagnetic evidence that supported the mobilist solution to the distribution of the Permo-Carboniferous glaciation, and concluded (1964a: 121), "a general theory of continental drift ... must now be accepted even if any special theory is premature."

Harland also arranged for inclusion of a companion paper by Rudwick in the NATO conference proceedings. Rudwick did not attend the meeting. Rudwick, a paleontologist at the Sedgwick Museum, considered the paleontological implications of Harland's worldwide Infra-Cambrian glaciation, the question of the explosion of new and relatively complex life forms at the beginning of the Cambrian. He suggested that the worldwide glaciation would have created a world unfavorable to the development of life because of adverse climates and loss of shallow seas conducive to the development of marine life. With the recession of the glaciation and milder climates, the newly flooded shallow seas would have become a new frontier for pioneering biota. Thus, the infra-Cambrian glaciation and the changes it brought about in the environment "could have 'triggered off' a phase of evolutionary change on a scale never again repeated" (Rudwick, 1964: 154). Rudwick (1964: 155) argued that his solution "is perfectly actualistic, even though it is not strictly uniformitarian," sheltering his view against a methodological difficulty. Harland also had to consider the second-order problem, namely, what would have caused such worldwide glaciation. He (1964: 124), citing S. E. Hollingworth, suggested a decrease in the level of solar radiation. "In the absence of an adequate terrestrial explanation for ice ages (cf. Hollingworth, 1962), this general cooling must be attributed to a fall in the level of solar radiation." Rudwick also noted this possibility. He referred to Harland, and non-gradualistic accounts of evolution that had been proposed to account for the Cambrian explosion of novel life forms.

This evidence of at least one world-wide glacial period in the late Precambrian time (Harland, 1963), itself produced perhaps by an extraterrestrial event, suggests that the latter hypothesis (non-gradualistic evolutionary change) may be nearer the truth than the gradualistic hypotheses which attempt to explain away the sudden appearance of the Cambrian fauna.

*(Rudwick, 1964: 154; the Harland reference is to the paper which appeared*
*in the conference proceedings)*

Rudwick said nothing about continental drift or its paleomagnetic support. He had been supportive of continental drift, however, since his undergraduate days at Cambridge, and had argued in its favor in his Presidential Address to the Sedgwick Club.

I and my undergrad contemporaries (Tony Hallam was one year junior to me, for example) were keen mobilists, but well aware that it was frowned on by most of the Geology faculty, with the well-known exception of Harland. I think I gave a paper enthusing about continental drift to the undergrad geology club (the long-established Sedgwick Club) during my final year 52/53, and it may have been my presidential address! Certainly we all knew of Holmes's advocacy in his Physical Geology book, and I know I bought myself a copy of Du Toit's book (a major financial investment for an impecunious undergraduate at that time), so I was certainly committed to thinking seriously about drift. I think I probably set out the palaeo arguments rather than anything (geo)physical.

*(Rudwick, August 28, 2005 email to author)*

Harland's paper was not well received. Harking back almost forty years, Harland (2007: 637) said that his idea of a Great Infra-Cambrian (or Eo-Cambrian) Ice Age had been "ridiculed" at the 1963 NATO conference.[20] Perhaps he had Crowell in mind, who had cavalierly on the basis of a brief study doubted the glacial origin of the Bigganjarga "tillites," while Harland (1964a: 137) had described them as "now classic tillites."

Recently I have visited briefly several localities of presumed tillites, although at no place have I undertaken extensive study. It nevertheless seems appropriate here to make a few comments concerning them ... A very profitable few hours were spent along the north shore of the Varangerfjord, and in visiting the famous Bigganjarga "tillite," commonly referred to as "Reusch's moraine." Without wanting to appear unduly skeptical, I nevertheless came away from the brief visit with the feeling that several important questions needed to be answered before glacial origin could be accepted without reservations. On the basis of lithology and outcrop relations, "Reusch's moraine" could as well have been formed by a subaqueous mud-flow or slump as by deposition from a glacier ...

*(Crowell, 1964: 94–95)*

Crowell (1964: 95) admitted that the "striae on the underlying bed" were "puzzling," but argued that the bed had probably then been soft sand and slumping rocks had marked its surface. Harland immediately responded, and even added the following note to his NATO conference paper.

A collection was made from this area in the summer of 1963 for mineralogical and sedimento-logical investigation to be reported elsewhere. It may be noted that although the striations below Reusch's tillite would appear to be the result of subaqueous sliding rather than subgla-cial abrasion, associated and more extensive horizons of tillites with good evidence of ice-rafting make it unnecessary to doubt a glacial origin for the clumped mass of Reusch's moraine.

*(Harland, 1964a: 121)*

In a fuller response, Harland (1964b) again argued that Crowell's interpretation of the Bigganjarga tillite as non-glacial was mistaken; he also expressed his strong support for mobilism and its paleomagnetic evidence, discussed his own paleomag-netic studies which suggested that Greenland and Norway had been close to the equator during the Infra-Cambrian, and further defended his idea of low latitude Infra-Cambrian glaciation. Turning his attention to Bucher, Harland charged:

Such scepticism [about Infra-Cambrian glaciation] was readily shared (e.g., W. H. Bucher, in press), partly because of these criticisms [about whether or not Infra-Cambrian tillites are glacial], and possibly also because an acceptance of evidences of glaciations had been an embarrassment, first because late Palaeozoic tillites might indicate continental drift, and then because Infra-Cambrian tillites conflict with a simple palaeomagnetic and palaeoclimatic picture.

*(Harland, 1964b: 47; Bucher reference is to his proceedings paper of the NATO conference*
*discussed in §1.10; my bracketed additions)*

Harland's first charge against Bucher is correct. Bucher attempted to weaken mobilism's advantage over fixism in explaining the distribution of Permo-Carboniferous glaciation by extending to it the skepticism that had been voiced about Infra-Cambrian glaciation. But I am less sure about Harland's inclusion of paleomagnetism in his second charge; surely the notion of low latitude Infra-Cambrian glaciation did conflict with a "simple ... paleoclimatic picture." However, at the time, Bucher had no faith in any paleomagnetic picture, simple or otherwise, engaged as he was in questioning all of paleomagnetism by attacking the GAD hypothesis (§1.10).

A year later in *Scientific American*, Harland and Rudwick (1964) revisited the question of Infra-Cambrian glaciation and the Cambrian explosion of faunal life, and continued to strongly favor continental drift and its paleomagnetic support. Rudwick thinks (but is not sure) that Harland approached *Scientific American*. He does remember that Harland thought that a paper in *Scientific American* "would help spread the idea and bypass the troublesome referees that a more specialist paper might have had" (Rudwick, December 18, 2009 email to author). Harland also invited Rudwick, the paleontologist, to be a co-author because he wanted to link the Infra-Cambrian glaciation with the burgeoning of animal life during the Cambrian.

I'm almost certain it was his idea to bring me in, as a sympathetic palaeontologist, to put the case (very briefly) for the reality of what became known as the Cambrian explosion, and thereby to get people thinking about a possible causal link with the "Infra-Cambrian Ice Age." (Until I read his posthumous paper very recently, I didn't realise, or had forgotten, that the linkage was not a new idea.) He knew that at that time I was much concerned with the analogous issue of the reality of the great Permo-Triassic extinction (which knocked out most of my favourite groups of brachiopods), on which I felt that other palaeontologists were more concerned to explain AWAY the evidence rather than to try to explain it.

*(Rudwick, December 18, 2009 email to author)*

Their paper was an amalgam of Harland (1964b) and Rudwick (1964) written in a popular style; they presented criteria for distinguishing tills formed by glaciation from those formed by rockfall, by slumping and sliding, by mass flow, or by turbidity currents, and argued in favor of a great widespread Infra-Cambrian ice age with glaciation at low latitudes, which they thought responsible for, or at least linked with, the Cambrian explosion of animal life. They also argued in favor of continental drift, and singled out the importance of its paleomagnetic support. After introducing continental drift and its solution to the distribution of Permo-Carboniferous glaciation, which they called the "Gondwanaland hypothesis" and which they characterized as "one of the strongest arguments in favor of the hypothesis that the continents have drifted across the surface of the earth," they described its paleomagnetic support as "decisive" and noted the consilience between paleomagnetic and paleoclimatic findings.

> Support for the Gondwanaland hypothesis has come from many other lines of evidence, but the paleomagnetic evidence is decisive. It indicates that the rocks associated with tillites were formed at high latitudes, that is, at latitudes near poles. Thus a reconstruction of the continental positions in Permo-Carboniferous times would place the South Pole somewhere in Gondwanaland and the North Pole in an enlarged Pacific Ocean. Paleomagnetic evidence also shows that both Europe and North America were then situated in lower latitudes. The extensive coal beds and salt deposits formed in those times in both regions confirm that the climate there was warmer.
>
> *(Harland and Rudwick, 1964: 29)*

Paleomagnetism also played a key role in their defense of Infra-Cambrian low latitude glaciation. Even though there were examples of Infra-Cambrian glaciation at places presently in low latitudes, Harland and Rudwick were mobilists, and therefore needed an example of Infra-Cambrian glaciation at low latitudes during the Precambrian to make less plausible the possibility of rapid polar wandering. Harland and Bidgood had found such examples in Greenland and Scandinavia (Harland and Bidgood, 1959, 1961a, 1961b), and Harland and Rudwick noted them.

> To check this point magnetic determinations have been made for sedimentary rocks closely associated with the tillites in Greenland and Scandinavia. These too yielded readings that pointed toward an equatorial position.
>
> *(Harland and Rudwick, 1964: 33)*

The dispersion of magnetization directions was high but reversals were said to be present so the results had some credibility. Their ideas were certainly of much interest to the general mobilism question.

## 1.12  Responses of some biogeographers to the paleomagnetic case for continental drift

During the first half of the 1960s biogeographers who were already mobilist-minded welcomed the paleomagnetic support for continental drift. Their fixist-minded counterparts ignored or dismissed it. However, there were a few fixist biogeographers who changed their minds in the light of the new paleomagnetic evidence, coming out in favor of mobilism, or at least acknowledging that it might be correct.[21]

W. G. Chaloner, a British botanist, who likely was the first biogeographer to write knowledgeably and sympathetically on paleomagnetism, argued that it had turned continental drift into a respectable hypothesis, and urged biologists to examine anew the biological support for it. Chaloner's 1959 essay appeared in the Penguin Books

Series entitled *New Biology*, which was designed to introduce new developments in biology and related areas. Fairly and clear-mindedly, Chaloner summarized the state of the mobilist controversy prior to the rise of paleomagnetism. He cited Köppen and Wegener (1924) and carefully explained du Toit's modifications of Wegener's theory (I, §6.7). Noting the absence of an acceptable mechanism as the key difficulty facing mobilism, he claimed that Holmes' mantle convection provided one (I, §5.8).

> The essential feature of this [Holmes'] mechanism is that the sial blocks are carried along on the moving sima beneath them; they do not move through the sima as Wegener suggested. The difference is that of a ship frozen into, and being carried along by, drifting sea ice, compared with an ice-breaker cutting its way through it. This suggested convection mechanism for drift removes the objection that the opposing force of the sima could never be overcome by the softer sial.
>
> *(Chaloner, 1959: 23; my bracketed addition)*

He noted the previous strong geophysical opposition to mobilism and emphasized that the new paleomagnetic findings were themselves geophysical.

> A few years ago it might have been said that the controversy of continental drift had reached something of a deadlock. The biological sciences, particularly the study of fossils, offered a number of problems which could only be explained on the assumption of drift; some more purely geological evidence also seemed to favour this explanation. But none of this evidence ... is in itself conclusive. On the other hand geophysicists were generally opposed to drift on the grounds of the lack of any satisfactory mechanism. A new development in the last decade, the study of rock magnetism, has now broken the impasse. It seems that under some circumstances the residual magnetism of rocks has preserved the direction of the earth's magnetic field as it was at the time of their formation. This study of "fossil magnetism" has revealed changes both in position of the poles with respect to the continents, and of the continents with respect to one another. This evidence is of particular interest as it comes from a field which has previously seemed to offer the strongest opposition to drift.
>
> *(Chaloner, 1959: 23)*

He then gave a concise description of the paleomagnetic method. Citing Cambridge (Creer, Irving, and Runcorn, 1957; Runcorn, 1955b) and Australian (Irving and Green, 1957a) work, he described the remanent magnetization of igneous and sedimentary rocks, the GAD hypothesis, and APW.

Chaloner first examined results from the Northern Hemisphere, describing the divergence between the APW paths from Great Britain and North America. He examined results from the Gondwana continents, describing the latitude changes of Australia and India, and accented the consilience between du Toit's postulated movement of the continents and the paleomagnetic results (RS1).

> The paleomagnetic data are therefore consistent with India and Australia having moved from positions relatively close to the Antarctic continent, into their present locations, during Tertiary times. Now these are exactly the same direction and order of drift postulated by du Toit on the basis of geological data unconnected with paleomagnetism ... It is this agreement

between totally unrelated sources of evidence which has given a new air of plausibility to this interpretation of all the other facts pointing to the occurrence of continental drift.

*(Chaloner, 1959: 29)*

Making sure that biologists and biogeographers would not miss his point, he assured them that they could interpret their data in terms of continental drift without fear that they were using a hypothesis of little worth.

The physical evidence just considered has, in one sense, a particular interest for biologists. In the past there has been a tendency to set aside biological evidence for drift in the face of seemingly irrefutable contrary evidence from the physical sciences. Now, if the contribution of palaeomagnetism to the controversy cannot be said to have "proved" the occurrence of drift, it can fairly be claimed to have made it a more respectable hypothesis. At least, it has helped to create an atmosphere in which the biological evidence may be examined anew on its own merits.

*(Chaloner, 1959: 29)*

Edna Plumstead, the South African paleobotanist who had added much to the understanding of *Glossopteris* flora and its distribution (I, §6.3), welcomed the consilience between occurrences of fossil plants and paleomagnetic results from the Gondwana continents.

Paleomagnetism has now been tested for rocks of several different ages in all continents, including Antarctica and, although the study is still in its initial states, the evidence strongly supports a considerable movement of continents. In the case of Antarctica and other parts of Gondwanaland this evidence dovetails neatly with that of the fossil plants. The causes of such movements and their mechanism have been much debated and are not yet fully understood but the evidence continues to accumulate until what was once a mere flight of the imagination is now not far from being generally accepted as an established fact.

*(Plumstead, 1961: 180)*

J. B. Hair, a New Zealand paleobotanist, whose work on podocarps (primitive conifers) provided evidence for continental drift, took paleomagnetic results seriously, emphasizing the general agreement between them and du Toit's reconstructions of the motions of continents.

Recent geological, geophysical, and oceanographic studies also reflect the new readiness of physical scientists to "look again dispassionately" at the evidence (Runcorn, 1960). The most significant information comes from paleomagnetic studies ... A new analysis of all available rock magnetism by Blackett and his colleagues (1960) strongly indicates that the divergence observed is systematic rather than random throughout geological time. The authors assume continental drift as the most likely explanation. In a later study, Blackett (1961) has tested this hypothesis by comparing ancient latitudes, as deduced from rock magnetism (on the assumption of an axial dipole field), and ancient climates, as deduced from geological data. The close correspondence found between ancient magnetic and ancient geographic latitudes supports the reliability of the magnetic data and in turn the theory of continental drift. Further paleomagnetic data from southern lands are needed, particularly in Africa and South America, so that correlations with other sources of information appearing to favor continental drift may yet be tested. One suggestive comparison can already be made. The paleomagnetic evidence is consistent with India and Australia having moved from positions relatively close to Antarctica into their present positions (Irving and Green, 1957), at a rate of 2–8 centimeters per year, over a period of the order of $2 \times 10^8$ years (Blackett and others, 1960). This is the order and direction of drift postulated by du Toit from geological data.

> *(Hair, 1963: 408; the reference to Irving and Green (1957) is to my (1957a),*
> *their letter in* Nature; *other references are the same as mine)*

Lucy M. Cranwell, another New Zealand paleobotanist, whose work on the fossil pollen of the genus *Nothofagus* added new support for mobilism, welcomed the paleomagnetic findings. Attempting to explain the distribution of various types of *Nothofagus*, she appealed to Irving's work.

Thus, if we bear in mind fresh evidence for appreciable drift, as urged recently by E. Irving in particular, we can envisage a node of Cretaceo-Tertiary development somewhere between the Andean Province of Antarctica and New Zealand (since the oldest fossil traces are found there), and almost certainly in mid-latitudes, in order to account for the consistently temperate facies of the flora. As I have stated in another section of this Congress, it is difficult to imagine the growth and regeneration of these "massive and light-demanding forests ... in an Antarctica characterized by a long, dark, and inevitably cold polar winter ..." and further, that the genus "may well have succumbed in the south as the Antarctic masses came to rest under the unfavorable conditions of the Miocene."

> *(Cranwell, 1963: 396)*

The British botanist J. D. Lovis stressed the importance of the paleomagnetic support for mobilism at the 1960 Royal Society meeting on the biology of the southern cold temperate zone. He was distressed that paleomagnetism had yet to be mentioned at the symposium and chastised his co-symposiasts. He stressed the importance of Irving and Green's Australian work, and went so far as to say that biogeographers would soon have to develop solutions that accord with paleomagnetic results.

Although in the course of this meeting two such eminent phytogeographers as Professors Skottsberg and Du Rietz have both rejected the hypothesis of continental drift, I am still very surprised that throughout the two days of this Discussion the word palaeomagnetism has not once been mentioned. As a single example of the type of palaeomagnetic evidence in favor of continental drift, and one which is particularly relevant to this Discussion, I would cite the information presented by Irving & Green (1957a) which indicates that although in the late Tertiary Australia and Antarctica occupied virtually the same positions relative to one another that they do today, in the early Tertiary these two land masses were adjacent! If, as seems likely, further palaeomagnetic evidence in favour of drift accumulates, and this type of conclusion is substantiated, then whether we like it or not, biologists will have to accept continental drift, and our theories concerning the origins of plant and animal distributions will have to conform to the geological facts.

> *(Lovis, 1960: 669–670)*

The British botanist Ronald Good, a longtime mobilist and former student of Seward's (I, §3.5), was greatly impressed with developments in paleomagnetism. As Professor Emeritus of Botany at the University of Hull, he came out with a new and revised third edition (1964) of his scholarly monograph *The Geography of the Flowering Plants*. Here is the opening paragraph from his preface.

In this, its third edition, *The Geography of the Flowering Plants* had again been revised throughout, and a good deal of new material, including some figures and many plant names added. These additions are mainly concerned with the notable development that has taken place in the last few years, chiefly as a result of investigations into paleomagnetism and related matters, in the general attitude towards theories of continental drift. A brief formal account of this recent work is given in Chapter 20 but it had also been the chief motive for adding to the book an entirely new chapter, designed to give a more integrated and concise review than is readily available in other chapters of the facts, and some of the hypotheses about those floras of the Southern Hemisphere which are of special interest in relation to theories of continental displacement, and most often mentioned in discussions about them.

*(Good, 1964: v)*

Good was thoroughly familiar with the methods and results of paleomagnetism, referencing about a dozen works, including those by Collinson, Cox, Doell, Green, Hospers, Irving, and Nairn. He touched on methods of paleomagnetism, and proposed displacements of every continent. He referred to Irving, Nairn, Opdyke, and Runcorn on paleoclimates, and discussed Gold's work on polar wandering. After discussing specific results, he added:

These at least are some of the many references to subjects too technical to be covered any more fully here, and though it may be that some of the conclusions will have to be modified in the future, there can be no doubt that this work has given most important fresh support to the idea of continental movement, and, in particular, has done much to remove some of the previous difficulties.

*(Good, 1964: 411)*

The North American paleobotanist Henry W. Andrews, who had favored mobilism in 1947 (I, §3.5), reaffirmed it in his 1961 general textbook in paleobotany. He emphasized the Late Paleozoic glaciation of Gondwana, listed relevant biotic disjuncts across the Atlantic, and devoted a section to the distribution of *Glossopteris* flora, which he took (1961: 418) to be the "most impressive of all positive evidence" favoring mobilism. He appealed to paleomagnetism and the paleowind studies of Opdyke and Runcorn in defense of "polar wandering." As discussed in II, §5.14, the evidence he referred to was not direct evidence of polar wander in the classical sense but a change of latitude or APW likely caused by continental drift. After arguing that the Arctic had enjoyed a much warmer climate in the past, he discussed paleomagnetism and paleowind studies.

A new and somewhat controversial approach to an understanding of fossil climates comes from paleomagnetic studies. Recent investigations of the magnetization of rock formations indicate that some sedimentary deposits may retain their direction of magnetization over long periods of geologic time and the direction perpetuates that of the earth's field at the time they were formed. For example, certain Carboniferous strata of Derbyshire and Yorkshire, England, are said to be magnetized with a low angle of dip which suggests an original magnetization in a field fairly close to the equatorial region. Paleomagnetic measurements in western United States also indicate a closer proximity to the equator in the late Paleozoic. In both western United States and in Great Britain studies of windblown sand deposits correlate with the paleomagnetic evidence in suggesting that these areas were within the belt of the northern trade winds, assuming that the present width of this belt is typical. Thus the evidence from this direction supports the view that the axis of rotation of the earth, also referred to as polar wanderings, may have undergone significant changes in the past.

*(Andrews, 1961: 393)*

Andrews cited two non-technical papers both appearing in *Endeavour*, Runcorn (1955b) and Opdyke and Runcorn (1959). He also (1961: 417), however, directed readers to "Chaloner's well-formulated summary."

Indeed, I go full circle, and return to Chaloner *via* the Canadian botanist N. W. Radforth. Radforth, then a member of the Department of Biology, Hamilton College, McMaster University, took part in a symposium on continental drift in June 1964 at the annual meeting of the Royal Society of Canada. Participants included J. T. Wilson, MacDonald, and Deutsch. The proceedings edited by G. D. Garland were published in 1966. Radforth gave an extensive defense of mobilism in terms of its solution to distributional problems in paleobotany, and took Chaloner's advice that biologists pay attention to the paleomagnetic case for mobilism.

Chaloner (1959), one of our most active British palaeobotanists, attempted the analysis and assessment of recent geophysical evidence in relation to the plant world. Despite the fact that the mechanics of drift as conceived by Wegener involved the application of forces of irrational size, Chaloner was impressed by the accumulation of palaeomagnetic evidence, for both polar wandering and continental drift ... Chaloner (1959) encourages the view that it is time to ascribe a realistic value to these dynamic features of the earth and states: "Now if the contribution of palaeomagnetism to the controversy cannot be said to have 'proved' the occurrence of drift, it can fairly be claimed to have made it a more respectable hypothesis." He reaches this decision through several considerations, one of these being that: "The palaeo-magnetic data are therefore consistent with India and Australia having moved from positions relatively close to the Antarctic continent, into their present locations, during Tertiary Times." He shows that these new positions involve "exactly the same direction and order of drift postulated by Du Toit on the basis of geological data unconnected with palaeomagnetism." This conclusion is comforting to palaeobotanists, for the Du Toit concepts were basic in accounting for distributional phenomena among the ancient flora.

*(Radforth, 1966: 53)*

Radforth heeded Chaloner's advice, for when discussing what he called "the legacy of paleomagnetism" he even appealed, as Chaloner had done, to Creer, Irving, and Runcorn (1957), and Irving and Green (1957a).

According to Irving and Green (1957) to bring palaeomagnetic directions into conformity in the Southern Hemisphere the continents would have to be brought together to make a supercontinent. If this situation did in fact obtain, during pre-Tertiary times there would have been a supercontinent (Gondwanaland) at the position of the present South Pole. The components of the Glossopteris flora at present separated by oceans would then be in reasonable apposition and place in proximity. In post-Miocene times on palaeomagnetic evidence there has been no appreciable shift in the position of the South Pole. In the Carboniferous, the Equator, according to palaeomagnetic studies, would have crossed southern Europe and struck the North American coast south of Charlottetown to traverse the United States. The Carboniferous North Pole (Creer *et al.*, 1957) would have been slightly north of the Cathaysia flora and in the Angara flora. This is plausible paleobotanically.

*(Radforth, 1966: 60; reference to Irving and Green (1957) is to my (1957a),*
*their short letter in* Nature*)*

Thus botanical biogeographers sympathetic to mobilism welcomed the results from paleomagnetism, and favorably compared them with their data (RS1). Mobilist-minded botanists used the same strategy (RS1) in reverse as had paleomagnetists who enhanced the reliability of their results by comparing them and establishing consilience with paleoclimatological and paleontological data (§1.18; II, §3.12, §5.10–§5.15).

However, in response to paleomagnetic results, only a few fixist-minded biogeographers actually switched to mobilism, or began to think mobilism worthy of serious consideration. Of the few, W. D. L. Ride is the clearest example (I, §9.5); initially believing in a fixed Asian source for Australian marsupials, he switched to mobilism and argued that marsupials invaded there from South America via Antarctica because of paleomagnetic findings.

In 1963, Wilma George, a vertebrate zoologist at Oxford University, who explicitly acknowledged the early influence of the staunch fixist G. G. Simpson, nonetheless came to regard mobilism as a legitimate possibility. Formerly a fixist, citing biogeographical difficulties against continental drift, she began to consider drift seriously because of the new paleomagnetic findings. She argued that it was too early to tell whether they would prevail, and suggested that any proof would come from physical not from biogeographical evidence. She specifically referred to Irving and Green's paleomagnetic results indicating a shift in position of Australia relative to Europe. Here is what she said.

[Continental drift] is an attractive idea. South America and Africa would fit together very neatly, and there is no reason to reject the Antarctic connexion out of hand. But evidence for the drift theory is conflicting. The Atlantic and Indian Oceans may date from Jurassic days at least, and this would make the drift theory necessarily a much earlier happening than Wegener estimated. Doubts have been cast on the sial-sima theory, since parts of the floors of the ocean are now known to be sial like the continents. On the other hand recent work on rock magnetism has brought up the whole problem of continental drift in a new light and further information must be awaited (see Irving and Green 1957) ... If this modern work proves the drift theory to be correct, continental drift might be the answer to some of the difficult problems of discontinuous distribution. It might account for the distribution of some of the frogs and fish. But in its turn it will raise a host of new problems ... And yet physical evidence is necessary before a decision can be reached. So far, however, the modern work on rock magnetism has not been able to show whether the continents were actually closer to one another in an east-west direction than they are today ... But it seems easier to make animals move round the world and fill permanent continents than to make the continents move round to collect them ... If geologists find that either drift or land bridges can be proved on physical evidence, then the zoological facts will have to be looked at again. But while the biological facts are the main evidence there is not conclusive reason why the theory of permanence of the continents should be rejected.

*(George, 1962: 95–96; my bracketed addition; Irving and Green (1957) is my (1957a))*

G. G. Simpson (I, §3.6–§3.8) and P. J. Darlington Jr. (I, §3.8) both took notice of the new paleomagnetic evidence for mobilism with very different consequences. Darlington was an entomologist at the Museum of Comparative Zoology, Harvard University. Although the paleomagnetic results were not the only reason for his change in attitude, they influenced him significantly. He announced his change of mind in summer 1964 in an address to the National Academy of Sciences (NAS). He did this after spending almost two years in the Southern Hemisphere, where he saw for himself some of the entomologic and paleoclimatic evidence for mobilism. In his address, he summarized the reasons for his conversion and previewed his *Biogeography of the Southern End of the World*, which appeared the next year. He confessed:

As a result of 19 months spent in Tasmania and eastern Australia and 7 weeks at the southern tip of South America, followed by a year of reading and writing on the biogeography and history of the southern end of the world, I have gradually become convinced that continental drift is not just an intellectual exercise but a probable reality. I was not easily convinced of this. I am a conservative, sixty-year-old biologist and also a professional biogeographer (if there is such a thing), and I have been and still am repelled by the exaggerations and unintentional misrepresentations of biogeographic "evidence" by Wegener and his followers. Nevertheless I am now a Wegenerian, of sorts. I think that some continents probably have drifted, although not in quite the usual Wegenerian pattern. I give this fragment of my own history to justify reviewing a subject which, if there is any reality in it, is of great interest and importance to many persons ... Convincing evidence – evidence that has convinced me – of movement of continents comes from the matching shapes of Africa and South America and their relation to the Mid-Atlantic Ridge, from new evidence of convection currents in the earth's mantle that might move continents, from the distribution of glaciation and floras in the late Paleozoic, and from the new (and still very incomplete) record of paleomagnetism.

*(Darlington, 1964: 1084)*

Darlington had read works in paleomagnetism (referenced below), and understood what he read. While discussing the separation of Africa from South America, he cited Creer (1959), and also remarked that paleomagnetism does not determine longitude (see II, §5.16 for exceptional circumstances).

Paleomagnetism does not show longitudes and therefore does not show how far apart Africa and South America were in the past. However, if basic assumptions are correct, paleomagnetism does show rotations of continents. If Africa and South America were united, they have not only separated but have also rotated so that once-parallel coast lines now diverge southward at an angle of about 45°, and paleomagnetic data seem to show that the divergence was already about 22° early in the Jurassic.

*(Darlington, 1964: 1084–1085)*

Darlington considered Australian paleomagnetic work and the Permo-Carboniferous glaciation there.

Paleomagnetism shows Australia almost motionless, with its Tasmanian corner near the South Pole, from the Carboniferous to the Jurassic. [Reference to the secondary source Runcorn, 1962b included here as an endnote.] During most of this time even Tasmania was ice-free and forested. But during part of this time ice sheets formed not only on Tasmania but across the whole southern half of Australia, and then the ice sheets disappeared again. If paleomagnetism is evidence of movement of continents, it is also evidence that Australia moved little during this time and that the glaciation of this and presumably of other southern continents was brought on and ended by revolutions of climate, not (as often claimed) by movements of land. Actually, glaciation such as occurred in the southern hemisphere probably required both a more-southern position of the land and a climatic revolution comparable to that of the Pleistocene.

*(Darlington, 1964: 1089; my bracketed addition)*

Darlington launched into a summary of paleomagnetic results. He remarked repeatedly on the indeterminacy of longitude, and recognized that there remained huge gaps in the paleomagnetic record. Although sensitive to issues raised by Cox and Doell in the 1960 *GSA* review, he thought the paleomagnetic case for mobilism was probably correct by reason of the general consilience among the paleomagnetic, glacial, and paleobotanical data (RS1). Because Darlington was such an important and conservative figure, and had actually changed his mind substantially because of the paleomagnetic case for mobilism, it is worth repeating what he wrote.[22]

*Paleomagnetic Evidence.* – If certain assumptions are correct, as I think they probably are, a paleomagnetic determination from a given rock formation established an arrow pointing at the position of the North or South Pole when the rock was formed, like a man pointing an arrow at a mark on the ground. (This is a gross oversimplification, but true for present purposes.) The dip of the paleomagnetic arrow shows the distance to the pole, and the direction of the arrow shows the orientation (not the east-west position) of the continent concerned. But the continent carrying the arrow may be anywhere around the earth at the distance from the pole and with the orientation indicated, just as a man may place himself anywhere on a circle around a mark on the ground and keep his arrow always pointing at the mark. Therefore, longitudes cannot be determined from the paleomagnetic data. This fact is not always emphasized by Neowegenerians.

Practically, paleomagnetic evidence [Reference to Cox and Doell, 1960 included here as an endnote] is further limited by enormous gaps that still exist in the known record before the Tertiary. Determinations of the (mid-) Mesozoic are scattered and not very satisfactory. They show Africa and South America near their present latitudes, Australia and India far southward, and Antarctica slightly off the South Pole. Determinations for the Carboniferous and Permian are satisfactory apparently only for Europe, North America, and Australia. Australia, including Tasmania, is therefore the only place where the paleomagnetic record can be compared directly with the record of glaciation and of floras in the southern hemisphere in the Permo-Carboniferous. The comparison gives exciting new information about the limits and tolerances of southern coal floras and about the revolutions of climate that apparently accompanied glaciation. It is unfortunate that, although paleomagnetism places India far south of its present position, about halfway between the Equator and the South Pole, probably in the Jurassic, no rocks have yet been found in India suitable for paleomagnetic determinations between the Cambrian and the Jurassic.

Nevertheless, in spite of the limits and gaps, the paleomagnetic record does agree in general with the joint record of glaciation and of distribution of floras. Both records strongly suggest that most continents lay farther south in the past than now, and that northward movements have occurred. Why should the two independent records suggest this, unless in fact northward movements have occurred? I think they probably have.

*(Darlington, 1964: 1089–1090)*

However, he did not change his mind solely because of paleomagnetism. The almost two years he spent in Australia and South America had enabled him to see at first hand and at leisure the biogeographical support for mobilism. He was impressed with the fit of South America and Africa (Carey, II, §6.11) with each other and with the Mid-Atlantic Ridge, and also with the new work on mantle convection.

George Gaylord Simpson, a longstanding fixist (I, §3.6–§3.8), remained one. The paleomagnetic evidence lessened his antagonism to continental drift, admitting, citing Cox and Doell (1960), that the new paleomagnetic data "raise serious doubts" about the fixity of the continents before the Cenozoic.

It may now also be taken without serious doubt the three southern continents had approximately the same geographic relationships to each other, to the northern continents, and to the poles and equator throughout the Cenozoic. That statement was often disputed in the past, and there is a large literature both geological and biogeographical claiming the contrary. Specifically, the various conflicting theories of continental drift, land bridges, and transoceanic continents have involved intercontinental land connections in the Southern Hemisphere during the Cenozoic. If so, the mammals themselves peculiarly took no advantage of them, and indeed the land mammals provide conclusive evidence against any such connections. Even the new paleomagnetic data, which raise such serious doubts as regards earlier times, confirm that the southern continents have been at least near their present positions throughout the Cenozoic (e.g., Cox and Doell, 1960). Whatever may prove to be true for other organisms or earlier times, it should really now be taken without argument that there has been no direct connection of terrestrial mammalian faunas between any two southern continents during the Cenozoic.

*(Simpson, 1965: 212–213)*

But he should have gone much further. The positions of Australia and India during the Cenozoic as determined by the Canberra and London groups, and given in the 1960

*GSA* review he cited, were in flat contradiction to his first sentence above. The "new paleomagnetic data" for Australia and India, far from confirming "that the southern continents have been at least near their present positions throughout the Cenozoic," showed them to be two and five thousand kilometers respectively south of their present positions in the early Tertiary. Simpson should have actually studied the original papers on Australia (Irving and Green, 1958; Green and Irving, 1958) and on India (Clegg, Radakrishnamurty, and Sahasrabudhe, 1958), that were cited in the *GSA* review (1960: 759); instead he relied solely on Cox and Doell's interpretation of them.

Van Steenis, the doyen of Dutch botanists and strong proponent of landbridges, did not accept mobilism until the 1970s (I, §3.4; I, §8.14). In the 1960s, he rejected its paleomagnetic support, in part because, like Billings (II, §7.12) and Bucher (§1.10), he erroneously imagined that the paleomagnetic method requires coincidence of the geomagnetic and rotational axes at every instant in time, whereas it calls for a time-average of the field. Clearly he had not read the papers of his fellow countryman, Hospers, or subsequent reviews of his work (II, §2.9). Citing Opdyke and Runcorn's non-technical 1959 paper in *Endeavor* (II, §5.14), van Steenis examined this

new branch of research of geophysicists who study the "fossil" magnetic record of ancient lavas ... Under this theory it is assumed that the magnetic axis of the Earth is at least approximately correlated with the geographical axis and that this correlation also existed in the past, allowing for a deviation of not more than c. 20–30° latitude. Another axiom is that the remanent magnetism in basalts and red earths has remained unchanged through the ages. If applied to paleogeography its implications are very wide-going, even in the Tertiary. Besides shifts of the crust in relation to the rotation axis it has led to a renewed consideration for continental drift of no mean magnitude. A curious, unexplained fact is of course that the magnetic axis of the dipole magnetic field of the globe does not coincide with the geographical axis but is eccentric for c. 17° latitude ... and does not run through the centre of the Earth. Properly it is not known what causes the magnetism of the Earth: is it terrene or cosmic? Research with this methodology, which yields figures in palaeolatitude, not longitude, has started only in recent years and many more facts must be known, especially in the southern hemisphere, to gain confidence that there was indeed in the past a persisting, approximate correlation between the two axes which is axiomatic for palaeographical deductions. Provisionally the course of the shift of the magnetic North Pole cannot be reconciled with what is known from palaeontological data at least for the Cenozoic. Besides, an uncertainty of 20°–30° latitude, which has an enormous significance in plant geography, is too coarse for phytogeographical purpose.

(*van Steenis, 1962: 320*)

By 1962 much work had been done in the Southern Hemisphere, the self-exciting dynamo hypothesis had become the predominant theory, and the GAD had acquired strong empirical and theoretical support, but the "Billings" misapprehension died hard.

To close this review of the opinions of biogeographers, I shall begin to monitor Daniel Axelrod's attitude toward mobilism and its paleomagnetic support. Axelrod, a highly influential botanist from the Department of Geology at the University of

California, Los Angeles (UCLA), was one of the last fixist holdouts. He, like others at the NATO meeting, completely rejected the paleomagnetic support for mobilism, his intransigence matching Stehli's. Like Good, he worked on the geographical distribution of flowering plants, but unlike him, he believed it best explained by fixed continents. In "Fossil floras suggest stable, not drifting continents," written in 1962 and published the next year in *Journal of Geophysical Research* (*JGR*), Axelrod attacked both paleomagnetism and biogeographers sympathetic to mobilism. He rejected paleomagnetic results because they were at variance with the fixed paleogeography required by his analysis of paleobotanical data.

There apparently is no paleobotanical evidence that unequivocally supports paleomagnetic data which ostensibly show that the continents have drifted as much as 60° to 70° of latitude since the Carboniferous. Several lines of evidence – climatic symmetry, local sequences of floras in time, distribution of forests with latitude, evolutionary (adaptive) relations – all seemingly unite to oppose the paleomagnetic evidence for major movement.

*(Axelrod, 1963a: 3262)*

The evidence mobilist paleobotanists such as Plumstead and Chaloner appealed to was either irrelevant or could be explained without drift. Moreover, studies of the biogeography of ancient floras should be left to those competent to undertake them.

To establish more firmly the relations which the fossil floras appear to indicate, it is apparent that broad regional studies of the older floras by paleobotanists competent in paleobiogeography are urgently required.

*(Axelrod, 1963a: 3262)*

Just in case paleomagnetists or even incompetent paleobotanists missed his point, he repeated yet again, "To establish firmly the relations which the flora evidently illustrate, we urgently require broad regional studies by paleobotanists competent in paleobiogeography" (Axelrod, 1963b: 364).

Perhaps to Axelrod's dismay, Hamilton (1964b) and Radforth (1966) wrote rebuttals. Hamilton was brief. He was one of the very few American geologists to argue strongly in favor of mobilism before confirmation of the Vine–Matthews hypothesis (I, §7.6–§7.9, and the next section). Hamilton was not a paleobotanist, but noted that he had had insightful discussions with several, including Plumstead, and his firsthand experience in Antarctica gave him an appreciation of the importance of *Glossopteris* in the mobilism debate. He stressed the consilience among paleomagnetic, paleoclimatological, and paleontological data (RS1), and argued paleontological and paleoclimatological data agreed rather than disagreed with the paleomagnetic data – fixists had more difficulties to deal with than mobilists (RS3).

Paleolatitudes as deduced from the climatological evidence of marine invertebrates, land plants, and sedimentary environments, and from most paleomagnetic data, are remarkably consistent with each other and with complex continental drift. They are equally incompatible both with present geography and with crustal shift without both drift and complex internal deformation of continents by bending and strike-slip faulting. Doubting specialists attempt to discredit one aspect or another of the paleolatitude determinations, but the mutual consistency of the independent criteria is strong evidence for their general validity.

*(Hamilton, 1964b: 1666)*

Axelrod, categorized by Hamilton as nothing more than a "doubting specialist attempt[ing] to discredit one aspect or another of the paleolatitude determinations," wrote a rejoinder (Axelrod, 1964), which I leave to the reader to follow up.

Radforth devoted two pages to Axelrod. Unlike Hamilton, he could not be dismissed by Axelrod with the *ad hominem* that his views were not worth considering because he was not a paleobotanist. Radforth admitted that Axelrod's objection that paleobotanical evidence does not support a very large northward drift of Africa since the Carboniferous needed to be "overcome."

Axelrod (1963) does not feel that there is any evidence derived from paleofloristic study to support a contention that continents have drifted. He claims that, for example, the proposed shift of South Africa, which was centered as claimed, over the South Pole in the Carboniferous, through 60° degrees of latitude in the proposed Permo-Triassic continental migration northward would induce considerable floristic change which is not actually in evidence . . . It would take strong geomorphic evidence to overcome Axelrod's objections.

*(Radforth, 1966: 64; Axelrod (1963) is my Axelrod (1963a))*

Here the issue was not the lack of intercontinental disjuncts, which, Radforth thought, provided strong support for drift, but the apparent lack of intracontinental "floristic change" over time that would likely be induced by substantial latitudinal changes. Radforth dismissed Axelrod's claim that the Permian coal forests in Africa imply that the continent could not have been located far from the equator because such forests were formed under tropical conditions. Radforth (1966: 64) argued that there is no reason to believe that the African Permian coal forests "must be tropical" for they could have just as easily been temperate, just as "most of our present-day organic deposits are temperate." Radforth disagreed with Axelrod's basic assumption "that major floras are disposed in accordance with successional symmetry conforming to latitude." Other factors besides latitude determine the distribution of floras.

One of Axelrod's basic assumptions in his objection to the drift hypothesis is that major floras are disposed in accordance with successional symmetry conforming to latitude. It is difficult to accept this *in toto*. The culminating flora of the organic terrain in Canada shows major structural and floristic differences often reflected in depth; these do not necessarily conform to latitude. Arnold (1947) emphasizes the importance of geomorphic inference, not latitude, in accounting for floristic distribution. A high water table and proximity to the sea make for broad constitution in related flora. Also, environmental combinations overshadow latitude implications often significantly to the point where primary morphological entities remain faithful to type as in sedge peat-cover, which is similar from 400 miles south of the present geographic North Pole to position 40 degrees north in latitude.

*(Radforth, 1966: 65)*

Radforth also showed that Axelrod himself realized there are exceptions to his symmetry assumption.

> I am hopeful that Axelrod will qualify his views on the primary significance of his insistence on this symmetry factor in relation to latitude, particularly when in another account he claims (Axelrod 1952, pp. 51–52): "Cain (1944, p. 75) has pointed out that breadfruit (*Artocarpus*), which is now strictly tropical, and which has been recorded at higher latitudes, may have been represented in Cretaceous and Early Tertiary floras by one or more temperate to warm temperate species as in the case of certain genera today, such as persimmon (*Diopyros*), avocado (*Persea*) and magnolia (*Magnolia*)."
>
> *(Radforth, 1966: 65)*

Radforth easily demolished another one of Axelrod's obstacles against drift, namely that large-leafed flora are tropical and small-leafed temperate.

> Axelrod, in his claims supporting the permanence hypothesis, emphasizes the importance of large leaves for the characterization of tropical plants in contrast to small leaves for temperate ones. This observation has questionable application to modern times. Plants with large leaves at the tropics today are predominantly angiosperm, whereas in the Carboniferous angiosperms were non-existent. Thus, the production of large-leaved plants is fundamentally a genetic function with selection following. A genetic mechanism, despite climate, produced the Glossopteris large-leaved types. They too were secondarily selected. Climate did not "induce" their evolution.
>
> *(Radforth, 1966: 65)*

Despite the difficulties raised by Axelrod, Radforth remained in the mobilist camp, again championing Chaloner and his progressive approach.

> In fairness to those faithful to the idea of continental drift, I must side with Chaloner and more with the paleobotanists who welcome the new support for this theory, which encourages us to look again at the organization of our plant fossil record.
>
> *(Radforth, 1966: 66)*

## 1.13 Hamilton welcomes paleomagnetism's support of mobilism

> Re Cox and Doell 1960 – I was by then already an actively-writing drifter, so I was disgusted with the paper. They compiled all that beautiful data, recognized that it defined polar-wander curves, and tiptoed away saying that it could not really mean anything. I don't recall any measure of its impact on those with whom I was in contact. The American geoscience world was populated overwhelmingly by people who knew that drift was impossible, and Cox and Doell reassured the dinosaurs who wanted to be told that they could ignore the p-mag. My hunch is that the paper changed few minds in either direction, but that it might have delayed the revolution by deepening the rut the dinosaurs walked in.
>
> *(Hamilton, July 8, 2002 email to Irving, author copied)*

By the early 1960s Warren Hamilton had become an active drifter whose opinions had developed independently but mindful of advances in paleomagnetism or oceanography

(I, §7.6–§7.9). Hamilton was certainly unimpressed with Cox and Doell's rejection of fellow paleomagnetists' pro-drift solution to the problem of widely divergent APW paths. Hamilton's first two appeals to paleomagnetism were brief. The first was in a paper he presented in September 1963 at a symposium in Cape Town sponsored by the Scientific Committee on Antarctic Research, the meeting at which Mirsky read Long's paper (I, §7.5). He (1964a: 678) argued that the S-shape of the Palmer Peninsula was partially caused by "folding in plan," and argued: "The general 30° swing of palaeo-magnetic declinations in the granitic rocks of the northern two-thirds of the peninsula (Blundell [1962]) is direct evidence for this interpretation." (See II, §5.7 for early paleomagnetic work on Antarctica.) His second appeal (1964b) to paleomagnetism was his response to Axelrod described above (§1.12).

Hamilton made a more extensive appeal in a paper he wrote in 1963 which was not published until 1965. In it he devoted a section to paleomagnetic results, arguing that they agreed with his own interpretation of Antarctic geology (RS1). In it he showed that he had read the Cox and Doell *GSA* 1960 review carefully, found their references useful, but did not agree with them that a pro-drift solution to the intercontinental scatter of poles faced substantial difficulties and should not be generally accepted. He also demonstrated that he understood paleomagnetic methods and was familiar with paleomagnetic studies on continents other than Antarctica.

Published studies by workers in different laboratories agree on the position of the paleomag-netic pole indicated by remanent magnetism of the diabase sheets intrusive in the Beacon Sandstone in several parts of Antarctica (fig. 23). Turnbull (1959, p. 155) reported on 57 samples from 5 localities near the upper Ferrar Glacier; all were magnetized normally and were grouped about a magnetic south pole at lat 58° S., long 141° W., in present Earth coordinates. One sample from the Queen Maud Range was magnetized about 12° from this direction, but its plot was well within the scatter to be expected of single observations so that the difference cannot be considered significant (Turnbull, 1959). Bull and Irving (1960a, b) made 12 measurements on samples from several localities in the basement and peneplain sills near Lake Vanda in Wright Dry Valley and found a close grouping around a pole at lat 51° S., long 132° W. Blundell and Stephenson (1959) reported measurements on samples from seven diabase sheets in a broad region in the Theron Mountains, Shackleton Range, and Whichaway Nunataks and found mostly normal polarities but some reversed; their measurements showed an average south paleomagnetic pole at about lat 54° S., long 136° W.

These three independent studies show remarkably close agreement on a paleomagnetic pole near lat 54° S., long 136° W., which falls in the southwest Pacific about 1500 miles east-southeast of New Zealand ...

Despite the separation between the present magnetic and geographic poles and despite the nondiametric opposition of the present north and south magnetic poles, measurements by many workers on rocks from all parts of the world show middle and late Cenozoic paleomag-netic-pole determinations to cluster about the present geographic poles (Cox and Doell, 1960, pp. 736–739). "The earth's average magnetic field throughout post-Eocene time was that of a dipole parallel to the present axis of rotation" (Cox and Doell, 1960, p. 739). Although some geophysicists have speculated that the Earth's field has been nondipolar in ages farther past, there is little basis for this hypothesis in the data available for testing it. The close agreement between the different groups of Antarctic determinations cited above suggests strongly that the continent lay in lower latitudes in Early Jurassic time than it does now and that it has moved 35° relative to the rotational axis in the intervening period. As Blundell and Stephenson (1959), Turnbull (1959), and Bull and Irving (1960b) emphasized, these determinations from Antarctica when considered with the very different poles indicated for each of the other continents having approximately correlative diabase sills and basalts (India, Australia, southern Africa, and South America), are consistent with continental drift along the general lines suggested by Du Toit (1937) and Carey (1958a). The data seem irreconcilable with the assumptions of either a stable Earth grid or of a shifting crust upon which the continents are fixed in relative position. Although the geology of Antarctica was long so little known that it was inadequate for use in arguments either for or against continental drift, the relatively abundant new data obtained during the past few years provide strong support for drift.

... Paleomagnetic studies on other continents normally show marked changes in paleomagnetic-pole positions within this interval [i.e., Permian to Jurassic], and it is likely that a similar actual change occurred in Antarctica. (Hamilton 1960, 1961, 1963).

> *(Hamilton, 1965: B28–B29; my bracketed addition; Hamilton's references are the same as mine except Carey (1958a) is my Carey (1958))*

Because Hamilton was, by 1965, such a committed mobilist, it might seem that he should have begun appealing to paleomagnetism much earlier. Not necessarily so, he did not become an *active* mobilist until 1960 after he found his *own new evidence* for it in Antarctica (I, §7.7).

## 1.14 Kay and Colbert reassess mobilism because of its paleomagnetic support

Kay and Colbert wrote *Stratigraphy and Life History* (1965), an introductory textbook finished in 1964; Colbert's were the lesser contributions, and were primarily concerned with vertebrate paleontology. They devoted one chapter to the mobilism controversy, which I shall consider in some detail because it reveals how two of North America's firmest fixists now began in the early 1960s to struggle with paleomagnetic findings. It was more balanced than Kay's account twelve years earlier (I, §7.3) and than Colbert's recent paper given at the 1963 NATO meeting (§1.9). Now they had come to acknowledge that, all along, mobilism had had some support; they admitted that paleomagnetism offered important new and independent support, although their discussion was several years out of date. They did not favor mobilism, but had become open to the possibility that it might be correct; their new attitude was one of wait-and-see. They distinguished Wegener's theory from other versions; some of his solutions, such as the formation of island arcs and mountains, they argued, were clearly wrong, but this did not mean that mobilism was wrong because some other version of mobilism might explain them satisfactorily. These nuances signaled a new openness to mobilism in fixist ranks, for which I believe the new paleomagnetic findings were primarily responsible.

I first consider their treatment of the classical arguments. Beginning with the South Atlantic, they (1965: 458) went so far as to categorize the "similarities of the geology on the two sides of the South Atlantic in Late Paleozoic and Early Mesozoic" as "remarkable." They considered the Permo-Carboniferous glaciation, the *Glossopteris* flora, and *Mesosaurus*, three of the stronger points of comparison, noting that they did not necessarily require mobilism and repeated familiar fixist alternatives. Referring to a map they compiled from du Toit of the distribution of late Carboniferous glacial deposits "of the theoretical Gondwanaland," they acknowledged:

Late Paleozoic consolidated glacial deposits, tillites, are widely present in South America – Argentina, Brazil, and the Falkland Islands – and in South Africa … Similar tillites are present also in the peninsula of India and in western Australia (Fig. 19–4 [after du Toit]). Marine beds associated with the glacial rocks show them to range from lower Carboniferous into the Permian, though there are some differences in age assignments in the several regions. Several of the glaciers that formed the tills seem to have flowed from beyond the borders of the present lands as shown by the orientations of grooves and striations. The glaciers are generally thought to have been broad ice sheets rather than restricted valley glaciers. The presence of the glacial deposits in the scattered regions, of course, shows that the climates were quite alike in each.

*(Kay and Colbert, 1965: 458; my bracketed addition)*

Highlighting Plumstead's discovery of *Glossopteris'* seed-like reproductive organ (I, §6.3), they favored open trans-oceanic seed dispersal in a fixist world.

The fact that the plants are remarkably similar in the latest Paleozoic and Mesozoic in these southern areas – Australia, south India, south Africa, and southeastern South America – has led to the belief that they are relics of forests on parts of the great continent of Gondwana (Fig. 19–6 [which showed reconstructions of Gondwana by Maack, Ahmed, and Ma]). There must have been environmental similarities, but plants on separate continents would be similar if climates were comparable and if they had means of traveling across intervening water barriers. Seeds can be drifted by sea and spores carried by winds. There were no birds in the late Paleozoic to transport seeds or plants. Though the chance of a seed crossing the South Atlantic seems remote, the dominant winds, tremendous number of individuals available, and enormous duration of geologic time make such a crossing conceivable if not quite possible.

*(Kay and Colbert, 1965: 459–460; my bracketed addition)*

The case for *Mesosaurus* as evidence for drift was, they thought, much stronger.

The small, fresh-water, fish eating reptile, *Mesosaurus* is found in South Africa in white sandstones directly above the late Carboniferous tillites and an overlying tongue of marine shales; it is in sediments in a comparable stratigraphic position in Brazil. The fact that the plants are similar may not seem compelling evidence for drift. And marine animals such as those in associated strata could have drifted across a sea or followed its circuitous shores. But it does seem unlikely that small reptiles could swim or be carried across such a great expanse of ocean as the South Atlantic. They might have traveled by roundabout land routes across thousands of miles, or they might have crossed the ocean on an island belt. But forms like

*Mesosaurus* have not been found in intervening regions that would be possible land routes; hence it has been thought that they did not migrate along the course of present lands. The factors of the accidents of preservation and of discovery cannot be completely discounted in explaining their absence from regions other than South Africa and Brazil. And it is quite impossible that the forms on the two sides of the Atlantic that are so alike could have developed independently.

*(Kay and Colbert, 1965: 460)*

Regarding the North Atlantic, they (1965: 462) characterized similarities on opposite sides as many but "not strong." Concentrating on the structural and stratigraphic disjuncts between Newfoundland and northwest Europe, Kay's longstanding research interest, they raised difficulties with mobilist reconstructions (RS2) and recycled fixist alternatives (RS1) much as he had done earlier (1952a, b) (I, §7.3).

The evidence may be compatible with drift ... But are there alternative hypotheses that will explain the facts? Though there is similarity between the stratigraphic sequences and strong events on the two sides of the North Atlantic, there is not identity. For instance, the Cambrian and Ordovician of northeastern Newfoundland is as similar to that along the coast of Norway as to that in Scotland, and also has close similarity to successions in northwestern Scotland, and also has close similarity to successions in northwestern Argentina. The Cambrian section in eastern Newfoundland is not very like that of Wales, resembling more that of Sweden. Possibly all are along a belt of similar character, an eugeosynclinal belt of the early Paleozoic that was deformed by orogenies in the later Paleozoic; perhaps this belt continues beneath the present North Atlantic, in which case the ocean north of the belt might differ from that to the south, being essentially a depressed part of a continental block rather than truly a simatic ocean basin. Islands in the northern-most Atlantic, Spitsbergen and Bear Islands have Ordovician sections like those found on the continents. There are other alternatives.

*(Kay and Colbert, 1965: 464, 466)*

After describing stratigraphic disjuncts between Newfoundland and Scotland, they labeled mobilism's explanation of them "quite speculative," and once more reminded readers of fixist alternatives.

Just as in the case of the stratigraphic evidence, there are many interesting points of similarity, but few that might not have developed in distant lands having similar histories. Any attempt to match these records on the two sides of the Atlantic Ocean as evidence of a once continuous continent is quite speculative, as our knowledge is limited to lands that are widely separated not only by sea, but by broad continental shelves.

*(Kay and Colbert, 1965: 469)*

Kay and Colbert (469) rightly wanted evidence that was "more definitive" and "more than suggestive." They wanted to know what lay beneath the continental shelves of Newfoundland and northwest Europe, and they hoped to find out from magnetic and seismic surveys – just what Ewing and his colleagues at Lamont were then doing. If mobilism is correct, truncated structures should be found.

In time, the nature of rocks beneath the continental shelves of Newfoundland and Ireland will be known. For instance, the rocks in a small and perhaps representative area in northeastern central Newfoundland are in fault-separated belts having Ordovician with frequent lava flows contrasting with Silurian conglomerates and argillites that are volcanic-poor. The different rocks affect the earth's magnetic field so that their distribution sometimes can be recognized from records of flights with airborne magnetometers, just as has been discussed for the iron ranges of northern Minnesota ... In the future, similar data will be gained from flights over the shelf, permitting better interpretation of the structural nature of the submarine rocks; perhaps some of this information ultimately will be gained by direct observation. Critical seismic surveys may determine the trend lines of the structures. And having established something of the character of the shelves of Newfoundland and Ireland, there should be abrupt truncation of the structures at the edge of the continent if there has been drift, for these two areas are exceptional in that the oceanic margins cut across the continental structures rather than parallel them. Thus, the present is one of promise and opportunity rather than of conviction from the stratigraphic and structural knowledge along the North Atlantic.[23]

*(Kay and Colbert, 1965: 469–470)*

The two longstanding fixists did acknowledge that mobilism had paleoclimatic support. They even referred to work on Late Paleozoic and Early Mesozoic aeolian sandstones of the Colorado Plateau. They did not mention Opdyke and Runcorn's work, but only that of Poole, who initially (1957) extended the belt of trade winds further north without invoking mobilism (II, §5.14), and who later at the 1963 NATO meeting expressed no opinion on continental drift (§1.9). They again declared that there was insufficient evidence to decide between mobilism and fixism.

Unfortunately, rocks of similar age are rarely of such wide distribution, and preserve enough records of these climatic indicators to permit unequivocal conclusions as to the relations among continents. Distributions that seem to some to demonstrate that continents and poles in the late Paleozoic were as today, seem to others to show constant relations among the continents with a shift of poles ... and to still others to represent changing relative position of continents.[24]

*(Kay and Colbert, 1965: 470)*

They briefly considered possible horizontal displacements along what were then considered transcurrent faults. They instead presented a diagram, a simplified version of a map by Carey,[25] showing lateral movements along the Robeson and De Geer megashears in the Arctic, and the Great Glen Fault in Scotland. Kay, I suspect, liked Carey's diagram because it connected movement along the Great Glen Fault with transcurrent faults in Newfoundland, which he himself had suggested, regardless of whether mobilism had occurred.

As so many had done, Kay and Colbert stressed mobilism's mechanism difficulties (RS2) (I, §4.3).

The nature of the earth's crust beneath the oceans should have a bearing on theories of drift. Drift is conceived of as rather resistant sial masses floating through a weak sima layer that underlies the continental sial plates as well as the floors of the ocean basins. Only in the past decade or two have we come to know the nature of oceanic crust, which is quite thin compared to that of the continents. Such a weak suboceanic crust must have means of gaining excessive depth, for the deepest troughs of the oceans like the Puerto Rico Deep north of that island, Tuscarora Deep off Japan, and Mindinao Deep of the Philippines, are along the borders of ocean basins.

*(Kay and Colbert, 1965: 470)*

Thus, even though Wegener's conjecture that oceanic crust was thin had recently been confirmed, they thought the existence of oceanic troughs ruled out that it was also weak. They thought the mechanism difficulty was real and could not be dismissed on the grounds that if continental drift can be shown to have occurred then suitable forces must exist.

Though it has been argued that the strength of earth materials is too great to permit the slight forces to produce drift, the validity of such objections is not acceptable to those who are convinced of drift – they can but say that the forces are of some additional cause or have been misjudged. If the continents did drift, there were sufficient forces to move them!

*(Kay and Colbert, 1965: 474)*

Like many, they ignored Holmes' now old theory of mantle convection. They also ignored Hess's and Dietz's versions of seafloor spreading (§3.14, §4.9), published three and four years before.

Kay had his own geosynclinal theory for the formation of island arcs and mountain belts, and it is not surprising that he and Colbert severely criticized Wegener's explanations of them (RS2). They (1965: 470–471) repeated the old chestnut that Wegener's drift mechanism and mountain-building theory were incompatible, requiring simultaneously weak and strong oceanic crust (I, §4.3). They repeated Kay's (1952a) previous objections (I, §7.3); on Wegener's theory, mountains and island arcs occur on both leading and trailing edges of purportedly drifting continents, but mountains should not be on the latter or arcs on the former; continental drift does not explain the formation of mountains within continents, mountains that are similar to those at the edges of continents.

Such folded, eugeosynclinal volcanic belts are found all along the Pacific shores of both Asia and the Americas, as well as southerly into New Zealand, presumably in the front of the drifting blocks. But similar, older belts also lie on the opposite sides of North America, where the Atlantic Coast northeastward to Newfoundland is on the "lee" of the continent; but the similar belt is the "front" of the European continent from Scotland to northern Norway and Lapland. If the structure of the West Coast of the Americas must be attributed to drift westward, those of the East Coast must be attributed to drift eastward at an earlier time – in fact long before the supposed beginning of drift of Wegener; but westward drift would be required at the same earlier time in northwestern Europe. The orogenic belts of the coasts of the continents are apparently of quite the same character and presumably, therefore, of similar origin as those found within the continents, not only in the Paleozoic of such belts as the Urals but also in the Precambrian going far back toward the beginning of earth history. If we attributed the drift at time earlier than Wegener's dating as of medial Mesozoic, the very evidence that was thought most significant in the comparison of the southern continents becomes anachronous. However, this simply shows that the mountains were not formed on the fronts of drifting continents, it does not disprove drift.

*(Kay and Colbert, 1965: 472)*

As in 1952, Kay made no mention of Argand's as yet untranslated idea of a proto-Atlantic, of drift having occurred more than once (I, §7.3, §8.7). Although he was sure that existing mobilistic explanations of mountain-building were wrong, and mountains do not necessarily form at the leading edge of drifting continents, he and Colbert acknowledged that some version of mobilism could be correct.

Turning now to Kay and Colbert's 1965 treatment of the new science of paleo-magnetism, they explained its methods and, unlike Bucher, Billings, van Steenis, Munk, and MacDonald, declared them sound. They did not think that paleomagne-tists had made silly mistakes.

The techniques are somewhat sophisticated, but the results are so consistent in rocks of the same age within a region that they permit the determination of the positions of the magnetic and approximate rotational poles when there are many readings.

*(Kay and Colbert, 1965: 474–475)*

Recounting geological history from a North American perspective, they described its APW path first. They reproduced figures from the 1960 *GSA* review.

But when observations are made of rocks in the distant past, it develops that the position of the magnetic pole has changed through time as measured on a single continent, relative to the present rotation pole (Fig. 17-6) [which showed pole positions from rocks of Jurassic age from North America, Europe, South America, and Africa, and was essentially Fig. 29 of Cox and Doell, 1960: 752]. Thus for North America the relative position of the pole in past times on the present globe is progressively southward in Asia through the Mesozoic and then eastward into the Pacific from southern China in late Paleozoic time and going toward the center of the Pacific at the equator in early Paleozoic (Fig. 19-17). [Fig. 19-17 is actually two figures. Fig. 19-17a was Fig. 33 of Cox and Doell, 1960: 758, which showed in Kay and Colbert's words "successive positions for Europe and North America; map centered in the south Pacific Ocean."] This does not mean that the pole of the earth's rotation has changed, but it may mean that the crust of the earth has moved over the mantle and core through time.

*(Kay and Colbert, 1965: 476; my bracketed additions)*

They next introduced the possibility of continental drift, adding the (earlier con-structed) European APW path. Again they reproduced figures from the *GSA* review (1960: 758).

Another anomaly appears when observations are extended to Europe. From observations in North America, the pole for the late Paleozoic should have been in southeast China; but we find on making a similar study in Europe (and presumably the same would be true of studies made in Asia) that during the same time the pole was somewhere in eastern Siberia (Fig. 19-17)! So, when the earth's pole relative to North America should have been in southeastern China, that area was not itself in that position! This can be explained on the assumption that the poles for the time were the same, but the relative positions to each other of the rocks of the crust of the two continents have changed.

*(Kay and Colbert, 1965: 476)*

Although they gave no hint that they had studied original papers, their summary was, up to this point, informative and reasonably accurate, bearing in mind it was four years out-of-date. They could now have gone on to discuss the APW paths from Australia (from Late Carboniferous on), India (from Jurassic on) and Japan (from Cretaceous on) because they were included in their Fig. 19-17. They also had in the literature available results from South America, Africa, and Antarctica; they could have declared that the positions of the APW paths from the Southern Hemisphere and peninsular India were dramatically different from those from Europe and North America and were consistent with the theory of continental drift. But they did not. What they did do, however, is most curious. Having displayed the markedly divergent APW paths, they then, as Hamilton said Cox and Doell had done (§1.13), "tiptoed away" from them as if they did not exist. They wrote as if none of the work on continents other than North America and Europe had ever been done; it was as if their 1964 account of the state of the paleomagnetic support for mobilism had been written in 1955 before results became available from Australia, India, South America, and Africa; they simply ignored evidence from Gondwana. Theirs is an astonishing absence of mind. They continued, remarking:

When the records have been made and analyzed, the present opinion, by no means unanimous, is that they will show that there has been separation of continents in the Western Hemisphere from those in the Eastern and some separation among Australia, Africa, and southern Asia. It remains to be seen whether the pieces will be found to form a single block as envisaged by Wegener; the present impression is that they will not. Moreover, the present judgment is that some of the continents have rotated relative to others. And there should be some changes in relative position as an effect of orogenic movements if we interpret them correctly.

*(Kay and Colbert, 1965: 476–477)*

This passage is highly misleading. What records did they have in mind, records from the Southern Hemisphere and peninsular India, results that had already been obtained and which they had displayed in their figure? Whose "present impression" did they have in mind? What judgment were they referring to? The context indicates they meant that of the paleomagnetists. But paleomagnetists were almost unanimously in favor of mobilism. By 1964, other than US (Cox, Doell, and Graham) and Japanese (Nagata, Kobayashi, Takeuchi, and Uyeda) workers, there were no active paleomagnetists who worked on obtaining poles who did not favor mobilism. Paleomagnetists from the USSR, Britain, Holland, Australia, South Africa, India, and South America all wrote favorably of it. Kay and Colbert prophesied that future results would not confirm Wegener's Pangea, yet this is precisely what (by 1964) paleomagnetists had just done; they did not even mention that the paleomagnetic results already obtained were in broad agreement with the reconstructions of Wegener, du Toit, and Carey.

Their last two sentences in the above quotation are vague but not dismissive of mobilism. Indeed, paleomagnetists outside North America had already obtained many, many results, published in scores of papers, which indicated that continents had rotated relative to each other, and work in the western Mediterranean region had already indicated the presence of large local rotations. It was as if Kay and Colbert, linked so firmly to North American regional geology, and still imprisoned by their earlier staunch opposition to mobilism, were not yet willing to face up to results from too far away and to bring themselves to rethink their fixist world. They made it seem as if paleomagnetists had only just begun to advance arguments favoring mobilism, whereas they had been doing so for over a decade on the basis of evidence they had obtained from seven continents. By freezing paleomagnetism in the mid-1950s, perhaps they thought they could postpone the day of judgment when they would have to fully embrace mobilism. Indeed, D. L. Jones (1965: 489), a reviewer of their book, aptly described their discussion of continental drift as "chary." Perhaps they still hoped that the paleomagnetic results would prove utterly incorrect.

This was not all. Kay and Colbert continued their strange account, disconnected as it appears to have been from any serious study of original papers, by describing the resistance to drift in this way.

The objections that were raised against the drift hypothesis seem to have been valid, in that drift is not responsible for some of the earth features that were attributed to it. Arguments that seemed to oppose drift were really criticisms of the effects that were attributed to drift, such as having it form folded mountains and other earth features. Some of the observations that relate to climate seem to be consistent with paleomagnetic results (Fig. 19-17).

(*Kay and Colbert, 1965: 476–477*)

So Kay, Colbert, and others who had strongly criticized mobilism had not really done so, they had only criticized its alleged effects without acknowledging that this is in fact how science works. Kay had not opposed mobilism *per se*; he was just opposed to its solution to the origin of fold mountains and island arcs. Kay and Colbert understated the coherence between paleoclimatic and paleomagnetic results, only "some" paleoclimatic observations "seem to be consistent with paleomagnetic results." Surely they should have squared with the reader and made clear that there were only minor uncertain exceptions, such as the Squantum Tillite whose age and glacial interpretation Dott, one of Kay's former students, had questioned (§1.2); they should have made clear that there was a remarkably general consistency which strongly fortified the paleomagnetic case for mobilism.

Kay and Colbert closed their discussion of tectonics with an idiosyncratic version of Earth expansion, and looked forward to new discoveries in marine geology.

One of the recent postulates is that the Mid-Atlantic Ridge has a central rift or graben (Fig. 19-20) that represents the widening gulf between the continents. Earthquakes are frequent along the rift, and volcanic rocks have been recovered from bottom dredging there. Such rifts could develop not only if the continents were drifting from an original single mass, extending and separating the ocean area, but also if they retained their size and the whole earth were to expand. The latter would not produce the relative change in positions shown in the paleomagnetic data. The present is a time of many new discoveries that are giving better understanding of the relations of oceans to continents.

> *(Kay and Colbert, 1965: 477; they attributed Fig. 19-20, a section across the*
> *Mid-Atlantic Ridge showing its central rift, to Heezen)*

They were right about one thing: the mid-1960s was a time when many new discoveries were being made about oceanic crust and continent–ocean relationships. Because they paid little attention to these discoveries, not even acknowledging seafloor spreading proposed several years earlier (§3.14, §4.9), the softening of their attitude to continental drift compared with their earlier rigid fixism was likely because of the land-based discoveries by paleomagnetists.

Kay and Colbert's discussion of paleomagnetism, although flawed, was an advance on those of other fixists, most of whom ignored it. In their 1965 summary, they recognized that the paleomagnetic evidence could not be ignored and required that they should at least temper their former strong anti-mobilism sentiments. Had they offered a similar analysis earlier, in 1959, 1960, or even 1961, it would, in contrast to the overwhelming fixist ambience in North America, have been progressive, forward-looking; it would have shown that they had been monitoring paleomagnetism's growing support for mobilism, had generally understood it, and were willing to reevaluate their earlier assessment of mobilism. But in 1965, with their almost complete reliance on a single, secondary source (the 1960 *GSA* review), their assessment was outdated by half a decade. Finally, their discussion of the paleomagnetic evidence was even less up-to-date than the diagrams they included; this disparity between narrative and diagrams made it seem as if they themselves did not realize or did not want to admit the full strength of paleomagnetism's support for mobilism; they may even have thought it was a bad dream that would eventually go away, something on which they should not dwell. Since they apparently did not read the original papers, they underestimated both the strength of support for continental drift and near unanimity outside North America among paleomagnetists.

In Volume I, Chapter 1, I presented the thesis that, prior to the later 1960s, regionalism was the principal reason for the dominance of fixism in North American thought, and in Chapters 6 through 9, I defended it. This discussion of Kay and Colbert (1965) provides, at first blush, reason to resist this thesis. The title of their book, *Stratigraphy and Life History*, was misleading. But it placed them in a tradition in which authors wrote books primarily about the geology of one region but whose title made it seem as if they had written about the whole world; titles promised more than texts delivered. Their book epitomized this not uncommon habit. They were quite explicit; their book was not about stratigraphy and life history in general but mainly about North America. "*Stratigraphy and Life History* presents the principles, as well as giving a summary of some of the main events in the history of North America" (1965: 1). They then referred to the provincial nature of historical geology, expecting that students would be much more interested in the geology of their chosen region than in other regions.

> For each span, such as of a system, the details are provincial – interest particularly to those who are near to them. What is of greatest interest to the student in one region will be little appreciated by someone in a distant place, where geographic separation reduces the pertinence of the material and ability of the individual to comprehend it. There are reference articles that will be most useful for students living in each of the many parts of North America; the instructor will be familiar with these.
>
> *(Kay and Colbert, 1965: 1)*

There could hardly be a clearer acknowledgment of the degree to which regionalism pervaded geological literature of North America, and a clearer statement of the unwillingness of influential educators to even attempt to broaden readers' outlook, a narrowness of outlook that was accepted, even catered to, pandered to. Under such circumstances, it is little wonder that students of Earth science, who went on to work in North America whose geology, according to dominant opinion, did not require former tectonic connections with other continents, gave mobilism little thought. Such a regionally centered regime must have made it very difficult for the student to strive for a global approach to tectonics and thus be able to comprehend the morphological, biological, and geological intercontinental disjuncts that are central to the classical case of continental drift.

### 1.15  Japanese rock magnetists avoid accepting the paleomagnetic case for mobilism

As already noted (II, §5.8) T. Nagata, S. Akimoto, Y. Shimizu, K. Kobayashi, and H. Kuno (1959) constructed an APW path for Japan but remained uncommitted to continental drift even though, as they noted, their path was very different from the others (II, Figure 5.11). Nagata and some of his colleagues and former students at the University of Tokyo revisited the issue in 1961 in the revised edition of their book *Rock Magnetism*, originally published in 1953. Nagata got Akimoto, Uyeda, Kobayashi, and Shimizu to help, and Nagata and Kobayashi added two sections in which they examined the paleomagnetic case for continental drift. Some years later Takeuchi, Uyeda, and Kanamori wrote a book, *Debate About the Earth* (1967), assessing the evidence for continental drift up to the mid-1960s. I comment on these two publications.

In *Rock Magnetism* (1961), they updated their 1959 figure of the APW paths for Japan, Europe, North America, Australia, and India, and added the APW path for Antarctica. They (1961: 292–293) noted the paleomagnetic support of the GAD hypothesis "for nearly the past 30 million years," that poles from each place differ "further from the present pole accordingly as" their "geological age increases," and that poles "estimated from rocks of the same geological period but in different continents do not coincide with one another for times prior to the Palaeogene age." They also noted that poles from the same geological period and from the same landmass "are generally in fairly good agreement with one another, even if the circumstances in which the rocks were generated are quite different." They (1961: 293) added that such agreement "is evidence that the deflection of the ancient pole positions from the present one is not caused by the physico-chemical processes in rocks, such as pressure effect."

They rejected polar wander as the exclusive cause of these results (1961: 294) because "there are definite differences among the traces for different landmasses." They also dismissed Hibberd's idea of rapid polar wandering (II, §7.9) because it depended on incorrectly supposing that rocks used in paleomagnetism "have an appreciable amount of the secondary component of magnetization (e.g. chemical remanent magnetization (CRM), VRM, etc)." This left them with continental drift, with or without polar wandering, or a geomagnetic field that is neither dipolar nor axial. They claimed that a non-dipole field could not be eliminated as a possible interpretation of the paleomagnetic data, citing Runcorn (1959a), which is curious, as he had rejected geocentric axial multipoles as an alternative to Wegenerian drift (II, §5.18). Nonetheless, they (1961: 294) still viewed continental drift, with or without polar wandering, to "be the most plausible explanation of the paleomagnetic results." Thus far, their descriptions and arguments followed closely those of British, Soviet, Australian, and South African workers.

They next examined Irving's and Blackett's analyses of the paleomagnetic data. They (1961: 295) appreciated but rejected Irving's 1958 attempt to reconstruct Gondwana, claiming (not necessarily incorrectly) (II, §5.16), like Blackett and others, that the relative ancient longitudes of landmasses can never be determined by paleomagnetic data alone. They reviewed approvingly Blackett and company's 1960 analysis (II, §5.19).

It is concluded from the figure [reproduced from Blackett *et al.* (1960), and reproduced as my Figure 5.23 in Volume II] that four land masses, Europe, American, Australia, and India have been moving steadily northwards with velocities between 0.2 and 0.8° of latitude per million years and that Europe has rotated about 50° clockwise relative to North America during the last 300 million years.

*(Nagata* et al., *1961: 296; my bracketed addition)*

Because Blackett and company's rates of latitudinal change of each continent differ by a factor of four, it seems that Nagata and Kobayashi had come to accept that the idea of a non-dipole field was not a reasonable possibility and that continents had moved relative to one another. They also spoke favorably of the agreement between paleomagnetic and paleoclimatic results, citing Irving (1956), Nairn (1961), and the work on paleowind directions by Opdyke and Runcorn (1960), Shotton (1937, 1956), Laming (1958), signaling, or so it would seem, their acceptance of the GAD hypothesis and mobilism. This, however, was not so; in their final section Nagata and Kobayashi retreated from acceptance of mobilism.

They singled out Blackett and company's 1960 analysis establishing that the paleomagnetic data were statistically significant and remarked:

Therefore, we *may* be able to believe, at least, that the magnetic poles of the earth were situated near the equator in old geologic times ($10^8$ years ago in order of magnitude) with respect to the present distribution of continents.

*(Nagata and Kobayashi, 1961: 305; emphasis added)*

There was no mention of continental drift. They then reminded the reader of the resistance to continental drift among geologists who emphasized the stability of continental regions and vertical tectonics, referring to Beloussov (I, §8.5), and the lack of a viable mechanism for polar wandering, contrary to comments by Green (II, §5.3, §5.17), and A. E. Scheidegger (IV, §1.6).

However, it must be noted that many geologists have shown geological evidence that the shield areas of old continents have been extremely stable ... showing no evidence of movement since they were formed, and consequently they are strongly against the hypothesis of the drifting of these old continents. On the other hand, no convincing geodynamical theory has yet been established about the possibility of the polar wandering hypothesis. It might be possible, as suggested by some authors ... but still the hypothesis will meet many contradictions from geodynamical view point of the creation and revolution of the earth's figure and structure.

They (1961: 309) next cited as an obstacle the absence of a thorough understanding of Earth's geomagnetic field and its reversals.

As for the origin of the geomagnetic field and its change, concrete theoretical view has not yet been sufficiently established, though the theory of the self-exciting dynamo system within the core is exclusively promising in affording and in explaining these phenomena. For example, a theoretical possibility for the reversal of the total magnetic moment has been demonstrated only for the simplified and ideal case of two ideal dynamos, coupled with each other ... This result is promising in explaining the reversal of the geomagnetic field, but the model adopted in the theory is still far from the actual physical condition within the earth's core. Mathematical treatment in the theoretical calculations, even for the above-mentioned simplified cases, is extremely complicated and difficult.

Were they saying that one must know everything before accepting anything, that progress could not be made in the interpretation of paleomagnetic results until there was an actualistic mechanism for the field and its reversals? Yet, earlier, such considerations had not hindered their seeming acceptance of field reversals when they (1961: 302) found it "very difficult to interpret" the many examples of "reversed NRM in both the baking igneous rocks and the baked sediments only in terms of the self-reversal mechanism."

Finally, and this was truly strange, they expressed full-blown doubt regarding the reliability of paleomagnetic data despite their earlier endorsement of the paleomagnetic case for mobilism made by Blackett and colleagues, acknowledged that Du Bois *et al.* (1957) had shown that physico-chemical processes, including effects of pressure, were not responsible for oblique directions, and had spoken (1961: 271–274) favorably of Stacey's work on magnetostriction (II, §7.5).

Finally, there still remains the fundamental uncertainty regarding the reliability of the NRM of old rocks (say older than $10^6$ years) for the purpose of palaeomagnetic research, especially because CRM and PRM [pressure remanent magnetization] are now believed to be very common in the old rocks. One may still raise a number of inquiries about the origin and stability of the NRM of natural rocks. For instance, nothing has been known about the NRM of rocks under a very long time effect of stress in a magnetic field. Summarizing all the remarks mentioned above, we may say that the results of palaeomagnetic research still contain a number of problems to be solved in the future and some of them look like being serious ones, from the viewpoint of physics. Nevertheless, the palaeomagnetic data obtained hitherto are of great significance in the field of the earth's physics described in the preceding section, and they are of great value as one of the main tools for solving through physics the great problem of the "history of the earth."

*(Nagata* et al., *1961: 307–308; my bracketed addition)*

There seems to be no reason for their sudden about-face. It was as if in the middle of writing their account they suddenly remembered that they ought not to accept the paleomagnetic case for mobilism. Perhaps Nagata and Kobayashi were restrained by the deep conservatism of Japanese society. But this is not apparent in their acceptance of geomagnetic field reversals even though there was no realistic explanation of their origin. Whatever the cause, their failure to accept the paleomagnetic support of mobilism as essentially difficulty-free on the basis of the evidence as they had reviewed it, was, I believe, unreasonable. By 1960 the GAD hypothesis was well-founded, they had endorsed Blackett and colleagues' answer to the non-dipole hypothesis and certain rocks were known to preserve records of the ancient field. There were difficulties, but paleomagnetists had learned a lot about how to avoid them.

This ambiguous attitude of Japanese rock magnetists toward the paleomagnetic case for mobilism seems to have continued unresolved into the mid-1960s. Several years later, H. Takeuchi and Uyeda reassessed the case, devoting much of their book (co-authored with H. Kanamori), *Debate About the Earth* (1967), to it. The book was based on Takeuchi and Uyeda's 1964 *Chikyu no kagu* (*Science of the Earth*), and the same or very similar figures are used in both. There are two references after 1964, one to the 1965 Proceedings of the Royal Society of London's 1964 symposium (IV, §3.3), and the other to T. Rikitake's book (1966), *Electromagnetism and the Earth's Interior*. *Debate About the Earth* represents the authors' assessment of continental drift up to and including the London symposium. They also referenced Irving's *Paleomagnetism*, which provided an account through 1963. Takeuchi and Uyeda were responsible for the assessment of the paleomagnetic support for mobilism. Their presentation is ambiguous, perhaps losing something in translation. Reproducing the same figure from Blackett and colleagues that Nagata and Kobayashi had used, they accepted continental drift because of its paleomagnetic support.

It is clear from this figure that all the continents have drifted northward throughout geologic history. The speed of drift is about 0.2°–0.8° in latitude or about 20–90 km per million years. Particularly for the last 300 million years, Europe and North America have moved northward abreast. They have rotated, however, in opposite directions. Europe has rotated about 50° clockwise relative to North America. The movement of Australia is somewhat complex. India has moved the farthest of all ... Clearly it is impossible to explain all these facts by polar wandering alone.

*(Takeuchi* et al., *1967: 188–189)*

This is a clear statement of drift accepted. Then they raised the possibility of a non-dipolar geomagnetic field, and like their fellow compatriots, offered the ever-popular blanket opinion that paleomagnetism provides no information about relative changes in longitude (II, §5.16). Perhaps they considered these not significant difficulties but only clarifications.

Therefore, what we can say with certainty at the present stage is this: unless we bring in the hypothesis of non-dipolar field, some kind of continental movement is certainly needed to explain paleomagnetic data. However, as to the question of the direction in which the continents have moved, rock magnetism can provide no information about longitudinal movement.

*(Takeuchi* et al., *1967: 188–189)*

They did not discuss the support for GAD provided by studies of dispersion of the geomagnetic field as a function of latitude (II, §5.18), although they had raised the question of the form of the geomagnetic field.

They then turned to the question of reliability.

There are problems concerning the reliability of the paleomagnetic method. For instance, the direction of remanent magnetization can sometimes be affected by various stresses acting on naturally occurring rocks. Sometimes, in sedimentary or metamorphic rocks, the ferromagnetic mineral grains or their crystal orientations are aligned in a certain direction. In such a case, the direction of remanent magnetization is influenced not only by the external geomagnetic field direction but also by the direction in which the grains or the crystals are aligned. Such a rock would not be a faithful fossil of the past geomagnetic field.

*(Takeuchi* et al., *1967: 189)*

But magnetostriction had already been shown to be irrelevant in rocks used in the drift test (II, §7.5), and in the sedimentary rocks relevant to the drift test thermal cleaning was becoming widely used (II, §5.5). What is unclear is whether they considered that these difficulties had already been avoided by judicial selection of rocks, by the use of field stability tests and by magnetic cleaning, whether they thought them endemic, ever-present and unavoidable. Their concluding statement was, "Of course, rock magnetists are doing their best to overcome these difficulties." Did they think rock magnetists were succeeding? They did not say.

They also discussed the new work in marine geology, but without mentioning the ideas of Hess, Dietz, or Vine and Matthews. They spoke favorably of Wilson's presentation of mantle convection, motions of the seafloor, and continental drift. Indeed, they (1967: 240–243) thought that descending convection currents explain the distribution of heat flow and the occurrence of deep focus earthquakes and volcanoes in and around Japan. But their concluding remark indicates that they were not at all committed to continental drift. Alluding to Newton's remark about his being a little boy collecting ever-smoother pebbles or prettier shells beside the undiscovered ocean of truth, they conclude:

Will the theory of continental drift and the debate about it prove to be merely a pebble? Or is it a shell significant of the fact that we are on the shore of an ocean of truth? No one yet knows. Nor is it as important that we answer that question, as it is important that we continue to debate and search for truth.

*(Takeuchi* et al.*, 1967: 276)*

Taking them at their word, I suggest that it was not reasonable, first to describe approvingly the paleomagnetic case for continental drift and then to repeat old worn-out difficulties without making clear that they had already been answered in situations relevant to the drift test. Perhaps they wanted to assure the reader that they had considered all angles and followed all procedures, but in doing so they left uncertain their position in the mobilism debate.

## 1.16  Further poles from Australia, 1958–1964

In the mid-1960s, the impetus in the mobilism debate shifted from continents to oceans, caused by an avalanche of results produced by the ever-increasing number and scale of surveys of the oceans; so far as geomagnetic studies were concerned, there was a switch in emphasis from paleomagnetism of continental rocks and APW paths to the paleomagnetism of oceanic crust, marine magnetic anomalies, and reversals of the geomagnetic field. A little earlier in the first half of the 1960s, there was a strong surge of continental results, featuring the newly developed magnetic cleaning techniques, and taking advantage of burgeoning radiometric dating. Paleomagnetic pole studies continued into the 1970s, and are still active, but their role in the global intercontinental drift test was essentially completed by the mid-1960s when substantial results from all major continents had been obtained. The next three sections describe examples from this final stage of the global drift test. Results from elsewhere for this period have already been noted (II, §5.7, §5.9).

First there is the paleomagnetic work at Australian National University (ANU). Irving recalled:

Jaeger must have been pleased with what Ron Green and I had done, as he supported new students, allocated funding for laboratory extension and for construction of AF and thermal demagnetization equipment. Students Peter Stott and Bill Robertson arrived in 1958, Don Tarling in 1959, and Jim Briden in 1961. Neil Opdyke came as a Fulbright Fellow for one year in 1959. Importantly I had also met Martin Ward, a chemistry graduate from Melbourne University who now wanted to take a degree in mathematics and needed a part-time job to keep him afloat. Jaeger agreed and Martin worked as my assistant until he graduated with a second degree. He became adept at computing and eventually wrote an elegant paper on Earth expansion [described in IV, §3.10]. After Martin left, Alan Major was taken on as a full-time technical assistant for the next several years.

> *(Irving, January 2011 note to author; my bracketed addition)*

The work was on the Silurian through Early Tertiary of southeastern Australia. Bill Robertson (1963), Robertson and L. Hastie (1962), and Peter Stott (1963), who also studied stress effects with Stacey (II, §7.5), worked mainly on the Mesozoic, and Irving (1966) completed the study of Late Carboniferous and Permian that he had started in 1955 (II, §3.12). Briden (1965, 1966) extended Green's work (II, §5.3) on Siluro-Devonian volcanic rocks, and made progress in understanding the role of old remagnetizations and paleoclimates (§1.18). Don Tarling continued his work on Pacific islands and the reversal timescale with McDougall (II, §8.15).

The geographic south poles for Late Carboniferous to the present are given in Figure 1.2. They cluster to the south and southeast of Australia in the Late Paleozoic and Mesozoic, and move to the present pole in the Cenozoic. Although there remained very many complexities to work out,[26] the general picture was by now clear; Australia was in high southern latitudes from Late Carboniferous through Cretaceous, and it moved about 45° north during the Cenozoic, in excellent general accord with the maps of Köppen and Wegener (1924).[27]

There were three Late Carboniferous poles from localities spread over 400 km. They were from volcanigenic sediments of glacial origin and were supported by a definitive tilt-test; folding is pre-Permian. There were Permian poles from lavas and dykes at three localities spaced over 500 km. All had near vertical magnetization, all were reversed. That there are throughout the Late Carboniferous and Permian of Australia sequences of glacial origin was no longer in dispute (§1.2); these are interbedded with sequences deposited in milder (but never hot) climate under which the Permian *Glossopteris* flora flourished.

Mesozoic rocks also had steep magnetizations. There were two Triassic poles from localities 700 km apart, four Jurassic poles (including the 167 Ma Tasmanian dolerites (II, §3.12) and the 168 Ma Prospect Dolerite near Sydney) spread over 1000 km, and three Cretaceous poles (the 140 Ma Noosa Heads Intrusive Complex, the 104 Ma Cygnet Alkaline Complex, and the 93 Ma Mount Dromedary Igneous Complex) spanning 1800 km. All the intrusions are post-tectonic. The Early Cretaceous rocks at Noosa Heads had reversals making any regional over-print unlikely, and there were positive baked contact tests in the other three

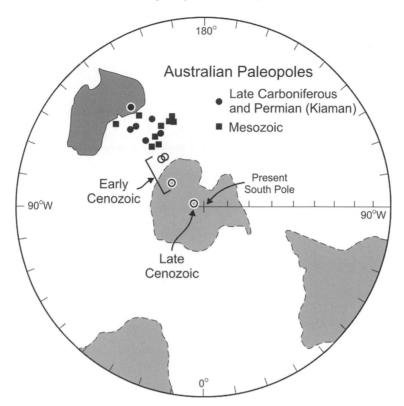

Figure 1.2 Paleomagnetically determined South Poles from Late Carboniferous and younger rocks of southeastern Australia. Redrawn from Irving, Robertson, and Stott (1963) Figure 1 and Irving (1966) Figure 26. Three poles of uncertain age are not included; they all fall with the others close to southeastern Australia. Ages are indicated by the symbols, details in text.

mid-Cretaceous intrusives. It was very likely that Australia remained in high southern latitudes throughout much of the Mesozoic. However, there is no evidence of glaciation in Australia in the Mesozoic. In fact there is little or none anywhere on Earth; the Mesozoic Australian climate was temperate or cool temperature throughout – no coral reefs (Figure 1.7), no evaporites, and no desert sandstones. As Brooks (1949) recognized, Earth was then in a non-glacial (greenhouse) state, and so Australian paleomagnetically determined latitudes and climatic evidence are in good accord.

There were four poles from Cenozoic lavas, notably two (the intertwined open circles in Figure 1.2) from Eocene basalts, one from the state of New South Wales and one from eastern Victoria; the latter was by Gus Mumme (1962a, b, 1963) at the University of Adelaide, where for a short time he had an excellent paleomagnetic facility under the direction of David Sutton. Reversals are present. Mumme's work on the magnetic stability of the Older Volcanics of Victoria was a most

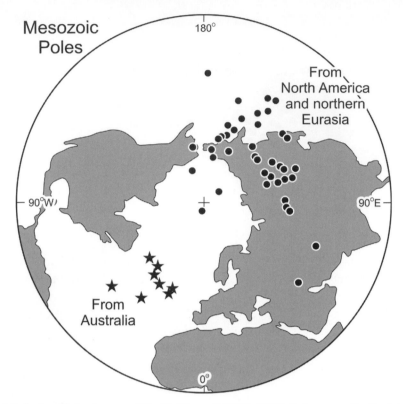

Figure 1.3  Irving, Robertson, and Stott's Figure 3 (1963: 2316). Paleomagnetically determined North Poles from Mesozoic rocks of Australia (stars) compared with those of North America, Europe, and Siberia (dots).

informative study, and fully justified these basalts as accurate recorders of the geomagnetic field. This work on Eocene basalts demonstrated lateral agreement over 1000 km. The Plio-Pleistocene Newer Volcanics of Victoria (over 4 Ma) as we have seen (II, Figure 5.4) gave a pole coincident within error with the present geographic pole. Later work on these basalts figured in the development of the reversal timescale (IV, §5.3).

The results of Figure 1.2 are very different from other continents whether presented as APW paths (II, Figures 2.1, 3.6, 5.1; colored version visible at www.cambridge.org/ frankel3) or as individual poles (II, Figure 5.9). In Figure 1.3 Mesozoic poles from Australia are compared with those from North America, Europe, and Siberia all lumped together. Northern poles are more dispersed because no account had been taken of the opening of the North Atlantic and data had been obtained by many laboratories from a variety of localities. Notwithstanding, there is a vast undeniable difference between them and Australian poles. Right from the outset, Pre-Neogene (older than 20 Ma) Australian poles were not even remotely close to those from the northern continents. This was confirmed by the 1958 APW path (II, §5.3). Now it was surely a certainty that there was a vast difference between the Australian poles and those of the northern continents, and that they were in excellent accord with the Australian paleoclimatic evidence and with Köppen and Wegener's (1924) paleogeographic reconstructions.

### 1.17 Further poles from Africa: the Salisbury (Harare) Group and further work at the Bernard Price Institute, Johannesburg, 1959–1964

It was proving difficult to get poles from Africa that were directly applicable to the paleomagnetic test of Wegenerian drift. Required were samples from the Karroo System (Upper Carboniferous through Triassic), Cretaceous and Early Tertiary rock formations. But except for the Early Jurassic basalts and dolerites that mark its end, the Karroo contains few mafic lavas or red sandstones – few ideal paleomagnetic targets. Outcrops had commonly been struck by lightning and were deeply weathered. Stratigraphic correlations were often uncertain. Getting suitable samples was an uphill task (II, §5.4).

K. W. T. Graham and Hales at Bernard Price Institute and Nairn working from Newcastle upon Tyne in the UK, had made a good start (II, §5.4). At the end of the 1950s an all-out effort to obtain them was initiated by Ian Gough, who in 1958 moved from the Bernard Price Institute to the Physics Department at the University College of Rhodesia and Nyasaland (now the University of Zimbabwe) at Salisbury (now Harare) in Southern Rhodesia (now Zimbabwe), and this, together with continuing work at the Bernard Price Institute, resulted in great progress. Gough's renewal of African work began in earnest in 1959, the same year Nairn published the summary of his earlier work (II, §5.4). He installed a spinner magnetometer, and persuaded Michael McElhinny, already at the University College of Rhodesia and Nyasaland (hereafter, UCRN), to switch his research from ionospheric physics to paleomagnetism; McElhinny eventually became one of its leaders. They were soon joined by two other staff members, physicists Dai Jones and Andrew Brock, and they too opted to work in paleomagnetism. Gough needed a geologist, preferably one with some training in paleomagnetism. With support from Walter Munk at Scripps Institution of Oceanography (hereafter, Scripps), he successfully obtained the first grant given outside the United States by its NSF; it was to build up the APW path for Africa and to fund a Research Fellow. He recruited Opdyke, who arrived from Australia in January 1961. He was a perfect fit. He had worked on paleowinds and paleoclimates with Runcorn, had spent a year with Irving in Canberra learning paleomagnetic work, and had become familiar with magnetic cleaning. An alternating field (AF) demagnetizer, together with a drill to obtain samples to a depth of twenty feet, was built, and, after Opdyke arrived, a thermal demagnetizer was added. The group was unique, all its members had Ph.D.s in various other subjects. Gough (1989: 24) described things in this way.

Our move to Salisbury, Rhodesia (now Harare, Zimbabwe) was motivated less by our strong dissent from the apartheid policy than by the hope of doing something positive for inter-racial collaboration in the (then) more favourable Rhodesian milieu … When I joined the University College of Rhodesia and Nyasaland (UCRN, now the University of Zimbabwe) it had a mainly white student body, which has seen gradual change to a dominantly black one … I set up the spinner magnetometer I had brought from Johannesburg, the workshop made us a drill after Ken Graham's design, and Michael McElhinny moved from ionospheric research to join me in palaeomagnetism. Through the kind offices of Walter Munk I secured the first grant ever given outside the USA by the National Science Foundation, and this enabled me to bring Neil Opdyke to join us for three years on a fellowship. Dai Jones and Andrew Brock joined the group, and we were five, or six with our field assistant Pascal [Khame]. McElhinny and I began with a study of the Great Dyke, which again gave a pole position in north-eastern Africa near the Bushveld and Pilanesberg poles [McElhinny and Gough, 1963]. Neil Opdyke joined me in a study of Cretaceous volcanics in Mozambique, which gave some of the best palaeomagnetic data I have seen [Gough and Opdyke, 1963]. In the five years, 1958–1963, I spent at UCRN our group made strides in defining Precambrian polar wander relative to the Zimbabwean craton [Gough and van Niekerk, 1959; McElhinny and Opdyke, 1964]. McElhinny built us an alternating-field demagnetizer and Opdyke a thermal demagnetizer, the latter for his classic study of Permian redbeds from Tanzania [then Tanganyika], which showed that Africa had been in Antarctic latitudes at that time. Opdyke and I left at the same time, at the end of 1963, and to mark our departure the group published six papers in one issue of the *Journal of Geophysical Research*. One of these papers [Gough, Opdyke, and McElhinny, 1964] summarized the rest, and gives references to them.

*(Gough, 1989; my bracketed additions)*

Opdyke also reminisced about the group and its accomplishments.

When I joined the Salisbury Group they had just begun to do paleomagnetic research. Ian Gough had obtained an NSF grant and proposed to study rocks from Rhodesia. There was a lot of sediments available but the lithologies were all wrong. The proposal was supported by Walter Munk and was funded. McElhinny later received another grant. Ian was a true pioneer and had been doing paleomagnetism on the Pilansburg Dikes in South Africa where he ran into the lightning problem. He had to go into mines to get a result. He had produced a spinner magnetometer that was very sensitive and we used this instrument in Salisbury. He had a strange agreement with Anton Hales that the Salisbury group would not cross the Limpopo into South Africa. When I first arrived in Africa I went on a field trip to Victoria Falls with Geoffrey Bond – who was the newly appointed Chairman of the Geology Department. I looked at the Karroo in the Wanki Basin and Victoria Falls. The lavas were good but the sediments were all the wrong color, no red! We all knew that the redbeds and lavas would give good results but that drab sediments were a losing proposition. I therefore went into the library and studied the Karroo of central Africa. I realized that if I couldn't go south I had to go north and east. No pole existed from the Cretaceous of Africa so Ian and I decided to sample the Lupata lavas in Mozambique. One of my jobs was to develop a drill that would drill to depth (20 ft) to try to get away from lightning effects and we took this drill with us. The study was successful; the fieldwork was done in 1961 and 1962, and we published it in 1963. McElhinny had got the AF demagnetization going and we were off. I pushed to go north to study the Karroo. This eventually happened. Gough and McElhinny are two of the smartest people that I know and we were all marching in the same direction. We all wanted to get a good polar wander curve for Africa. They were both physicists and I added the geological dimension. Ian as principal investigator of the NSF project was worried that we would expend all the funds on travel. However it all worked out. The Triassic collections were after 1961 in 1962 and 1963.

*(Opdyke, September 5, 2007 email to author)*

When Gough (1956) was at the Bernard Price Institute for Geophysical Research he studied the Pilansberg dykes of South Africa and found steep downward magnetizations giving a Precambrian pole (7.5° N; 42.5° E) in the horn of Africa (II, §5.4). While there, Gough and C. B. van Niekerk (1963) worked on the Bushveld gabbro, and obtained another Precambrian pole (23° N; 36° E) nearby. At UCRN McElhinny and Gough (1963) continued to work on the Precambrian, this time on the Great Dyke of Southern Rhodesia (Zimbabwe) and, trying out McElhinny's new AF demagnetizer, they removed secondary VRM and located a pole at 21.5° N, 61.5° E. Nairn (1963) from Newcastle had also sampled the Great Dyke, obtaining a pole of 10.5° N, 69° E, yet again in the same region, and Gough *et al.* (1964) emphasized their agreement. McElhinny and Opdyke (1964) found that the Precambrian Mashonaland dolerites and the Umkondo dolerites, the former intruding the latter, had very different poles. These five poles were sufficient only to construct in rudimentary fashion a short segment of the Precambrian APW path for Africa (Gough *et al.*, 1964); it was not a direct test of Wegenerian continental drift but it showed once again that geomagnetic field directions in the remote past were very different from those in the Phanerozoic.

But it was results from the Carboniferous to Late Triassic Karroo System and from younger rocks that were still needed to test Wegenerian drift. Gough and Opdyke (1963) began by sampling the Mesozoic, post-Karroo, Lupata volcanics of Mozambique, which were very stable magnetically. They located a pole in the Antarctic Ocean south of Africa (Figure 1.4). In the meantime, van Zijl, K. W. T. Graham, and Hales (1962a, b) at the Bernard Price Institute sampled in detail two sections each over 1000 m thick and 100 km apart in the Stormberg (Karroo) lavas of Basutoland (now Lesotho). The lavas overlie the Triassic Cave. After magnetic cleaning they found the lower one-third was reversed, followed by a ~30 m transition zone, and the upper part was normally magnetized. Baked sandstones in the reversed and in the transition section agreed with their overlying lava: excellent baked contact tests. There was also the earlier work of Nairn and Graham and Hales (II, §5.4) on Karroo intrusive sills and dykes, the hypabyssal equivalents of the Stormberg lavas. Gough and Opdyke were now in a position to compare Jurassic and somewhat younger results. They noted:

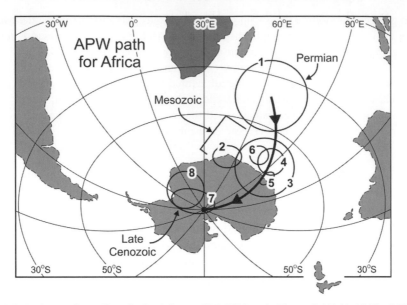

Figure 1.4  Redrawn from Gough, Opdyke, and McElhinny's Figure 3 (1964: 2513). APW path relative to Africa from Permian to the present. 1, Ecca red beds; 2, Triassic red beds, Zambia; 3, Shawa Ijolite; 4, Marangudzi and Mateke Hills; 5, Karroo lavas; 6, Lupata volcanics; 7, Miocene volcanics; 8, Plio-Pleistocene volcanics.

While the age of the Lupata Series is subject to some uncertainty, it is certainly considerably younger than the Karroo basalts. It appears that the pole may not have moved greatly relative to Africa during the Jurassic and Cretaceous Periods.[28]

*(Gough and Opdyke, 1963: 467–468)*

The "Salisbury" group then turned to older rocks. In 1961 Opdyke collected Triassic red sandstone from the Karroo System in the Lusitu River valley in what was then Northern Rhodesia (now Zambia). The results were encouraging, and he returned in August 1963. Earlier Nairn had obtained results from samples from these beds, samples sent to him by the Northern Rhodesia Geological Survey, but before demagnetization became available (II, §5.4). Opdyke magnetically cleaned his samples and found "the direction of magnetization remained unchanged in temperatures at high as 650° [which] demonstrated that only a single direction of magnetization is present" (Opdyke, 1964b: 2496). He calculated a pole just off the coast of Antarctica, "similar to that found for igneous rocks of Mesozoic age ... in central Africa." Next there was work on the Shawa Ijolite by Gough and Brock (1964) and on the Marangudzi and Mateke Hills intrusions (Gough *et al.*, 1964) dated at 209 Ma and 182–196 Ma (Early Jurassic–latest Triassic) and likely closely related to the Karroo lavas and dolerites. Their work demonstrated the excellent lateral consistency of magnetizations of the Karroo volcanics and intrusives over 600 km – what is now called the Karroo Large Igneous Province. Nairn (1964c) also obtained poles from Late Cenozoic volcanics close to the present geographic pole (Figure 1.4).

Opdyke recognized that it was critical to obtain Late Paleozoic results to fix the latitude of the Permian *Glossopteris* flora, which had such a central role in the drift debate (I, §3.3–§3.9, §6.3–§6.5, §6.12, §7.4). Nairn had already obtained Permian results from two localities in Kenya that were somewhat inconsistent but without demagnetization (II, §5.4). In Tanganyika, Upper Ecca (Lower Permian) contains red siltstones sandwiched between coal horizons with *Glossopteris* flora, and Opdyke planned to sample them. Both Nairn (1959: 402) and K. W. T. Graham (1961) had earlier tried unsuccessfully to get results from Ecca-aged strata in South Africa; the strata do not contain red beds, being in higher paleolatitudes. Nairn had also not been able to obtain satisfactory results from the Songwe-Kiwira coalfield, a place Opdyke planned to sample in Tanganyika.

Opdyke, his wife Marjorie, their two sons, Scott, aged two-and-a-half years, and Bradley, aged six months, and Khame, a Zimbabwean, left Salisbury for Tanganyika in July 1962. He and his wife kept a journal.[29] On August 30 they wrote:

We rose early and started to work our way up the river, and successfully reached the scarp near the gorge about 9 AM. We walked up the trail and down the gorge and found our *lovely hard red* siltsone.

On September 6, they drilled six cores. The Opdykes wrote:

In the afternoon we drilled six long cores. We made camp by the Land Rover and after dinner while I was drinking coffee I heard a crash in the bush and the elephants had arrived. They passed about 150 yards from camp. Later we heard a lion. I didn't get much sleep but lay there waiting for the crashing to start again.

They returned to Salisbury on September 15, 1962, with over 100 oriented cores from the Galula (six sites), the Songwe-Kiwira (two sites), and from the Ketewaka-Mchuchuma (four sites) coalfields, spread over several hundred miles (Opdyke, 1964a). The NRMs were scattered and thermal demagnetization reduced them spectacularly. All magnetizations were reversed. Opdyke obtained a pole for the Upper Ecca in the southwestern Indian Ocean, locating southern Africa in mid to high latitudes (Figure 1.4). Turning to the "paleoclimatic implications of the Permian pole position for Africa," he first discussed Dwyka glaciation, arguing that there was agreement between the paleomagnetic and paleoclimatic data from southern Africa, which supported the GAD model of the geomagnetic field.

It has long been maintained by southern hemisphere geologists that during the Ecca stage of the Karroo system southern Africa lay in high latitudes. The reason for this contention is that the Ecca is the rock unit which deposited on the glacial tillites, varves, and glacial pavements of the Dwyka glacial horizon. The reality and intensity of the Dwyka glaciation in Africa has seldom been questioned and is well documented ... The Dwyka glaciation is remarkably extensive, and glacial tills and varvites have been recorded from the Congo, Southern

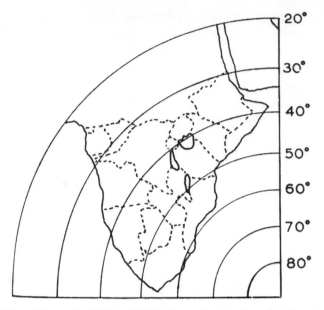

Figure 1.5 Opdyke's Figure 12 (1964a: 2485). The paleolatitudes of Africa during the lower Permian.

Rhodesia, Northern Rhodesia, Tanganyika, and South-West Africa, as well as from the well-known localities of the Republic of South Africa. African geologists have been forced by this evidence to postulate that their deposition occurred at higher latitudes than those in which they are found. Since many of the localities now lie well within 15° of the equator this is easy to understand. Because of this close stratigraphic association with Dwyka glaciation, the Ecca has also been assigned a position in upper or middle latitudes [*du Toit*, 1954; *Flint*, 1961]. If the common assumption of the coincidence of the geographic and the mean geomagnetic pole is accepted, this work provides strong support to this hypothesis. The Ecca pole position (Fig. 12) [reproduced as my Figure 1.5] derived from the results presented in this paper would place those parts of Africa which were glaciated within 40° of the pole. During the Pleistocene of North America continental glaciers advanced farther than this from the geographic pole.

*(Opdyke, 1964a: 2484–2485; second bracketed item is mine)*

Opdyke then turned to the distribution of *Glossopteris* in the Permian of southern Africa, confirming du Toit's and Plumstead's arguments (§1.12; I, §6.3–§6.7).

The Permian of southern Africa was the great period of coal formation; consequently the flora of this era is well known. This flora is part of the *Glossopteris* flora which is present in all of the southern continents, including Antarctica and India. Northern hemisphere elements are missing or are only a minor constituent [*Plumstead*, 1962]. There are good reasons for believing that this flora grew in a temperate climate: 1. Remains of the *Glossopteris* flora have been found interbedded with or beneath glacial horizons in both South Africa and Australia [*du Toit*, 1954]. 2. *Plumstead* [1962] presents strong evidence that the *Glossopteris* flora of South Africa was deciduous and that many of the plant-bearing beds represent autumnal accumulations of leaves.

*(Opdyke, 1964a: 2485)*

Not finished, Opdyke took on Axelrod, who claimed that Late Carboniferous climate in southern Africa had been moist and tropical, that its moist coal swamps were similar to the Pennsylvanian/Westphalian (Carboniferous) coal measures of the Northern Hemisphere, and that this was inconsistent with Africa being near or within the Antarctic Circle (Axelrod, 1963: 3260). Axelrod had disagreed with du Toit, Plumstead, and other mobilists, including recently Opdyke. Opdyke was not amused. He raised three difficulties (RS2): (1) coal in southern and central Africa formed during the Permian, not the Carboniferous; (2) flora of the northern coal swamps were not the same species as the *Glossopteris* flora of Gondwana; and (3) contrary to what Axelrod seemed to maintain, the *Glossopteris* flora did not as at present have "a pole-to-equator distribution in the southern hemisphere." Axelrod (1964) claimed that the *Glossopteris* flora was a temperate upland flora that lived alongside the tropical flora said to be the *Gigantopteris* flora. Opdyke disagreed.

It should be pointed out that the *Gigantopteris* flora is unknown in Africa, and a mechanism has to be imagined whereby the highland temperate flora was selectively preserved while the lowland tropical flora was totally destroyed, a proposition which seems to be entirely untenable.

*(Opdyke, 1964a: 2485)*

Opdyke (1964a: 2486) disputed Axelrod's claim that the paleomagnetic data implied a large movement of Africa between the Late Carboniferous and Early Permian, suggesting that he had been misled by Nairn's Lower Permian pole, which "was of reconnaissance nature only," based on limited sampling and no magnetic cleaning. Opdyke (1964a: 2486) then noted that his own result is "in agreement with the botanical evidence," because it places southern and central Africa in mid- high latitudes in the Early Permian and unlike Axelrod's does not require a large shift in latitude between the "glacial" Late Carboniferous and Early Permian.

Gough, Opdyke, and McElhinny (1964) summarized the significance of their group's work in a sixth and concluding paper. They supported mobilism strongly, basing it "on the most reliable paleomagnetic results available."

Mesozoic paleomagnetic poles from the four southern continents are shown to form widely separated groups. Polar wander alone cannot account for both the divergence of the poles and the stability of the Mesozoic paleomagnetic poles relative to Africa and Australia. The results can be reconciled, however, by the supposition that relative movement had occurred between the southern continents since the Mesozoic ... These are based on the most reliable paleomagnetic results available.

*(Gough et al., 1964: 2509)*

There was a way to avoid continental drift – postulating that Earth's "magnetic field during the Mesozoic was not dipolar" – to which they had an answer: they showed that the angular dispersion of the geomagnetic field during the Mesozoic was similar to that estimated for the present field, and underscored the extreme position that those who still took seriously the alternative of a non-dipolar field must adopt in order to avoid drift.

> If the earth's magnetic field was not dipolar during the Mesozoic, its configuration must have been such as to produce the same distribution of angular dispersion with "latitude" as the present approximately dipolar configuration does. In such a case the "latitude" will have been calculated on the false assumption of a dipole field.
>
> *(Gough* et al.*, 1964: 2515)*

By the time the group's six papers appeared consecutively in *JGR*, Opdyke and Gough had left UCRN. Gough joined Anton Hales, who had moved to the Southwest Center for Advanced Studies, now the University of Texas, Dallas, in 1962. In January 1963 Opdyke moved to Lamont, which as Opdyke (1984 interview with author) remarked, "was a citadel of continental fixism." Irving, involved in the event leading to Opdyke's going there, recalled what happened.

> The second happening at the IUGG meeting September 1963 [in Berkeley] was that on a field trip to the San Andreas Fault I got to know Jack Oliver (head seismologist at Lamont, now Lamont Doherty Earth Observatory) who told me they wanted to get into paleomagnetic work and did I know of anyone suitable? I said yes, Neil Opdyke, whose term as an NSF post-doctoral fellow in Rhodesia (Zimbabwe) was coming to an end. He was a graduate in geology from Columbia (of which Lamont was a part) who had played football for the university. He was a local boy and Margie his wife was a local girl so there would be no settling-in problems. Jack took that message back to Lamont, to Jim Heirtzler in particular who was in charge of geomagnetic programs. In Dec 1963 Jim offered Neil a job which he took up in Jan 1964. It was at Lamont that Neil made his reputation especially on reversals in oceanic cores [see IV, §6.4].
>
> *(Irving, December 2008 email to author; my bracketed additions)*

## 1.18 Convergence of paleomagnetism and paleoclimatology at Canberra, 1959–1966

Irving continued to work at the intersection of paleomagnetism and paleoclimatology, combining forces with Tom Gaskell, his student Jim Briden, and David Brown.

While in England in late 1957, Irving arranged with Gaskell and Alan H. Cook, co-editors of the new *Geophysical Journal*, to publish lists of paleomagnetic results, to record their rapid growth and to ensure wide awareness of what had been done.[30] He talked with Gaskell about using them to determine the paleolatitude of oil deposits. Irving recalled:

If oil is derived from remains of abundant life then it should show latitude dependence. After I returned to Australia, Tommy sent me a global list of oilfields and I calculated latitudes and compiled histograms of them.

*(Irving, June 13, 2007 email to author)*

At the time there were nineteen oilfields for which paleomagnetically determined latitudes could be calculated and all had formed in low latitudes (Irving and Gaskell, 1962). Gaskell was a former Bullard student and an employee of British Petroleum.

From a study of the magnetism of a rock sequence the paleogeographic latitude at the time the beds were laid down may be estimated. Little work of this type has been reported from actual oil-fields but useful results can be obtained by extrapolating from magnetic data already recorded from nearby regions. The paleolatitudes of many pre-Tertiary oil-fields obtained by these methods are shown to be usually less than 20° although their present latitudes range from 25° to 59°. These results are consistent with the hypothesis that oil originated preferentially in low rather than high latitudes. They also suggest that the measurement of the magnetic inclination in rock sequences which are being explored for oil is a useful tool for deciding the likelihood of the occurrence of oil.

*(Irving and Gaskell, 1962: 54)*

At the time, most paleomagnetic data relevant to oil deposits were Late Paleozoic or Triassic. They had Triassic but no later Mesozoic paleomagnetic data to provide latitudes for the vast rapidly developing Middle-east oilfields. Nevertheless, their work showed, for example, the low latitude of the then very important Permian fields of the United States and USSR.[31]

When Irving (1956) first compared paleomagnetic and paleoclimatic data, he did so to determine if the GAD hypothesis was applicable for periods of time prior to the Miocene, and so determine if continents had drifted. He, Briden, and Brown now began using paleomagnetism tentatively to refine models of climate zonation. Briden studied geology as an undergraduate at the University of Oxford (1957–61), but, like Creer and Hospers, he took physics courses too. Versed in mathematics and physics, he saw geology as in need of both, and combining them as a means of avoiding a deskbound career.

I was a geology undergraduate at Oxford 1957–61 but, uniquely at that time, took 50% physics for the first 2 years. My determination to enter geology arose at school: I was looking for a non-deskbound career that would use my predominantly math and physics ability, and I was stimulated into geology via the subsidiary physical geography that I was studying around age 16. I resolved to enter geophysics prompted by finding the then leading applied geophysics textbook thoroughly unsatisfying: they seemed to address a geologically unreal earth; and the geology I read was more natural history than natural science. I felt that bringing a physics and maths background to this might build a career and the more I looked at it, it seemed that not many people were doing that. So with geology I could do that and see the world in the process.

*(Briden, June 21, 2007 email to author)*

Continental drift was not discussed very much at Oxford during his undergraduate days, and there was little geophysics.

The Oxford course was almost devoid of geophysics and continental drift barely mentioned in courses or in planned tutorials. Of course CD "theory" at that time lacked any physics and chemistry dimensions, though it did have a biogeography element.

*(Briden, June 21, 2007 email to author)*

Briden, however, was introduced to continental drift during a course "in global geology" and found himself wondering about the fit of South America and Africa and orogenic disjuncts on opposite sides of the Atlantic. Overall, he thought "continental drift had lots of appeal but no clinching data"; he "was not bothered about mechanism," and decided to "let the evidence lead" (Briden, June 21, 2007 email to author). Interestingly, in light of his future work, he also wondered (if the fixity of continents is assumed) about the apparent lack of latitude-dependency of past climates as indicated by climatic indicators.

I was highly stimulated by a course in global geology by the stratigrapher K. S. Sandford. It was a global tour de force, but left me utterly unsatisfied e.g. about the disappearance of orogenic belts at continental edges, no latitudinal pattern about likely paleoclimatically significant indicators, the tantalising possible fit of South America into west Africa.

*(Briden, June 21, 2007 email to author)*

Although Briden enjoyed his courses, he found "the strength of an Oxbridge education then as now was the function of tutorials to open the mind and explore the unknown" (Briden, June 21, 2007 email to author). During one of his tutorials, he wondered, unaware at the time of the work of Shotton, Opdyke, and Runcorn (II, §4.2, §5.10–§5.14), if studying past wind directions could be used "to determine paleolatitudes."

I do recall "discovering" paleowind directions and discussing with my tutor at the time (Harold Reading – distinguished sedimentologist) whether they might be used with the planetary wind system to determine past latitude. I did not know of the published work on these lines at the time.

*(Briden, June 21, 2007 email to author)*

He planned to become a geophysicist, and applied for a Shell scholarship in the Department of Geodesy and Geophysics at Cambridge, but changed his plans after hearing Blackett lecture on paleomagnetism. Briden recalled:

The seminal moment for me was a Friday night at a meeting of the Physics Society in the University Museum of Natural History (in the same lecture theatre as the famed Huxley/ Wilberforce evolution confrontation a century before) hearing Patrick Blackett talk about continental drift. He must have focused on the divergence of Europe/North America data (Runcorn, Graham and contemporaries) and no doubt Ernie Deutsch's work in India. He likely indicated work was beginning in Australia. This was it for me! Use my maths and physics in geology; see the world; work on the biggest imaginable problem in geology. I applied for the

Shell scholarship in Geodesy/Geophysics Dept Cambridge and was apparently in a shortlist of 2. But I was not impressed with the remaining paleomagnetist in Cambridge (Runcorn, Creer, Irving already having dispersed). Meanwhile one Sunday morning I saw an advertisement in *The Observer* for scholarships at ANU (which I had never heard of), including paleomagnetism. I applied and was interviewed by Irving who was traveling on sabbatical (We walked around the University Parks) and then with Jaeger (Sunday lunch at The Savoy; that was his London abode). The contrast was staggering, and the excitement of Australia irresistible. [Deciding to go to ANU] I withdrew from the Cambridge application. Bullard later said to Irving that it was the strangest reason he had ever heard for not going to Cambridge. But of course it made my career; Irving was the ideal supervisor for me; gave me lots of rope and only tugged on it when necessary. And in the little-known ANU geophys/geochem Jaeger had accumulated more than a dozen feisty young guys nearly all of whom would be academicians in their own countries within a decade.

*(Briden, June 21, 2007 email to author; my bracketed addition)*

In July 1961 arrangements were made for Briden to work with Irving, and he arrived at ANU in mid-September 1961. He read Blackett's *Lectures in Rock Magnetism* on his way to Australia. Irving was still on sabbatical when Briden arrived; Paterson, who had helped Carey (II, §6.8), became his supervisor until Irving's return in early 1962. He found "ANU was continually in healthy dialogue about CD," except for Ringwood who was "contemptuous of CD and above such frippery" and "implacably hostile to paleomagnetism and paleomagnetists ... I vividly remember him and Sid Clark, a visiting geochemist, sitting in the back row at a paleomagnetic seminar and talking right through it" (Briden, June 21, 2007 email to author; slightly altered October 2007). Briden's Ph.D. thesis, "Palaeolatitudes and palaeomagnetic studies with special reference to pre-Carboniferous rocks in Australia," completed in 1964, ranged very widely, but it is his work on paleoclimates that is relevant here.[32] Irving described Briden's contributions to paleoclimatology as "considerable. He tackled the whole range of sedimentary climate indicators" (Irving, June 11, 2007 email to author). Their first joint paper, "Palaeolatitude of evaporite deposits" appeared in late 1962.

Irving probably suggested the evaporite study; he was keen for me to achieve an early publication. Who-said-what after that is beyond my memory. He let me get on with it with timely joint brainstorming. Most important for me was the lesson in succinct and precise writing (though I reckon the Oxford tutorial grounding with a full essay every week must have made me a more advanced pupil than most). The attraction? I think it follows naturally from what I have told you above, starting with the Oxford experience [wondering about how paleoclimatically significant indicators appeared to show no latitudinal pattern of zonation].

*(Briden, June 11, 2007 email to author; my bracketed addition)*

They compared the present and past latitudes of evaporites for each period in the Paleozoic and Mesozoic. There were none from Australia (and Antarctica), itself a telling point. Evaporites indicate climates of "excess evaporation ... where temperature is high, at least at some season of the year, and rainfall fairly low."

Yet the present geographical distribution of Paleozoic and Mesozoic evaporites is asymmetrical about the Earth's equator ... the bulk of them being in the northern hemisphere in intermediate or high latitudes often in regions which nowadays experience both heavy rainfall and low temperatures.

*(Irving and Briden, 1962: 425)*

They gave three possible explanations: changes in location of evaporite deposition, northward movement especially of North America and Europe, or some combination of the two. Like Irving (1956) and Blackett (1961), they argued that evaporites "were laid down" in low latitudes and are now in higher predominantly northerly latitudes because North America and Europe had drifted northwards.

The present latitudes of evaporites are mostly intermediate or high and predominantly northerly. The full range is 31° S.–83° N. But 73 percent lie between 30° and 60° N. In contrast, the estimates of paleolatitudes are generally low, ranging from 59° S.–44° N., with 76 percent of the values lying within 20° of the palaeoequator. They are consistent with the view that evaporites were laid down preferentially in tropical latitudes and hence that areas of excess evaporation were in low latitude.

*(Irving and Briden, 1962: 427)*

However, it was not simply a question of latitude. The latitude distribution of evaporites was not constant: in the Mesozoic, although low latitude, they were prominently in the Northern Hemisphere whereas in the Paleozoic they were distributed about the equator, and this might reflect "the paucity of land and shallow seas in low southerly latitudes" during the Mesozoic (Irving and Briden, 1962: 428).

Their next paper appeared in the 1964 proceedings of the January 1963 Newcastle NATO conference, although neither was able to attend. They forcefully explained that in the absence of an independent and quantitative means of determining paleolatitudes little progress had been made in testing models of Earth's past climatic zonation (Briden and Irving, 1964). Without it, there was no reasonable way of deciding (1) whether and to what degree Earth's past climatic zones differed from the present, or (2) whether there had been changes in the climatic tolerance of the various indicators.

In the interpretation of specific instances one of the most intractable problems facing the geologist arises from the qualitative nature of these two arguments. Possible variations, difficult or impossible to assess, may occur in the tolerance of the chosen indicator to the palaeoclimates, and also in the model used for assigning the indicator to its palaeoclimatic zone. For example, at the present time, corals in bioherms will not tolerate continued minimum temperatures of less than 18 °C, but, because of biological changes in corals, it may not be safe to assume that this same isotherm has delimited the region of growth of biohermic corals since their beginnings in the Ordovician; further, the position of the 18 °C isotherm may have been well to the north or south of its present position, so that the allocation of a coral occurrence to a particular palaeolatitude is a qualitative procedure open to easy criticism. On the basis of assumed fluctuations in the model, some authors (cf. Hill, 1957, 1958) contend that biohermic coral growth at some periods extended almost to the North Pole.[33]

*(Briden and Irving, 1964: 200; Hill, 1957 is my Hill (1957b); Hill, 1598 is my Hill (1959); see I, §9.4)*

Nothing illuminates this problem better than coral reefs, which first occurred in the Ordovician. Globally and lumped together through their time range, reefs have these general characteristics:

The latitude distribution of modern coral reefs is symmetrical about the Equator. The maximum frequency is between 10° and 20°, and most occurrences are within 30° of the Equator ... Their present latitudes range from 40° S to 80° N, with a high concentration between 20° and 50° N: the frequency of occurrence of fossil reefs, in the latitudes in which modern reefs occur is low. The palaeolatitude spectrum, on the other hand is very similar to that of modern reefs, over 95% of values occurring within 30° of the paleoequator (Fig. 19).

*(Briden and Irving, 1964: 213; Figure 19 is reproduced as Figure 1.6)*

The regional example in Figures 1.6 and 1.7 shows the occurrence through time of coral reefs in Australia. Reefs were common in the Late Silurian through Early Carboniferous when Australia was in low latitudes as determined paleomagnetically; they were absent in the Late Carboniferous through Early Tertiary when latitudes were high; they reappeared with the commencement of the Great Barrier Reef in Late Tertiary as northern Australia moved into low latitudes. The striking cut-off of reefs during the time of rapid increase in latitude in the Carboniferous and their reappearance as latitudes became low again in the Late Cenozoic is in strict agreement with the GAD hypothesis and the paleomagnetically determined latitude drift of Australia.

The results of Figures 1.6 and 1.7 can hardly be coincidence. Fixists, in order to retain their beliefs, had not only to reject paleomagnetic findings, they also had to deny that there was little climatic latitudinal zonation in the past and/or maintain that massive reef-builders had been tolerant of cold conditions. The straightforward interpretation was that during and since the Ordovician Period climate has been strongly zonal and *approximately* the same as at present, that there has been no *gross* change in the climatic tolerance of reef-building corals, and hence there had to have been large relative motions of continents.

Briden and Irving expanded their consideration of climatic indicators to include carbonates, dolomites, red beds, desert sandstones, coals, and the Permo-Carboniferous glacial formations of Australia. Concerning carbonates, they emphasized the

rather striking general point that ... 90% of the palaeolatitudes fall within 40° of the palaeoequator, with maximum frequency within 30°. This is closely similar to the [present] distribution of shallow water carbonate deposition and reef growth.

*(Briden and Irving, 1964: 213–218; my bracketed addition)*

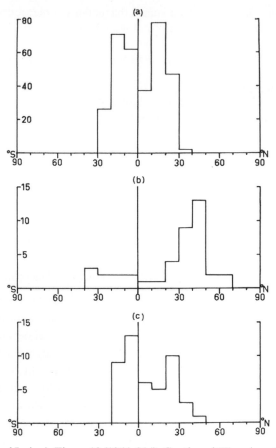

Figure 1.6  Briden and Irving's Figure 19 (1964: 214). Captioned "Equal-angle latitude histogram for organic reefs. (a) Present latitude of modern reefs …; (b) present latitude of fossil reefs; (c) palaeolatitude of fossil reefs."

The present-day latitude of occurrences of dolomites for which paleolatitudes were available was "markedly asymmetrical," their paleolatitudes were symmetrical "about the equator with 95% having values less than 30°." Red beds mimicked evaporites. Desert sandstones are now mostly found between 18° and 40°; their paleolatitudes ranged from 20° N to 30° S, and likewise are symmetrical about the equator. Late Paleozoic glacial beds of Australia, the only Gondwana glacial beds for which there were, at the time, reliable paleomagnetic data, yielded high paleolatitudes: 80° S to 68° S for the Late Carboniferous and 70° S to 80° S for the Middle Permian (Briden and Irving, 1964: 205, 218).

Coals were particularly interesting. They could not form in arid regions. They tend to form "where accumulation" of vegetation "greatly exceeds removal or decay" and are found in hot rain forests, where growth exceeds decay, and under cool conditions, where growth may not be as rapid, but where "decay is heavily inhibited by cold

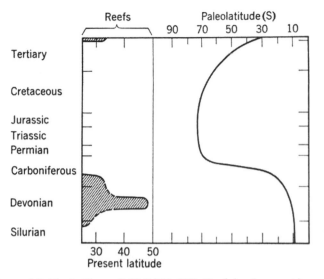

Figure 1.7 Irving and Briden's Figure 9.58 (1964: 222). Reef development in eastern Australia plotted as a function of present latitude on the left, compared with paleolatitude on the right.

winters." Tropical coals should be "associated with trees without rings, whereas ringed trees" should be "common" in coal measures formed under cool conditions (Briden and Irving, 1964: 216). The Carboniferous coals of the United States and Europe are associated with trees without rings; the Permian coals of Gondwana with well-developed rings. Post-Permian coal measures are generally associated with plants with well-developed tree rings, and occur in intermediate to high latitudes.

Bearing these considerations in mind, coal paleolatitudes revealed a pattern consistent with mobilism: Carboniferous coals of the United States and Europe fall between present latitudes of 45° N to 50° N, but their paleolatitudes are within 30° of the Carboniferous paleoequator. But this was just the beginning; the paleomagnetic studies also showed a

somewhat abrupt change from low palaeolatitude coal in the Carboniferous to intermediate and high palaeolatitudes in Permian and later times … Furthermore, the locations of this latter [Permian and later] group are consistent with present conditions, since nowadays peat is rare in the equatorial regions but abundant in latitudes of 50° to 70° N. Finally, the frequency minimum … at about 25° which separates the two distributions corresponds to the present arid zone, in which coal formation would be most unlikely to occur. This frequency minimum is very good evidence for supposing that since the later Palaeozoic there has been a persistent arid zone at about this latitude, but it does not preclude the possibility of extensions of this zone at particular times.

*(Briden and Irving, 1964: 222; my bracketed addition)*

Drawing together their findings (Figure 1.8), they claimed that past climates had been latitude-dependent during and since the Cambrian, that the consistency between the

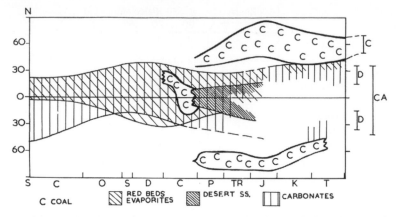

Figure 1.8 Briden and Irving's Figure 26 (1964: 223) showing variation through time of the sediment types they analyzed. Latitude is plotted against time. The range of present-day counterparts is given on the right: D, desert sands; CA, shallow marine carbonates; C, peat. Red beds and evaporites are grouped together because differences in their distributions were very negligible at this level of generality.

paleoclimatic and paleomagnetic evidence from the same region strongly supported the GAD hypothesis, and hence that there had been large-scale continental drift. They compared their combined paleomagnetic and paleoclimatic analysis with Köppen and Wegener's purely paleoclimatic analysis, noting similarities and differences.

> Consistency [between the paleoclimatic and paleomagnetic results] *only* arises when results from the *same* general region are taken together. This may be regarded as a statement analogous to the thesis of Köppen and Wegener that it is not possible to reconcile the past distribution of palaeoclimatic indicators with the present distribution of the continents, if it is assumed that the palaeoclimatic indicators are related to palaeoclimatic zones based on the modern analogy. But the difference is, that the result is obtained not by assuming some model of palaeoclimatic zones, but by assuming an axial geocentric dipole configuration for the palaeogeomagnetic field. *If* the palaeolatitude spectra described above are in fact true spectra, relative movements between the land regions in question would seem to be a necessary condition.
>
> (Briden and Irving, 1964: 221–222; my bracketed addition; their emphasis)

They then turned to the question of whether the latitudinal zonation of climate had ever changed in the past, a question that the independent and quantitative nature of the paleomagnetic information allowed them to address. They claimed that "the width and mean position" of the climatic zones has changed over time.

In conclusion we return to geological arguments about temperature. The palaeolatitude results show that deposits (such as organic reefs) laid down under what are considered to be "warm" conditions give low palaeolatitude estimates, and those (such as the glacial tillites of Australia) deposited under "cold" conditions give high palaeolatitude estimates. These estimates are based on magnetic considerations and not on palaeoclimatological arguments. This evidence suggests that there have always been latitude-dependent climatic zones. The time differences that occur between spectra from the same type of sediment are interpreted as reflecting changes in past land distribution, which have progressively influenced the development of arid and semi-arid belts. The present time appears to be typical, to the extent that the climatic zones are latitude-dependent; but the detailed variations in the width and mean position of these zones preclude direct and quantitative extrapolation from the present analogy.

*(Briden and Irving, 1964: 223)*

They went on to claim:

The present results make clear the potentialities of the palaeomagnetic method for defining these time variations in a numerical fashion. When sufficient results are available from other regions, it should be possible to build up a record, period by period, of the palaeolatitude distribution of each palaeoclimatic indicator, and thus to surmise the time variation in the palaeoclimatic zonation.

*(Briden and Irving, 1964: 223)*

They then presented their "tentative and highly generalized" scheme (Figure 1.8 below), singling out these noteworthy points: the low latitude distribution of carbonates, red beds, evaporites, desert sandstones and Carboniferous coals, the high paleolatitudes of post-Carboniferous coals, and the low (often near-equatorial) values for Late Palaeozoic and Early Mesozoic evaporites and desert sandstones (Briden and Irving, 1964: 224).

Although their findings were somewhat qualitative, Briden and Irving had given a new slant to conjoint studies of paleomagnetism and paleoclimates, adding, to their justification of the GAD hypothesis, new information about variations in past climate zones.

While working with Briden, Irving began a study with Brown of the distribution of the large terrestrial, warmth-loving Late Paleozoic and Triassic Labyrinthodont amphibians, and eventually suggested that climatic zones were wider during the Triassic than during the Late Paleozoic. After introducing Brown, I shall examine their work and their subsequent debate with Stehli.[34]

David Brown, a stratigrapher and paleontologist, was one of the very few supporters of drift in Australia before confirmation in 1966 of the Vine–Matthews hypothesis (IV, §6.6 and §6.7). He was appointed in 1959 to the Chair of the Geology Department at Canberra University College, an undergraduate college that had been affiliated with the University of Melbourne before becoming part of ANU in 1960. Brown received his B.Sc. at Auckland University College in New Zealand (1932–4), majoring in geology and chemistry, and worked as a geologist in New Zealand until the outbreak of World War II in September 1939. After the war, he earned his Ph.D. at Imperial College, London, studying New Zealand Cenozoic Bryozoa. He worked as senior geologist with the New Zealand Geological Survey, and then became reader in geology (1949–58) at the University of Otago, Dunedin, New Zealand.

As an undergraduate he never heard of continental drift (Brown, February 4, 2004 email to author). Reading du Toit's 1937 *Our Wandering Continents* while at the Survey turned Brown into a mobilist.

I was particularly attracted to Du Toit's conclusions in 1937 because they were backed up (a) by good stratigraphic evidence on both sides of the South Atlantic Ocean ... and (b) by the very graphic matching of the coastal outlines of the South American and African Continents, noted by 15th Century explorers and emphasized by Wegener in 1912 (*The Origin of the Continents and Oceans*). These opposing coastlines are now some 4500 miles apart, and there is still no evidence that a land-bridge of that magnitude ever existed between these continental masses.

*(Brown, June 1, 2007 email to author)*

He read Wegener and Carey, but mobilism did not seem relevant to his particular paleontological interests, and he paid it no attention (Brown, June 1, 2007 email to author).

On arrival in Australia Brown soon became conscious of strong anti-mobilism sentiment there.

Yet when I arrived in Australia in 1959, I was amazed to note the almost complete rejection of any sense of continental displacement by the geological and geophysical fraternity there. The only supporter of Du Toit, known to me was Prof. Sam Carey of the University of Tasmania.

*(Brown, February 4, 2004 email to author)*

He also encountered this strong anti-mobilism attitude in a distinguished English visitor, Harold Jeffreys, who gave a fixist seminar at ANU (§2.10). Brown questioned him about Carey's excellent fit of South America and Africa, and was met with frosty stares from the audience.

This episode, however, had a happy outcome. Irving, who also attended Jeffrey's seminar and understood the pertinence of Brown's question, approached him, and later they began working together.

You attended Jeffreys' lecture September 1959; I think he and Bertha his wife stayed a week or so. You asked the Carey question as you described. You encountered hostile stares from the audience that had to be predominantly from the ANU geochemistry group – Ringwood etc. I presumably recognized you as a welcome ally and (1) approached you (perhaps after the lecture) about looking at latitude distributions – this may have been the first time we met; I would know at the time that you were a paleontologist because the start up of a new department in the Canberra University College was understandably a topic of conversation in the Geophysics department. You then got back to me (2) with the proposal to do Labyrinthodonts for the excellent reasons you gave. We did the work, wrote it out (AJS 1964) and then had our tiff in the journal with Stehli.

*(Irving, June 15, 2002 email to Brown)*

Brown remembered it this way.

However, SHJ's [Sir Harold Jeffreys'] declaration [that he had not read nor intended to read Carey] did one good thing – it brought Ted Irving, a geophysicist, and me, a palaeontologist and stratigrapher, together to formulate a paper, involving two very distinct and independent disciplines (palaeomagnetism and palaeontology), which, we believe, clearly demonstrated the falsity of (a) the "land-bridge" concept, and (b) the "fixist" theory supported by SHJ and his ilk.

*(Brown, June 1, 2007 email to author; my bracketed addition)*

They decided to work on terrestrial rather than marine fauna as Stehli had done "because their remains usually occur in tectonically relatively stable regions from which nearby paleomagnetic results are available" while marine invertebrates "arc less well controlled paleomagnetically" and their remains are often found in "geosynclinal rocks" "subject to little known tectonic displacements" (Irving and Brown, 1966: 491). Brown suggested they study labyrinthodonts. They are a well-defined group, and have a "considerable fossil record" from the Carboniferous through Triassic periods "for which many paleomagnetic data are available" (Irving and Brown, 1964: 690). Being almost certainly cold-blooded like their modern counterparts, and often very large, sometimes four or five meters in length, they were unlikely to have survived cold winters in the open, and thus probably lived in warm climates, and "when considered over their whole time span may be expected to be dependent on latitude" (Irving and Brown, 1964: 691).

They compiled labyrinthodonts and calculated their paleolatitudes, restricting their attention to occurrences from North America, Europe and northern Asia, and Australia where Permo-Carboniferous and Triassic paleomagnetic results were reasonably abundant and reliable. They compared the present and past latitude distributions. On the basis of fixed continents (that is, present latitudes) the abundance and diversity of Triassic labyrinthodonts was not greatest in low latitudes. Almost all Late Paleozoic labyrinthodonts occur between present latitudes of 30° to 75°, and most Triassic forms in the 30° to 60° range. In stark contrast, for the whole time span, 80% of genera were located within 30° of the paleoequator, which strongly supported the GAD hypothesis and therefore favored drift. Results were particularly striking for the Late Carboniferous and Permian; most of the 122 occurrences fell "within 15° of the paleoequator, and [there were] only two known occurrences above 60°." Of the eighty-one Triassic occurrences seventy-five were "in the range 10° to 30°" and six were "above 60° indicating a wider latitudinal spread in the early Mesozoic as compared with the late Paleozoic" (Irving and Brown, 1964: 705; my bracketed addition). Noting the "greater confinement" to lower latitudes of Late Paleozoic to lower latitudes as compared to Early Mesozoic labyrinthodonts, they suggested that the difference "may reflect the establishment, at the beginning of the Mesozoic, of a milder, more uniform, non-glacial regime." The latter can now be seen to be an important observation, a specific paleomagnetic/paleontologic confirmation of the change in climate regime from the polar-glacial (icehouse) regime of the Late Carboniferous, Early and Mid-Permian to the non-glacial (greenhouse) regime of the Mesozoic.

Irving and Brown argued that the strength of their method lay in the consilience between independent methods (RS1).

In general terms, the strength of this method is that it is based on comparisons between results of independent studies. On the one hand, the paleogeographical distribution of a given fossil group and its comparison with the zoogeographical distribution of the closest modern relatives may suggest that the occurrences of the members of the group were confined to a certain habitat, which in its turn was subject to a latitudinal temperature control. On the other hand, the paleolatitude estimates are based on an entirely different set of observations and assumptions ... Therefore, any strong correlation found as a result of these comparisons is likely to be significant, whereas the absence of such correlation may imply either that there is no dependence on paleolatitude or that one or more of the underlying paleoecological or paleomagnetic assumptions is incorrect.

*(Irving and Brown, 1964: 689–690)*

Quietly confident, they (1964: 690) did not however maintain that theirs was the "final answer" but wished "only to give an illustration of a type of inter-disciplinary comparison we feel is likely to be of assistance in placing the study of paleozoogeography on a more objective basis than hitherto."

Irving and Brown's mobilist explanations were attacked in 1966 by Stehli (II, §7.10), then at Western Reserve University, Cleveland, Ohio. Irving and Brown (1966) responded and their papers were published together. Theirs was a telling encounter worth examining in detail.

Stehli (1966), for no stated reason, restricted his attention to the Late Permo-Carboniferous labyrinthodonts. He raised several difficulties, questioning Irving and Brown's weighting procedure. They took 300 mile diameter circles as their area size; if *n* occurrences of a particular genus were encompassed by one such circle, they weighted it as one, regardless of *n*'s value. Stehli insisted that they should have used the raw data. Irving and Brown countered by showing that using raw data gave almost identical results as they had obtained.

Our weighting procedure was an attempt to answer the general question, "What constitutes a locality for analysis purposes"? Is it one fossiliferous exposure? Is it a general area containing several exposures? We made the latter choice and gave circles of 300 miles diameter as our area size. In analyses of this type some assumptions about locality size and weighting are always made implicitly or explicitly. *However the matter is not important because our weighting hardly changes the primary data ... it is imperfect (like the weighting procedures) but is not so imperfect that its removal alters our case at all.*

*(Irving and Brown, 1966: 492; emphasis added)*

Claiming that the general noise associated with Permian labyrinthodont amphibians was particularly high, Stehli rejected, as unreliable, data from geographically isolated sampling localities in Australia, Siberia, and Africa, although these were just the places where Permian labyrinthodont amphibians exhibited low diversity, a fact central to Irving and Brown's argument. Having denuded the database of these particularly critical examples, he argued that the remaining data were insufficient to decide whether the distribution of Permian labyrinthodont amphibians was better explained by fixism or by mobilism. Hardly a surprise! But Irving and Brown were not to be fooled.

Our discussion with Professor Stehli is about continental displacement. He is hostile to our opinion that the distribution of labyrinthodonts suggests that the latitudes of the continents have changed in much the same way as the paleomagnetic results indicate. He argues that some labyrinthodont data are unrepresentative of past conditions. He identifies these data by means of his own specially designed reliability criteria, rejects them, and then claims that the remainder cannot discriminate between the present latitude frame and that derived paleomagnetically; Stehli rejects just those items that make the data interesting for paleolatitude studies.

*(Irving and Brown, 1966: 488)*

They proposed (1966: 493) "to take the opposite view and accept the fossil occurrences at their face value" and to see whether or not they are consilient with the paleomagnetic data; they rejected Stehli's arbitrary rejection of the data inconsistent with his fixist position.

Because of the arbitrary nature of Stehli's selection procedure it is to our way of thinking at least as reasonable to take the opposite view and accept the fossil occurrences at their face value. This is an example of the dilemma into which many geologists have been forced when considering the internal evidence of the sedimentary and fossil record in relation to the hypothesis of continental displacement, and this is precisely the point at which paleomagnetism is making its greatest contribution to paleogeography, because it is now possible to assess the usefulness of paleoclimatic indicators against an independent measure of paleolatitude.

Heeding their own advice they demonstrated consilience.

Applying the paleomagnetic results to the labyrinthodont record one finds that ... here high diversity stations ... are all in low paleolatitudes whereas the three low diversity stations are all in intermediate or high paleolatitudes. Six positive correlations out of six make a coincidence unlikely. The observed low diversity in Siberia, Africa, and Australia [i.e., areas Stehli eliminated as being unreliable] may be rationally explained as being due to their intermediate or high paleolatitude in the Permian. There is no need to reject data. *All* the data, including those Stehli regards as spurious, are entirely consistent with the independently obtained paleomagnetic data. It seems to us more correct scientifically to seek first (as we did successfully in our paper) for a single coherent explanation of the data *in toto* rather than to dismember them by an arbitrary selection.

*(Irving and Brown, 1966: 493–494; my bracketed addition)*

Using all the labyrinthodont data, and now including African Permian occurrences, Opdyke having just obtained reliable Permian paleomagnetic results from there (§1.17), Irving and Brown found agreement between the latitudinal distribution of Permian labyrinthodont amphibians and that of their closest living relatives, frogs, "remarkable."

Stehli asks if it is possible by returning to our primary data to test the rival hypotheses. He says not, but he first *rejects certain data*. We have taken up Stehli's suggestion, returned to the primary data, and performed the dispersion analysis, but *without rejecting data*. The result ... is a remarkably symmetrical distribution about the paleoequator. No significance is attributed to details, but the factor of 10 increase in dispersion within the 30° parallels compared with that beyond is felt to be significant. The new African data have a generic frequency and paleolatitudes intermediate between those from Australia and North America and are a group of low diversity in intermediate southern paleolatitudes balancing the Siberian faunas in comparable northern latitudes. Comparisons may be made with frogs as the closest living relatives of labyrinthodonts. Of 207 [extant] frog genera about 75 percent occur in the tropics and

25 percent outside (Darlington, 1957). The corresponding values for the [Permian] labyrinthodonts are also about 75 and 25 respectively. The agreement is remarkable.

> *(Irving and Brown, 1966: 492; my bracketed additions; reference to*
> *Darlington is the same as mine)*

Stehli also cavalierly remarked that the abundance of Late Paleozoic labyrinthodont
amphibian fossils found within the present-day continental areas between 30° N and
60° N, "the latitudes characterized by the highest levels of scientific technology," had
more to do with "the greater concentration of exploratory effort and study" found
there than elsewhere.

It is probably not accidental that, as Irving and Brown note (1964, p. 704), the present-day
latitude distribution of late Carboniferous and Permian labyrinthodonts is mainly northerly
with 87 percent of occurrences lying between 30° N and 60° N. There is certainly a remarkable
correlation between these fossil occurrences and the location of well-established centers of
scientific technology.

> *(Stehli, 1966: 482)*

Stehli was not the first Northern Hemisphere scientist to be so openly smug. Smuts
(1925) had asked Southern Hemisphere scientists to resist the hegemony of Northern
Hemisphere science (I, §6.3); du Toit (1924) opined that if geology had begun in
South Africa, continental drift would be orthodoxy (I, §6.5); Irving had encountered
it at the 1957 Toronto IUGG meeting when Rutten implicitly questioned the stratigraphic competence of Australian geologists (II, §8.12).

Irving and Brown were not going to let Stehli's arrogance go unchallenged. Characterizing his suggestion as another example of "a common northern hemisphere conceit"
and keeping their anger in check, they would have made Smuts and du Toit proud.

*The technological belt hypothesis* – In a footnote Stehli revives a common northern hemisphere
conceit ... that the concentration of known fossil localities in the latitude belt 30 to 60° N may
relate to the greater exploration activities in these latitudes "... characterized by the highest
levels of scientific technology." We shall refer to this as the *technological belt hypothesis* and
will discuss it because it is an argument commonly used against the application of southern
hemisphere evidence to the hypothesis of continental displacement.

We cannot accept the suggestion that the absence of an abundant Australian labyrinthodont fauna is simply a question of lack of study; the eastern Australian sections have been
intensively studied, quarried, and mined in several areas for over a century. Such a
contention is even more tenuous in the case of the small Permian faunas of southern
Africa, because it is from these same regions that a large Triassic labyrinthodont fauna has
been described (listed in Irving and Brown, 1964: table 2); it is highly unlikely that the
same geologists, working over the same area, would have noticed three times as many
genera in the Triassic as in the Permian sequence if they were not, in fact, more common in
the former. If South African geologists had found as rich a fauna in the Permian as in the
Triassic, then this would have been strong evidence against our hypothesis since in
paleolatitudes of 50 to 60° this would not be expected, but the fact is they have not, and

until they do, we contend that the occurrence of comparatively few genera in the Permian of the Karroo is strong support of our hypothesis.

*(Irving and Brown, 1966: 494–495)*

It is no wonder Stehli ignored Irving and Brown's analysis of Triassic labyrinthodont amphibians.

Irving and Brown proceeded to further nail Stehli's technological belt hypothesis. The first was particularly appropriate to the mobilist controversy.

If the technological belt hypothesis is true then it should apply to all fossils. Yet it fails completely to explain the *Glossopteris* flora found only *outside* the technological belt. It fails when such forms as the Cambrian Archaeocyathinae are considered – they have their richest development in eastern Siberia, Australia, and Antarctica!

*(Irving and Brown, 1966: 495)*

Edna Plumstead would have enjoyed that remark (I, §6.3). Perhaps Irving and Brown also were alluding to the misidentification, presumably by experts from the technological belt, of *Glossopteris* flora from Permian Russia (I, §3.4, §3.7). However, Runcorn, Brown, and Irving had not deterred Stehli from trying to discredit paleomagnetism and its support of mobilism (II, §7.10); for at the 1966 Goddard conference, he again attacked the paleomagnetic evidence and the GAD hypothesis, which he claimed did not hold during the Permian (IV, §6.13); but by then Stehli's was a lost cause. Later, in his 1968 Presidential Address to Section C of ANZAAS (Australian and New Zealand Association for the Advancement of Science) at Christchurch, Brown (1967) expanded their explanation to include the distribution of Late Paleozoic and Triassic reptiles.

This ability of paleomagnetism to make fundamental contributions to paleoclimatology is one of its durable contributions to Earth science. It is the ancient geomagnetic field through the study of paleomagnetism, not plate tectonics, that fixes paleolatitudes, and thereby enables paleoclimatologists to reconstruct past environments, to assess models of climatic zonation, and determine the degree to which latitude affects climate. Land-based paleomagnetism provided mobilism with its first, albeit generally unrecognized, difficulty-free solution. But it did, and continues to do much more. For example, it guides all current studies of past climate (e.g., Evans, 2006) and it guides attempts to track pre-Jurassic continental drift (Muttoni *et al.*, 2003), and when combined with plate tectonics provides the global reference frame for paleogeographic maps (Phillips and Forsyth, 1972; Besse and Courtillot, 1991; Kent and Irving, 2010).

Right from the beginning of the drift/mobilist controversy (I, §2.7, §3.10–§3.12), some of the most important evidence for continental drift is the occurrences of Carboniferous and Permian glacial beds spread over five continents across half the globe. The area, mobilists argued, becomes manageable climatically if these continents were then assembled around the south geographic pole at that time. However, the glaciated area is still large, approaching 93° of arc on du Toit's 1937

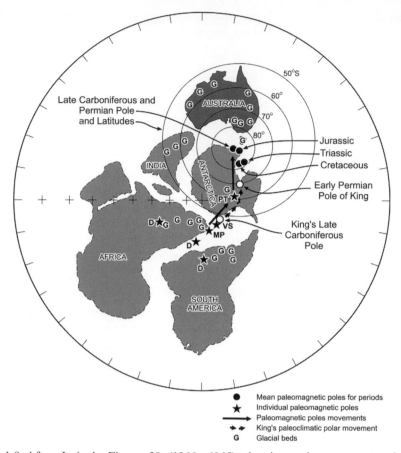

Figure 1.9 After Irving's Figure 28 (1966: 6045) showing pole movement relative to Gondwana. The base map is from King (1962) for the Late Paleozoic. The poles were observed from Australia (Irving, 1966). The mean poles for Kiaman (Late Carboniferous and Permian), Triassic, Jurassic, and Cretaceous. The Late Paleozoic shift is defined by paleomagnetic results from Late Silurian or Early Devonian. MP, Mugga Porphyry (Briden, 1966); VS, Volcanic Stage of what was formerly called the Lower Kuttung; PT, Paterson Toscanite (mid-Carboniferous).

reconstruction of Gondwana (I, Figure 6.4). Köppen and Wegener (1924) suggested a polar shift from southern Africa toward Australia between the Carboniferous and Permian. Du Toit followed this, as did Lester King (1962) who, elaborating on it, suggested that the motion was "from early Carboniferous until mid-Permian."

For comparison Irving (1966) plotted Australian poles on King's Gondwana reconstruction (Figure 1.9); he also marked glacial beds. All these results were based on demagnetization studies, except for the early Devonian results obtained by Green (1961). The Devonian through Late Carboniferous polar shift mirrors the trans-Gondwana trend of glaciations. This was an early attempt to track polar migration across Gondwana that was soon followed by others beyond the time-frame of this book (Creer, 1968; Briden, 1967a; McElhinny, 1973). It was, however, sufficient to show that not only did the Late Paleozoic glaciations of Australia fit the paleomagnetic latitudes very well but also that the extension backward of the Australia results followed the distribution of glaciations across Gondwana. The trend with time was consistent in general fashion with the trend with time of glaciations, confirming in a general way the speculations of Köppen and Wegener, du Toit, and King.

This essentially concludes my account of the continent-based paleomagnetic test of continental drift. Later chapters of this volume deal with mid-1960s responses to this test. At this time workers were on the move. Gough moved to the United States in 1962. Opdyke returned there in 1963, to Lamont where he worked on the magnetization of sediment cores, further developed the reversal timescale and helped confirm the Vine–Matthews hypothesis (IV, §6.4). In 1963 François Chamalaun, a student of Creer's, arrived at ANU and worked with McDougall on the reversal timescale (IV, §6.3). In 1964 Irving left ANU and moved to the Dominion Observatory, Ottawa, Canada. In 1967 McElhinny was appointed to Irving's position at ANU. McElhinny went on to do important work which falls largely outside the period dealt with in this book.

## Notes

1 R. H. Dott Jr. became a leading American sedimentologist, spending most of his career at the University of Wisconsin. He was President of the Society of Economic Paleontologists and Mineralogists (SEPM) for 1981–2. He received the Ben H. Parker Medal from the American Institute of Professional Geologists in 1993 and the W. H. Twenhofel Medal from SEPM and the Society of Sedimentary Geology in 1993. Dott has made substantial contributions too as an historian of the Earth sciences, and received the 1995 History of Geology Award of the GSA.
2 See also Wertenbaker (1974: 106–119) for an account of Ericson, Ewing, and Heezen's work on turbidity currents.
3 Dott (1961) cited Crowell and Winterer (1953), and Crowell (1957).
4 Dott's criticism could have applied to Upper Carboniferous glacial beds of southeast Australia, which are in an orogenic zone, are non-marine, and could have been deposited at some elevation, but it almost certainly did not apply to the glacial beds that occur intermittently through much of the overlying Permian there (David, 1950). These are interbedded with shallow-water marine and terrestrial sedimentary beds, indicating that they were deposited at or close to sea level.
5 Nairn correctly claims that Earth scientists first learned about recent glaciation before recognizing Paleozoic and Precambrian glaciation. Both earlier glaciations were known before Wegener's theory appeared, and geologists had already given up the idea that Earth's climate had always been mild up through the Tertiary. Moreover, Cambrian and Precambrian glaciation had been identified in other parts of the world. Coleman (1907) discussed Huronian (early Proterozoic) glaciation in Canada. Bailey Willis *et al.* (1907) identified what they described as early Cambrian or possibly Precambrian glaciation on the Yang-tse river in China; Strahan (1897) corroborated Reusch's 1891 discovery of an early Cambrian glacial deposit in Norway; David (1907) described ancient glacial deposition of Eocambrian age in Australia, and Cambrian or Precambrian glaciation was also discovered

in southern Africa and India; all these are now, like the Squantum Tillite, regarded as latest Precambrian (§1.11). The existence of Permian-Carboniferous glaciation was well known before Wegener proposed continental drift; in fact his theory arose in part to explain its origin and distribution (I, §2.7).

6 In an amusing aside Romer suggested that the Disney movie *Fantasia* played a role in directing research in the effect of changing temperature upon crocodilians.

A converse of this theory, that of the heat-death of the dinosaurs, suggested by Cowles, was adopted in the well-known moving picture, "Fantasia", by Walt Disney. In this film, to the harsh accompaniment of Mussorgsky's "Night on Bald Mountain," [Stravinsky's "The Rite of Spring"] the last of the dinosaurs are seen staggering to their death across a hot and arid plain. This suggestion led to an interesting study by Colbert, Cowles and Bogert (1946) on the effect of temperature changes on crocodilians – the closest living relatives of dinosaurs. To Colbert and Bogert the results do not appear to warrant the conclusion that over-heating was important in dinosaur extinction.

*(Romer, 1961: 197; my bracketed addition)*

Birds are closer to dinosaurs than crocodiles, but Romer was not to know that.

7 I say "publicly" welcomed, because I asked King when he began to have much faith in paleomagnetic reconstructions, and he (December 29, 1981 letter to author) wrote, "I have never had much faith in geomagnetics. You see both north and south magnetic poles have moved 500 km in my lifetime of 74 years. And there has been no such continental displacement in the land masses across which they have moved." His misunderstanding is the same as Billings expressed in his Presidential Address to the GSA (II, §7.12).

8 Nairn's three references were to Irving (1956), Opdyke and Runcorn (1960), and comments he had made and reported in Clegg (1956). Although I am not sure of the other references Irving had in mind, substantial contributions that Nairn did not refer to were found in Creer, Irving, and Nairn (1959), Irving (1957a), Irving and Green (1958), Laming (1958), Opdyke (1959), Opdyke and Runcorn (1959), Runcorn (1959c), Shotton (1956), Schove, Nairn, and Opdyke (1958), Stehli (1957, 1959).

9 Runcorn did not refer to Holmes in this paper. In his short and incomplete first discussion of his hypothesis, he mentioned Vening Meinesz and Hills; he (1962a) later added Griggs. Holmes was finally cited by him (1965: 229) at his presentation during the 1964 Royal Society Symposium on continental drift (IV, §3.3). The general lack of reference to and impact of Holmes' convection is remarkable, and lamentable, but true.

10 Reading this letter almost 50 years later, Irving noted:

My statement that APW paths are alike is incorrect, that is the poles that define them along their Late Paleozoic and Mesozoic segments is generally incorrect. It is correct for N. America and Europe, but not for the paths for Australia and India. To generally assume that they were gave these crazily fast rates of drift as discussed by me and Creer *et al.* in 1958. We discussed the alternative which implied APW paths were not well matched and which gave the lower estimates of drift which we preferred. We did discuss alternatives and stated our preference but we ought to have rejected the former outright. We were not always clear headed about this; Deutsch was (Irving, October 2007 note to author). See II, §5.16 for discussion of Creer *et al.* 1958.

11 Irving also appealed to simplicity in his letter to Deutsch. "I prefer to attribute this similarity to polar wandering which is the simpler hypothesis." Polar wandering is, in a straightforward sense, simpler than continental drift. Deutsch wouldn't have disagreed. But Deutsch took the combination of polar wandering and continental drift to be less simple than continental drift by itself. Deutsch was correct.

12 Runcorn's passing comments about Bucher ring true and reveal Bucher's narrow-mindedness about continental drift. If he thought that the APW paths for Europe and North America were well-founded, he should have thought that the APW paths for Australia and India were also well-founded, and therefore supported continental drift.

Much of the work that established Gondwana Permo-Carboniferous glaciation was old, but that is not a reason in and of itself to question it. Indeed, sedimentologists had reexamined it many times. Bucher was not a sedimentologist, and, as far as I know, had not personally observed the classic sites of this glaciation.Yes, turbidites and tillites may sometimes be confused. Dott, for example, would agree, but he also claimed that evidence for Gondwana glaciation was strong (§1.2). Also, Runcorn's passing comment about Wilson's anti-mobilism attitude may sound surprising because he was soon to become a committed mobilist. But Runcorn was correct; Wilson argued against mobilism as late as 1959 (IV, §1.7).

13 Runcorn's penchant for getting King's College, Newcastle upon Tyne, and later the University of Newcastle upon Tyne (which King's College became in 1963 when it separated from the University of Durham) to award honorary degrees to prestigious foreign Earth scientists once backfired on him. Runcorn got the University of Newcastle to award Beloussov, the great Soviet Earth scientist and longtime foe of mobilism, an honorary degree. Constantin Roman, a Romanian influential scientist, who Creer and Runcorn helped get out of Romania by offering him work in paleomagnetism at Newcastle, and who worked later with Dan McKenzie and Teddy Bullard at Cambridge, where he obtained his Ph.D. under Bullard, described what happened.

Runcorn, our roving ambassador, had just arranged, after some arduous behind the scenes work, for the University of Newcastle to award an honorary degree to a Soviet scientist, Professor Beloussov of the Soviet Academy of Science. The degree ceremony was scheduled in August 1968, just as the Russian troops marched in to Czecho-Slovakia: what a contretemps and Beloussov had just arrived in Newcastle ready to take his degree. Runcorn was embarrassed, but had to go through the motions as sponsor, so was the University Chancellor, Professor Bosanquet, who had to confer the degree of Doctor Honoris Causa on Beloussov. The University academic staff were on the warpath and voted with their feet in boycotting the ceremony. Hardly anybody was in sight, so, the resourceful Runcorn had to ask all the students, secretaries, and cleaning ladies in the University to fill the public gallery in order to furnish the hall. It was rather like Catherine the Great's Siberian villages, except that on this occasion, instead of bogus villages, we had a bogus audience. All rather low key, with poor Beloussov bearing the brunt of the invasion, but he was rather thickskinned, so he could take it.

*(Roman, 2000: 54)*

14 Yang and Oldroyd (2003) give an account of Ma's work.

15 Nor did Crowell question in print the evidence in support of Permo-Carboniferous glaciation on Gondwana continents. I asked Crowell (January 2008) if he questioned them during any conversations, especially with Bucher, at the NATO meeting. At the age of 91, he declined to answer.

16 Figuring out the nature and age of the Squantum strata proved to be quite difficult. Socci and Smith (1987) give an excellent summary. Before Dott's 1961 study the Squantum was viewed as evidence of late Paleozoic glaciation, and taken to be contemporaneous with the vast glaciation in the Southern Hemisphere. The date was established from two purported tree fossils. They were the only fossils. But investigators, believing that Earth had undergone general refrigeration to account for the massive Permo-Carboniferous glaciation in the Southern Hemisphere, were quite happy to accept a Permo-Carboniferous age for the Squantum strata. Dott failed to convince many that the Squantum was not glacial; however, a few researchers continued to argue that it was glacial but also involved the action of turbidity currents. All researchers continued to assign a late Paleozoic date to the Squantum. However, Thompson and Bowring (2000) obtained Precambrian radiometric dates for the Squantum and Lenk *et al*. (1982) found Precambrian microfossils. The current view is that the Squantum is late Precambrian (Vendian) and marine glacial. Once the Squantum was taken to be late Precambrian, it was associated with the vast glaciation that is now believed to have occurred during the late Precambrian – Harland was

right. Moreover, researchers matched the Squantum deposits with similar ones
in Newfoundland and Nova Scotia. With the advent of plate tectonics, the Boston area,
Newfoundland, and Nova Scotia are taken to be part of the Avalonian terrane. But it
should also be added that most workers believe that turbidity currents played some role in
forming the Squantum.

    The only strong support for the marine glacial interpretation of the Squantum is the
presence of dropstones. Dropstones, often a meter to several meters in diameter,
are embedded in icebergs. They drop to the bottom of the ocean when the iceberg melts.
Dropstones are distinguished from other stones by the surrounding sediment; they
have disrupted laminae below, and draped sediments above them. I also recommend
Billings (1982) for an interesting and entertaining summary of work in the Boston Basin
and Squantum. He probably devoted more time than anyone trying to figure out the
geological history of the Boston Basin. His students mapped most of the Boston Basin in
his field courses, and every new excavation for roads and tunnels gave him and his
students more to investigate.

17 Of course, until the Squantum rocks were determined to be Precambrian and not Late
Paleozoic they were not associated with any nearby continental glacial strata. Nearby Late
Paleozoic strata were all indicative of a warm environment.

18 Brynjolfsson actually reported:

Twenty-five postglacial lavas were investigated. The intensity of magnetization . . .
in lava-flows, which cooled down after A. D. 500, fluctuated between $1.2 \times 10^{-2}$ and $1.1 \times 10^{-3}$ e.m.u./g with a mean of $5.1 \times 10^{-3}$ e.m.u./g . . . The intensity in the older
flows was about 20% smaller.

Contrary to Bucher, the lava-flows were not weakly magnetized.

19 While working on his book, Popper visited ANU for some weeks and gave a course of
lectures, and Irving attended them (November 1999 comment to author). He recalled,
"I felt he described, in his lectures, very well what I was actually doing. I did not discuss
things with him, but just marveled from the back of the class, and then went off and read
some of his works." Irving obviously thought Popper's remarks were relevant to the
mobilism controversy, and the reception of its paleomagnetic support. He also placed the
following quotation from Popper's *Conjectures and Refutations* on the frontispiece of his
*Paleomagnetism*, which he obviously felt described the current situation in the Earth
sciences.

. . . a theory may be true even though nobody believes it, and even though we have no
reason for accepting it, or for believing it is true; and another theory may be false,
although we have comparatively good reasons for accepting it.

20 Harland died in 2003 before finishing what became Harland (2007). The editors of
*Geological Magazine*, with the help of I. Fairchild, lightly edited the manuscript and had it
published.

21 Reactions of biogeographers and paleontologists are discussed in Frankel (1981, 1984a, b).

22 Irving recalled that Darlington heard him talk about the Paleozoic data of Australia,
and the positive response of Darlington, the former fixist, made Irving feel that "all was
not lost."

Darlington came to my lecture on Paleozoic data of Australia at MIT winter 1964–5. We
chatted afterwards, and he gave me or sent me later a copy of his book, which I still
have. Meeting him made me feel that all was not lost.
*(Irving, November 1999 comment to author)*

23 How prophetic it is that they called for magnetic surveys of the ocean floor. They
did not have the Vine–Matthews hypothesis in mind, and they made no explicit reference
to it or to Hess's or Dietz's version of seafloor spreading. They had in mind a magnetic
survey similar to one over the Mesabi Range, which revealed formations rich in iron. They
had showed (1965: 45) results of such a survey in a previous chapter.

24  Kay and Colbert included among their mobilist reconstructions one of a Cambrian Pangea
    by Amadeus W. Grabau. Grabau spent much of his career in China. He was a
    paleontologist. He defended mobilism in his (1940) *The Rhythm of the Ages, Earth
    History in the Light of the Pulsation and Polar Control Theories*, which was published in
    Peking. He had little influence on other mobilists. Marvin (1973: 110–114) discusses
    his work.

25  Carey (Figure 9, 1958: 206–207) described the movement along these megashears and
    the Great Glen Fault as left lateral. Kay's Figure 19-10, shows the movement along the
    Great Glen Fault as left lateral, while the movement along the Hampden and Lukes
    Arm Faults in Newfoundland is shown as right lateral (Kay and Colbert, 1965: 465).
    However, he and Colbert reversed the movement along the Great Glen fault in the text
    where they wrote:

    The Great Glen Fault crosses central Scotland (Fig. 19-10); it is a transcurrent fault, one in
    which a more northerly block moved easterly relative to a more southerly one, called
    right lateral because as one stands on one block, the other block that he faces
    moved relatively to the right. In eastern Newfoundland, an important right-lateral fault,
    the Lukes Arm Fault, enters from the sea in central Newfoundland ...

    *(Kay and Colbert, 1965: 466)*

26  See for instance Wellman *et al.* (1969), Schmidt (1976), Schmidt and Embleton (1981), and
    Opdyke *et al.* (2000).

27  The Carboniferous map of Köppen and Wegener is Figure 3.7 of Volume I. For the
    Triassic through Cretaceous maps see du Toit (1937), Figures 39–42. Australia's position
    given by them is in excellent agreement with the paleomagnetic results of Figure 1.2 herein.

28  The Karroo basalts are now known from radiometric dates to be Early Jurassic (Jones
    *et al.*, 2001) and the Lupata Series Cretaceous (Briden, 1967b).

29  Their privately published journal contains separate entries by Neil and Marjorie Opdyke;
    Marjorie's the lion's share. She provided a straightforward description of journeying across
    southeastern Africa with their infant and two-and-a-half year old sons. I thank them for
    letting me quote from their journal.

30  Seven lists were published from 1960 through 1965. They were the beginning of what,
    through the efforts of M. W. McElhinny, later became the global paleomagnetic database.

31  Irving returned to the issue of oil latitudes in 1974.

    I revisited oil latitudes much later with Ken North from Carleton University and a
    summer student Rich Couillard (Irving, North, and Couillard, 1974). Now we had data for
    middle-east fields which dominated the picture. When we included them, it became
    clear once again that the vast majority of oil reserves had accumulated in very low
    paleolatitudes. Migrations into younger strata and in a few cases (the Algerian fields)
    into older strata complicated the picture but we made suitable corrections. It was a large
    task and our paper became widely used.

    *(Irving, June 10, 2007 email to author)*

    Irving has no direct evidence that his work on oil latitudes led to the discovery of new fields,
    but the citation for his 2005 Wollaston Medal says that it did.

32  Briden's other Ph.D. work included paleomagnetism of the Adelaide System and Cambrian
    of South Australia and of pre-Carboniferous rocks of southeastern Australia, studies of old
    secondary remanent magnetization, the intensity of the ancient geomagnetic field and of
    the magnetic inclination in bore cores. These other aspects of Briden's Ph.D. work assumed
    importance after 1970 (beyond the scope of this book) when a return was made to
    paleogeographic studies and to Paleozoic and Precambrian mobilism. With this strong
    background, Briden succeeded Opdyke as an NSF Fellow (1965–6) at the University
    College of Rhodesia (§1.17).

33  Briden and Irving were right to choose Hill as an example, but it did not seem to have done
    much good. Her response in the Clarke Memorial Lecture for 1971 (prepared the year

before) illustrates the strength of the resistance to mobilism among paleontologists. Even after the introduction of plate tectonics, she could (referencing their paper) still claim:

> We are forced to admit that either polar waters were warmer then to support such reef growth, or that the invertebrates involved in reef formation could proliferate in colder waters than those of today. Thus we could not say that the palaeomagnetic latitudes deduced for the Lower Carboniferous give better palaeoclimatic zonal arrangements than would the undrifted continents. This does not, of course, prove that continental drift has not occurred.
>
> *(Hill, 1970: 99)*

Indeed! One is forced to accept a grossly different latitudinal zonation or a significant increase of reef-building in colder waters, only if one is unwilling to accept mobilism. She was unwilling to admit that repositioning of continents as indicated by paleomagnetism gave palaeoclimatic zonal arrangement much as at present and did not require proposing that reef-builders thrived in much colder waters than today.

34 Stehli (1957, 1959) had already objected to paleomagnetism's support of mobilism by arguing that the distribution of brachiopods and fusulinid *foraminifera*, two groups of Permian marine invertebrates, is better explained by fixism and a *generally* mild Permian climate than by mobilism (§1.9). Runcorn (1959c), and later Irving (1964: 217, 224), argued that Earth's Permian climate was not mild for there was the well-established Permo-Carboniferous glaciation, and Irving proposed that fusulinid distribution was better explained by mobilism than by fixism (II, §7.10).

# 2

# Reception of the paleomagnetic case
# for mobilism by several notables: 1957–1965

## 2.1 Introduction

The paleomagnetic case for drift has been the main subject of the previous volume and previous chapter, and in this chapter I review opinions expressed about the case by six notable non-paleomagnetists during the mid-1950s up to the eve of plate tectonics. All six played important roles, either directly or through their influence, and so their attitudes need monitoring. I close with a summary of the case's general development and reception. I argue that it confirmed continental drift by providing a difficulty-free solution to the fixist/mobilist question, and show that this was recognized by thoughtful researchers who had worked through the complexities. It was not, however, recognized by the majority of Earth scientists, who generally had not seriously thought about and understood what paleomagnetists had done; they saw no evidence for drift in their own work, and they continued to doubt the validity of the paleomagnetic results long after the various difficulties raised against them had been removed.

Gutenberg, a longtime mobilist, thought the new paleomagnetic evidence for continental drift was strong and broadened drift's effectiveness. Holmes, the old mobilist, enthusiastically welcomed it. Both Bullard and Vening Meinesz came to favor mobilism during this period, being strongly influenced by the new evidence. MacDonald and Jeffreys rejected it outright and remained fixists. Unlike some of their contemporaries, these six, with the possible exception of Jeffreys, were in a position fully to understand the paleomagnetic case for mobilism. Jeffreys had the intellect, but did not keep abreast of developments. Here to begin is the seismologist Beno Gutenberg and his mobilist theory.

## 2.2 Gutenberg's career

Before coming to California, Gutenberg had never personally experienced an earthquake. On Friday evening, March 10, 1933, he attended a lecture on the California Institute campus, given by Albert Einstein. After the lecture, the two were walking together, discussing earthquakes and seismic waves, when one of the staff stepped up to them and abruptly asked whether they had noticed the earthquake. The response was "What earthquake?"

*(Charles Richter, 1962: 93–94)*

Gutenberg and Einstein were concentrating, and they knew how to. Gutenberg was a leading geophysicist and influential seismologist during the first half of the twentieth century.[1] He and Reginald Daly, both immigrants, were the most important Earth scientists in the United States supportive of continental drift before the rise of paleomagnetism. Born in 1889 in Darmstadt, Germany, where his father owned a soap factory, he was the elder by four years of two sons, who were expected to join the business. But Beno was interested in science. While attending the local *gymnasium*, he helped operate the meteorological station, and became interested in weather forecasting and climatology. Learning that Emil Wiechert offered a course at the Institute of Geophysics of the University of Göttingen on the use of geophysical instruments, Gutenberg moved there in 1908, and obtained his Ph.D. in 1911 at the age of twenty-two, learning from leaders in geophysics and mathematics. He described his education when accepting in 1953 the AGU's Bowie Medal awarded for unselfish cooperative research.

Originally I had intended to study mathematics and physics, although during school years I already had an interest in problems of weather forecasting and climatology. Thus, when as a sophomore, as it would be called here, in Göttingen in 1908 I found that a course on instrumental observations of geophysical phenomena was offered at the Geophysical Institute under Professor Wiechert, I took this course with three other students. We were introduced there into observational methods of meteorology, into the handling of seismographs, the reading of records, and the determination of exact time by astronomical means. In addition, I took three lecture courses by Professor Wiechert during that sophomore year, one on terrestrial magnetism, one on tides and one on geodesy. During my third year, I followed this up with a course on earthquakes, a geophysical seminar, and advanced laboratory work in seismology, all under Professor Wiechert, who then told me that I had learned practically all that was known in seismology, and was prepared to continue with research for a Ph.D. thesis … Most students had no difficulty in taking advanced calculus and advanced physics in their "freshman year" and got an early basis for advanced study in all fields of natural science. For example, during my second year at the university I took courses on principles of mathematics by Hilbert, mechanics by Klein, and algebra by Minkowski; during my third year I took theory of numbers by Hilbert, infinite series by Landau, potential theory by Wiechert, non-Euclidian geometry by Klein, elasticity by Born, and a year later theory of functions by Weyl, as well as mathematical exercises for advanced students by Landau. However, after receiving my degree, I was not yet ready for a full-time assistantship but took one more year of advanced courses and research during which I worked out and finished my paper on the Earth's core.

(Gutenberg, 1953: 354–355)

It was a very thorough education. His paper on Earth's core gave the first accurate estimate of the depth of the core/mantle boundary.

At the outbreak of World War I in August 1914, Gutenberg was inducted into the army. Serving in the infantry he was, like Wegener, almost immediately injured in the head by a grenade. His helmet saved his life. In 1916 he volunteered for the weather forecasting service, and was sent to various fronts where he attempted to predict, and thus avoid, the backward drift of poison gases onto German troops. He also may have been the first to use seismology as a military tool, estimating the position of enemy artillery from travel-times of the sound waves produced when fired.

After the war he took a position of Privatdozent at the University of Frankfurt-am-Main, and was promoted to Professor Extraordinarious in 1926. That year his father died, and Beno was obliged to run the family's soap factory, doing research in his spare time and on weekends. Gutenberg's old mentor, Emil Wiechert, died in March 1928, and Gutenberg was his natural successor. However, he was not chosen, and was unable to find a good academic position despite his high reputation; Max Born (Knopoff, 1999: 127) believed rising anti-Semitism responsible. The Carnegie Institution in Washington organized a conference in 1929 at its Seismological Laboratory in Pasadena, California. Gutenberg and Jeffreys were invited, and the two geophysicists, who had already begun corresponding, finally met. Gutenberg so impressed the attendees that he was offered a post by the California Institute of Technology (Caltech), where he would have the opportunity to work with staff at the Seismological Laboratory. He accepted and took up the new post the following year. In 1936 the seismological program of the Carnegie Institution was transferred to Caltech. Gutenberg directed the program, although he was not officially given the title of Director until 1947, which he retained until 1958. He retired in 1959. He made great contributions to seismology – his determination of the depth to Earth's core, made at age twenty-three, was only the beginning. Working with Charles Richter, Gutenberg calculated time-travel tables for seismic waves from distant earthquakes to rival those produced by Jeffreys and Bullen. Gutenberg also helped Richter develop his scale of earthquake magnitude. He confirmed Wadati's discovery of very deep foci earthquakes, and he pioneered the geographical mapping of them and of earthquakes of shallow and intermediate foci. He argued, correctly, for the existence of a low-velocity layer in Earth below the Mohorovičić discontinuity. In addition to the Bowie Medal (1933), he was awarded the Wiechert Medal by the German Geophysical Society (1956), was a member of the NAS, and President of the Seismological Society of America from 1945 until 1947. He died in 1960.

## 2.3 Gutenberg supports mobilism during the 1920s and 1930s

Gutenberg was an early supporter of mobilism, and many of the modifications he proposed to Wegener's theory were based on new seismological information. He stressed the importance of mobilism's success in solving problems in palaeoclimatology, climate being an early interest. In 1927 he first introduced his "*Fliesstheorie*" or theory of continental flow as an improvement on Wegener's theory.[2] He thought that seismological findings indicate that the present floor of the Pacific is entirely simatic, while the other oceans have an upper thin sialic layer. On Wegener's theory, all ocean floors were simatic. According to Gutenberg (1927), Earth's surface was completely sialic before the Moon was ripped away very early in Earth's history, leaving a scar that became the Pacific Basin. What remained of the original sialic layer was located primarily in the Southern Hemisphere. Gutenberg proposed that the sialic layer spread outward as a result of isostatic readjustment, and equatorwards because of *polflucht*. The originally east–west trending Caledonides formed as the spreading sialic mass reached the equator. As the sial continued to advance north-ward, the Hercynides formed at the equator and south of the earlier Caledonides. As northward migration continued, the Alps formed to the south of the Hercynides and also at the equator. Subsequent general northward migration moved all three orogenic belts into the Northern Hemisphere. Meanwhile the mass of sial, thicker under continents than under oceans, also spread to the east and west, resulting in its separation into the present continents. Isostatic adjustment kept the thicker sialic areas beneath continents higher than the thinner layer beneath oceans. In this way, Gutenberg argued that the timing of his spreading of the sial both northward and toward the Pacific Basin explained the formation of the three Phanerozoic east–west trending mountain ranges. He claimed that his positioning of the continents offered a better paleoclimatic solution than did fixism, especially for periods preceding the Carboniferous.

Gutenberg presented a revised version in 1936. Beginning with a summary of the reasons why he had proposed his *Fliesstheorie* in the first place, he then explained its advantages over Wegener's theory.

[Wegener's] work was a first attempt to explain a vast body of problems with one hypothesis, and it is therefore, not surprising that numerous objections arose against general or specific parts of this statement. One by the writer. According to Wegener's concept, the continents were blocks of sial, completely separated from each other by the sima forming the bottoms of the oceans. Seismological, as well as geological, observations show a decided difference in composition between the upper crust beneath the Pacific Ocean and that beneath the Atlantic and Indian oceans ... Further, the continents and the Atlantic Ocean bottom are found, from seismic as well as geologic data, to be continuous; the transition from the continents to the Pacific Ocean is abrupt ... To obviate these and other discrepancies in Wegener's theory, without affecting its advantages, the writer, in 1927, suggested a modification of the concept, to the effect that the continents did not break, but flowed apart, and that a sialic connection between them still exists across the bottom of the Atlantic and the Indian oceans.

*(Gutenberg, 1936: 1588; my bracketed addition)*

Although Gutenberg still kept his *Fliesstheorie* largely intact, even reusing the same diagram to display the former positions of the sial and continents, he did make several changes. He began by abandoning the idea that the Pacific Basin had formed by the birth of the Moon, realizing that he could not account for its momentum, as Wade (1935) had argued a year earlier (I, §8.13). This left Gutenberg without an explanation of the simatic layer beneath the Pacific Basin. Notwithstanding, he maintained, for seismological reasons, that the floor of the Pacific Basin was simatic, while the continents and the floors of the Atlantic and Indian Basins had a thin sialic layer. He was undecided about the Arctic, data being scanty. Finally, he argued that his theory explained the distribution of deep-focus earthquakes. In his 1936 version, he did not repeat his explanation of the origin of the Caledonides, Hercynides, and the Alps.

H. H. Turner had proposed the existence of deep-focus earthquakes as early as 1922. But their existence remained doubtful until 1927 when "proved beyond doubt" by the Japanese seismologist K. Wadati. Gutenberg argued that differential movements along "faults" caused deep-focus earthquakes, which, together with their distribution around the Pacific, would be expected from his theory.

The conclusion to be drawn from all these observations is that deep-focus earthquakes are neither produced by an explosion-like event nor by a collapse of material but by differential movements along "faults." In the regions from which there are sufficient data, these movements are in the same direction in deep and in shallow shocks. The "faults" are farther inland around the Pacific Ocean. This does not necessarily mean that they dip inland under angles of 30 to 60 degrees, but, more likely, that the faults at the greater depths are displaced in a direction away from the Pacific Ocean. If the Fliesstheorie be correct, this fact could be easily explained, for the sial of the continental crust, flowing at the surface toward the Pacific, must be replaced at depth by sima, flowing in the opposite direction.

*(Gutenberg, 1936: 1604–1605)*

As the sial spread outward it displaced sima. But where did the sima go? It flowed back under the advancing sial as indicated by the dip of the "faults." This was certainly an original way of looking at what are now called Benioff or Wadati zones.

Next, he argued that Earth's mantle possesses little strength. His argument was an old one; Wegener had used it.

Just below the melting point, and at higher temperatures, materials have no strength. For this reason, it has been assumed that the strength decreases rapidly with increasing depth and is practically zero, possibly at depths below 50 kilometers, but at least at depths of a few hundred kilometers. The fact that the slight stresses, remaining from the melting of the ice after the ice age, still produced a rising of the areas in the Great Lakes region and north of it in North America, and in Scandinavia, indicates that plastic flow is possible at relatively slight depths and that the strength must decrease rapidly with depth. Isostasy would be impossible, if there were noticeable strength at depths of a few kilometers.

*(Gutenberg, 1936: 1605–1606)*

A region of little strength so close to the surface gave Gutenberg the plastic flow he needed for sialic masses to spread laterally, and for continents to drift apart. However, if the mantle had little strength, how could deep-focus earthquakes occur?

The cause of deep-focus earthquakes seems due to one of two possibilities: either stresses originate suddenly, or there is great strength at depths down to 700 kilometers, which makes accumulation of stresses possible during the period of several years – the time interval between succeeding shocks in the same region. The first possibility seems to have been eliminated by recorded facts; the second possibility would mean that isostasy is impossible. The later hypothesis has, therefore, been adopted by only a few, but in all publications on this problem authors are clearly at pains to indicate that both possibilities meet serious objections.

*(Gutenberg, 1936: 1606)*

Gutenberg had a solution: an upper mantle devoid of strength, which would allow for mantle creep.

Apparently, it has been overlooked that there is a third possibility ... which avoids all difficulties: viz., that the material at these depths has no strength, but the viscosity is so great that plastic flow takes place slowly. This would enable the restoration of isostatic equilibrium, if it is distributed, in the course of hundreds or thousands of years, but would not prevent, on the other hand, the accumulation of stresses during a few years. The fact that the equilibrium in the Great Lakes region and in Scandinavia is not yet complete, is not only in favor of such an assumption, but even requires it ... One may safely conclude therefore, either that stresses, acting in the uppermost crust and producing shallow earthquakes there, induce almost as great stresses at depths of a few hundred kilometers, or that the stresses originate at these great depths, and that the stresses producing shallow earthquakes are only a consequence of these deep-seated stresses. Possibly, both types of mechanism occur.

*(Gutenberg, 1936: 1606–1607)*

He also referred to what had, by now, become the familiar mechanism problem. Although he (1936: 1609) acknowledged that the appeal to radiogenic heat "by Joly, Holmes, and others" may be of "possible importance," he concluded, "the problem – how the energy liberated in the radioactive process is finally used up in mountain-making – is still unsolved."

Gutenberg did not in 1936 pay much attention to paleogeography. As already noted, he illustrated the probable distribution of continents and oceans during Carboniferous, Cretaceous, Eocene, and Recent times using the same rather vague diagrams as he had in his 1927 papers. He did not consider it his place to fill in the details, and noted Wegener's unwise attempts to provide specifics based on studies he could not himself properly evaluate. He relieved himself of this burden, insisting that it was the job of geologists and paleontologists to provide the details.

All, including the writer, agree that it is impossible for one man – and especially a geophysicist – to discuss all details involved with the Fliesstheorie. It is the geophysicist's task to outline the general problem and to test the physical foundations, but the detailed mapping of the world and of the distribution of the sialic parts in each period must be done by geologists and paleontologists. They are familiar with the fundamental data and need not rely only, as Wegener and the writer had to do, on papers by others, with little possibility of independent judgment concerning the methods and accuracy of the findings. The writer, therefore, refrains from any attempt to improve his original maps or to add sketches for earlier periods.

*(Gutenberg, 1936: 1609)*

In the mid-1930s, little did he know that it would be continental paleomagnetists, and marine geologists and geophysicists who would be doing the mapping.

## 2.4 In the 1950s Gutenberg reconsiders mobilism
## and appeals to paleomagnetism

In 1951, in a lengthy essay entitled "Hypotheses on the development of the Earth," Gutenberg revisited his *Fliesstheorie*. This served as Chapter IX in the second edition of *Internal Constitution of the Earth* (1951b), which he edited under the auspices of the "Committee on the Interior of the Earth" of the US National Research Council. The first edition (1939) was out of print, and the Committee decided that instead of reprinting they would engage someone to assemble a new edition. Gutenberg obliged. Many essays were extensively revised, including his Chapter IX. Much new data and several chapters were added.

Gutenberg had now become a less confident mobilist than he had been during the 1930s, echoing the general decline of enthusiasm for it. In the introduction, he (1951a: 4) informed readers, "It is noteworthy that in 1949 our knowledge concerning Earth seems to be less definite in many respects than it seemed to be in 1939." Turning to hypotheses about Earth's major features, he began by noting, "Books and papers dealing with hypotheses on the development of the earth's crust are as the sands of the sea" (Gutenberg, 1951b: 178). He began by considering hypotheses, viewing convection more positively than contraction, and favored both polar wandering and continental drift, but did not see the last two as necessarily at odds with one another.

He (1951b: 211) then summarized his own ideas, and explained their advantages over Wegener's. He again labeled mechanism as "the most serious problem to be solved," regarded convection as promising, and referred to Holmes, Joly, du Toit, and others. But he did mention two new discoveries favorable of large-scale relative movement – displacement across the San Andreas and other faults in California, and paleomagnetism. Referring to the new paleomagnetic work at the Carnegie Institution of Washington, he thought the young science might prove useful in solving the problem of crustal movements.

Results on the secular movements of the magnetic poles can be expected to contribute to the problem of movements of the continental block. Preliminary results of investigations on remanent magnetism in rocks carried out under the supervision of M. Tuve at the Department of Terrestrial Magnetism of the Carnegie Institution of Washington indicate that there was no appreciable change in the average direction of the magnetic declination or inclination in various parts of the United States since the Jurassic (at least 50 million years ago). Several rock samples from the early Silurian show a magnetization with a direction almost opposite to that of geologically younger samples. While most of the data do not agree with the curves in Fig. 12 [a figure showing polar wander curves put forth by Kreichgauer, Köppen and Wegener, Milankovitch, and Köppen], it is too early to draw conclusions. A larger body of data and a critical discussion of their interpretation promises to aid greatly in solving the problem of crustal movements during the earth's history.

(*Gutenberg, 1951b: 204; my bracketed addition*)

Events proved Gutenberg was correct about the potential of paleomagnetism. Indeed, both Hospers (II, §2.9) and Creer (II, §3.6) used his Figure 12 or its equivalent in their initial discussions of polar wandering based on paleomagnetic findings.

It is notable that he did not explicitly mention relative displacement of continents. Nor did he mention paleomagnetism in his discussion of continental drift, but only when considering polar wandering. It seems he had yet to realize that it would be helpful in discriminating between them, but thought of it only in terms of what it said about motions relative to Earth's axis of rotation. At the time, the rise of paleomagnetism in Britain had after all only just commenced (II, §1.1–§1.6) and it would be three years before Creer drew the first APW path (II, §3.6).

Eight years later, a year before his death, Gutenberg (1959) again reviewed his *Fliesstheorie*. His summary remained unchanged. He added further examples of horizontal movement along fault systems in New Zealand and Japan and "fracture zones" in the northwest Pacific, but his major addition was a well-informed review of paleomagnetism, which he now thought had progressed sufficiently to offer strong support for mobilism (RS1).

The foregoing discussion [see also Irving, 1959] shows that the recent results concerning movements of the magnetic poles relative to the continents agree well with the older conception about continental movements based on paleontological data. This is an example of an hypothesis which has been strengthened by the addition of many new data of a type different from that on which it had been based originally. It seems to be more probable now than at the time when Wegener's theory had been formulated that, during the geological history, continents or portions of them have moved considerably relative to each other and relative to the earth's axis.

(*Gutenberg, 1959: 220; reference to Irving is my 1959*)

For him it was not a matter of coincidence that paleogeographic reconstruction based on paleomagnetism accorded with paleontological and paleoclimatological data. As he put it,

Thus, there are now so many areas of which the findings derived from climatological evidence and those from paleomagnetic data are in good agreement that it is very difficult to consider this a coincidence.

(*Gutenberg, 1959: 219*)

The overall argument in favor of mobilism had been strengthened by paleomagnetism and especially by its consilience with paleoclimatic evidence (RS1). He concluded his discussion of mobilism and paleomagnetism in this way:

Finally, we must realize that all the findings and hypotheses to which we have referred in this section are only rough qualitative outlines which certainly will change considerably in the future. As Graham (1956) and others have pointed out, the data are consistent with various versions of polar wandering and continental drift, but they do not reveal which processes have actually taken place. To find a better and more reliable approximation to the history of the earth during the last few hundred million years and to its present structure

THE DATA "MUST BE GREATLY AMPLIFIED AND STRENGTHENED."

This might be called the motto of this book.

*(Gutenberg, 1959: 221; upper case emphasis is Gutenberg's)*

He still did not think mobilism warranted complete acceptance. Unfortunately, Gutenberg died six years before the data were sufficiently amplified and strengthened to the point where most Earth scientists would switch to mobilism.

Gutenberg had a particularly good and thorough understanding of the new science. His twenty-two citations went up to 1959. He found Irving's 1956 paper in *Geophysica Pura e Applicata* particularly helpful. He was impressed with the great deviation between APW curves for Australia and India and those from North America and Europe. Nor did he neglect the systematic difference between the polar wander paths for North America and Europe. He cited Stott and Stacey's work countering Graham's appeal to the possible effects of magnetostriction (II, §7.5). He appreciated Irving's appeal to paleoclimatology. But he did not hide the fact (1959: 216) that "Graham ... and others point out that great caution is required in the drawing of conclusions as the observations could be interpreted differently." Perhaps he had not forgotten that Wegener had got into trouble by citing work somewhat uncritically. Unlike almost every non-paleomagnetist at the time, Gutenberg thoroughly appreciated the paleomagnetic case for mobilism. Already inclined toward mobilism, he welcomed but stopped short of recognizing it as difficulty-free.

## 2.5 Vening Meinesz reconsiders mobilism

The old fixist, Vening Meinesz, became supportive of mobilism during the early 1960s. Although it is difficult to determine how much effect the paleomagnetic evidence had, it certainly was, as I shall show, one of the more important factors that led him to change his mind. Beginning with the revision of his theory of convection, I shall describe how he believed that the new discoveries about the seafloor increased the number of problems that mantle convection could solve; I also shall argue that this did not require him to embrace mobilism fully.

By 1960 Vening Meinesz's rejection of mobilism had begun to soften. He was not in favor, but no longer rejected it outright. He first raised the possibility of mobilism after modifying his earlier account of the breakup of his proposed, very old, proto-continent to explain the origin of mid-ocean ridges. He now argued that ridges formed above the rising convection currents that had fragmented the proto-continent; originally he had believed that ridges were bits of sial left behind as fragments of the early proto-continent journeyed to their present positions. He argued that ridges floated higher than the ocean floor because they had sialic interiors. Convection accounted for their central rift valley, the prevalence of volcanoes, and their median position in the Atlantic and Indian oceans. This modification allowed him now to introduce the possibility of mobilism through the splitting apart of his old proto-continent, which he now called "the urcontinent."

> When after the period of rest the cooling at the surface again brought about instability, a new system of currents must have originated in the mantle, which distributed itself according to the thickness of the mantle. It tore the urcontinent apart in the way it was discussed above and transported the parts more or less towards the places now shown by the continents; we may leave it an open question whether this distribution was reached at once or whether this took more than one period of convection-currents. We may perhaps even assume that still the relative positions of the continents are not stable. During the first part of the earth's history, however, the transportation of the continents through the unconsolidated mantle's surface must have been a simpler phenomenon than more recent changes of relative position.
>
> *(Vening Meinesz, 1960: 417–418)*

He now began appealing to paleomagnetism. In a lecture at Princeton University, December 1958 (1961a), he invoked it as support for polar wandering and mantle convection without mentioning its support for continental drift.

> A strong argument in favor of mantle-currents is provided by the evidence given by geomag-netism which points to great shifts of the earth's rotation axis with regard to the crust. According to Runcorn, the North Pole, since Carboniferous times, must have described a curve from the middle of the Pacific via Japan towards its present site. This movement cannot have been a shift of the rotation axis in space; with regard to space, it must have remained stable but it must have meant a movement of the earth's crust relative to the earth's interior. If we assume systems of convection currents in the mantle, such a relative movement can easily be explained; it is unlikely that the drag exerted by such a current system on the crust would have no resulting moment causing such a relative movement.[3]
>
> *(Vening Meinesz, 1961a: 507)*

In a second paper, he (1961b) referred to Hess's now famous preprint "Evolution of ocean basins," in which he proposed seafloor spreading (§3.14), singling it out as especially influential, but made no mention of paleomagnetism and said nothing about seafloor spreading itself, noting only that Hess and Menard had suggested that mid-ocean ridges are short-lived.

> In long and much appreciated discussions with my friend H. H. Hess of Princeton University he gave me several arguments for the view that the mid-ocean ridges are ephemeral features, probably present during an orogenic period. One of the strongest arguments is the existence, derived by Hess from his great many soundings of atolls and guyots, of a Mesozoic ridge extending from the Mariannes arc towards Chile. It has largely disappeared since middle cretaceous time, leaving a belt of atolls and guyots which have subsided 1 to 2 km.
>
> *(Vening Meinesz, 1961b: 526)*

Vening Meinesz had been corresponding with Hess about ridges, and he wrote to him in May 1961, about two weeks before presenting these two papers (1961a, b) to the

Royal Netherlands Academy of Arts and Sciences, asking him if he could refer to his preprint. He also summarized his own views, and asked for comments.

With great appreciation I studied the paper you gave me. Have you already sent this paper around and can I mention it in a paper I am preparing for the proceedings of the Academy at Amsterdam?

As far as the mid-ocean ridges are concerned, I gradually came to the supposition that in the beginning of the Earth's history low mid-ocean ridges did already come into being because of some sial being left by the rising limbs of the mantle convection. But, following your arguments and your most important discovery of a Mesozoic Pacific ridge which [has] since disappeared, these mid-ocean ridges only became as high as they now are when rising currents originated below them, which also in our period is the case (rise because of the high temperature of the rising convection-current). The sialic core [or ridges] may, because of its higher content of radio-active matter, cause higher temperature below the ridges and thus bring about that the rising convection-currents are usually found below them. Though [this is] not necessarily so (see Pacific ridge which disappeared). The sialic core also [is] responsible for the low seismic speed.

I should highly appreciate your views on these ideas!
*(May 19, 1961 Vening Meinesz letter to Hess; my bracketed additions)*

Hess quickly replied, permitting him to quote his preprint, and commented on his old friend's sialic cores of ocean ridges.

As for your question about possible sialic material under mid-ocean ridges: (1) if there is any liquid magmatic phase coming up it would be sima (basalt); (2) it would presumably be dispersed again by moving laterally with the convection cell; (3) it seems now, having looked a little further into the matter, that [Frank] Press' suggestion of decrease in mantle velocity because of higher temperature is adequate to account for the low velocity by itself.

P.S. I see no reason why you may not quote from the preprint of the chapter to appear in "The Sea."
*(May 29, 1961 Hess letter to Vening Meinesz; my bracketed addition)*

Because of Hess's and Menard's work, Vening Meinesz abandoned his belief that the low-density ocean ridges were sialic remnants of urcontinent's breakup. Listening to Hess, he changed his mind, but remained loyal to his idea that ridges were ancient features formed during his second stage of convection (§2.5). Even though they were ancient features, formed at the same time as the breakup of urcontinent, they were only intermittently elevated as convection currents arose beneath them. Their present high elevation and the prevalence of earthquakes arose because it was a convecting phase.

Turning to continental drift, Vening Meinesz admitted that displacement of continents relative to each other may be occurring currently, but much less extensively than during the early breakup of his old urcontinent.

As the cooling continued, new convection-currents must have originated in the mantle, and we may expect these currents to have drawn the rigid urcontinent apart and to have transported the parts over the mantle till they come to a stop above areas of subsidence of the currents. A group of independent continents thus came into being, showing similarities of shape on both sides of the rift oceans dividing them. We may expect that at some time of the Earth's history the mantle surface between the continents also solidified; further relative movements of the continents must then have become more difficult. It is not easy to form an opinion at what time this occurred. As we shall deal with afterwards, the relative movements of continents still continued and probably even go on in the present period, but the velocity of movement must have been diminished.

*(Vening Meinesz, 1961b)*

Thus, by 1961, he had begun to entertain mobilism. He thought he had a solution to the origin of oceanic trenches and coastline similarities that did not require Phanerozoic continental drift. With the early Precambrian breakup of urcontinent, he thought he could explain coastline similarities and the origin of ocean ridges as a consequence of his convection currents without needing to entertain Phanerozoic drift. Perhaps he was waiting to see what new paleomagnetic work would bring. If paleomagnetism's support for mobilism did not grow, perhaps he still had solutions to two problems; if it did continue to grow, he could add subsequent phases of mantle convection and thereby accommodate Phanerozoic episodes of continental drift.

### 2.6  Vening Meinesz becomes favorably inclined toward mobilism because of its paleomagnetic support

The following year, 1962, Vening Meinesz came out in favor of Phanerozoic continental drift. Why did he change his mind? I think he did so because of the strengthening argument from paleomagnetism. Although I am not sure that this was his only reason, I suspect that once he realized that paleomagnetic support was increasing and that it was likely to continue to do so, he came to accept its consequences. I also suspect that Runcorn, who had his ear, kept him informed about this increasing paleomagnetic support for mobilism. Vening Meinesz then found a way to incorporate drift into his theory of mantle convection. Moreover, as later explained (§3.12), Hess also told him that mobilism's paleomagnetic support was compelling. It is also possible that he changed his mind primarily because he had found a way to subsume it into his theory, but this is very unlikely because he did not present his final convection solution until a year after he changed his mind. Notwithstanding, he came to accept the paleomagnetic support as strong.

What did he actually say about this support? In the 1962 anthology *Continental Drift*, he recognized two main evidences for mobilism. There was paleomagnetism and there was Permo-Carboniferous glaciation.

There are two indications that the relative position of the continents is still slowly varying [in post-Precambrian times]. The first group of indications, obtained by Runcorn and others, is provided by the determination of rock magnetizations. It forms an important part of the subject of this book and so will not be discussed here. The second indication in favor of relative movements of continents is founded on the data on the Permian glacial period. As is well known we find evidence of Permian glaciation in Southern Brazil, South Africa, the Indian

Peninsula and the southern half of Australia. In their present relative positions it is impossible to combine them in a polar cap of the dimensions we may imagine these caps to have had during a glacial period. Only relative movements of the continents can have brought about this divergence. Such movements can easily have taken place in such a way that the arctic basin was originally large enough to explain the absence of Permian glacial evidence in the northern hemisphere. The evidence in favour of relative continental movements thus obtained seems fairly reliable, although we must recognize that the whole phenomenon of glacial periods is still unexplained; the background of our evidence is, therefore, uncertain.

*(Vening Meinesz, 1962: 175; my bracketed addition)*

Of course, Vening Meinesz knew all about this glaciation evidence many years before he came out in favor of mobilism, but it had not then been enough for him to change his mind. He had invoked polar wandering, but only before the Permian Period. Thus, it appears that it was either the very new paleomagnetic evidence that prompted him to take mobilism seriously or this combined with the evidence of widespread Permian glacial strata.

He delivered a short paper to the Royal Netherlands Academy of Arts and Sciences in October 1962, entitled "Relative movements of the continents." He again stressed the importance of paleomagnetism.

As it is well known, there are many arguments in favour of this hypothesis, which have e.g., been advanced by Taylor, Wegener, Runcorn … we shall not mention them here except that Runcorn has recently given strong reasons for accepting them based on the magnetization of rocks of different periods.

*(Vening Meinesz, 1963: 3)*

The following year Elsevier published his book *The Earth's Crust and Mantle*, as the first in its series entitled *Developments in Solid Earth Geophysics*. In it he expanded his convection theory. He discussed problems that he believed his theory explained, including continental drift. Again, he regarded the paleomagnetic support as particularly strong, but he now appealed to very recent work in oceanography that indicated horizontal movement of oceanic crust, as well as to the much earlier work of Taylor and Wegener.

Likewise see that this conclusion is right, notwithstanding the rigid character of the earth's crust below the oceans, which is shown by many straight fault planes through this crust, of which evidence is given by the Menard escarpments and by a considerable number of straight rows of volcanic islands and atolls in the Western and Middle Pacific. Because of numerous arguments in favour of continental drift, given by Taylor (1910), Wegener (1929) and more recently by Runcorn (1963), this is a welcome conclusion. To the older arguments, Runcorn added important and strong arguments based on the magnetization of rocks, which has such a stability that a disagreement of their magnetic properties with regard to the magnetic field now present at the places where these rocks are located, provides us with a strong indication of the mobility of the earth's surface.

*(Vening Meinesz, 1964: 52; Runcorn (1963) is my Runcorn (1962b))*

Later in his book, he stressed the paleomagnetic support and its consilience with Wegener's results, even adding paleoclimatic arguments, but made no further mention of new oceanographic findings.

So, the result of these considerations is that the effect of the mantle currents on the crust leads to only slight deformations of the continents in the geosynclinal belts, but that the oceanic crust is more malleable and yields nearly indefinitely. These results fit in well with the great number of facts pointing to continental drift. To all the arguments given by Wegener (1929), Runcorn (1963) in recent times has added the geomagnetic arguments, based upon the fact that magnetized rocks indefinitely retain their magnetization, and that, therefore, rocks showing a magnetization that does not fit the geomagnetic field of their present position, must have undergone a transition. Runcorn's studies comprise extensive and numerous research data along these lines, and these data are completely in agreement with Wegener's results which Runcorn, moreover has further developed, especially by new studies on climatic changes. The mobility of the continents with regard to the poles is obviously included in these views.
                    *(Vening Meinesz, 1964: 76; Runcorn (1963) is my Runcorn (1962b))*

Although giving few details, there is no question that paleomagnetic evidence played a crucial role in Vening Meinesz becoming favorably inclined toward mobilism. It was his manner to devote little attention to details unless discussing his own work on mantle convection. Perhaps he was short on details because he had not read paleomagnetic papers, becoming informed about paleomagnetic work primarily through talking with Runcorn and from Runcorn's review, itself largely the work of others. It certainly was not through careful study of the original papers that Gutenberg had made in his much more thorough account of the paleomagnetic case for mobilism (§2.4). Hess also told him in no uncertain terms that he should pay attention to the paleomagnetic evidence, saying that it made acceptance of continental drift "almost compelling" (Hess, July 6, 1959 letter to Vening Meinesz; see §3.12 where I quote this important letter).

   Vening Meinesz first came out in favor of continental drift in 1962. By this time, he had come to view the mantle as crystalline, arguing that it underwent "pseudo-flow," which roughly amounted to solid-state creep. He argued that this created a drag on the rigid oceanic crust. The oceanic crust downbuckled, forming an oceanic geosyncline or trench brought about by descending convection currents. These currents upwelled along mid-ocean ridges. Because oceanic crust has nearly the same mean density as the mantle, he suggested that the shortening of oceanic crust brought about by downbuckling is much more extensive than that of continental crust. However, it was not until the following year that he fleshed out his view, and proposed that oceanic crust may actually begin to descend deep into the mantle where it melts – a prescient thought. As a result, shortening of the oceanic crust could become extensive so long as convection continued to draw the crust down.

As we suppose that the mantle-currents occur in a crystalline mantle, these considerations cannot prevent us from assuming a strong shortening over thousands of kilometers of some deep ocean-basins. We may think here of the ocean-basins, which should have to give way, if we assumed that great relative movements of the continents occur. These basins must be situated over subsiding mantle-currents. The bottom relief of the basins does not give indications about such a strong shortening, but we must be conscious of the possibility that sedimentation may hide them. Considering the rigid properties of the ocean-crust, it appears probable, that we must look for the explanation of our problem in the same direction as we did for continental crust. In the same way we find that if in a belt of relative weakness the compressive stress exceed the elastic limit, the crust will buckle downwards. In our case, however, the crust has nearly the same density as the mantle, while for the continental case it is clearly lighter. As a consequence of this the downward movement of the crustal matter can proceed to much greater depth, and the shortening of the crust can be correspondingly larger. We might perhaps even suppose that at greater depth the downbuckled matter, because of the greater heat around, would itself assume higher temperature and would mix up with mantle. We may perhaps put it in this way that in this belt the crustal matter is swallowed up and that thus great parts of the oceanic crust would disappear. In this way we might understand shifts of the continents over indefinitely large distances. We might in fact imagine that the displacement would more or less equal the displacement of the mantle-current below the crust.

*(Vening Meinesz, 1963: 6–7)*

Like the good scientist he was, Vening Meinesz changed his mind because of evidence, primarily because he found that the paleomagnetic support for Phanerozoic continental drift was, by the early 1960s, sufficiently compelling. He altered his convection theory to account for these new results (RS1), and in doing so, thought he had found a way to explain geologically late and current mobilism.

### 2.7 MacDonald denies mantle convection and Runcorn responds

Unwary readers should take warning that ordinary language undergoes modification to a high-pressure form when applied to the interior of the Earth; a few examples of equivalents follows:

| High-pressure form | Ordinary meaning |
|---|---|
| certain | dubious |
| undoubtedly | perhaps |
| positive proof | vague suggestion |
| unanswerable argument | trivial objection |
| pure iron | uncertain mixture of all three elements |

*(American geophysicist Francis Birch, 1952: 234)*

In the earlier years of the continental drift debate, the sternest critic of the mechanism proposed by Wegener, Joly, and Holmes was Jeffreys (I, §4.3, §4.11, §5.5). In later years, this role was increasingly assumed by Munk and MacDonald (II, §7.11) and MacDonald (§1.6). In the early 1960s MacDonald elaborated his objections, which I deal with now. Jeffreys returned to the fray in the mid-1960s (§2.10).

To introduce MacDonald's criticism of convection, which was directed personally at Runcorn, I start with Jeffreys' *The Earth*, to see what was known about Earth's gravitational field prior to the acquisition of data derived from perturbations in the orbits of artificial satellites. Ideally Earth's figure could be approximated by an ellipsoid of revolution, whose equatorial axis exceeds the polar by 43 km, very slightly flattened at the poles and raised at the equator. The geoid is the equipotential surface that approximates to mean sea level; it is less regular and is slightly elevated or depressed relative to the ellipsoid. These deviations are called geoid anomalies or geoid undulations. Geoid anomalies can be calculated from surface or, more recently, from satellite measurements. Jeffreys produced a map (1952: Figure 21, facing p. 177) of Earth's gravitational field obtained from harmonic analysis of Earth-based measurements. Superimposing it onto a relief map, he found no correlation between locations of geoid anomalies and elevations or depressions of Earth's topography. He (1952: 199) argued that the variations must correspond to uncompensated inequalities of density, to difference in load, and therefore to differences in stress. These differences extend far into Earth's interior, and could not, he argued, be maintained unless the mantle had great strength.[4]

Artificial satellites provided another method. By careful tracking, geodesists identified perturbations in their orbits. Factoring out effects of atmospheric drag, solar radiation pressure, and perturbations by the Sun and Moon, they mapped geoidal undulations. By the early 1960s, satellite geodesists had obtained reliable results that were in general accord with Jeffreys' Earth-based determination. Munk and MacDonald noted this agreement in a paper (1960b) that appeared shortly after the publication of their book *Rotation of the Earth*. They noted (1960b) the lack of correlation between the positions of the continents and the geoid anomalies, which they attributed to density variations in the underlying mantle; positive (negative) anomalies indicated greater (less) than average density. They also noted that some might interpret these in terms of convection, matching less dense areas with hot rising currents and denser areas with sinking cooler currents, but claimed that lack of correlation between positions of the continents and geoid anomalies spoke against convection. Although they did not elaborate on this difficulty, they could have done so by remarking that the convective pattern already hypothesized by mobilists to explain continental drift (currents rising under oceans and sinking beneath continents) was not consistent with this lack of correlation.

It is hard to escape the conclusion that density variations in the mantle, unrelated to the distribution of continents, are the important factor (aside from rotational deformation) in the gravitational coefficients of low order [i.e., the gravity anomalies]. This conclusion will be interpreted by some as favoring mantle convection. It should be added that this lack of correlation between the large features in the gravitational field and continentality was quite apparent from Jeffreys' global chart of free-air gravity anomalies [Jeffreys, 1959, Figure 21], but the satellite values have served to emphasize the discrepancy still further.

(*Munk and MacDonald, 1960b: 2171; my bracketed addition*)

If the regions of low and high density required by undulations of the geoid could not be understood in terms of rising and sinking convective currents, then they must be very long-term features, which, as Jeffreys claimed, meant great mantle strength. MacDonald closed his review of *Continental Drift* by referring to the new satellite measurements of the geoid (they were not mentioned by Runcorn), and their general accordance with Jeffreys' determination of the geoid from surface measurements as further support for Jeffreys' inference that the mantle was too strong to allow large scale convection.

Unbeknownst to MacDonald, Runcorn was already studying the new satellite data, and his first paper on "Satellite gravity measurements and convection in the mantle," appeared in *Nature* five months after MacDonald's review (Runcorn, 1963). Runcorn recalled retrospectively how he first came to hear of the new "satellite geoid."

In 1961 there was this Virginia Polytechnic Institute conference … on satellites around the Earth, and I would have heard for the first time about the first satellite geoid. The person who gave the talk was Dr. R. Kershner, and he was a group leader at the scientific laboratory at Baltimore. He gave a talk about his group's work and this was the first time I heard that the geoid had been discovered [from satellites]. I immediately thought that we have to explain these gravity anomalies in a different way from Jeffreys, and that was the origin of that paper [Runcorn, 1963]. I went on to write a number of other papers on the problem of explaining gravity fields in terms of convection, and the rather definitive first one [Runcorn, 1964d] was in 1964 in *JGR*. When I sent that one into *JGR*, it was sent to a referee who wrote a rather abusive review, and I can remember the words: "Runcorn has written several papers recently about convection, and hope that this is the last." He then went on to say that convection in the Earth's mantle was quite inadmissible on grounds of the fact that we know the mantle is rigid. I think that he was probably a seismologist. At that time in seismology, people were saying that the continents had some sort of roots down to 600 km. That was the view of Gordon MacDonald, and Gordon may have been the reviewer because he was very prominent in discussions on this matter, and he was very anti-convection. These papers interpreting the geoid in terms of convection really fell on very unsteady grounds and it has only been until relatively recently that it has been seen as evidence for convection.

*(Runcorn, 1984 interview with author; my bracketed additions)*

Even if Runcorn's surmise that MacDonald had been the reviewer is incorrect, he correctly identified MacDonald as a vehement and influential opponent of mantle convection.

The NSF sponsored the conference Runcorn had attended. R. Kershner, the scientist mentioned by Runcorn above, worked at the Applied Physics Laboratory, Silver Springs, Maryland. Runcorn (1963: 630) noted, "he was stimulated to think about the problem presented by the gravity data discussion" at the conference and singled out Kershner and his colleagues. Runcorn knew about Munk and MacDonald's early discussion of the satellite-based measurements, seemingly "a strong argument against convection."

The discovery of non-axial and odd axial harmonic terms in the Earth's gravitational field from surface gravity observations and, more recently, from the orbits of satellites has been widely interpreted as showing that the Earth's mantle possesses a finite strength and that in consequence the occurrence of convection currents in the mantle is excluded. The argument is that if the mantle acts as a fluid over periods of millions of years and convection currents are occurring in it, the continents should lie over the descending currents. Munk and MacDonald compared the coefficients of the spherical harmonics which represent the distribution of the oceans and continents with those of the gravity observations and found that no correlation exists, so that a strong argument against convection seemed to have been found.

(*Runcorn, 1963: 628*)

But he already had a counterargument (RS1); Munk and MacDonald, he argued, were mistaken in claiming a lack of correlation between topographic features and geoid undulations. There was a correlation, albeit not immediately obvious, and it indicated mantle convection upwelling beneath oceanic rises. As Runcorn put it a year later:

In fact, inspection reveals no simple correlation between the undulations of the geoid, the maps of which are groups of oval curves surrounding gravity highs and lows, and the ocean ridges which are linear features, presumably related to the convection pattern. However a new approach was made by Runcorn [1963b], who showed that while the ridges could be lines from which flow diverged they could not be lines of zero horizontal flow, as had been thought. He therefore argued that the points of zero horizontal flow would occur where the rises bifurcated or sharply changed direction. He showed that there are five such regions, and these are close to the gravity lows. As these will be over the main upwelling currents, this correlation gives support to the convection hypothesis.

(*Runcorn, 1964d: 4389; my bracketed addition*)

He was right about the obscurity of the correlation he was claiming. If the geoid could be explained in terms of convection, then gravity lows should correlate with ridges. Consider Figure 2.1, which is a slightly modified version of the figure Runcorn presented (1963) showing the highs and lows of the geoid relative to continents, oceans, and ocean ridges. There is no obvious correlation. Runcorn argued that given his convection hypothesis, with its postulation of five convective cells ($n = 5$), there should be five lines of longitude where convection currents diverge and create central rifts along oceanic rises. Next, he claimed that, contrary to what had been assumed, purely rising convection currents, without any horizontal component, would occur only at junctions of oceanic ridges; he identified five referring to his Figure 1 (Figure 2.1).

Fig. 1 shows the distribution of the ocean rises after Menard. There are five clear cases where the ridges bifurcate or sharply change. (1) the intersection of the Pacific Antarctic rise with the rise running northwards into New Zealand; (2) the branching of the mid-Indian Ocean rise; (3) the join of the north and south Atlantic rises, associated with the east-west fractures near the equator; (4) the join of the East Pacific rise with the Pacific Antarctic rise; and (5) the northern extremity of the East Pacific rise in the Gulf of California.

(*Runcorn, 1963: 630*)

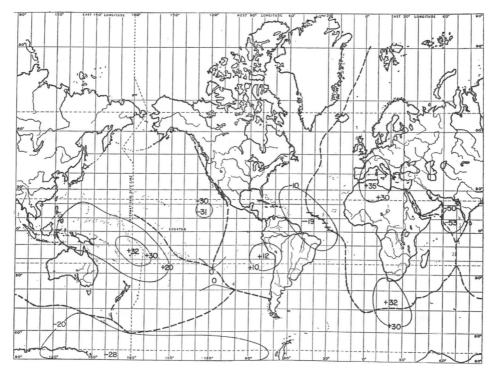

Figure 2.1 Runcorn's Figure 1 (1963: 628). The dotted lines are the ocean rises. The thin lines are contours of the geoid surface determined from satellite observations. Positive and negative numbers given in centers of contours show positions and values of gravity highs and lows in meters.

Turning to the undulations of the geoid, Runcorn identified five points of minimum and four of maximum gravity. The five minima correlated with his five bifurcating ridges: the −20 mgal and −28 mgal contours correlate with (1), the −50 mgal and −53 mgal correlate with (2), the −10 mgal and −19 mgal with (3), the 0 mgal and 0 mgal with (4), and the −30 mgal and −31 mgal with (5). He also claimed a correlation between three of the four places of maximum gravity and descending convective currents: the fourth showed no correlation, which he categorized not as an anomaly but as an unsolved problem.

On this view we would naturally associate the positive anomalies with the descending currents and therefore with zones of compression. The two zones of positive anomaly in south Europe and the Andes agree with this expectation very well. The elongated positive anomaly between the mid-Pacific and the East Indies covers a region of strong compression discovered by Vening Meinesz and the Tonga trench. The positive anomaly south of South Africa is of considerable interest, for little is known of the topography of the ocean floor in this region and points the need for investigation there.

*(Runcorn, 1963: 630)*

Runcorn also argued that the heights of the geoid anomalies roughly agreed with the rates of continental drift given his assumptions about the viscosity of the mantle and the temperature and density differences between the columns of rising and falling mantle streams. Hence he declared that mantle convection explained geoid undulations.

> The explanation for the low degree harmonics of the Earth's gravitational field ... may more plausibly be sought in thermal convection [than in the view advocated by Jeffreys, Munk, and MacDonald]. An order of magnitude argument established the reasonableness of this solution for the Earth. I showed [in Runcorn, 1962a] that if a value for the viscosity of the mantle is taken of $10^{21}$ poise, and the velocities of the convection currents are inferred from the rates of displacement of the continents, the temperature differences between the rising and falling streams on the equipotential surfaces are of the order of $1/3\,°C$. Taking a volume coefficient of expansion for the mantle of $3 \times 10^{-5}$ per $°C$, the deficiency of mass per unit surface in the uprising volume compared with the downgoing one is about $5 \times 10^3\,g/cm^2$. This gives a gravity anomaly of the order of $10^{-2}\,cm/sec^2$, or $30\,m$ in geoid height, which are the values given by Jeffreys and those who have reduced the satellite gravity observations (See Fig. 1).
>
> *(Runcorn, 1963: 629; my bracketed additions)*

In Runcorn's next paper, the one that had provoked a referee to write "[I] hope ... his last," he (1964b) went even further in an attempt to explain the geoid anomalies in terms of his convection hypothesis. If MacDonald actually was the referee as Runcorn suggested, then MacDonald might have come to regret remarking in his review of *Continental Drift*, "The editor and several of the contributors prefer to ignore the new data." Runcorn was very adept at exploiting new data. He took full advantage of the satellite data, aiming (1964d: 4389–4390) to use them as an "analytic method of determining the nature of the flow patterns in the earth's mantle." He displayed his results in the diagram shown in Figure 2.2.

Runcorn discussed the divergence of the currents in the east Pacific, the Atlantic Ocean, and the middle of the Indian Ocean, associating them with oceanic ridges and rising convection, and emphasized his findings about zones of compression.

> Turning our attention to the zones of compression indicated by the strong convergence of arrows in Figure 2, see that the clearest examples are off the Andes and the associated ocean trench, off the west coast of South Africa, off the Asian mainland, over the Japan trench, and in southern Europe. All these are associated with mountain building or ocean trenches with the exception of Africa, for which there is as yet no explanation. There is a suggestion of a convergence of arrows around the Tonga trench, though this is displaced to the east, but smaller-scale features such as the East Indies Trench and the African rift system do not emerge in Figure 2, but this is to be expected for the data are based on low harmonics.
>
> *(Runcorn, 1964d: 4393–4394)*

Runcorn thought he had converted the difficulty raised by Munk and MacDonald into a solved problem by showing his solution to be superior to that proposed by Jeffreys and supported by them. He pressed his case, arguing that their solution was

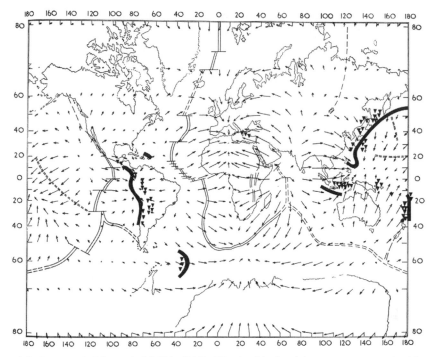

Figure 2.2 Runcorn's Figure 2 (1964d: 4393). The double-lined features are oceanic ridges, the thick dark lined features are ocean trenches, and the solid triangles are deep-focus earthquakes. The arrows are convection vectors. They were determined from a more recent geoid based on satellite measurements than Runcorn had used the previous year. They are associated with the five minimum gravity points. Incoming convective currents surround the points of convergence, which are associated with the five minimum gravity points.

improbable because it required supposing that Earth's mantle is made of materials whose creep rates are very much less than those observed under laboratory conditions, where the stresses and temperature conditions are not nearly as extreme as in the mantle (Runcorn, 1963: 629; 1964d: 4389). Runcorn summarized with sarcasm to match MacDonald's.

It is remarkable that our geophysical colleagues have not made widely known to engineers their technological breakthrough in discovering a material which, at elevated temperatures, has a creep rate only one millionth millionth of those encountered in the laboratory.

*(Runcorn, 1963: 629)*

The following year, Runcorn drew even more sharply the contrast between their competing solutions.

But *Jeffreys* (1962) has argued against this view [of convection] by interpreting the low harmonics of the earth's gravity field as evidence for the earth's mantle having finite strength, by which inequalities of density may be retained. Their origin, on this explanation is obscure, but can hardly be more recent than the last few hundred million years and might more plausibly be associated with the formation of continents. The retention of these anomalies requires that creep rates in the mantle be less than $10^{-12}$ of those observable in laboratory experiments [*Runcorn* (1963)], although the stress differences are of the order of 10 to 100 bars and the temperatures are elevated. Before postulating such extreme properties it is desirable to estimate whether the density variation could result from convection.

> *(Runcorn, 1964d: 4389; first bracketed addition mine, and references are to the fourth edition*
> *of Jeffreys'* The Earth *and Runcorn, 1963)*

It was not only mobilists who had mechanism difficulties. According to Runcorn, people who maintained that Earth's mantle was sufficiently strong to maintain the load differences implied by undulations in the geoid also had to invoke materials, he thought, with unbelievably small creep rates.

### 2.8 MacDonald renews his attack on the paleomagnetic case for mobilism

MacDonald's hostile review of the anthology *Continental Drift* was only a prelude to his later, untethered attacks on mantle convection. Later in the year, he (1963b) wrote an extensive paper that appeared in *Review of Geophysics*, a new quarterly journal of the AGU that he edited. He expanded his attack against large-scale mantle convection and its employment by Hess (§3.14), Runcorn, and Vening Meinesz as a solution to the cause of mobilism. He proposed a fixist theory of the origin and evolution of continents and oceans that included no role for convection, and he continued to question mobilism's paleomagnetic support. Indeed, MacDonald declared that he had developed his theory because he wanted one without convection.

The studies presented have been stimulated to a large extent by the emphasis of recent speculations about the mobile nature of the earth. In these speculations continents move relative to each other, the poles tip, and convection currents constantly stir the interior. It seems that a fluid earth gives the investigator a great deal of flexibility in interpreting geophysical and geological features, since all limitations on convection are then minimal. I have been impressed by the strong evidence, which I present here, that there has been relatively minor horizontal transport of material. To develop a theory of continents and oceans in which convection plays no part is the purpose of these studies.

> *(MacDonald, 1963b: 590)*

MacDonald (1964) also wrote an abridged version in *Science*, and a third for the Royal Society's symposium on continental drift. For now, I shall briefly consider only his attack on paleomagnetism.

MacDonald raised the well-worn old theoretical difficulties: the geomagnetic field may not have always been dipolar; the origin of the field was not yet established; it might not remain dipolar while undergoing reversal; and he seemed to think that the field itself, or perhaps its dipolar nature, may not be, geologically speaking, all that ancient. He appealed again to Axelrod's fixist paleobiogeography, fellow fixists seeking mutual support.

The paleomagnetic data discussed by Runcorn [1962a], Cox and Doell [1960], and others can be interpreted in terms of motions of the continents relative to each other. This interpretation presumes that the field has remained dipolar over geologic time. The theory of the causes of the earth's magnetism is far from being well established, and it can be argued that the observed reversals in the field suggest a rather short time scale for the field and intervals during which the internal field may not be dipolar. Further, there is accumulating paleontological evidence that argues strongly against continental drift [Axelrod, 1963].

(MacDonald, 1963b: 611; Runcorn reference is to his own (1962b) contribution in Continental Drift, *other references same as mine*)

MacDonald's attack on paleomagnetism did not have quite the spirit of his attack against large-scale mantle convection; of course it was not his expertise, but that had not previously hindered him. His attack was milder than what Munk and he had mounted in *The Rotation of the Earth*; he did not play off Cox and Doell against Runcorn, and did not, as he had previously, emphasize Cox and Doell's reservations about the incompleteness and variable reliability of paleomagnetic data. Whether or not this notable change in tone reflected any subtle change in MacDonald's attitude toward the paleomagnetic support for mobilism, he certainly remained adamantly opposed to mobilism and mantle convection.

## 2.9 Harold Jeffreys, his career

[Harold Jeffreys] was a solitary thinker and, like Newton, he thought by himself to some purpose.

(Alan Cook (1990: 304), Biographical memoir of Harold Jeffreys)

In 1961 the RAS sponsored a symposium entitled "The Earth Today" to celebrate Jeffreys' seventieth birthday. Some of Jeffreys' former students and colleagues presented papers, which were published two years later as Volume 4 of the *Geophysical Journal*. The volume sold well, and the RAS, as noted in the report of its Council at the Society's 143rd Annual Meeting (see Anonymous, 1963: 152) used the proceeds to honor Jeffreys once again by establishing "the Harold Jeffreys Lectureship for an annual lecture on some subject of geophysical interest to the Society." This was quite an honor since the only other lectureship the society had established was named for George Darwin, Jeffreys himself delivering the first lecture on October 25, 1963. He expressed his appreciation to the society, paid homage to Darwin, and recalled that Darwin's work was important in getting him interested in geophysics.

I should like to make it clear that although I am greatly honoured by having my name given to the lecture, I felt great reluctance in agreeing to it. The Council must have thought of Sir George Darwin, but his name is pre-occupied. However, as I have been asked to give the first lecture I think it is appropriate that I should say something about the extent of his influence on modern geophysics. I became interested in problems of cosmogony and geophysics through reading his popular book on the tides. I never knew him personally and he died during my third year at Cambridge. But I had often seen a man practicing archery when I was taking walks on Coe Fen, and I found out much later that Darwin was an archer and his home was near by. So I must have seen him without knowing who he was.

*(Jeffreys, 1964: 10)*

Jeffreys proceeded to speak about Earth's strength, and argued against continental drift and its paleomagnetic support. In fact, his attack on paleomagnetism was more vigorous than that he had written two years before in the fourth edition of *The Earth* (II, §7.13). However, before examining his critique, I want to use the occasion of the establishment of the Harold Jeffreys Lectureship to comment briefly on Jeffreys' brilliant career. It is a vast mistake to view Jeffreys only in terms of his unbending rejection of mobilism, for doing so not only ignores the fact that he was one of the principal geophysicists of the first half of the twentieth century, it also makes difficult understanding why his continued rejection of mobilism before the 1960s was taken so seriously by so many for so long.[5] Indeed, none other than Runcorn had advised geologists on both sides of the Atlantic to read Jeffreys, if they thought geophysics was nothing more than "continental drift, seafloor spreading, plate tectonics and mantle plumes ..."

Geologists on both sides of the Atlantic who suppose – perhaps understandably as the highways of earth science are presently congested by careering band wagons – that continental drift, seafloor spreading, plate tectonics and mantle plumes is the whole of geophysics might do well to read again Sir Harold Jeffreys' *The Earth* ...

*(Runcorn, 1974: 794)*

Jeffreys has already appeared in earlier volumes (I, §4.3, §4.11, §5.3–§5.6, §5.9, §8.13; II, §7.13) and it is now time to say more about him and his very long involvement in the drift debate. Jeffreys was almost ninety-eight when he died in 1989.[6] The future Sir Harold Jeffreys was born in April 1891, at Fatfield, County Durham, he was the only child of Elizabeth Mary and Robert Hall Jeffreys. His parents were schoolteachers. As a young boy, he was interested in botany. Jeffreys attended school at Fatfield, where his father was headmaster. Winning a Durham County scholarship in 1903 to Rutherford College in Newcastle upon Tyne, he remained there until 1907 after which he went to Armstrong College, Newcastle. Armstrong College, the forerunner of the University of Newcastle, was part of the University of Durham. He took courses in mathematics, physics, chemistry, and, for one year, geology, graduating in 1910 with distinction in mathematics and first class marks in his other subjects.

Encouraged by his mathematics teachers, Jeffreys applied to Cambridge, and in 1909 was awarded a £60 entrance scholarship at St John's College as one of four mathematics scholars. At first he found mathematics at Cambridge much harder than at Armstrong, but still won a prize for an essay on precession and nutation of the Earth, a subject to which he often returned. His work improved. In his third year, in Part II of the Mathematics Tripos in 1913, he was a Wrangler, and was awarded one of two Hughes Prizes for undergraduates who had done best in his college in any subject. Jeffreys' financial support was increased during this fourth year, and he began research. He was elected a research fellow of St John's College in November 1914, and remained one for the rest of his life.

From 1915 until 1917 Jeffreys worked part-time in the Cavendish Laboratory on wartime problems. In 1917 he joined Sir Napier Shaw of the Meteorological Office in London and, like Wegener and Gutenberg, worked on meteorological problems during World War I. It was during this time in London that Jeffreys met Arthur Holmes, who was at Imperial College. They became lifelong friends, despite their strong differences over mobilism. Jeffreys returned to Cambridge in 1922 as College lecturer in mathematics, became reader in geophysics in 1931, and was elected to the Plumian Professorship of Astronomy and Experimental Philosophy in 1946. He was elected fellow of the Royal Society in 1925, and Foreign Associate of the NAS in 1945. He received numerous awards, and the societies giving them indicate his range of interests. They include the Buchan Prize (Royal Meteorological Society, 1929), the Gold Medal (RAS, 1937), the Murchison Medal (Geological Society of London, 1939), Royal Medal (Royal Society, 1948), the Bowie Medal (AGU, 1952), The Copley Medal (Royal Society, 1960), the Guy Medal (Royal Meteorological Society, 1963), Wollaston Medal (Geological Society of London, 1964), Harry Fielding Reid Medal (Seismological Society of America, 1978), and the Vetlesen Prize which he shared with Vening Meinesz (Columbia University, 1962). He retired in 1958 to remain active in research and continued to publish.

Jeffreys made very important contributions in fluid dynamics, seismology, dynamics of the solar system and Earth, geophysics and physics of Earth's interior, probability and scientific inference, and mathematics. I consider a few. Jeffreys became interested in fluid dynamics while working at the Meteorological Office. He dealt with problems concerned with why fluids resist motion through them, and these included studies of the dynamics of winds in the atmosphere, waves in streams, shallow seas and oceans, and convection in a variety of conditions and media. Jeffreys' contributions to seismology rivaled those of Gutenberg. The tables he and K. E. Bullen, one of his students and eventual fellow of the Royal Society, constructed for the travel times of seismic waves generated by distant earthquakes, rivaled those of Gutenberg and Richter. He was the first to demonstrate that Earth's core was essentially liquid, his analysis involved seismology, and planetary dynamics. Although it already was known that Earth's core failed to transmit S waves, implying that it was liquid, Jeffreys (1926) hypothesized a liquid core to reconcile the conflict between the required rigidity of the whole Earth as determined from precession and nutation, and its rigidity as determined by the velocity of S waves within the mantle, the latter value being much higher. He realized that this would be done by hypothesizing a liquid core. (See Brush, 1996a for details.) He did extensive work on gravity, calculating the anomalies in Earth's gravity field. His analysis of the geoid was the best available before the launching of artificial satellites (§2.7). During much of his career Jeffreys defended the view that the solar system had formed through tidal disruption of the Sun by a passing star. However, in 1948, he found all theories about the origin of the solar system flawed and thereafter supported none. He addressed issues concerned with the evolution of Earth's surface, championing contractionist theories of mountain building, learning much from his friend Holmes. But Jeffreys remained a contractionist long after Holmes switched to mobilism, and became mobilism's sternest and most enduring critic. His work on probability and scientific inference was very influential. He developed and modernized Bayes' idea that probability is a degree of belief in a statement given other information, and his work on it remains highly respected. Cook (1990: 327) reports that Jeffreys' *Theory of Probability*, first published in 1939, and reprinted three times through 1967, "was [again] reprinted in 1983 by popular request of statisticians, not all of whom adopted Bayesian principles."[7] The honor the RAS bestowed on Jeffreys by naming its second lecture series after him was surely well deserved. He is generally acknowledged to be the premier mathematical geophysicist during the first half of the twentieth century, and the range of his accomplishments is truly amazing. Even to a non-geophysicist like myself, his book *The Earth* is a model of clarity, conciseness and wit.

Jeffreys participated in three controversies: mobilism, probability, and the origin of the solar system. It is his involvement in the first that is my concern. He argued with R. A. Fisher, a frequency theorist, about probability theory; they were friends, but their friendship, like the one between Jeffreys and Holmes, did not prevent argument. Doris Reynolds, Holmes' second wife, recalled that Jeffreys and Holmes used to send rude postcards to each other. As a proponent of tidal theories, Jeffreys disagreed with the planetesimal theory, which was championed by T. C. Chamberlin and F. R. Moulton during the first decade of the twentieth century (Brush, 1996).

Although Jeffreys wrote clearly and concisely, he did not talk very much. Cook (1990: 328) describes him as shy but at heart a sociable man, and remarked that his lack of oral communication was sometimes misinterpreted.

## 2.10 Jeffreys renews his attack on mobilism in the first Harold Jeffreys Lecture

Returning now to the first Harold Jeffreys Lecture of the RAS in which he attacked mobilism, at first sight his criticisms appeared unchanged. He repeated his old argument against convection. He ignored Runcorn's hypothesis and the difficulties that he had raised against Jeffreys' rejection of convection (§1.5). He ignored Carey (II, §6.11), and repeatedly dismissed as fiction the excellent fit of opposing continental shelves, even the fit between South America and Africa was poor.

One writer called the agreement [between the coastlines of South American and African continental shelves facing the Atlantic] a coincidence; I simply deny that there is agreement.

Another said that the fit would be better if the boundaries were taken at the 1000 fathom line instead of at the coast; but this would only alter the angles by a few degrees, in the same direction, and leave the disagreement much as it was.

*(Jeffreys, 1964: 16; my bracketed addition)*

Jeffreys not only ignored Carey in print, but had no intention of ever reading him. Some years before his Sir Harold Jeffreys Lecture, Jeffreys gave a talk at ANU in 1959. David Brown (§1.18), who had recently become chair of the Geology Department at Canberra University College, attended Jeffreys' lecture and asked him about Carey's fit.

I well recall an interview with the visiting applied mathematician Sir Harold Jeffreys, who stated that if one studied a simple map of the South American – South African situation (presumably Mercator's Projection) closely, one could readily recognize a 15-degree "misfit." When I asked him whether he had examined Sam Carey's three-dimensional (almost perfect) "fit," Jeffreys said, "I have never read any of Carey's papers, and I have no intention of doing so." Ted Irving was at that meeting and vouches for that statement.

*(David Brown, February 4, 2004 email to author; Irving, February 4, 2004 email to author*
*confirmed Brown's recollection)*

Sure of himself, Jeffreys would not reconsider what he thought he already had disproved.

Regarding biotic disjuncts, Jeffreys (1964: 17) cited G. G. Simpson, Charles Darwin, and others to dismiss the "biological arguments ... based on similarities of fauna and flora on land masses now separated by oceans."

Jeffreys thought the paleomagnetic case for mobilism had weakened. He began by repeating, as Munk and MacDonald had done and as paleomagnetists had long recognized, that the paleomagnetic pole method does not provide unique positioning of the continents.

Further, it [the paleomagnetic support for mobilism] cannot be checked. Munk and MacDonald point out that if there are $n$ continents the data yield $2n$ equations. If there is polar wandering but no drift, there are 2 unknowns. If there is drift but no polar wandering there are $2n$. If there are both, there are $2n + 2$. There is no satisfactory solution in the first case. In the second, we can fit any data whatever and can get no check on consistency (except so far as one continent might turn out to be on top of another). In the third, the data can be fitted in an infinite number of ways. If we could get any check on the consistency of the interpretation it might tend to show that the hypotheses are fairly reliable, but there cannot be one.

*(Jeffreys, 1964: 18; my bracketed addition)*

But he went one better: he asked Bullard about the status of paleomagnetic support for mobilism, perhaps seeking reassurance having become somewhat alarmed by its favorable reception in some quarters. Bullard, who was by then favorably inclined toward mobilism, told Jeffreys, presumably not wanting to disappoint him, that there was yet another degree of freedom. As Jeffreys (1964: 18) put it, "Bullard points out to me that if the continents rotate there are $n$ more unknowns."

In his lecture, Jeffreys also raised theoretical difficulties by questioning the three key assumptions behind the paleomagnetic method.

Many determinations of magnetizations of rocks have been made recently. It is assumed that the rocks were magnetized by the Earth's field when they were formed, and have retained their magnetism ever since. On the assumption that the original field was not far from a dipole field along the axis of rotation, as at present, the direction of magnetization gives the direction of the axis of rotation relative to the place where the rock was found. Now these directions are found to be far from the present axis. To begin with, this suggested that the axis of rotation has moved relative to the continents and people spoke of polar wandering. Then rocks of the same period in different continents were found to give different directions of the axis – on the supposition that the continents were then where they are now. Hence it was inferred that the continents instead have moved relatively to one another, and there is a new type of evidence for continental drift, which is spoken of as a fact. It is nothing of the sort; it is an inference based on at least three shaky hypotheses.

*(Jeffreys, 1964: 17–18)*

As far as Jeffreys was concerned the ideas, that rocks could accurately record the geomagnetic field, that they retain this record, and that the axis of the mean geomagnetic field lay close to the axis of rotation, were all shaky.

Jeffreys' criticism of paleomagnetism continued unabated; he quoted Axelrod, who was fast becoming fixists' favorite biogeographer.

Evidence for large displacements in latitude has been derived from palaeomagnetism. Such displacements were formerly inferred from distribution of vegetation. A recent analysis by D. I. Axelrod contains the following passage. "There apparently is no paleobotanical evidence that unequivocally supports paleomagnetic data which ostensibly show that the continents have drifted as much as 60° to 70° of latitude since the Carboniferous. Several lines of evidence – climatic symmetry, local sequences of floras in time, distribution of forests with latitude, evolutionary (adaptive) relations – all seemingly unite to oppose the paleomagnetic evidence for major movement."

*(Jeffreys, 1964: 18)*

Thus Jeffreys, now perhaps taking his cue from the youthful MacDonald's allegiance to Axelrod, appealed to him (1963b) to confirm the absence of agreement between paleobotanical data and mobilist interpretations of paleomagnetic data.

Jeffreys closed with a touching appeal to R. A. Fisher. According to Jeffreys, Fisher had been wrong about probability, and he was wrong about continental drift. Unfortunately Fisher was no longer alive to respond to his old acquaintance and sparring partner's disapproval of continental drift, which he had come to support (II, §2.7).

I do not try to explain the palaeomagnetic data; my subject is mechanics. But it is relevant to quote a remark that R. A. Fisher once made to me. "At any time in the last hundred years there has been a piece of evidence appearing to support the inheritance of acquired characters that had not been explained otherwise." That is the position with regard to continental drift.

The Africa–South America fit was bad from the start; the climatic evidence has disappeared; the palaeontological evidence has been explained otherwise and might have been suspect from the start if any advocate of continental drift had read *The Origin of Species*. Even if we take the evidence as it stands it does not hang together. To bring North America and Europe into contact needs a displacement of 40° since the Cretaceous. Palaeomagnetism gives only 20°.

*(Jeffreys, 1964: 18)*

Jeffreys thought the overall case for mobilism had weakened; even with the new evidence from paleomagnetism, which, he argued, gave mobilists only half of what they needed.

There is no question that Jeffreys thought he had learned a great deal about paleomagnetism from reading Munk and MacDonald. But, I believe that his familiarity with paleomagnetism was cursory and that he had made no serious study of the original papers. He certainly cared and knew less about the subject than Gutenberg, Bullard, or Holmes. Consequently, as brilliant a scientist and skillful a writer as he was, almost total reliance on "mechanics" and his misconceptions regarding paleomagnetism placed him at a disadvantage when he attempted to evaluate the status of the mobilism debate in the early 1960s.

## 2.11 Bullard's journey to mobilism: his early career

Bullard has already entered this story but has not been properly introduced. To locate the source of the geomagnetic field he had worked on the dynamo theory of the field (II, §1.4) and on the origin of secular variation (II, §1.6), had suggested to Runcorn the mine experiment (II, §1.5), had chaired a critical session on paleomagnetism at the IUGG meeting in 1957 (II, §8.12), and had, as just seen, discussed paleomagnetism with Jeffreys. Also to be noted (§1.2), is Bullard's pessimism regarding paleoclimatology's ability to offer acceptable solutions, although it was not within his expertise.

However, Bullard merits special consideration at this point in the mobilism debate, for several reasons. There was his work on geomagnetism and on marine geophysics, which included pioneering work on heat flow. There were his assessments of mantle convection. There was his work with Everett and Smith on computer reconstruction of the continents, using Euler's point theorem (IV, §3.4). There was his encouragement of young workers, Matthews, Vine, Everett, McKenzie, and Parker in the Department of Geodesy and Geophysics at Cambridge, who were destined soon to make critical discoveries. There was his easy and good humored manner with students and colleagues. He knew many of the key participants on both sides of the controversy on both sides of the Atlantic, and he arranged for several of them, most notably Menard, Fisher, Hess, and Wilson, to make highly creative visits to Cambridge during the mid-1960s.

During the 1950s Bullard was not strongly inclined toward any particular global theory, but he was willing to discuss their failings and merits as he saw them, and to suggest fruitful ways to test them; he was a well-informed, broadly based geophysicist with a strong interest in geological matters and he was not afraid to good humoredly discuss major hypotheses and throw out fruitful ideas to test them. He had a very wide range of experimental and theoretical skills related to work both on land and at sea. He was well and broadly educated, wrote and spoke eloquently. And what is more, he wrote the most informed and detailed review endorsing the paleomagnetic case for mobilism offered by anyone working outside the field, including Gutenberg (§2.4). It is time to get better acquainted.

The future Sir Edward Crisp Bullard was born in Norwich, England, in 1907, sixteen years after Jeffreys, a decade after Blackett, and fifteen years before Runcorn. His was a prosperous family, brewers of Bullard's Ales. He was the oldest child and had three sisters. His childhood was not a happy one. His relationship with his father was often strained. McKenzie provides an interesting anecdote about Bullard, later known to colleagues as Teddy.

When Teddy was in his late teens his father gave him about £1000 of brewery debentures and pointed out that, whatever happened, the interest would be sufficient to buy a good meal every day. Even when Teddy was appointed to a university post his father continued to ask 'What use is this work? Why do people pay you to do it? Only when Teddy began to talk frequently on the radio after 1943 did he accept that Teddy had somehow established himself in a respectable career.

*(McKenzie, 1987: 68)*

Bullard's early experiences in school were not completely cheerless, but mostly so.

He first went to Norwich High School for Girls, which he quite enjoyed, especially the arithmetic lessons. At nine he moved to Norwich Grammar School, where he was extremely unhappy and thought of killing himself. After two years of misery he was taken to see a psychiatrist, who recommended a boarding school. Though Teddy was horrified at the idea of going away to school he was sent to Aldeburgh Lodge in Aldeburgh, Suffolk, just before he was twelve. This change had the desired effect: he started to work hard and to take an interest in the world around him. He took up photography and started to read the newspaper. On entry he was put in the bottom but one class with eight-year olds. He rapidly came top of the class. He attributes this change in his fortunes to the two headmasters of the school, Mr Sturgeon and Mr Wilkinson, who took great interest in his progress. The only remaining signs of his emotional disturbance were an inability to spell and a dislike of team sports, both of which persisted throughout his life.

*(McKenzie, 1987: 68–69)*

He went to Repton School in 1921. At first he was unhappy, but later enjoyed himself. Although he continued to have unhappy periods, such as his time in Toronto, his adult life was much happier than his childhood.

McKenzie considers Bullard the leading British geophysicist of his generation.

Edward Crisp Bullard ... was the most distinguished and best known British geophysicist of his generation, whose experimental and theoretical work contributed to every aspect of the subject. He will be remembered as one of the major figures in the development of the Earth sciences during the 20th century, both for his own contributions and for his influence on his colleagues and students.[8]

*(McKenzie, 1987: 65)*

He went up to Clare College, Cambridge, in 1926 to read natural science. Although he was somewhat poorly prepared before entering Cambridge, he eventually earned a first, and became a research student in the summer of 1928 at the Cavendish, where he remained until 1931. Although Blackett was his supervisor, Bullard saw more of Rutherford because Blackett was often in Germany. While at the Cavendish, Bullard gained confidence in his abilities to undertake difficult experiments and obtain reliable results. He co-authored several papers with his fellow research student, H. S. W. Massey, on electron scattering in gases.

Bullard took a job in what was then the new Department of Geodesy and Geophysics at Cambridge in 1931.

Teddy often told the story of how he became a geophysicist. He had had enough of electron scattering and was thinking of the possibility of developing an electron microscope. Then one day at tea in the Easter term of 1931 Norman Feather told Teddy that he had been offered a job helping Sir Gerald Lenox-Conyngham teach surveying and geodesy to people who were to be sent out to the colonies. He said he had refused it and asked Teddy if he was interested. Teddy then talked to Rutherford who said, "There are no jobs and there are a lot of people just in front of you. If I were you I would take any job I could get." Teddy accepted the post and became a demonstrator in the newly formed Department of Geodesy and Geophysics, which at the time of its formation consisted of only one person, Sir Gerald Lenox-Conyngham.

*(McKenzie, 1987: 71–72)*

Bullard was hired in 1931 as the demonstrator, along with a sixteen-year-old boy named Leslie Flavill as assistant. Flavill became a very talented instrument maker; recall the role he played in helping Creer build the Cambridge astatic magnetometer (II, §2.11). Bullard held Flavill in particularly high regard. According to Walter Munk (1978: v), Bullard used to maintain, "Harold Jeffreys, Johnny von Neumann, and Leslie Flavill [were] the three most intelligent people he [had] ever met." Bullard began to apply his expertise as an experimentalist to geophysics. With Flavill at his elbow, he perfected a pendulum apparatus to measure gravity on land, describing his work in his 1932 Ph.D. thesis. The following year he took his pendulum to the East African rift valley region to measure gravity, and from the results he attributed (wrongly) the rift's origin to compression; but, he had embarked on his long career obtaining and using geophysical data to try to answer important geological problems. On the strength of his work, he was awarded the Smithson Research Fellowship of the Royal Society in 1936, and he resigned his position as demonstrator to be replaced by B. C. (Ben) Browne, who also had been trained at the Cavendish. (See II, §1.12, §1.19, §1.21, §2.10, §3.4 for Browne and Cambridge paleomagnetists.) During the next few years, Bullard worked in seismic profiling and the measurement of heat flow on continents. Dick Field at Princeton, who also encouraged Hess (§3.2) and Ewing to begin geophysically exploring the seafloor (§6.2), also inspired Bullard (Bullard, 1980). Bullard, Browne, and T. G. (Tom) Gaskell began seismic profiling of the continental shelf south of Ireland, just as Ewing started similar profiling on the other side of the Atlantic.

During World War II, Bullard first worked on the detection of magnetic mines, and later joined Blackett as Associate Director of Naval Operational Research. After the war Bullard returned to Cambridge. It was not to his liking.

When the war with Germany ended in May 1945, Bullard returned to Cambridge. He found a shambles; the door of the laboratory had not been opened for years, most of the equipment had been removed, and what was left was covered with rust. There was no money and no help; his first task was to scrub the floor.

*(Munk, 1978: vi)*

Much needed to be done: the university had few physical resources, but, as it turned out, substantial human resources. Bullard was offered a job by the University of Toronto as Head of the Physics Department. He was tempted, accepted, and went there in spring 1948. Runcorn was soon hired to fill the position he vacated (II, §1.15). Although Bullard accomplished much at Toronto, working on the origin of secular variation and geomagnetic field, learning about computers, and getting the department going, he and his family were unhappy and wanted to return to England. He was hired as the director of the United Kingdom's National Physical Laboratory. Before returning, however, he visited Scripps, where he later spent many summers. During this first summer (1949) he built a heat probe to measure heat flow through the ocean floor. Munk, who was at Scripps, recalled what happened, and stressed the importance of Bullard's use of O-ring seals in ocean equipment.

He and Art Maxwell, then a graduate student, built the long-postponed equipment for measuring heat flow at sea. He shared my office, and I remember the enthusiasm with which he and Art spent long days in the workshop building this quite complicated machine with their own hands. Watertight equipment for use on the floor of the deep sea was at that time a novelty, and, so far as I know, the heat-flow equipment was the first to use the now indispensable O-ring for its watertight joints.

*(Munk, 1978: vii)*

As a result, they obtained the first reliable observation of heat flow beneath oceans, finding higher than expected values (Hess and Maxwell, 1953).

Bullard remained at the National Physical Laboratory for five years, and by all reports was an excellent director. He continued to work on the origin of the geomagnetic field, using digital computers to develop more fully his dynamo model. He continued his experimental work in marine geology, obtaining heat flow results from the Atlantic and Pacific. But the National Physical Laboratory was not Cambridge, which is where he wanted to be.

Finally, in 1956, he got his wish, regaining his old position that Runcorn had held and then vacated upon moving to Newcastle (II, §4.1). Leaving Cambridge, each had opened their position for the other. The Department of Geodesy and Geophysics at Cambridge benefited greatly. Both made very important contributions to the Earth sciences, and helped their students and colleagues do likewise.[9]

### 2.12  Bullard considers mantle convection and measures ocean floor heat flow

Like many people I had been uncertain about the reality of such movements [continental drift] ever since I was an undergraduate. The thing that convinced me was the work of E. Irving on the magnetization of Australian rocks. The discrepancy with the European and American results was so great that there were only two possibilities: either Australia had moved by 40° or so relative to the northern continents or the earth's magnetic field during the Mesozoic was grossly different from that of a dipole. There is some direct evidence that the field was not far from a dipole field, and it should be possible to put the matter beyond doubt; in the meantime the assumption of a non-dipole field seems a quite arbitrary and unattractive hypothesis ...

*(Bullard, 1968: 231; my bracketed addition)*

The first results to suggest strongly that the continents did not all follow the same track were those of Runcorn (1956) for Europe and North America ... In a statistical sense the difference is clear and is in the expected direction but at the time many, including myself, were not fully convinced. I feared that there might be unknown systematic errors which were different for the two continents. For me full conviction came when Blackett *et al.* (1960) collected the world-wide data and presented it in a way that clearly indicated the reality of continental drift. A little later Irving (Runcorn 1962) showed that the Permian and Carboniferous poles derived from Australian rocks were about as far as they could be from the European and North American poles. The agreement of the results from lavas and sediments was also important.

*(Bullard, 1975a: 14; Runcorn (1956) is my Runcorn (1956b), Runcorn (1962) is my Runcorn (1962b), and Blackett* et al. *(1960) is the same as mine)*

Bullard said in the above autobiographical comments that he became "convinced" of mobilism in general because of its paleomagnetic support. He singled out the great discrepancy between poles from Australia and those from Europe and North America (§1.16; II, §5.3), and Blackett and company's (1960) presentation of the paleomagnetic evidence for mobilism (II, §5.19) as particularly important.

Having followed Bullard's return to Cambridge, I want to look more closely at some of his seafloor work, and his attitude toward mantle convection just prior to his above positive assessments of mobilism and its paleomagnetic support. I shall examine Bullard's papers from his first accounts of measuring the heat flow of the ocean floor until he became favorably inclined toward mobilism and show that they concur with his retrospective comments.

Bullard (1951) recognized right from the beginning the relevance of heat-flow work, how it could help to decide whether contractionism or mantle convection was responsible for mountains and island arcs. However, he initially considered convection as a cause of orogeny and not of continental drift.

> There are two widely held views on the cause of the deformation of the Earth's crust. The first, which has been supported by Jeffreys, is that the crumpling observed in mountain ranges is due to the adaptation of the cold exterior to the contraction of the cooling interior. The other view supposes that the stresses which cause mountain building are the shear stresses associated with convection currents in the material immediately below the crust.
>
> *(Bullard, 1951: 520)*

There is no evidence that he gave much thought to continental drift during the early 1950s.

Bullard raised two difficulties with contractionism (RS2), the first as old as the hills, and he noted previous attempts to remove it. The second was relatively new, but important enough for him to look elsewhere for a solution to the origin of mountains and island arcs: it turned him to convection.

> There has been much controversy as to whether Jeffreys' mechanism is adequate. Until recently the geological estimates of the shortening required to produce a major mountain range has been of the order of 300 km and far in excess of what thermal contraction can provide. These estimates are, however, contradicted by the estimate of shortening obtained from the volume of the roots of the mountains revealed by gravity survey. The geological estimates for the shortening are much reduced if it is supposed that the isolated blocks that were previously thought to be the remains of continuous nappes have slid downhill to their present positions. It is of fundamental importance to inquire whether a reasonable adjustment on these lines can bring the geological and gravimetric estimates of shortening into agreement and whether the gravimetric estimate is consistent with the contraction hypothesis.

> The contraction hypothesis is much less attractive now than it was ten years ago, since most cosmologists now require an initially cool Earth that subsequently gets hotter. If this is so, there is no contraction and we are forced to seek some other cause for folding. The suggestion that has been most considered is that of convection currents. Vening Meinesz, Griggs, Hess, and others have given an attractive account of the genesis of island arcs and mountain ranges from this point of view.
>
> *(Bullard, 1951: 520)*

However, Bullard then noted that some sort of independent evidence was required to render mantle convection plausible.

> The argument [in favor of convection] would, however, be more convincing if there were strong evidence for a system of forces tending to produce convection and some independent reason for believing in their existence. The currents are usually supposed to be driven by the action of gravity upon a density difference between the matter beneath the oceans and the continents. This difference in density is supposed to be due to the material beneath the oceans being cooler than that beneath the continents, as a result of the lesser radioactivity of the oceanic rocks.
>
> *(Bullard, 1951: 520; my bracketed addition)*

What might provide that independent check? His answer: measurements of heat flow through the ocean floors.

The work of Pettersson and of the Scripps Institution of Oceanography has shown that it is possible to measure the rate of flow of heat through the floor of the ocean. On the hypothesis described above this should be markedly less than for the continents. At present we have only preliminary results. Their verification and extension is one of the most important problems of experimental geophysics.[10]

*(Bullard, 1951: 520)*

Reporting next year on the early heat-flow measurements, Bullard found them surprising. They were roughly the same as those through continents, which being sialic should be higher and more radioactive.

The preceding communication by Revelle and Maxwell gives a result which is completely unexpected, and demonstrates again how little we know of submarine geology. The observations do, I believe, demonstrate that the heat flow is roughly the same under the oceans and continents.

*(Bullard, 1952: 200)*

Bullard threw out a possible solution involving continental accretion or expansion. The idea was common currency at the time, and Bullard drew on the work of J. Tuzo Wilson.

It seems almost certain that the heat found by Revelle and Maxwell must be generated by radioactivity in the rocks beneath the oceans, and therefore that the total amount of radioactivity beneath unit areas of continent and ocean is the same when summed down to a depth of a few hundred kilometers ... This would be very surprising if the continents were formed from a primitive sialic layer not present under the oceans, and are underlain by material which is the same under continents and oceans. It would, however, be natural if the continents are continuously expanding by a process of differentiation in which radioactive material is concentrated vertically. [Bullard here referenced a paper by Wilson.] The rocks beneath the oceans would then have the same total amount of radioactivity as those beneath the continents, but spread through a greater range of depth; this would give the same heat flow as beneath the continents, but higher temperatures at depth.

*(Bullard, 1952: 200; my bracketed addition)*

Continuing to speculate about the vertical migration of radioactive material, Bullard turned to the problem of the origin of continents, and recommended further work.

A possible interpretation of the results therefore appears to be that when the earth solidified most of the radioactivity was concentrated in the upper 150 km. of the mantle; under the oceans this distribution still exists, but under the continents a further concentration has occurred into the top 10 or 20 km. On this view it would be expected that the oceanic ultrabasic rocks would contain more radioactive material than the continental ones. There are few reliable measurements; but those that do exist do not show such a difference. This matter should be further investigated.

*(Bullard, 1952: 200)*

Although its stock had plummeted, Bullard was unwilling to dismiss convection.

Other explanations can be suggested. It might, for example, be supposed that at some not too remote time a convection current rose under the Pacific and brought hot material near the surface, or that the horizontal limb of a convection current had transported material from beneath the continents to the central Pacific. Such suggestions are pure speculation, and there is no other evidence in their favour.

*(Bullard, 1952: 200)*

The next results, obtained from both the Atlantic and Pacific Oceans, confirmed that heat flow beneath the oceans truly was roughly the same as on continents and greater than expected (Bullard, 1954; Bullard, Maxwell, and Revelle, 1956). The lowest values were in the region of the Acapulco Trench (now the Middle American Trench) in the Pacific, off the western coast of Mexico. Even though the new findings agreed with earlier ones, Bullard realized that his previous explanation, based on conduction only, would not work; it would not get rid of enough heat and the mantle would melt.

It is usually supposed that there is no granite and about 5 km of basalt above the oceanic Mohorovicic discontinuity. This would yield $0.08 \times 10^{-6}$ cal/cm$^2$ s. There remains $0.6 \times 10^{-6}$ cal/cm$^2$ s to be accounted for in the Atlantic and over $0.8 \times 10^{-4}$ in the Pacific. The former would require the radioactivity of 2000 km of ultra-basic rock. This is impossible if the heat is to be removed by conduction, as there has not been time for heat to reach the surface from so great a depth. Moreover as Revelle has pointed out, melting would then occur in the mantle.

*(Bullard, 1954: 428)*

Just as Holmes had done, Bullard realized that there was too much heat for contractionism to be valid, and he turned to convection to remove the excess (I, §5.3). The difference between them was that Holmes promptly connected convection and mobilism (I, §5.4), while Bullard did not do so until about six years after he had first invoked it. He introduced convection in this way.

There appear to be only two ways of escape from the difficulty. Either the rocks in the first 100 km beneath the oceans are much more radioactive than has been assumed, or heat is transported from deep in the mantle by convection. If the radioactivity is distributed through a greater range of depth under the oceans than under the continents, the temperatures under the oceans will be higher, and there will be a tendency for convection currents in the mantle to rise under the oceans and sink under the continents. This is the reverse of what has usually been assumed.

*(Bullard, 1954: 428)*

Like Holmes, Bullard began to wonder about the pattern of mantle convection. However, unlike Holmes, Bullard had the prospect of obtaining more data. He had a heat probe that he and others at Scripps used.

There are clearly large variations of heat flow from place to place, and it may well prove that these form a significant pattern. Whether this would be a pattern of radioactive content, of convection currents in the mantle, or of igneous activity is at present obscure, but the method does seem to provide a new experimental approach to the study of the material underlying the ocean basins.

*(Bullard, 1954: 428)*

Two years later, Bullard and colleagues reconsidered the higher than expected heat flow beneath the oceans. The mantle's ability to transmit seismic shear waves indicated that in the short term it was solid, and the problem was to find a process that would remove heat without it melting. Convection was a possibility; another was unusually high thermal conductivity so heat escaped without causing melting.

The rough equality of the oceanic and continental heat flows can scarcely be a coincidence. The simplest explanation would be that the radioactivity originally in the upper part of the mantle has been concentrated in the crust, allowing the continental heat to escape by conduction, while beneath the oceans the same amount of radioactivity is still distributed through the mantle and the heat is brought to the surface by convection or by unexpectedly high thermal conductivity.

*(Bullard* et al., *1956: 178)*

They had new measurements. Again they found low heat flow in the Acapulco Trench, but they now found that the two highest readings were associated with the Albatross Plateau, which was only later identified as part of the East Pacific Rise. They suggested that convection currents might sink beneath the Acapulco Trench (Bullard *et al.*, 1956: 177).

They noted that others had proposed convection to account for orogenesis, but once again there was no mention, or even hint, of them connecting it with continental drift. That continents had remained fixed relative to each other and to the mantle was implicit.

An alternative method of bringing heat to the surface without excessive temperature gradients is by convection in the material of the mantle. Very slow convective motions suitable for this purpose have been suggested by many authors as a mechanism of orogenesis.

*(Bullard* et al., *1956; their "many authors" were Griggs, Vening Meinesz, and Perkeris)*

As already noted, they believed that radioactive material originally located in the upper mantle beneath continents became concentrated in continental crust and its heat escapes by conduction, while radioactive material beneath the oceans has remained in the upper mantle where it likely escapes by convection. If continents had moved relative to each other or the mantle, it would be difficult to explain why concentrations of radioactivity in the mantle beneath continents and oceans were different. In fact, as I shall later explain, Bullard raised just this difficulty with his own view several years after he came out in favor of mobilism (IV, §6.13).

Bullard and Alan Day gave a talk in 1960 at the meeting of the RAS commemorating Jeffreys' seventieth birthday. Their paper (Bullard and Day, 1961) was submitted in May 1960. They discussed the new measurements of heat flow from ocean ridges, which were generally higher than elsewhere in oceans, leading them to the important conclusion that ocean ridges are sites of rising convection currents. Moreover, following suggestions made independently by Richard von Herzen (1959) who was at Scripps working on heat flow and Ron Girdler (1958) whose work also influenced Runcorn and who had obtained his Ph.D. in 1958 from the Department of Geodesy and Geophysics at Cambridge, Bullard and Day proposed that uprising convection currents might explain the origin of mid-ocean ridges.

The origin of the Mid-Atlantic Ridge has been discussed by Hess (1954) and Heezen, Tharp & Ewing (1959). It forms part of a worldwide system of connected ocean ridges, one branch of which runs into the Red Sea. Since the Red Sea has a central valley which is believed to be a tensional crack and to be floored by a dyke (Girdler 1958), it is natural to suggest that the central valleys or other parts of the system, and in particular the central valley [of the mid-Atlantic Ridge] in which [a heat-flow value 4.7 times larger than elsewhere is obtained] are similar features. Such a world encircling tensional feature might be due to convection currents in the mantle rising under the ridges, spreading out sideways and sinking again. The horizontal lines of the currents would produce the forces that open the tensional crack and the rising current would bring hot material near the surface and produce volcanism and a high heat flow ... If the rising current under the ridge turned down again on each side, a low heat flow would be expected over the descending limbs ... Von Herzen has found [very low values] in the basin 1600 km to the west of the crest of the East Pacific Rise and a value [almost as low] 1000 km to the east. He suggests that the pattern he has found is due to convection in the mantle. Clearly much remains to be done to elucidate the meaning of these high and low heat flows which were entirely unexpected.[11]

*(Bullard and Day, 1961: 290–291; my bracketed additions)*

Perhaps out of deference to Jeffreys who, it should be remembered, was at the time being honored, there was still no mention of continental drift. Perhaps Bullard and Day could not agree. Perhaps at the time they were not in favor of mobilism. Whatever their reason, they certainly thought that the hypothesis of mantle convection with rising currents at ridges was worth serious consideration, but, and I say it again at risk of repetition, they made no connection of it to continental drift: on the evidence of the printed record, Bullard favored mantle convection long before he expressed any opinion about or even mentioned drift. Like Vening Meinesz (§1.5), Bullard's acceptance of mantle convection, and even his acknowledgment of its likelihood as a cause of oceanic ridges, was not the reason he began to view mobilism favorably.

## 2.13  Bullard begins to consider mobilism seriously

Bullard must have started wondering about mobilism by 1956, long before he publicly expressed sympathy for it, for he served as a referee for the series of papers in the *Philosophical Transactions of the Royal Society* detailing the paleomagnetic findings by Collinson, Creer, Irving, and Runcorn while they were together at Cambridge. The papers were received at the Society on April 17, 1956. Creer remembered Bullard's referee's report.

Bullard was one of the referees of the series of papers (Collinson, Creer, Irving, and Runcorn) that were published by the Royal Society in 1957. His review raised no objections to continental drift in principle, nor to the validity of our paleomagnetic conclusions. But, he did not come out until the mid-1960s.

*(Creer, March 2000 comment to author)*

Moreover, Bullard had thought well enough of Irving's presentation at the 1957 IUGG meeting in Toronto to offer him encouragement (§8.12), and remarked at the time, as Irving recalled (November 1999 comment to author), "God would not be so bloody-minded as to put these magnetizations there just to fool us."

When did he first record or publicly express his sympathy for mobilism? I know that he and Day did not mention continental drift in the paper they gave at the meeting honoring Jeffreys, which was submitted in May 1960, and in which, as I have just explained, they may have refrained from doing so to spare the old fixist embarrassment. The two relevant publications recording when he did express sympathy are Bullard (1961) and (1962). The first was based on an invited lecture he gave at the International Oceanographic Congress, held in New York, August 31 to September 12, 1959; there was no pre-meeting abstract, no recorded date of receipt of his manuscript, and a long publication delay, so this does not date precisely when he actually first became willing to express public support for mobilism, but does provide an early limit. The second paper (Bullard, 1962) was based on a talk that he presented at a meeting in November 1960, which was sponsored by the Royal Society and entitled "A Discussion on Progress and Needs of Marine Science." It was received at the society in July 1961, providing a likely later limit. Thus, he began to examine the question of mobilism sympathetically with a mind to committing his views into print sometime between the end of 1959 and mid-summer 1961.

In the first of the above papers, Bullard began by commenting on its evidential support and describing his ideas for further testing. He suggested that it would be helpful to see if the fold-belts of eastern North America and Western Europe extended beneath the Atlantic. He also claimed, referencing Blackett *et al.* (1960), that paleomagnetism provided a strong enough case in favor to merit further testing.

If the fold lines of western Europe and of Newfoundland are cut off between the shore and the edge of the continental shelf, it would strongly suggest that the ocean floor was not present when the folding took place and would be consistent with the hypothesis of continental drift. New evidence on this has come from paleomagnetism; the whole material has recently been reviewed by Blackett *et al.* (1960), who make a strong case for relative movements of the continents. The evidence seems strong enough to justify the effort needed to obtain some independent check; a continuation of Caledonian or Hercynian structures from Europe well out into the Atlantic would go a long way to show that the ocean had not been formed by a post-Paleozoic westward movement of America.[12]

(*Bullard, 1961: 48*)

Strangely, Bullard cited only Blackett's work, saying nothing specific about the truly pioneering work done at Cambridge by Creer, Irving, and Runcorn, even though he had recently acted as reviewer for their work, had rejoined his old department, and was well aware of what they had done.[13]

Asserting (incorrectly) that paleomagnetism is in all circumstances blind to changes in longitude,[14] he stressed the need to find other ways to detect east–west movements of continents, and summarized recent work on the seafloor, especially, mid-ocean ridges, which suggested a widening of the Atlantic and the Red Sea. But, again, he thought such findings only suggestive.

Paleomagnetic studies cannot detect a movement in longitude; particular interest therefore attaches to any evidence that can be found for east–west movements from a study of the oceans. Many of the arguments that have been suggested are inconclusive. The observed general similarity of crustal structure in the Atlantic and Pacific is perhaps not to be expected if the Atlantic is a recently opened gap. The mid-Atlantic ridge and particularly its central valley might be regarded as the place where the Atlantic is at present widening. The continuation of this feature round the south of Africa and up the Indian Ocean to the Red Sea suggested by Rothé (1954) and by Ewing and Heezen (1956) is consistent with the movement of India away from Africa and with an incipient splitting of Africa from Arabia (Girdler, 1958). Such an explanation is not, however, available for the ridges of the Pacific which are supposed to be connected with the ridges of the Indian Ocean and thus with the mid-Atlantic ridge. It is important to be sure of the various suggested connections and to know how far the ridges are really similar. In particular, do the Pacific ridges often or usually lack a central valley as has been suggested by Menard (1958)?

*(Bullard, 1961: 48–49; Ewing and Heezen (1956) is my 1956b, Menard (1958) is my 1958a)*

He rounded off this brief initial foray into continental drift by claiming mantle convection offered the only plausible mechanism and might also account for mid-ocean ridges.

If the continents are moving, there must be horizontal forces, and the only plausible suggestion that has been made is that these forces are associated with convection currents in the mantle. The high heat flows found on the mid-Atlantic ridge and on the east Pacific rise suggest that, if the currents exist, their rising limbs are under the midocean ridges. The rising current would be of lower density than the material on each side and would explain the rough isostatic compensation of the ridge.

*(Bullard, 1961: 49)*

He considered his views tentative. He saw them only as hypotheses that should be seriously entertained, leaning toward them but not strongly. He also argued, and would later push more forcefully, that if the facts point to them, then mobilism and convection should not be dismissed on theoretical grounds, because not enough is known about long-term processes, and, to make matters worse, the processes cannot be directly observed.

Obviously, views about the existence of convection currents and continental drift must be tentative. The mechanisms, if they exist, are concealed so effectively and have such long time scales that they cannot be directly observed. Theory cannot assert or deny the possibility of such processes. If the facts show a high probability that they have occurred, we must make the best theory we can, but should not place too much reliance on it or let it obscure the very complicated facts.

*(Bullard, 1961: 49)*

So this is where Bullard first began publicly to express sympathy with mobilism, and to regard it as worthy of serious further discussion. He did so because of its paleomagnetic support, and because convection offered a solution to the first-order problem about the origin of ocean ridges and was, he thought, the only viable approach to the second-order problem regarding the forces responsible for drifting continents.

In his next talk entitled "The deeper structure of the ocean floor," Bullard (1962) again discussed convection and wondered whether the Appalachian, Caledonian, and Hercynian belts extended into the Atlantic, but made no mention of paleomagnetism. Instead, he turned to three other avenues of support: the evidence of large horizontal displacements of the ocean floor in the eastern Pacific off California and Oregon, the lack of thick sediments on the seafloor, and the fit between Africa and South America.

First there was Vacquier's work (1962), which was soon to be published in the anthology *Continental Drift* (§1.6), then matching of magnetic anomalies (discovered by R. G. Mason and A. Raff in 1954–5 at Scripps (§5.9; IV, §2.3)) across great fracture zones that strongly suggested a displacement of seafloor of approximately 1200 km (§5.6; IV, §2.3; IV, 2.13; IV, §3.8).

An examination of the magnetic profiles taken parallel to the escarpments suggests that they are faults along which a large transcurrent movement has occurred ... The profiles have been displaced in longitude so as to match as well as possible. Across the Pioneer fault the fit is so good as to leave no reasonable doubt that there has been a displacement of about 2.8° of longitude. The fit for the more easterly part of the Mendocino fault is also excellent and suggests a displacement of about 14° or 1200 km. The fit at the western end is not as good as it is to the east, but is nonetheless convincing.

*(Bullard, 1962: 392)*

Like Vacquier, Bullard argued that these displacements weakened mechanical arguments against mobilism; whatever theory may say to the contrary, displacements of the seafloor are possible because they have been demonstrated to occur.

Marine seismic surveys had revealed that seafloor sediments were only ½ to 1 km thick. If oceanic crust was very old and given present rates of sedimentation, this was much less than expected. Bullard agreed with Hess (1960d, 1962) and Dietz (1961a, 1962a) that the ocean floors might not be ancient features, although all three noted that there were alternative solutions such as the presence of an underlying layer of consolidated sedimentary rock.[15] In addition, like Hess, he urged that project Moho be undertaken to core and to sample through oceanic crust to the Mohorovičić discontinuity, in part to determine if consolidated sedimentary rock underlay the upper unconsolidated sediment.

Over most of the oceans unconsolidated sediments account for ½ to 1 km of the material above the [Mohorovičić] discontinuity. We have no direct evidence as to the nature of the material below the sediment. The seamounts that emerge from it are almost invariably composed of basalt and it is usually supposed that most of the material is basalt; it is possible, however, that consolidated deep-sea sediments occur between flows of basalt. If they do not, it is difficult to understand why there is so little sediment in the oceans. The present rate of accumulation is doubtless very variable but is probably of the order of 1 cm in 1000 years or 1 km in 100 million years. What has happened to the Palaeozoic and pre-Cambrian sediments? The most direct approach to this problem is to drill through the whole sequence in a number of places and find what range of age is represented. Information of this kind should be one of the earliest fruits of the Moho project. It is fruitless to speculate as to what will be found. Perhaps the Atlantic did not exist before the Cretaceous; perhaps the organisms living in earlier times dissolved and were deposited elsewhere as inorganic precipitates.

*(Bullard, 1962: 393)*

Finally, Bullard introduced Carey's (1955b) fit of South America and Africa in his discussion of ocean ridges (II, §6.11); unlike Jeffreys, he was impressed. He reproduced Carey's map, and related it to the mid-oceanic position of the Mid-Atlantic Ridge. Within a few years Bullard would return to the fit of the circum-Atlantic continents, and ask J. E. Everett, one of his Ph.D. students, to design a computer fit of continents (IV, §3.4). Bullard also suggested a causal link between the formation of oceanic ridges, growth of ocean basins, and continental drift. Characteristically, he offered a way to test the idea: determine if the age of volcanoes increases the further their distance from ridge axes.

Clearly this system [of mid-ocean ridges] is one of the main features of the earth's surface and any acceptable history of the earth must describe how it came into being. At present theories can only be tentative and are useful more to suggest what should be looked for than as serious accounts of earth history. A linear system of volcanoes suggests a crack up from which lava comes and this idea is reinforced by the occurrence of a faulted central valley, itself sometimes obstructed by later volcanoes. Perhaps the ridges mark out a continually widening and self-healing crack in the ocean floor. If this were the true explanation, the most recent volcanoes would be those along the central valley, where the crack is at present widening, and those on the flanks of the ridge would have been formed earlier. Such a view involves a widening of the Atlantic and Indian oceans and is closely connected with the hypothesis of continental drift. The incredibly accurate fit of the continental edges of South Africa and South America can hardly be fortuitous [see II, §6.11)] and the occurrence of a widening crack, which also fits both continental edges and is half way between them, would greatly strengthen the probability that the separation of the continents and the outpouring of the lavas of the ridge are part of the same system.

*(Bullard, 1962: 388–389; my bracketed addition)*

Just as Runcorn took up Bullard's idea of testing Blackett and Elsasser's competing theories of the origin of the geomagnetic field (II, §1.5), Wilson (IV, §3.2) would soon carry out Bullard's test to see if oceanic islands increased in age as their distance from ridges increased; I do not know if Wilson came up with this independently, but he could have read Bullard's paper.

### 2.14 Bullard recognizes that all obstacles to the paleomagnetic case had been removed and becomes a mobilist

In a lecture given in June 1963 ... I at last came out in favor of continental drift without reservations and without balancing of probabilities for and against.

*(Bullard, 1975a: 16)*

Bullard (1964) publicly came out unequivocally in favor of mobilism on June 19, 1963, at a lecture he gave before the Geological Society of London. The editors of the *Quarterly Journal of the Geological Society of London* received the written version on October 11, 1963; in it he showed that he had now acquainted himself with much of what happened during the classical stage of the mobilism debate. He described how geologists had been unable to resolve the controversy, and emphasized the continuing concerns over the reliability and interpretation of paleoclimatological data, as I have already described in relation to the distribution of Permo-Carboniferous glaciation (§1.2). Although not his expertise, he himself was distrustful of paleoclimatic interpretations. He singled out the distribution of *Glossopteris* flora and the fit between the continental edges of Africa and South America as two particularly good mobilist solutions, but suggested (1964: 3), "Clearly it is necessary to break away from this well-trodden circle of ideas." This is precisely what paleomagnetists, several of whom got their start in his own department, had been doing in the previous decade, and I want now to explain why I believe it was at this time that Bullard in effect recognized that the objections that had been raised against their arguments had, during the later 1950s and early 1960s, been disposed of: that is, for him the paleomagnetic case for mobilism had become difficulty-free.

Bullard explained in a very positive way the paleomagnetic case, remarking (1964: 7) that it has been the "main cause of the revival of interest in continental drift." He could not cover everything and referred the reader to reviews, leaning heavily on that by Cox and Doell.

It is not possible to review here all the methods and results in detail. An admirable and balanced discussion had been given by Cox and Doell (1960) and one from the point of view of a convinced believer in continental drift by Runcorn (1962). The most complete tables of results are those of Irving (1959, 1960–3). Blackett, Clegg & Stubbs (1960) have summarized the results for each continent separately.

*(Bullard, 1964: 7; Runcorn (1962) is my 1962b; Irving references are my 1959a, 1960, 1961,*
*1962a, b, c)*

Bullard's characterization of the 1960 *GSA* review seems surprising because, in stark contrast to its authors, he thought that paleomagnetism

has provided strong evidence for movement, the only alternative being to suppose that before the Eocene the Earth's magnetic field was nothing like a dipole and was not related to the axis of rotation.

*(Bullard, 1964: 1)*

Although Cox and Doell thought the geomagnetic field probably had had the form of a geocentric dipole throughout its history, they chose rapid polar wander with or without associated motion of the rotation axis, not continental drift, to explain the dispersion of Mesozoic and Early Tertiary poles (II, §8.4). Even though he made abundant use of the *GSA* review in defending the paleomagnetic case for mobilism, he never mentioned that they had stopped very far short of endorsing mobilism, had "tip-toed" away from it as Hamilton aptly remarked (§1.13). His use of their review might perhaps be seen as a subtle reproach for their reticence to endorse mobilism. Also it might seem odd that he should have given so much prominence to a review that took a negative view of drift rather than using the original papers that spearheaded the paleomagnetic case for drift, and which, to a very large extent, had their root and origin in his department. On the contrary, for his purposes Cox and Doell's review was a singularly appropriate source because other reviews were by mobilists: by using the *GSA* review, he could never be accused of selecting data biased in favor of his new-found belief in mobilism.

Before directly examining mobilism's paleomagnetic support, he acknowledged that there were complicating factors.

Interpretation of the results is made more difficult by the possible occurrence of three controversial processes: polar wandering, continental drift, and reversals of the field. There is also the well-established process by which certain uncommon rocks become magnetized in a direction opposite to that of the magnetic field in which they are placed.

*(Bullard, 1964: 7)*

Perhaps his mention of the first possibility of both continental drift and polar wandering was a nod to Munk's, MacDonald's, and Jeffreys' complaints about the apparently endless paleogeographic games they thought could be played with paleomagnetic results. However, I speculate, and Bullard did not mention their attacks on paleomagnetism.

Bullard began by identifying the two main assumptions: that the remanent magnetism of rocks records the direction of the geomagnetic field at the time of formation or at some determinable later time; and, the GAD hypothesis. Bullard defended the first by describing the conglomerate test (II, §1.8), citing an Australian study by Irving (1963) which showed:

Many samples from Tasmanian dolerite sills give concordant directions, but samples of the same dolerite in a conglomerate give random directions. Clearly the pieces of lava [he meant dolerite] in the conglomerate have been able to retain their magnetization through the processes that broke up the rock and formed the conglomerate and through the ensuing 60 million years.

*(Bullard, 1964: 7; my bracketed addition)*

He then turned to the igneous contact test of stability (II, §1.11) and described a wonderful example from Iceland observed by Einarsson and Sigurgeirsson (1955).

However, instead of citing their original work, as he had in the Tasmanian example, he (1964: 10) again referred to Cox and Doell's "admirable and balanced discussion," and reproduced their diagram (1960: 736) illustrating what Einarsson and Sigurgeirsson had found. Bullard also discussed the fold test (II, §1.8). He recognized that not all rocks are magnetically stable, but noted that paleomagnetists had designed techniques for cleaning samples; they also had learned through long experience which rock types were most likely to possess stable remanent magnetization, that basalts and red sandstones generally provide reliable results, and that slates and schists do not. Bullard closed his defense of the assumption of magnetic stability by dismissing Graham's magnetostriction difficulty (II, §7.4), referring (1964: 11) to the important work of Stott and Stacey (1960, 1961) (II, §7.5).

Although there are these difficulties at some sites, it seems to be established the magnetization of a suitable rock can usually be taken to represent the direction of the field at the time it was formed.

(*Bullard, 1964: 11*)

Bullard began his defense of the GAD assumption by admitting that it raised serious questions.

The relation between the direction of the field at a point and the position of the magnetic and geographical poles raises more serious questions. It is customary to work out a pole position by taking the mean of the measurements from a number of specimens from each site and then averaging the means for a number of sites. The position of the pole is then calculated on the assumption that the Earth's magnetic field is similar to that of a dipole and the dipole axis coincides with the axis of rotation.

(*Bullard, 1964: 11*)

Drawing on his own work on secular variation, which causes the geomagnetic field at any instant in time to depart somewhat from that of a geocentric axial dipole, he argued that it created no difficulties if samples were taken spanning thousands of years. Turning to the geographical half of the assumption, he said that historical observations of the geomagnetic field fail to show that the axis of rotation and that of the geocentric dipole field coincide. As he put it:

It is not clear from observations whether the difference between the dipole axis and the axis of rotation will also average out. Since 1600 the north magnetic pole has remained to the northwest of Hudson Bay and the axis of the dipole has, presumably, been not far away. The direction of the dipole axis was first determined in 1929. Since then it has shifted by less than 0.3° in latitude and only about 5° in longitude. From this it is uncertain whether it is bound in some way to the neighborhood of its present site or whether it can wander about and will, in the long run, have its mean direction along the Earth's axis.

(*Bullard, 1964: 12*)

To settle this question, he reproduced a diagram from the *GSA* review, remarking:

Fortunately this question [about the coincidence of axes] can be decided by the palaeomagnetic observations themselves. Fig. 7 [after Fig. 17 from Cox and Doell (1960: 734)] shows the pole positions found from rocks ranging from late Pleistocene to Recent (see Cox & Doell (1960) for details). The positions form a group about the geographical pole and 10 out of 13 lie within 5° of it. From these results it is clear that during the past one or two hundred thousand years the field averaged over a few thousand years has been close to that of a dipole with its axis along the rotational axis of the Earth.

*(Bullard, 1964: 12; my bracketed additions)*

This got him back to the late Pleistocene. To get himself into the Tertiary, he appealed to one of Runcorn's diagrams.

Observations on Tertiary rocks show a similar behaviour, but with a large scatter as shown [in a diagram, which is taken from Runcorn's review in *Continental Drift*]; the magnetic poles seem to be distributed at random within about 15° of the geographical pole. The reason for the increased scatter is not known; some of it may be due to errors in allowing for post-depositional tilts and some to real movement of crustal blocks.

*(Bullard, 1964: 12; my bracketed addition)*

Although at the end of his discussion on paleomagnetism Bullard found an interesting way to return both to the *GSA* review and to the dipole hypothesis, at this mid-point he simply asserted that the GAD hypothesis was reasonable.

It is not possible to say with certainty that because the field has resembled an axial dipole since the Eocene it must have done so for all time, but it does make this a reasonable hypothesis.

*(Bullard, 1964: 13)*

Having satisfied himself that the two key assumptions were no longer dubious, Bullard turned to the paleomagnetic findings that supported mobilism. He first reported the results for Europe that indicated that the pole had moved relative to Europe (APW). (I find it surprising that Bullard did not take this and other opportunities to tout work of his own department by mentioning Creer's pioneering and game-changing construction of the first APW path (II, §3.6).) He described the deviation between pole paths from North America and Europe, which indicated continental drift, although critics could suppose some systematic error or maintain that the field had not been dipolar.

On going further back in time it is found that the pole positions calculated from measurements on European rocks no longer lie around the present geographic pole. The results for Mesozoic rocks are neither so numerous nor so concordant as could be wished but suggest that the pole was in the north Pacific. Numerous results for the Permian and the Carboniferous form a group in the north-west Pacific Ocean. The results from North American rocks form a similar group whose centre is about 30° to the west of the centre of the European group. If the centres of the two groups had agreed one could have supposed that the magnetic pole in Permian and Carboniferous times was in the north-west Pacific and, on the evidence from the Tertiary and Recent rocks, one would naturally, but not inevitably, suppose the geographical pole also to have been there. The separation of the European and North American groups of poles shows that this hypothesis is too simple and that one must either suppose that the field had a substantial non-dipole component persisting for millions of years or that America was some

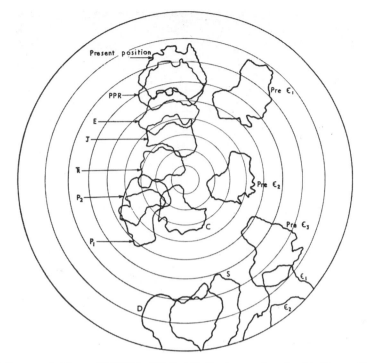

Figure 2.3  Runcorn's Figure 22 (1962b: 26) that Bullard reproduced as his Figure 10 (1964: 15). Latitudes of Australia through time. T, Tertiary; K, Cretaceous; J, Jurassic; Ŗ, Triassic; P, Permian; C, Carboniferous; D, Devonian; S, Silurian; O, Ordovician; Є, Cambrian; Pre-Є, Pre-Cambrian. The data came from Irving and Green (1958). Runcorn's key was sloppy. He lists positions of Australia during the Ordovician and Cretaceous in his key, but neither appears in the figure. The figure shows positions of Australia during the Eocene (E) and during the Pliocene, Pleistocene and Recent (PPR) but neither is identified in the key.

30° nearer Europe 200 million years ago than it is now. If this were all, perhaps one could accept the former alternative, or suppose there to be some systematic error in one set of measurements or the other; the scatter is comparable with the distance between the mean poles and the argument is not as clear-cut as one would wish for so important a conclusion.

*(Bullard, 1964: 13–14)*

Following in the decade-old footsteps of several earlier members of his own department, Bullard sought results from other landmasses. Using two figures adopted from Runcorn's paper in *Continental Drift*, he singled out Irving's findings from Australia (Figure 2.3), reported those from India, and showed schematically the various APW paths from various continents (Figure 2.4).

Data from other parts of the world considerably strengthen the case for relative motions of the continents. In particular, the careful and thorough work of Irving (1963) has shown that the Australian pole positions are completely incompatible with the European and North America ones. Positions of Australia since the Carboniferous are shown [in Figure 2.3]. Only latitudes

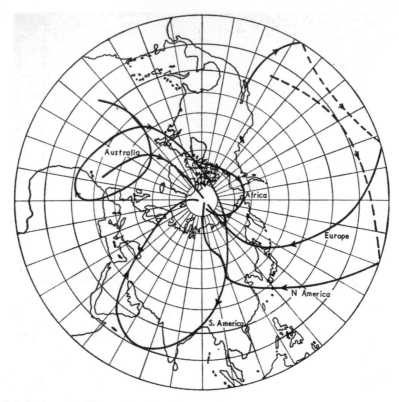

Figure 2.4 Runcorn's Figure 20 (1962b: 24) that Bullard reproduced as his Figure 11 (1964: 16). Schematic APW paths from various continents.

and the orientations are determined by the palaeomagnetic results; the longitudes are arbitrary and have been chosen to give a continuous track.

These and other results are summarized [in Figure 2.4] in the form of [apparent] polar wandering curves showing the calculated positions of the pole relative to the present positions of the continents. Results from India, which are not included in [the] figure, suggest that India has moved about 70° northward since the Jurassic. The differences between the curves are so large that they cannot be due to unexpected systematic errors or small departures from a dipole field. Either Australia or India have moved relative to the northern continents or the Earth's magnetic field was nothing like a dipole field and had no obvious relation to the rotational axis for tens of millions of years.

*(Bullard, 1964: 14; my bracketed additions; Irving (1963) is the same as mine)*

This passage may not have met with spontaneous applause from Bullard's audience, but Almond, Blackett, Clegg, Creer, Deutsch, Gough, K. W. T. Graham, Green, Hales, Irving, and Runcorn must have been happy with it, as well they might be, because he had adopted without reservation their general mobilistic interpretation. The contrast between Bullard's progressive attitude to paleomagnetism and the attitude of Cox and Doell and of Munk and MacDonald could not have been starker. Why is this? I consider Munk and MacDonald first. Bullard agreed with them, that unique reconstruction of the previous position of the continents based solely on paleomagnetic data was not possible; like them, he thought (erroneously, see II, §5.16) that the paleomagnetic method could not, in any circumstance, pin down ancient relative longitudes. Whereas Munk and MacDonald saw the added complexity brought about by the lack of coincidence of APW paths from different landmasses as reason to reject the method, Bullard, like the paleomagnetists listed above, saw it as supportive of mobilism, and viewed the idea of a rapidly changing geomagnetic field as unreasonable.

Bullard made no mention of Cox and Doell's worries about the scarcity of data for North America and Europe. He said nothing about the aberrant pole from the Eocene Siletz volcanics, which had led them to doubt continental drift and favor a rapidly changing magnetic field, with or without wandering of the rotational pole; perhaps he simply judged it, as Irving did (II, §8.11), to be a local side-show, an incomplete experiment, requiring confirmation. The poles, especially of Australia compared to Europe and North America during the Permo-Carboniferous and Mesozoic (Figures 9.2 and 9.3), convinced Bullard that Australia had continued to drift relative to them. He wanted more data, but there was sufficient he thought to accept that continental drift had occurred. He had jumped down from the fence. As he saw it, the drift solution to the widely divergent APW paths was essentially difficulty-free.

Bullard used Cox and Doell's (1961a) attack on Earth expansion as furthering support for the geocentric dipole hypothesis. They based their claim on the angular distance between Permian rocks in Europe and northern Siberia being the same as at present, and consistent with the Permian magnetic field being dipolar (II, §8.14).

Either Australia and India have moved relative to the northern continents or the Earth's magnetic field was nothing like a dipole field, and had no obvious relation to the rotational axis for tens of millions of years. In principle, detailed measurements over a large block that has behaved as a rigid whole, such, for example, as Africa south of the Sahara, can show how far the field at any epoch departed from a dipole field over that area. The only investigation of this kind is a comparison of the European and Siberian results for the Permian by Cox & Doell (1961). The poles agree, which shows that northern Siberia has not moved by a detectable amount relative to Europe since the Permian and that the fields in Europe and in Siberia were then related in the way that a dipole field would be (or, alternatively and improbably, that relative motion has occurred which has just obscured a departure from a dipole field).

(Bullard, 1964: 14–15; Cox and Doell (1961) is my Cox and Doell (1961a))

Bullard closed his discussion of paleomagnetism and support for mobilism.

In the absence of a world-wide net of measurements we have to make a tentative choice of a hypothesis. Either we must accept the mobility of continents or suppose, without any support-ing evidence, that the magnetic field in the past was quite different from what it is today and demonstrably has been since the Eocene. An extension of Cox and Doell's work to a wide geographical range of Permian and Carboniferous rocks in Eurasia and Africa would be of great value.

*(Bullard, 1964: 15)*

Would Cox and Doell have agreed with Bullard's last statement? Likely they would, but unlike Bullard, they would simply have preferred the second option. Bullard did not follow their piecemeal approach to the data; basically he saw as a whole, the vast and systematic difference between Late Paleozoic, Mesozoic, and Early Tertiary poles from Australia and the northern continents spanning over 200 million years, and saw it as compelling evidence of drift (Figures 3.2 and 3.3). Bullard gave his British audience no reason to suspect that Cox and Doell did not support mobilism. It was a masterly essentially unanswerable performance. Using as its main basis the *GSA* review that was far from favorable to mobilism, he turned it on its head into an account unequivocally favorable to it.[16]

Bullard also considered comparisons of paleomagnetic findings with paleobiogeography and paleoclimates. Although he did not think such studies could ever resolve the mobilist controversy, he wanted to make clear that they marched together, which MacDonald, relying not on his own expertise but on that of Stehli and Axelrod, had denied (§2.8). Bullard countered by appealing to work of Plumstead (§1.12; I, §6.3), Blackett (§1.7) and Opdyke (II, §5.12–§5.14). Choosing the best known paleoclimatological and paleontological evidence for mobilism accumulated during the classical stage of the controversy, Permo-Carboniferous glaciation and distribution of *Glossopteris* flora, he claimed that they too agreed with paleomagnetic results.

The recent evidence is sufficiently striking to suggest that we should look again at the older work to see how far it is consistent with the palaeomagnetic conclusions. It is remarkable that the palaeomagnetic results put the Permo-Carboniferous glaciations of Africa, Australia, and India in the polar regions. Other climatic indicators have been discussed by Blackett (1962) and by Opdyke (1962), who find substantial agreement with the palaeomagnetic results. Their conclusions agree with those of Plumstead (1962) but disagree with those of Stehli (1957) and Axelrod (1963). Opdyke has been accused by MacDonald (1963) of "careful selection of data". It is extraordinarily difficult to reduce the diverse and often dubious information about climates, depending on the largely hypothetical habits of long-extinct organisms, to any kind of order and there seems no prospect of agreement, even among the specialists. The trouble may lie partly in the occurrence of drastic changes of climate in a short time. It is easy to imagine the confusion that would occur were the present English flora to be found fossil associated with a glaciation only 10 000 years older and perhaps also with one a few thousand years later. Presumably some rapid change of this kind is needed to account for the association of the *Glossopteris* flora with glaciations.[17]

*(Bullard, 1964: 17–18; references correspond to my Axelrod (1963), MacDonald (1963a),*
*Blackett (1961), Opdyke (1962), and Plumstead (1962))*

Bullard clearly had become an important ally of paleomagnetists and their friends who had developed, expanded, and defended their case for mobilism. His was a prestigious endorsement. It is quite fitting that Bullard was soon to combine forces with Blackett and Runcorn in organizing the Royal Society's 1964 meeting on continental drift (IV, §3.3).

Bullard concluded his overall defense of mobilism by emphasizing its usefulness as a focus for future research.

Past experience suggests that it is too much to expect a hypothesis concerning a major geological problem to be wholly correct. It is sufficient if it is not clearly absurd either geologically or physically and if it suggests relations and other hypotheses that can be tested. The idea of continental drift is exceptionally fruitful in this way; it suggests large programmes of investigation on land, at sea, and in the laboratory. If these are pursued vigorously for ten or twenty years it is probable that general agreement will be reached. From what we know at present, it seems likely that the decision will be that the continents have moved.

*(Bullard, 1964: 24)*

Fifteen years before Bullard wrote these lines, Hospers had arrived in Cambridge, and began to plan his Icelandic studies, which culminated in his demonstration of the geocentric axial model of the geomagnetic field as well as sequential reversals of the field. Thirteen years before Bullard wrote these lines, Irving realized the connection between paleomagnetism and continental drift, and, with help from R. A. Fisher and the Indian Geological Survey, had procured samples from India. A dozen years before Bullard wrote these lines, Khramov had begun magnetostratigraphy surveys in western Turkmenia. Creer produced the first APW path ten years before Bullard wrote these lines. Eleven years before, Blackett recruited Clegg, and they, along with Stubbs and Almond, began their work. Du Bois, Opdyke, Deutsch, Nairn, Green, Gough, Hales, K. W. T. Graham, and others soon became involved, and a global paleomagnetic program to test continental drift was launched. By the time Bullard wrote these lines, seafloor spreading had been introduced, and Vine and Matthews, members of his own department at Cambridge, had just sent off their paper introducing their eponymous hypothesis. Bullard thought it likely that ten or twenty years of vigorous work would lead to general agreement that the continents have moved. Little did he or anyone else know that ultimately persuasive arguments were only three or four years away.

## 2.15 Bullard squabbles with geologists about the contributions of geology and geophysics to the mobilism debate

In spite of this deficiency [that there was almost nothing in it about the ideas that were to lead to seafloor spreading and plate tectonics] the talk [I gave before the Geological Society of London where I came out in favor of mobilism] had, I think, some influence in moving opinion in England towards acceptance of continental movement. I was surprised by the amount of support shown in the subsequent discussion.

*(Bullard, 1975a: 16; my bracketed additions)*

I am nearly 80, ten years younger than Arthur [Holmes] was, but to my knowledge continental drift has always been accepted by some geologists. At the present time [1979] geophysicists, a rather recent breed of earth scientists, are inclined to think that they have discovered everything about moving continents. The reaction of geologists to E. C. Bullard's lecture on the fit of the continents to the London Geological Society in 1964 was very amusing. The published discussion in the *Quart. Jour. Geol. Soc. London,* 1964, *vol. 120,* pp. 1–33 is toned down, but the geological audience was very irritated that a physicist should come and tell them what they had known for a long time, and one after the other rose to tell Bullard, in essence, that all he had said was "old hat."

*(Doris L. Holmes, February 4, 1979 letter to author; my bracketed additions)*

It is difficult to determine whether and to what extent Bullard's 1963 talk persuaded Earth scientists in Britain to look more favorably on continental drift. It appeared in January 1964, three years before transform faults and the Vine–Matthews hypothesis were confirmed, and was not often cited. Among participants at the Royal Society 1964 meeting on continental drift – held just two months after its publication of Bullard's 1963 talk but ten months before all manuscripts from that meeting were received for publication – only Harland, already a mobilist, referred to Bullard's 1963 paper (§1.11; II, §1.16). It is even difficult to tell whether his talk had any appreciable effect on the opinion of those in attendance. If Doris Holmes (Reynolds professionally) is correct, and listeners thought Bullard's talk was "old hat," then his talk changed few minds. Mobilists already believed, and fixists would have remained skeptical, despite what he had to say. Doris Holmes is probably right, Bullard simply managed to irritate geologists because they believed he had greatly underestimated the extent and strength of the geological support for mobilism before the rise of paleomagnetism and marine geophysics. But much of what Bullard described *was* new.

Bullard believed that even during the classical stage of the controversy the strongest arguments in favor of mobilism were inconclusive, that the mechanism difficulties had a very significant impact, and that, at the time, fixists were faring better than mobilists. Here is Bullard's summary.

The idea that the continents may have moved can be traced far back in geological thought, but first became a serious subject of speculation after the publication of Wegener's well-known papers (1912) and book (1924). During the nineteen-twenties and thirties there was a prolonged and, at times, acrimonious controversy (Waterschoot van der Gracht *et al.* 1928) which can fairly be said to have been inconclusive, in the sense that neither side convinced the other and that there remained well-informed and eminent supporters of both the fixity and the mobility of the continents. Perhaps on the whole the believers in fixed continents had the best of it, at any rate so far as teaching and textbooks were concerned (though Holmes's *Principles of Physical Geology* [1944] was a notable exception). By 1939 most people had become wearied of inconclusive argument, and the battle was not renewed with any vigour when the war ended. Gradually, however, new evidence accumulated, interest revived, and some conversions were announced, largely among geophysicists (Runcorn 1956; Blackett 1956; Bullard, Maxwell & Revelle 1956; Bullard 1961).

*(Bullard, 1964: 1–2; Runcorn (1956) is my Runcorn (1956b), other references are the same as mine)*

Bullard's view is substantially but not entirely correct. Oddly, Bullard's only outright mistakes are about himself. He, Maxwell and Revelle did not come out in favor of mobilism in their 1956 paper; they did not even mention continental drift! Moreover, his statement that he announced his conversion in 1961 is several years premature!

I turn now to comments made immediately after Bullard's address as recorded anonymously. Twelve geologists responded to Bullard's talk (Anonymous, 1964). Seven respondents (Sir Edward Bailey, S. E. Hollingworth, R. B. McConnell, E. R. Oxburgh, N. Rast, M. J. Rickard, and R. W. R. Rutland) were inclined toward mobilism; three (W. D. Gill, J. E. G. W. Greenwood, and B. Webster-Smith) favored fixism; O. T. Jones remained neutral; and J. V. Hepworth expressed no opinion. Seven (Gill, Hollingworth, Jones, Oxburgh, Rast, Rickard, and Rutland) mentioned or at least indirectly referred to the new evidence. Five (Hollingworth, Gill, Greenwood, Rast, and Rickard) disputed Bullard's view about the relative importance of the geological and geophysical contributions to the overall debate. I shall deal only with those who addressed the new support (especially paleomagnetism) for mobilism, and will take Hollingworth as representative of those who thought Bullard had under-estimated the geological support for mobilism.

O. T. Jones (Anonymous, 1964: 27), one of the few geologists at Cambridge who had encouraged Runcorn, was undecided about mobilism.[18] After noting his opposition to mobilism, he turned to its new support. He thought the paleomagnetic work important, but was troubled by Axelrod's claim that paleontological evidence did not support it. He found interesting both the relative displacement of ocean floor in the east Pacific as indicated by the magnetic anomalies, and Hess's suggestion about the youthful age of sediments on the ocean floor, but wondered why the shifts in the magnetic anomalies did not appear in the adjacent continental area.

Oxburgh (later Lord Oxburgh), a research student at the University of Oxford, favored mobilism. He thought drift's solution for India's Permo-Carboniferous glaciation more probable than fixism and a vastly different climate regime. However, he was unimpressed with paleomagnetism's contribution; he worried about the GAD hypothesis. He favored the use of oxygen isotope studies rather than paleomagnetism to determine past latitudes.

The speaker [Oxburgh] also suggested that paleoclimatic evidence supported by oxygen isotope data might well provide a more reliable indication of latitude during the earlier geological periods than palaeomagnetic evidence. The relationship between the Earth's axis of rotation and its climatic belts, although less precise, was better known than that between the axis and the Earth's magnetic field. It had been suggested by a number of workers that in the past the Earth's magnetic field might have been neither axial nor dipolar. If this were so, palaeomagnetic observations made for periods before the present axial dipole field became established could, at the moment, give little information on palaeolatitude.

(*Anonymous, 1964: 29; my bracketed addition*)

Bullard had emphasized how very closely Late Cenozoic poles conformed to the present geographic pole; he had discussed the difficulties that had been raised against the axial and dipolar nature of the main geomagnetic field, and found them wanting. Oxburgh thought otherwise, but he was correct about one thing: the relationship between "Earth's axis of rotation and its climatic belts" is certainly "less precise" than that between the "axis and the Earth's magnetic field." But did he realize how imprecise a measure of latitude temperature is? Mean temperature of 20 °C could correspond to latitude of 20° in the Quaternary but 50° in the Eocene; at best, oxygen isotope estimates of past temperatures provide only imprecise information about past latitudes.

Gill, a fixist, resented Bullard's remarks about the inability of arguments from structural geology to settle the mobilism question. He dismissed paleomagnetic findings as ludicrous.

In any objective effort to explain the Earth's orogenic belts it must be emphasized that the theory of continental drift was not a meaningful concept, and in many cases the geological facts adequately demonstrated the ludicrous nature of many palaeogeographic reconstructions based on paleomagnetism.

(Anonymous, 1964: 28)

Rutland and Rickard, who favored mobilism, seemed to prefer Earth expansion to convection: Rickard definitely did, Rutland may have. Citing Heezen (§6.9), Rutland perceptively asked how Africa could have simultaneously drifted away from both the Indian and Mid-Atlantic Ridges, if convection currents rising under oceanic ridges propelled continents sideways: with ridges on two sides, Africa would be unable to move toward either of them; this difficulty did not diminish the likelihood of mobilism, but only the likelihood of mantle convection. He did not go on to speak directly of expansion, but he thought something more plausible than convection was needed to explain mobilism. He thought mobilism should be treated as a working hypothesis, and its paleomagnetic support needed to be coordinated with the timing of various geological phenomena.

Rickard favored Earth expansion, declaring that it enjoyed paleomagnetic support. Writing up his comments, He claimed that Carey had come up with a coherent mechanism for continental drift, and that Bullard, a geophysicist, had made a grave omission. To him it seemed

incredible that a discussion on drift could take place, apparently, without any serious consideration of this work ... Carey's basic assumption is that the marked trend changes of the major orogenic belts may be simple bends, which he calls oroclines, formed during drift. By empirically unbending all the oroclines (there are some 30 of them) on a globe rather than on map projections, he finds that the continents fit back into place in much the same position as in the classical reconstructions of Wegener and Du Toit ... The theory thus neatly co-ordinates and explains the relationships between deep ocean basins, mid-oceanic ridges, tensional rifts, large wrench-faults, compressional mountain belts, and curving island arcs, all of which are considered to have developed during the process of continental drift ... The rotational movements of the oroclines all have a sinistral sense, and rotation of land blocks indicated by this analysis accounts for some of the apparent anomalies in palaeomagnetic pole determinations; the rotations of Spain to open the Bay of Biscay, and of India away from Africa have been independently confirmed by palaeomagnetic studies ... The tensional expansion calculated for the continental blocks is the same as that calculated for the Pacific margin, strongly suggesting the possibility that drift occurred as a result of *expansion* of the earth as a whole. More geotectonic and palaeomagnetic evidence is required to prove or disprove the validity of Carey's orocline concept but the preliminary results are promising enough to warrant further studies along those lines and the reconsideration of geophysical data in relation to a possible expanding Earth.

*(Anonymous, 1964: 32–33)*

Rickard did not explain that Carey introduced continental drift several years before appealing to expansion. He was also mistaken about paleomagnetism supporting Earth expansion. Paleomagnetic findings did not prove expansion, and Cox and Doell (1961a) had shown that paleomagnetic results were inconsistent with Carey's rapid expansion (II, §8.14). Nevertheless, Rickard's comments illustrate the tension alluded to by Doris Holmes between Bullard, the geophysicist, and his geological audience.

Bullard replied to Rickard. He named Egyed and Heezen (a geophysicist and a geologist/geophysicist) and others as proponents of Earth expansion, but surprisingly delegated Carey to "the others." He also raised an interesting theoretical difficulty against expansion, arguing that it was inconsistent with the mobilist solution to the origin of the Himalayas, and cited the paleomagnetic attack on Earth expansion.

In the interests of brevity and clarity I ignored the possibility that the Earth has expanded, as suggested by Egyed, Heezen, and others. Expansion of the Earth might pull the continents apart, and much of the evidence for continental drift could be interpreted in terms of expansion. It is to be hoped that a study of the movements will lead to a decision between the two views (for example, India has been thrust northward against Asia, this can hardly be due to expansion). It is difficult to believe that in the last few per cent of geological time the Earth has expanded sufficiently to open the Atlantic and Indian oceans. Cox & Doell (1961a) have produced palaeomagnetic evidence that no such expansion has occurred, though van Hilten (1963b, *Nature, Lond. 200*, 1277–81) claims that other palaeomagnetic evidence suggests that it has. Deutsch (1963b, in *Polar wandering and continental drift* [ed. Munyan, A. C.], Society of Economic Paleontologists and Mineralogists Special Publication 10) has reviewed the widely ranging rates of expansion that have been suggested.

*(Anonymous, 1964, 33; my bracketed addition)*

Thus, of the five geologists who actually discussed the paleomagnetic support for mobilism, Gill thought it "ludicrous," Jones viewed it as impressive but noted Axelrod's criticism, Oxburgh questioned the GAD hypothesis, Rutland thought paleomagnetism worth considering but accented the need for it to be coordinated with geological evidence, and Rickard thought it supported Earth expansion.

What about Bullard's assessment of the geological support for mobilism? Hollingworth, who had favorably reviewed Holmes' *Principles of Physical Geology* (I, §5.9), let Bullard know that he had underestimated the mistakes of geophysicists and the contributions of geologists. His comments are the most extensive among those who felt that Bullard had slighted geology. He began by recalling the controversy about the age of the Earth, about Kelvin's estimate being too short, and geologists being closer to the mark.

> Professor S. E. Hollingworth said that the vicissitudes of the continental drift hypothesis recalled the divergence of views between the physicists and the geologists in the controversy concerning the age of the Earth in the second half of the nineteenth century. Kelvin, using available physical data, was successful in persuading most geologists to cut their estimate of the age of the Earth to a fraction of the figure arrived at through geological evidence, but he did include in his argument an escape clause covering possible, but as yet unknown, factors.
>
> *(Anonymous, 1964: 27)*

Hollingworth criticized Bullard's characterization of the mobilist controversy as being at a standstill during the second quarter of the century. Although he considered Bullard's view as "admirable and balanced," he clearly disagreed with it.

> Should we accept the suggestion in the lecturer's admirable and balanced exposition that continental drift was dropped by "general consent" in the second quarter of this century – a view that is rather widely held by our geophysical friends? When the speaker began teaching physical processes in 1946, he found the available English textbooks, such as Holmes's *Physical Geology* (1944) and Wooldridge & Morgan's *Physical Basis of Geography* (1937), strongly supporting continental drift; and, somewhat later, Read's little book *Geology* (1949) at least fully sympathetic. This seemed to have remained the position pending, as the lecturer cogently remarked, positive evidence of movements of this type. Surely it was at least considered a stimulating and controversial topic throughout this period?
>
> *(Anonymous, 1964: 27)*

Hollingworth was being unfair to Bullard. Bullard actually said that the drift controversy had been lively during the 1930s, and that it was only at the end that participants became "wearied of inconclusive argument," and added that "the battle was not renewed with any vigour when the war ended." But Bullard mentioned Holmes' 1944 *Principles* as "a notable exception" to the general rule that most geology texts published in the United Kingdom during the forties were fixist in outlook. Although Bullard did not mention Wooldridge and Morgan's or Read's books, he did not claim that Holmes was the *only* exception but *a notable* one. In addition, Wooldridge, one of the authors Hollingworth had singled out, might have agreed with Bullard; in an address at a symposium on continental drift in 1950, Wooldridge, who had come to favor sinking landbridges more than continental drift, remarked that little had changed since the 1930s (I, §5.9).

Turning to Jeffreys' role in the controversy, Hollingworth thought his influence as negative, persuading geologists to disregard mobilism.

Let us accept that the weighty adverse opinion of Jeffreys and others influenced and indeed convinced many, and that a majority of structural geologists, the world over, whether regional or continent-wide in their interests, have found little need to invoke continental drift in discussing orogenic processes. The "anti-bias" even led one palaeoclimatologist not only to accept equatorial glaciation on a continental scale for the Permo-Carboniferous but to produce a dubious explanation.

*(Anonymous, 1964: 27)*

But the difficulties Jeffreys raised were substantial. Was Hollingworth suggesting that the palaeoclimatologist (perhaps he meant C. E. P. Brooks (I, §3.11)) or structural geologists rejected mobilism because they feared Jeffreys and geophysical arguments? Or was he suggesting that Jeffreys duped them, that they were spineless or stupid? Hollingworth admitted that geophysics, by providing new evidence, was now atoning for retarding the study of continental drift for twenty years. He closed by recalling the important support for mobilism put forth by two geologists before the "dawn" of geophysics.

If allowed that geophysical information retarded progress in the study of continental drift for twenty years, there had, at least in the past decade, been ample atonement for this lapse. In addressing these remarks to a predominately youthful audience the speaker could not refrain from remarking that before the geophysical dawn geologists in this room continued to make important contributions. Sir Edward Bailey was stressing the significance of the Hercynian and Caledonian belts across the Atlantic and Professor Shotton's classic contribution on Permo-Triassic wind directions came to mind. As the lecturer had remarked, the function of a theory was to stimulate its verification.

*(Anonymous, 1964: 27)*

Perhaps prompted by Hollingworth's mention of his work, Sir Edward rose to speak, not about his own work, but on G. W. Lamplugh's (1859–1926) idea of continental drift. He recalled that it was presented in a public lecture in 1910 at the Old Vic Theatre in London.

Sir Eward Bailey gave a personal explanation of how he came to be the first to lecture on the subject of drifting continents. It was in 1910, and the occasion was a request-lecture to a non-geological audience at the Old Vic. Lamplugh, at this time, was newly returned from South Africa ... An outstanding glacialist, Lamplugh had been immensely impressed by what he saw of the Permo-Carboniferous glaciation. The speaker [Bailey], for his part had then recently visited, under Collet's guidance, the wonderful thrusts of the pre-Alps. Lamplugh was interested, and [I] invited him for a week-end to talk things over. After the Alpine discussion, Lamplugh unfolded his theory of drifting continents, to which he had been led by the extraordinary present-day distribution of lands glaciated in Permo-Carboniferous times. Hence, the lecture mentioned above. Unfortunately, Lamplugh never published his idea. On returning from the First World War, the speaker [Bailey] asked him [Lamplugh] whether he was the source, direct or indirect, of Wegener's theory; but he was assured that it was a case of [an] entirely independent approach.

*(Anonymous, 1964: 28; my bracketed additions; see I, §8.13 on Bailey's seeing pre-Alps in 1909 with Collet)*

Bailey's remarks are of historical interest because they recount Lamplugh's version of continental drift to explain the origin and distribution of Permo-Carboniferous glaciations, and because they indicate that Bailey himself thought continental drift worth discussing at a public lecture he gave at least fifteen years before he himself supported it in print.[19]

What about Bailey and Shotton's support of mobilism, how strong was it? Although Shotton (1956) spoke in terms of continental drift and polar wandering in his second paper on determining the prevailing wind directions during the formation of the Bunter Sandstones, he was mute on mobilism in his much earlier study (Shotton, 1937). Bailey's support for mobilism, however, was consistent and strong, developing, as he did (1927), an original argument based on the crisscrossing of the Hercynian and Caledonian belts on both sides of the Atlantic (I, §5.4), by matching and detailing similarities among Caledonide fragments in Scotland, Scandinavia, and Greenland (Bailey and Holtedahl, 1938) (I, §8.11, §8.13), and continuing to defend it (with J. Weir) in their *Introduction to Geology* of 1939 (I, §8.13).

Although Bullard did not respond directly to Hollingworth, he did repeatedly answer the charge that he had underestimated the impact of the geological support for mobilism.

> The LECTURER, in reply, said that he did not mean to suggest that one need not consider the facts of structural geology; merely that they had been considered at great length for forty years and that there were still apparently irreconcilable differences of opinion among the experts as to which way they pointed; it therefore seemed more profitable to talk about other things.
>
> *(Anonymous, 1964: 29)*

He was entirely correct about the "irreconcilable differences of opinion among experts" as evidenced by the responses to his talk, for example, Webster-Smith's evaluation of mobilist and fixist explanations of the Permo-Carboniferous glaciation.

> Mr. B. Webster-Smith commented on the stress often laid in discussions of continental drift upon the Permo-Carboniferous glaciations of India, Southern Africa, and Australia and the implications that at that time the Earth's polar axis had wandered. The entire fossil record, and also the detailed study of ice fluctuations in the Pleistocene glacial stages, seemed to the speaker to prove the contrary and show that minor changes in solar radiation would explain all these matters without calling in such complications.
>
> *(Anonymous, 1964: 29–30)*

What about Doris Holmes' recollection that "the geological audience was very irritated that a physicist should come and tell them what they had known for a long time, and one after the other rose to tell Bullard, in essence, that all he had said was 'old hat'." Some of the geologists' anger surely was misplaced. If they really thought that the support they had garnered in favor of mobilism was more conclusive than Bullard suggested, then perhaps they were angry at themselves for not having offered mobilism the support it deserved in its time of need. Moreover, the new evidence in favor of mobilism from paleomagnetism and marine geology and geophysics that Bullard discussed was not "old hat." Old fixists might have found it ludicrous, fence sitters might have been troubled by Axelrod's alleged contrary results, and old mobilists might have viewed the geological record of drift as equal or better than the paleomagnetic evidence, but the latter was completely new, and perhaps that irritated geologists. Of course, Doris Holmes herself knew that work in paleomagnetism and marine geophysics and geology was new; she remarked on it and explained why her husband had been unable to undertake the sort of empirical studies needed to test drift.

As a physicist Arthur could not have investigated drift because the instruments now used by marine geologists and geophysicists only became available after the last war, nor was money made available for such investigations; one had to pay out of one's own meagre earnings for any research one wished to follow, at least in Britain.

*(Doris L. Holmes, February 4, 1979 letter to author)*

And about her husband, what did Holmes himself think about the new paleomagnetic evidence for mobilism?

### 2.16 Arthur Holmes' attitude to the paleomagnetic case for mobilism

Here are a heading and four quotations from Holmes' notebooks he evidently thought might prove appropriate to preface his discussion of continental drift.[20]

CONTINENTAL DRIFT

To be uncertain is to be uncomfortable; but to be certain is to be ridiculous.
JOHANN WOLFGANG von GOETHE 1749–1832

The man who will not alter his opinion is like stagnant water which breeds reptiles of the mind.
WILLIAM BLAKE 1757–1827

If a man will begin with certainties, he shall end in doubts; but if he will be content to begin with doubts, he shall end in certainties.
FRANCIS BACON 1561–1628
Advancement of Learning, I, v, 8.

The latitude's rather uncertain,
And the longitude also is vague.

WILLIAM JEFFREY PROWSE 1836–1870

Holmes finished the second and fully revised edition of his *Principles of Physical Geology* in 1964; it was published the following year. He had spent seven years, expanding it from 532 to 1288 pages. Although his commitment seems to have weakened somewhat during the first half of the 1950s, as evidenced by his 1953 review of the 1949 symposium on Mesozoic landbridges (I, §5.11), he never gave up supporting mobilism, and as he wrote, he looked ever more favorably on mobilism. In 1959 he prepared a revised geological timescale, adding sections favoring Earth expansion, proposing pre-Wegenerian drift, and praising the use of paleomagnetic surveys to speculate on Precambrian paleogeography. He announced that mobilism had finally become respectable.

The importance of such work [on Precambrian paleogeography], especially if combined with palaeomagnetic surveys, also of carefully dated materials, cannot be over-emphasised, now that "polar wandering," continental drift and global expansion have at least become respectable subjects for serious research.

*(Holmes, 1959: 211; my bracketed addition)*

The Goethe quotation suggests that Holmes thought rigid opposition to continental drift ridiculous; he used it in his speech accepting the 1956 GSA's Penrose medal (1957: 75). For his chapter on continental drift and paleomagnetism he chose the quotation from Prowse. I shall consider here only his ringing endorsement (1965: 963) of paleomagnetic support. "The study of palaeomagnetism has brought about a major revolution in the attitude of most scientists toward polar migration and continental drift." He thought drift is "as real as" erosion, and claimed that many geophysicists had been won over to it because of paleomagnetism.

Continental drift, now embracing crustal separation and ocean floor dispersal and renewal, is known to be in operation at the present time from the movements still continuing along the greater transcurrent faults or "megashears". The many forms of crustal drift – whatever their explanations may be – are activities of the earth as real as those of erosion. And geophysicists have been won over by the growing testimony of palaeomagnetism, i.e., by the results of investigating rocks that can be regarded as magnetic compasses: rocks which have retained a record approximating to north and south as these directions were at the time when the rocks received their magnetisation.

*(Holmes, 1965: 1203)*

In a section "Palaeomagnetism: rocks as fossil compasses," Holmes explained different ways that rocks acquire their remanent magnetization and introduced the GAD time-averaged model. The old drifter explained the common practices of collecting enough samples from a site to secure statistically reliable results and of gathering samples with sufficient "vertical" span to average out secular variation. He explained the igneous contact test (II, §1.11). He focused on the work of the Manchester/London, Cambridge/Newcastle, and ANU groups and referred to nine of their papers. He never himself got to know any of the active workers and there is no record of him visiting any of their laboratories or attending any meetings at which their work was explained.

Apart from some earlier work in France, which was not systematically followed up – probably because of the World Wars – the present rapidly accelerating study of palaeomagnetism was begun about 1950 by P. M. S. Blackett, who soon attracted an enthusiastic team of co-workers. Some of these, notably S. K. Runcorn, have started teams of their own. Because of the obvious bearing of the results on the geological problems of continental drift and the physical problems of the origin of the earth's magnetic field, the subject has rapidly spread throughout the scientific world as one of the most fertile and actively developing branches of geophysics.

*(Holmes, 1965: 1206)*

Holmes placed strong reliance on van Hilten's work, which likely resulted from their mutual interest in Earth expansion. Van Hilten was, I think, the only paleomagnetist who in 1964 still thought that paleomagnetism did not rule out rapid Earth expansion, and in his new edition Holmes expressed mild support for slow expansion coupled with mantle convection. Holmes also discussed work on paleowinds (II, §5.11–§5.15), and work comparing paleomagnetic and paleoclimatological findings (§1.7).

Holmes next discussed the many difficulties that had been raised against the straightforward reading of the paleomagnetic results as evidence for continental drift, and argued that they were insufficient and had been answered. After explaining the relationship between paleomagnetic inclination and paleolatitude, and the determination of poles from paleomagnetic directions, he discussed the dispersion of poles from the same time period from the same continental block, and argued that individual poles may generally be considered reliable to within about 15°; he concurred with paleomagnetists that dispersion in such poles could be attributed to age differences between samples, local displacements along transcurrent faults, or orogenic movements. He repeated an earlier suggestion (1959: 211) that dating uncertainties might be lessened through the use of radiometric dating of paleomagnetic samples.[21]

With present techniques, which are constantly being improved, individual pole determinations may be correct within about 15°, and possibly less if based on a dozen or more reliable samples. It should also be noticed that considerably greater variations may be found in the pole positions relative to a given continent and recorded for a given geological Period. These differences are not necessarily errors. Many of them may represent genuine changes in the position of the continent during the very long interval of time involved, or in the positions of certain regions that have been displaced by great transcurrent faults, or have gone through the severe disturbances of orogenic movements. The effects of some of these difficulties will be reduced as increasingly detailed work is accomplished year by year, and especially as it becomes more generally possible to combine radiometric dating with magnetic measurements.

*(Holmes, 1965: 1207)*

Turning to the use of paleomagnetism to construct paleogeographies, Holmes noted the uncertainty in longitude. He reproduced several paleomagnetic reconstructions, and concluded that, even though there was not yet paleomagnetic information to fix the position of *every* continent during *every* geological period, the current results were consilient with the findings obtained from geological and paleoclimatological studies (RS1).

As yet there is insufficient paleomagnetic information to carry out this programme [of reconstructing continental positions from paleomagnetism] in full – but data are rapidly accumulating. Meanwhile, it can be done for two or three periods or three continents at a time.

The impressive fact has already emerged that, despite all the very broad margins of error, the results are consistent with those that have been already inferred from the evidences of glaciation and other geological criteria to which appeal was made before the data provided by palaeomagnetic researches became available.

*(Holmes, 1965: 1212; my bracketed addition)*

Holmes briefly discussed polarity reversals, which he agreed represented reversals of the geomagnetic field. He did not consider the hemispheric ambiguity arising from reversals (II, §5.16) a serious problem because geological evidence could generally decide, but he admitted that there might be uncertainties when latitudes are very low.

Fortunately for our present purposes it does not seriously matter whether or not the magnetisation of a given rock is normal or reversed. Numerically, the magnetic dip remains approximately the same, and whether the calculated latitude is north or south of the Equator of the time can usually be decided by other relevant evidence, with which it has to be consistent ... Some ambiguities may arise when latitudes are low, i.e. near the Equator of the time; but in the present limited state of our knowledge it is a great achievement to have firm evidence, satisfying physicists as well as geologists ...

*(Holmes, 1965: 1214)*

Returning to the question of paleomagnetic support for mobilism, Holmes (1965: 1214) quoted Blackett *et al.* (1960): "on a broad scale, and allowing for vicissitudes in direction from time to time, the older the rock the greater is the departure of its original latitude from its present latitude" indicative, as they had shown, of a general but not uniform northward drift of landmasses (II, §5.19; II, Figure 5.23). He then thoroughly examined the paleomagnetic method and the two underlying assumptions, covering much the same ground as Bullard (§2.14), and like him thought them reasonable (RS1). In defending the assumption that certain rocks record the geomagnetic field at about the time of formation, he appealed in this elegant passage to actualism.

The observed fact that the stable magnetisation acquired by rocks in recent centuries is in line with the earth's field at the time of magnetisation is assumed to apply equally in earlier times. This is an application of actualism against which no geological objections can be brought. If the assumption were not valid, it would be difficult, if not impossible, to account for the fact that neighbouring rocks of the same age but of widely different kinds (e.g. igneous, sedimentary, and metamorphic) yield consistent results, although they received their magnetisation in very different ways. And it would be equally difficult to account for the manifest relationship between the age of a rock from a given region and the direction of its acquired magnetism.

*(Holmes, 1965: 1214)*

Actualism, akin to uniformitarianism, means "that the same processes and natural laws prevailed in the past as those we can now observe or infer from observations" (1965: 44). Uniformitarianism is usually taken to imply that they did so at the same rate. Blackett *et al.* (1960) had invoked the same defense without distinguishing between them. To deny this assumption raised difficulties (RS2). If rocks fail to acquire a remanent magnetization along the geomagnetic field, then consistency of magnetization should not be observed among neighboring rocks of similar age and their poles would not agree. The argument is greatly strengthened if consistency is observed among different sorts of rocks in different structural settings from the same region. If this should sound familiar, it is because Irving had used it at the 1957 Toronto IUGG meeting to contradict Graham's assertion that stress effects (magnetostriction) had seriously distorted the Australian APW path (II, §8.12) and because Du Bois *et al.* (1957) and Blackett *et al.* (1960) had raised in print similar objections to Graham's assertion that effects of magnetostriction had seriously impaired the record of the geomagnetic field and invalidated the first assumption (II, §7.5).

To defend the second assumption, the GAD hypothesis, he appealed to the consilience between paleomagnetic and paleoclimatological results. Again, this will sound familiar, being essentially that invented a decade earlier by Irving (1956) (II, §3.12) and later utilized by others, including Blackett (1961) (§1.7), which is probably where Holmes found it. Holmes added his own gently mocking twist, pointing to the two absurd difficulties that arise if the GAD hypothesis is denied (RS2).

It is assumed that in the past the average earth's field has not widely deviated for long periods from that of an axial dipole. Had there been such long-period deviations they must have been due to powerful non-dipole fields, each associated with a particular continent and each varying systematically with time in such a way as to present the illusion of continental drift (as implied in Fig. 866 [showing the APW paths for Europe, North America, Australia, India, and Africa]), and also to bring about the very real changes of climate to which the geological history of every continent bears irrefutable witness. These are fatal difficulties. Stated more simply, the general agreement between the climatic zones of a given time inferred from geological observations and the corresponding latitudes inferred from palaeomagnetic measurements suffices to justify the extension of the axial dipole assumption to ancient times.

*(Holmes, 1965: 1214–1215; my bracketed addition)*

Having disposed of the difficulties raised against the two key assumptions, he proceeded to apply them. Considering first polar wandering, he informed his readers:

Until about 1956 most geophysicists seem to have favored polar wandering (without continental drift) as a sufficient explanation for the changes of latitude disclosed by the palaeomagnetic data then available.

*(Holmes, 1965: 1215)*

Perhaps he had Runcorn in mind; he would not know of Irving and Creer's earlier support of both continental drift and polar wandering that they expressed in their Ph.D. dissertations (II, §3.4, §3.6–§3.7); if he knew about the Triassic work and similar sentiment of Clegg *et al.* (1954a, b) (II, §3.5), he did not say so. Without specifically discussing any APW paths, Holmes made the point, which was inescapable to him as well as paleomagnetists from Britain, Australia, and southern Africa, and to which Graham, Cox, and Doell, the leaders of paleomagnetic research in North America, seemed so blind, that the gross lack of agreement between APW paths from different continents made polar wandering without continental drift untenable.

If this [i.e., polar wandering] were an adequate explanation by itself, then the geographical positions of the poles at any given time would be the same for all the continents. Assuming the continents to have remained fixed relative to one another, there would be a single "polar-wandering curve." But later work soon revealed that this is far from being the case. Every continent has its own "polar-wandering curve" and these are so widely different that continental drift on a grand scale is obviously implied.

*(Holmes, 1965: 1215; my bracketed addition)*

Having established continental drift, Holmes now inquired whether it alone or polar wandering was also required. He knew about Bradley's paper (II, §4.8–§4.10, §4.12), and referring to it and Deutsch (1963a) he agreed with them that it was difficult to assign a precise meaning to "polar wandering." Was it the idea of the entire crust slipping over the mantle or the whole Earth toppling through its axis of rotation? He did not reject polar wandering outright. He thought drift had happened and polar wander might have. Following a suggestion by Blackett and quoting van Hilten, Holmes ended in this way:

The only remaining explanation is *continental drift*, which is abundantly demonstrated by both geological and palaeomagnetic evidence, whereas "polar wandering" is not. D. van Hilten has put the present position very clearly: "According to our present knowledge, both processes may have played their part in the geological history of the Earth. We are, however, completely at a loss about their relations. Are both processes occurring contemporaneously or in some succession, and is there any causal relation at all? Paleomagnetic evidence is not (yet) sufficiently accurate to produce the answers." In these circumstances it seems wise to adopt Blackett's practical suggestion that the term continental drift should be provisionally allowed to include any polar-wandering component that may eventually be recognized in such a way as to become separable from the total combined movements.

*(Holmes, 1965: 1216; the van Hilten quotation is from van Hilten, 1963a: 200)*

The developments must have pleased Holmes. He and Bullard had described how success had been achieved by applying physics and geophysics to a major geological problem. They, and others too, had come to agree that the absence of a known mechanism was not a legitimate difficulty, and despite its longevity, could now be recognized for what it had become – a phantom difficulty. He, like Bullard, believed that the mobilist solution to the divergent APW paths implied continental drift on a vast scale and that all the difficulties that had been raised against it had been satisfactorily disposed of: for them, it had in effect achieved difficulty-free status.

## 2.17 Mobilism's solution to divergent APW paths, its difficulty-free status

In these next three sections I want to comment on the wider reception of the paleomagnetic case for continental drift. I begin, however, with a very particular, much earlier incident: Neil and Margie Opdyke's arrival at Canberra airport, Australia, in late 1959.

We met Margie and Neil at the airport and Neil tells the story, which I have no reason to believe incorrect, that I greeted him in a totally unexpected way, which he found disconcerting. He says I said "It's all over, Neil, continents have drifted," or words to that effect. Neil remembers this vividly because he wondered why the hell he had come all this way simply to be told it was game over! I think this story means that it confirms that I personally must have by then been convinced, because of our new results, because of the paleoclimatic evidence, and because of Stacey and Stott's work [physically disposing of Graham's magnetostriction difficulty], that we now had a pretty water-tight case. I did not mean the task of converting the fixists was complete, or what Neil was about to do was not going to be important. I mention this because it dates, by means other than my own testimony, that at this time I personally had become convinced of drift; it is not just vague retrospective thinking on my part.

*(Irving, October 2010 email to author; my bracketed addition)*

When the Opdykes arrived in Canberra there were APW paths for Europe (II, §3.6), North America (II, §3.10), and Australia (II, §5.3), a rudimentary path for India (II, §5.2), and spot poles from South America (II, §5.6), southern Africa (§1.17; II, §5.4), Antarctica (II, §5.7), and Japan (II, §5.8) (II, Figure 8.1). Now, five years later, old paths had been confirmed and augmented, and others obtained and new cleaning techniques applied. Each path was continuous, each was different, and all converged to the present geographical pole. Paths for Australia and India, especially, diverged widely from those of North America and Europe. This, together with the markedly better agreement among poles of a given period from the same continent compared with that between continents (II, §5.17), and the differing rates of latitude change, especially the fast northward movement of India, required continental drift: alone, polar wandering was utterly insufficient. There were independent contributions from Soviet paleomagnetists, notably by Khramov and colleagues who in the 1950s, despite the predominance of fixism among Soviet geologists, made an important contribution to the paleomagnetic drift case based on their own observations and their knowledge of work internationally (II, §5.9). Their straightforward reading of the APW paths, like that of workers in British, Australian, South African, and Rhodesian paleomagnetic groups, contrasted strongly with the piecemeal approach of the then active US groups (II, §8.7). Although more poles were needed to clarify particular aspects of drift history, now, a decade later, poles based on reliable magnetizations observed from well-dated rocks had been obtained from all major landmasses confirming continental drift and warranting its acceptance. Let me justify this statement.

Paleomagnetists looked for fine-grained red beds and mafic lavas and intrusions because they generally provide reliable results. Igneous contact, fold including slump, conglomerate, and reversal tests became standard. Commonly poles were based on many samples yielding consistent results spaced through a substantial sedimentary or igneous sequence to average out secular variation, and spread laterally over a substantial area to test for local aberrations. Fisher's statistics were used universally to summarize directions of magnetization and corresponding poles were reported.

Data lists were published and freely circulated. Experiments on the settling of sediments determined how well the inclination of the magnetic field is recorded. Paleomagnetists sought contemporaneous sedimentary and igneous rocks, and rocks in differing structural environments because agreement between magnetizations acquired by different processes and preserved under different conditions enhanced reliability. Magnetic cleaning had been introduced as required routine (II, §5.5); their application allowed the use of rocks with substantial secondary but removable VRM overprints. Poles derived from rocks that were dated radiometrically began to be obtained in increasing numbers. Next, paleomagnetists validated empirically (II, §2.9) and theoretically (II, §2.13, §5.18) the GAD model of the geomagnetic field on which the method depended. Poles, extending in age in some cases as far back as Miocene, from Iceland, North America, Australia, South America, and Antarctica centered around the present geographic pole, confirming globally the GAD model.

Crucial support came from the demonstration of the consilience of APW paths and the paleogeographic reconstructions of Wegener, du Toit, Carey, and King based on entirely different data and methods. Crucial support also came from the agreement in all manner of situations between paleoclimatic and paleomagnetic findings from the same region, bolstering support for the GAD hypothesis (II, §3.12) as far back as the Paleozoic (II, §5.10–§5.15). Blackett, taking a phenomenological approach, argued in favor of continental drift, he claimed, without having to appeal to the GAD hypothesis (II, §5.19). Irving, influenced by Popper, argued that those who objected to the GAD hypothesis were replacing it with some unspecified non-dipole hypothesis that was unfalsifiable and thus non-scientific (§1.10). Paleomagnetists pushed relentlessly onward and, I would argue, conclusively answered the difficulties that had been thrown in their way (RS1): Graham's appeal to magnetostriction (II, §7.4, §7.5), Cox's troublesome Siletz River Volcanics (II, §7.6, §7.7) and Hibberd's single spiraling APW path (II, 7.9) were disposed of. Stehli's warm climate Permian brachiopods and fusulinids were shown to be beautifully consistent with mobilism (II, §7.10).

The attacks by Jeffreys in his fourth edition of *The Earth* (II, §7.13) and by Munk and MacDonald in their (1960) *The Rotation of the Earth* (II, §7.11) petered out, and the censures by Cox and Doell in their *GSA* 1960 and *AG* 1961 reviews (II, §8.7) fell by the wayside. Paleomagnetists did not, as Jeffreys supposed, use rocks whose magnetization changed when beaten with a hammer. Munk and MacDonald's complaint that paleomagnetists could not provide unique past continental positions was irrelevant because the mobilism case did not depend on there being a unique solution. Cox and Doell's piecemeal analysis of the paleomagnetic data was not generally adopted, soon lost by the wayside, and their neglect of its consilience with paleoclimatic evidence (§1.7, §1.18; II, §3.12, §5.12–§5.15) and of the spectacular agreement between paleomagnetic results and the mobilistic paleogeographies of Wegener, du Toit, Carey, and King, (II, §8.7) left a void in their arguments.

## 2.18  On the general failure to recognize the difficulty-free status
## of the paleomagnetic case for mobilism

Despite their successes, paleopole studies did not provide all that was needed to change fixism's deeply entrenched status. Why was that?

As expected, old-time drifters or drift sympathizers welcomed the new paleomagnetic results (RS1). Holmes thought its solution to divergent APW paths secured drift (§2.16). Harland and Rudwick declared the paleomagnetic case for mobilism decisive (§1.11). Gutenberg thoroughly familiarized himself with the paleomagnetic support, and stopped just short of fully accepting mobilism (§2.4). Carey (II, §6.15), Hamilton (§1.13) and Schwarzbach (§1.3) welcomed the paleomagnetic findings, and King, even if he misunderstood a key aspect of them, still acknowledged their support for mobilism (§1.3). Paleobotanists, especially those concerned with the Southern Hemisphere such as Plumstead, Good, Cranwell, Chaloner, Hair, Andrews, and Lovis welcomed the paleomagnetic results (§1.12), and geologists from South America such as Maack, Bigarella, and Salamuni were also pleased (§1.9). Westoll, who jointly supervised Opdyke, and got after Runcorn for clinging to polar wandering without drift, applauded the paleomagnetic results (§1.9). D. A. Brown, a New Zealand paleontologist, later at Canberra, welcomed mobilism's paleomagnetic support and worked with Irving (§1.18). All agreed that paleomagnetism had given mobilism new and independent support, and that the paleomagnetic, paleoclimatological, and paleobiogeographical results were consilient (RS1).

A few fixists changed their minds, partly or entirely because of mobilism's paleomagnetic support. Hales, one of the minority of South African Earth scientists who was opposed to mobilism, became a mobilist because of paleomagnetic work, including his own (II, §5.4). The Dutchman Vening Meinesz (§2.6) accepted mobilism because of the paleomagnetic results, and his countryman Martin Rutten, after arguing with Irving at the Toronto 1957 IUGG meeting (II, §8.12), declared, seven years later (1964), that paleomagnetism proved continental drift (IV, §3.3); both were formerly staunch fixists. Most importantly, Hess, as we shall see, changed his mind substantially because of the paleomagnetic evidence (§3.12), and Dietz was also impressed with it (§4.6). The US entomologist and biogeographer Darlington became disposed toward mobilism, in part, because of paleomagnetic findings (§1.12). Raasch, co-organizer of the symposium on polar wandering and continental drift sponsored by the Alberta Society of Petroleum Geologists, became strongly inclined toward mobilism because of its paleomagnetic support, and consequently began his own research into paleowinds (II, §5.15). Fischer, from Princeton, was impressed with the paleomagnetic results, and expressed his guarded preference for mobilism in his review of Nairn's *Descriptive Palaeoclimatology* (§1.4). The Australian biologist Ride became a mobilist because of paleomagnetism (I, §9.4). Bullard, who originally sat comfortably on the fence, recognized that the paleomagnetic case was fully justified, and accepted mobilism (§2.14). Longwell (II, §6.16), formerly a mild fixist, became sympathetic to mobilism because of its paleomagnetic support.

Yet fixism remained the majority mind-set, and most active fixists did not change their minds. Neither did the vast majority of "conforming" fixists, those who had simply accepted fixism as the ruling dogma without actively advocating it or involving themselves in it in any way. Orthodox fixism remained the safe refuge for those anxious at all costs to avoid mobilism and its implications; they attacked (RS2) or ignored mobilism's paleomagnetic support while arguing their case. As already shown, Jeffreys (UK), Billings (USA), Bucher (USA), Munk (USA) and MacDonald (USA) repeatedly attacked paleomagnetism (§2.10 and II, §7.13; II, §7.12; §1.10; §2.8 and II, 7.11, respectively), and they often succeeded in obscuring the issues from those not closely acquainted with paleomagnetic work and who were too incurious or too lazy to find out for themselves. Waters (USA) and Gilluly (USA) at first ignored the paleomagnetic case, although Gilluly later (1963) accepted mobilism, in part, because of it (IV, §3.4, §3.5). Ewing long ignored the paleomagnetic support (USA) (§1.9). Stehli (USA) continued to attack mobilism into the middle 1960s even though his earlier critiques had been fully rebuffed (§1.9, §1.18). Van Steenis, the doyen of Dutch paleobotany, and Axelrod (USA), diehard fixists, rejected paleomagnetic arguments (§1.12). Durham (USA) and Arkell (UK), both aware of mobilism's paleomagnetic support, remained fixists (II, §3.10). Despite Irving and Green's strikingly different APW path for Australia, Australians generally remained unmoved. E. S. Hills, Ringwood, and W. R. Browne were unconvinced (I, §9.3). Immigrant European geologists, Teichert, Glaessner, and Öpik, who became very influential in the Australian Earth science community, were unmoved by mobilism's paleomagnetic support (I, §9.4). Dorothy Hill was impervious to paleomagnetism (I, §9.3). Biogeographers working in Australia, such as E. Le G. Troughton, Riek, Paramonov, Balme, McMichael, and Iredale ignored it (I, §9.5). Keast, who remained undecided about mobilism even after confirmation of the Vine–Matthews hypothesis, mentioned mobilism's paleomagnetic support but offered no evaluation and remained a fixist (I, §9.5). Burbidge spoke favorably of mobilism's paleomagnetic support but remained inclined toward fixism (I, §9.5). At the 1963 NATO meeting Chaney, Barghorn, and Dorf, all from the USA, ignored paleomagnetism and Lotze (West Germany) also said nothing about it (§1.9). At the same meeting, Colbert (USA) did not mention paleomagnetism, and Lowenstam (USA) and Shirley (UK) restated their fixist arguments explicitly rejecting the paleomagnetic case: Lowenstam relied solely on the outdated Cox and Doell *GSA* review, and Shirley did not accept the Triassic paleomagnetic pole for Africa (§1.9).

Later, Kay and Colbert (1965) did acknowledge that paleomagnetic evidence could not in all fairness be ignored (§1.14). Had they done so five years earlier in 1959, 1960, or even 1961, it would, relative to the overwhelming fixist ambience in North America, have been most progressive; it would have shown that they had been monitoring the ongoing paleomagnetic support for mobilism, had understood it, and were willing, in the light of it, to reevaluate their earlier fixism. Nonetheless, theirs was a small improvement, indicative of how, in the early 1960s, active fixists had remained blind to the paleomagnetic case for mobilism. G. G. Simpson, prominent US vertebrate paleontologist and longstanding foe of mobilism, admitted that the new paleomagnetic data "raise serious doubts" about the fixity of the continents before the Cenozoic (1965: 212), but there is no evidence that he studied primary sources (§1.12). The British paleontologist George (1962), who acknowledged Simpson's influence, gave serious consideration to mobilism's paleomagnetic support, but remained a fixist (§1.12). By the mid-1960s the foregoing fixists admitted some softening of their positions, but all except George, who commented on paleomagnetism in some detail, relied entirely on the 1960 *GSA* review for paleomagnetic information, an outdated secondary source.

This short account of those who wrote unfavorably about mobilism during the late 1950s and early 1960s indicates (1) that few of them changed their minds about mobilism because of its new paleomagnetic support, (2) that resistance to mobilism remained strong in Australia and North America, (3) that all of them relied heavily on the 1960 *GSA* review, even when years out-of-date, rarely referring to primary sources, (4) that a few less rigid fixists or "neutralists" who came to favor mobilism because of its paleomagnetic support did not rely on this review but read original papers. Under point (1), most fixists were strongly prone to dismiss work outside their area of expertise if it clashed with their own, even if, as in the present instances, their own work was less relevant to the drift debate than the findings of paleomagnetism so evidently were. Specialization continued to retard the acceptance of mobilism; few specialists, to use Feynman's metaphor, appreciated the gems researchers in other fields were digging up just over the hill. Paleomagnetism was not taken seriously by most workers in other fields; not only by geologists and biogeographers but also by geophysicists such as Birch, Jeffreys, and MacDonald, and geochemists such as Ringwood who claimed mobilism was impossible – antithetic as it was to their own worldview. The strong pervasive presence of regionalism among Australian and North American Earth scientists underlies point (2). Apart from a few such as Carey, Jaeger, Ride, D. A. Brown, and Evans, Australian geologists and biogeographers were not paying enough attention, few cared much that Australia's APW path differed hugely from paths from other continents; it was not going to change their way of thinking. Points (3) and (4) speak directly to the delaying effect of the 1960 *GSA* review; reinforced as it was by the imprimatur from the dominant Geological Society of America, it gave geologists not conversant with paleomagnetism an excuse to remain fixist or at least not to take mobilism seriously.

However, I suspect there were some fixists, perhaps many more than I have been able to find, who in the early 1960s stopped defending fixism and took a wait-and-see attitude. They began to believe that the paleomagnetic case for mobilism might be correct. Dott, who lived through it all, believes that he was just such a person, one among many.

There were a lot of others who were thinking hard about the issue – I was one among the many. I was on the fence through the fifties, was "converting" in the early sixties, and was surely won completely by Vine and Matthews.

*(Dott, January 2001 email to author)*

Dott, however, may not have been at all typical, for he had been fortunate enough to witness the debate at Columbia University between King and Bucher, which by all reports King had won, and had been impressed with the paleomagnetic results after hearing a Runcorn lecture (§1.2). Charles Bentley, who like Dott had been a student at Columbia University and had witnessed the King–Bucher debate, may have been another "one among the many."

I was a graduate student at Columbia in the early 50's with Marshall Kay and Walter Bucher among my professors and Maurice Ewing as my thesis advisor. I thought continental drift was an intriguing idea, but, as you can imagine considering the environment, I was more impressed by the apparent lack of any viable mechanism. What sticks most clearly in my mind, however, is a debate on the subject between Bucher and the visiting Lester King. It was eye-opening – we graduate students pretty much agreed that King won the debate, i.e. had the better case because his evidence was so strong, although I (and others also, I believe) remained troubled by the mechanism problem. But I also thought the paleomagnetic results that were coming along then were powerful evidence (yes, the supporting physics was clear to me); all in all I was ready to become a full believer as soon as the mechanism problem could be solved. I think it is fair to say that, based on the evidence presented by Southern Hemisphere geologists and Northern Hemisphere paleomagnetists, I was leaning enough towards drift that if I had been forced to say "yes" or "no" I would have chosen "yes."

In late 1956 I went off to the IGY [International Geophysical Year] traverse program in Antarctica, not returning until early 1959, when I went to the University of Wisconsin to work on my two years of mostly glaciological data. Soon I was in charge of the continuing Antarctic oversnow traverse program, so I didn't have much time to think about basic solid-earth geophysics. But I was aware of the breakthrough Vine & Matthews work and I remember being pleased that the long-missing mechanism had finally been found. However, I was not active in professional debate or publication on the subject, because I had moved too far out of the field.

*(Bentley, January 14, 2008 email to author; my bracketed addition)*

Like Dott, Bentley had an illustrious career at the University of Wisconsin. A glaciologist, he returned many times to the Antarctic. Were Dott and Bentley part of a silent group that was impressed with the paleomagnetic case for mobilism? It is possible that paleomagnetism had a somewhat subterranean greater effect on Earth scientists than is indicated by the number of Earth scientists who commented favorably about mobilism and its paleomagnetic support prior to the confirmation in 1966 of the Vine–Matthews hypothesis. Or were Dott and Bentley just among the few who were freed from the legacy of North American antagonism toward mobilism by, at least in part, witnessing some special event, such as King's victory over Bucher or a lecture from Runcorn, that caused them to reexamine seriously the new evidence from paleomagnetism without prejudice? Regardless, paleomagnetists, by confirming continental drift, provided mobilism with its first difficulty-free solution. Albeit generally unrecognized as such, it rekindled interest in mobilism, gave hope to old-time drifters, and provided globe-wandering paleomagnetists with gainful employment and interesting fieldwork.

## 2.19  Unreasonableness of fixist responses

It goes without saying that if scientists do not accept a solution which they recognize as acceptable then they are behaving unreasonably. What, however, if scientists do not recognize that such a solution is acceptable when all the difficulties that have been raised against it have been answered? It depends. If they are in a position to recognize that the solution is acceptable and have the requisite knowledge to do so but nonetheless do not do so, they behave unreasonably. I think this happened with continental drift and its paleomagnetic confirmation. I have argued in this context that paleomagnetists when they limited themselves to one aspect of the debate – Cox, Doell, Graham, Nagata, Kobayashi, Takeuchi, and Uyeda – behaved unreasonably. What about those who specialized in other fields? Some certainly were fully capable of understanding the physics behind paleomagnetism but still actively rejected the mobilist solution. Munk and MacDonald come to mind: I believe that they behaved unreasonably. All these workers should have better acquainted themselves with mobilism's geological and especially paleoclimatic support, and recognized that the paleomagnetism case, like all other lines of evidence in the mobilism debate, did not stand alone but had been shown after detailed study to be remarkably consilient with these other lines of evidence. MacDonald, in particular, cherry-picked the work of paleontologists who rejected drift. He should have begun wondering if his continued arguments against mantle convection were invalid or depended on questionable data about the mantle's viscosity; as the paleomagnetic evidence piled up, his continued strong rejection of it became more and more unreasonable. What about those who did not understand or had not properly read the paleomagnetic case for mobilism? If they suspended judgment, and if they decided to wait and see, as Longwell did, then they behaved reasonably. Better yet if they began to consider whether their former fixist analyses could be reinterpreted in terms of mobilism, or whether their findings were less evidentially relevant than those from paleomagnetism. However, if they continued to argue against mobilism and did not recognize its paleomagnetic support because they took the 1960 *GSA* review at its face value, allowing it to think for them, and did not delve into original papers, they behaved unreasonably. The *GSA* review gave them a way out, and many accepted it uncritically. Kay, Colbert, and Simpson, for example, cited the *GSA* review long after its "expiration date," taking no account of advances in a rapidly developing field.

There were those like Billings, Bucher, and Van Steenis, and the mobilist King, who raised difficulties based on a misunderstanding of the paleomagnetic method, of the work of Hospers and others concerning the time-averaged GAD field. Bucher and Billings were structural geologists; Van Steenis, a botanist. They confused the time-averaged direction of the geomagnetic field and its direction at any given moment (II, §7.12; §1.12). They abused RS2. Theirs was a phantom difficulty. Runcorn found that geologists often shared Billings' misapprehension. Although Billings and Bucher admitted that paleomagnetism was beyond their expertise, they thought themselves competent to attack its support for mobilism. Completely misunderstanding Brynjolfsson's discussion about catching the geomagnetic field in the act of reversal, Bucher raised another phantom difficulty against the dipole hypothesis (§1.10). From what seems to me to have been a false sense of omniscience, he stepped outside his area of expertise to criticize mobilism's explanation of Permo-Carboniferous glaciation. He argued, erroneously, that because workers "sometimes" have difficulty distinguishing debris deposited from turbidity currents from true glacial deposits, then all claims of old glaciers were dubious; fortunately, Heezen did not let him get away with this extraordinary distortion of Dott's statement that it is "sometimes" difficult to distinguish glacial deposits from turbidites, but who noticed? Even Crowell did not question widespread Permo-Carboniferous southern hemispheric glaciation (§1.10). Then there was Bucher's insistence that proof of mobilism could only come from geology, and only then could paleomagnetism be shown to be an acceptable method for determining the past positions of continents. He was categorically predicting which field would spearhead progress. Prideful of what he imagined geology to be, Bucher apparently believed that he had a better understanding of specialties outside his own field than had those working within them. My argument about the unreasonableness of fixists' responses does not depend on my particular hypothesis about difficulty-free solutions being acceptable – on my argument that mobilism's solution to divergent APW paths was acceptable. Even if my account of what constitutes an acceptable solution is mistaken, mobilism was worthy of acceptance because its solution to divergent paleopoles was a far better solution than anything fixism had to offer, and not accepting mobilism by ignoring it or by raising repeatedly already answered or phantom difficulties was unreasonable.

## 2.20  Telling it like it was not, revisionist accounts of the paleomagnetic case for mobilism

The decision by the UK (Cambridge, later Newcastle, and Manchester, later Imperial College, London) and Australian and South African paleomagnetists to direct much of their research to testing mobilism was eminently reasonable. They had good ideas, developed efficient, practical plans and they worked with imagination, energy, zeal and rapidity. My account of the rise of paleomagnetism shows just how quickly and effectively these paleomagnetists built an impressive case for mobilism. Deeply mindful of methodology, concerned about the need to get reliable data, to get it quickly, and to get more and more of it, repeatedly finding new ways to justify key assumptions, and relentlessly disposing of the many obstacles that were thrown in their way, it is no wonder they were so successful, and were able to inextinguishably rekindle interest in mobilism.

My summary above of the contribution in the 1950s and early 1960s of mobilism-tolerant paleomagnetists differs radically from that held at the time by the majority of Earth scientists, and, to jump ahead a little, it differs radically from that offered for instance by R. Phinney in his introductory essay to the proceedings of the 1966 Goddard symposium, which I cover in Volume IV, §6.13. This very important meeting was intended to summarize the state of the mobilism debate in the mid-1960s. It was held at the Goddard Institute for Space Studies in New York in November 1966. Here is what Phinney, a seismologist, wrote:

It would seem that the more radical ideas, lying under the geophysical interdict against continental drift, did not achieve the status of serious proposals until, in the 1950's, the new field of paleomagnetism produced data which split the geophysical camp and brought the ideas of Wegener back into serious contention. Strong objections made to these ideas were based on modern data on the distribution of heat flow and the non-equilibrium component of the earth's gravity field (MacDonald, 1963; MacDonald, 1966). Many of the assumptions and techniques of paleomagnetism appeared to be on dubious grounds; attempts to demonstrate the fallacy of the paleomagnetic method, however, apparently stimulated the kind of definitive laboratory and field work required to clarify many of these questions (Cox and Doell, 1960).

*(Phinney, 1968: 5; MacDonald (1963) is same as my MacDonald (1963a);*
*other references are the same)*

Anticipating difficulties ahead, paleomagnetists had worked through most of the methodological issues well before their case for mobilism was attacked, and their every key move had been made years before the 1960 *GSA* review appeared. Phinney was simply wrong. Why was he so wrong? If he sincerely believed what he said, and if his referencing truly reflects his own reading, then the *GSA* review was likely the only paper Phinney had read on the paleomagnetic case for mobilism. Phinney, like so many others as I have amply documented, did not cite a single, original, mobilism-friendly paleomagnetic paper, only the *GSA* review with its avoidance of pro-drift analyses, "tip-toeing away" as it did (§1.13) from the broad, central, revolutionary issue of mobilism. Nor did he cite any such paper from the six active years that appeared after the publication of the 1960 *GSA* review; since then, as far as Phinney was concerned, paleomagnetism had been in the deep-freeze. But, if he had not read the original papers, then why was he so judgmental? Perhaps he was seeking reasons why Earth scientists from the United States were so slow to take seriously the paleomagnetic support for mobilism. It seems to me that Phinney, in his so very incorrect assessment of paleomagnetism, was merely stating what he and almost every other Earth scientist in North America thought about paleomagnetism. Group-think is very powerful, and difficult to recognize when so powerful. If so, then Hess and the few other North American Earth scientists who at least in part became mobilists because of its paleomagnetic support deserve special credit for being able to rise above the fixist fog, the group-think of North American Earth scientists on the mobilism issue.

Phinney also implied that MacDonald's arguments (§2.7), based on new heat-flow and gravity studies, precluded the possibility of convection; they could serve as a reason for not preferring mantle convection, but they did not make the paleomagnetic support for mobilism methodologically unsound: the two were unconnected. I suspect that many workers simply decided that it was not worth the effort of getting their minds around the novel paleomagnetic case for mobilism because they believed that convection or any other mechanism for continental drift was impossible; there was Gutenberg and his low-velocity zone, but his interpretation was considered questionable, and, of course, he was a mobilist. Perhaps most did not even imagine that collectively they might be wrong, and therefore saw no reason to examine the paleomagnetic work at its source. The *GSA* review could only have reinforced this view. So much for Francis Birch's warning to unwary readers that "certain," "undoubtedly," "positive proof," and "unanswerable argument" respectively meant "dubious," "perhaps," "vague suggestion," and "trivial objection" when applied to Earth's interior (§2.7). At the time, self-doubt seemed fairly rare among the cognoscenti of Earth's interior.

## 2.21 Presentation of the mobilist interpretation of the paleomagnetic results, Runcorn's tactical error

Although Phinney's characterization of the paleomagnetic support for mobilism is bad history, it points to a tactical mistake made by Runcorn in his advocacy of the paleomagnetic case. Runcorn decided that the Cambridge paleomagnetic work should be published together in the venerable *Philosophical Transactions of the Royal Society of London*, as they indeed were, but not until years later. I believe this was a major tactical error. If the Cambridge paleomagnetists, following Hospers' example – who was very much a free agent, Runcorn not being his supervisor – had presented their findings in support of mobilism in a more timely and incremental fashion, I think their case might have been more favorably received. Irving and Creer should have begun publishing their results in short notes in *Nature* or *Science* from 1952 as soon as they had finished each segment of their work. Except for Creer, Irving, and Runcorn (1954), which was in a conference proceedings, and which did not even give Creer's final thesis version of this seminal APW path for Britain (II, §3.6, §3.7) and Irving (1956), which included his Indian results, none of Irving's and Creer's British results from the Ph.D. theses were published under their own names until 1957, when their work appeared in the series of papers published as Runcorn had ordained in the *Philosophical Transactions*, which was read by few in North America. By 1955 Irving could have published short notes on his Torridonian results, and a note on his Indian results, and Creer could have published separate notes on each of his poles, and on the comparison of his British path with the pole from Graham's Silurian Rose Hill Formation. Runcorn, their supervisor, never encouraged them to do so, even though it was their work. Because of the delay, Earth scientists learned about paleomagnetic work in Cambridge during the mid-1950s primarily from listening to one of Runcorn's reviews and numerous presentations in Britain and North America. If they had learned of Creer's and Irving's rapidly emerging data and ideas from them rather than from Runcorn, with his lack of knowledge of and cavalier attitude toward geology, they may have paid more attention and later questioned Cox and Doell's critique. Of course, it may have made no difference, because many regionalist North American Earth scientists still might have ignored what was being done elsewhere. Perhaps, like Dott and Bentley, they may have also needed to hear someone as eloquent as King tell them in no uncertain terms of the strong geological support for mobilism from the Southern Hemisphere before they could even think of trying to escape the limitations of North American regionalism.

The delay in publishing the Cambridge results was a matter Blackett raised with Irving, as he recalls, after Fisher gave a talk probably late 1952 to early 1953 on continental drift at Cambridge (II, §2.7).

After the talk, Blackett walked with me a short distance as far as Great St. Mary's. It was a baffling few minutes. We discussed my work, and then, quite abruptly, he expressed his disapproval of the way Runcorn was handling the publication of things. He asked my opinion. I said I was in no position to comment. I had never written a scientific paper. I was so wrapped up in doing science that I can't recall ever giving any thought at the time to "writing up." Everything seemed to me to be in so unfinished a state. Also I was enjoying myself. He was I think disappointed.

As an administratively powerless student all I could do was to carry on my work, develop ideas, get as many achievements under my belt as possible, so that I could survive and hopefully emerge with something to show for my efforts. Both Blackett and Runcorn were now out to get recognition for their groups, and presumably, since they were human, for themselves. They gave lectures – Runcorn especially in the U.S.A. Blackett used the results of his group with more tact and authority than Runcorn did his. Blackett's lectures in Britain and around the world were enormously influential, and helped our cause immensely.

*(Irving, March 2001 note to author)*

Thinking later about what Blackett had said, Irving concluded that Blackett was right, and also explained why he (and Creer) did not publish their results as they came out.

In retrospect, the proper thing to have done at that time would have been for Runcorn and Blackett to have ensured that short accounts of our discoveries of 1951 be published. In my case Blackett should, of course, have been in the authorship. Although he played no part in the measurements themselves or the analysis, his role was vital. It was his magnetometer I used initially. Notes on the solution to the inhomogeneity problem, on the coherence and stability of the magnetization of the fine-grained red beds, the persistent oblique directions of the

Torridonian, and the confirmation of Indian drift by my Deccan Trap data should have been published. It was not until 1956 and 1957 that these were published under correct authorship ... Runcorn got his objective, a large set of papers in the *Philosophical Transactions of the Royal Society*, but by the time they came out (1957) the data had been reviewed, and lectured about across the world especially by Runcorn. Runcorn made his name. The result should have been published piecemeal, and then brought together in the occasional review. Instead we produced a set of papers the ideas in which were old-hat when published. Blackett was quite right to question Runcorn's management of the publication of our results.

*(Irving, March 2001 note to author)*

Runcorn wanted to have his students' results published as a single unit, causing several years' delay. This did not inhibit him from lecturing about his students' unpublished results, and even quoting them at length in his own reviews (Runcorn, 1955a). Timely publications by his students alongside his missionary lecture tours would have been a much more appropriate and effective means of bringing the word to North American Earth scientists.

Did Runcorn's decision adversely affect Creer's and Irving's careers? I think it definitely hurt Creer, and Irving's in the short run. Before explaining why, it should be emphasized that despite his mishandling of publications, Runcorn provided both Creer and Irving with excellent early leadership and the chance to become pioneers in a field that grew enormously when, and partly because, they entered it. Neither Irving nor Creer would have become paleomagnetists if it had not been for Runcorn's desire to work in the field, and his realization that he needed to bring together a team of researchers with diverse backgrounds. Moreover, Runcorn encouraged and supported them tremendously. Irving probably would not have become a researcher if it had not been for Runcorn, and Creer would have continued to work under other members of the staff. Runcorn was a great recruiter, and his success in getting good students who went on to have good careers is perhaps his greatest contribution.

Runcorn's decision to delay publication of Irving's and Creer's early work probably had little effect in the long term on their substantial accomplishments. But it did, I believe, obscure the fact that they were primarily responsible for the work they described in their Ph.D. dissertations. Because his talks were the only way outsiders learned about their work in the early 1950s, Runcorn received undeserved credit for their work. They were his students, but they soon outstripped him technically, which, together with his taking extended trips to the United States, freed them from detailed supervision. Obscured by Runcorn's growing shadow, Irving and Creer had to escape if they were to get the recognition they deserved. Irving emigrated, he escaped quickly. Creer remained in England and joined Runcorn in Newcastle; he took longer to escape.

When Jaeger hired him to start paleomagnetics at ANU, Irving got his chance, and he took it. Jaeger instructed him how to conduct his career. Irving, increasingly no longer viewed as Runcorn's student, formed his own group.[22] In the following years, Irving received ample recognition; he was elected to the Royal Society of

Canada in 1973, to the Royal Society of London in 1979, admitted as Foreign Associate of the US Academy of Sciences in 1998, and named to the Order of Canada (2002). He received eight medals, most notably, The Logan Medal (1975) of the Geological Association of Canada, the Walter Bucher Medal (1979) of the AGU, the Alfred Wegener Medal (1995) of the European Geosciences Union, the Arthur L. Day Medal (1997) of the GSA, and the Wollaston Medal (2005) of the Geological Society of London. But, Irving probably was hurt by Runcorn's decision in the short run because his Ph.D. examiners would have found it more difficult to fail him had he already published several letters to *Nature*.

Creer was less fortunate, but still rewarded for his work. He was elected a fellow of the Royal Society of Edinburgh (1985), of the AGU (1988), of Academia Europaea (1990), and a Foreign Associate of the Association Nacional de Geografia, Argentina (1985). He won the Fleming Medal (1990) from the AGU, the Gold Medal (1990) from the RAS, and the Prix Mondial Nessim Habif for Science (1987), which is awarded once every four years from the University of Geneva. He served as President of the European Geophysical Society 1992–4. But it seems to me that Creer's accomplishments exceed his accolades; it was years before he escaped Runcorn's shadow. Creer's medals came late in his career, even though they came in part because of the work he did at Cambridge while Runcorn's student. But in retrospect Creer made an unfortunate early decision that did not help: having finished his dissertation he had been awarded a "Commonwealth Fund" Fellowship to study in the United States and Verhoogen offered him a post-doctorate at Berkeley. For personal reasons he turned down the offer and joined the Brtitish Geological Survey. While this enabled him to acquire broad experience with geophysical field survey methods, it kept him out of active participation in paleomagnetism for two vital years. Creer poignantly reflected on what happened:

With hindsight, it is true what you say that Ted and I should have published earlier on. Ted had good reason – he had to go out to Australia and set up there. But I made a big mistake in not continuing in the field following on from my Ph.D., and in not taking up Verhoogen's invitation to join him as a post-doc. And I could have made a very nice little book out of my thesis. But it is no use fantasizing and I had to face the consequences of decisions I made at the time.

(Creer, February 8, 1999 email to author)

One can only wonder what might have happened at Berkeley if Creer had arrived while Doell was finishing his Ph.D., and had stayed while Cox began working in paleomagnetism. At the very least Doell and Cox would have had an experienced paleomagnetist to talk to, someone who had developed pioneering ideas about polar wander and continental drift, and Berkeley would most likely have got a high-sensitivity astatic magnetometer. If Creer had stayed at Berkeley for a few years, the development of paleomagnetism in the United States could have been quite different. There likely would have been greater early cooperation between the

ANU and Berkeley groups. Paleomagnetically, Creer wasted two critical years at the Geological Survey, before accepting Runcorn's offer of a staff appointment at Newcastle (at the same time as Frank Lowes and Raymond Hide). Those unfamiliar with the situation came to think of him as Runcorn's second-in-command and were unaware, and many still are, that it was Creer on his own who came up with the idea of constructing an APW path, synthesized the Cambridge work, and thus paved the way for the first successful physical test of continental drift. They know nothing of Creer's major role in getting Runcorn to come out in favor of mobilism. It was fitting that Irving, one of the few who knew firsthand what Creer had done, described his accomplishments when he received the Fleming Medal; here is part of what he said.

This is what Creer proceeded to do. Everything had to be put together more or less from scratch. Rocks had to be collected, instruments built, measurements taken, procedures honed, and data analyzed and written about . . . With the true experimenter's instinct, and informed by Fisher's newly-developed statistics, Creer realized that the average direction of the geomagnetic field drifted during some identifiable interval of time could be estimated to an accuracy of 10° or so from a score of the right sort of samples spaced through a rock-unit. Such a broad-brush approach carried with it certain risks. There would be fewer data on which the physical reliability of results could be based. A handful of spurious data could seriously prejudice the outcome. On the other hand, one could, with luck, obtain a quick preliminary glance at the very long-term motions of the continents relative to the time-average geomagnetic field. Creer's efforts paid off. In very short order he sampled widely through the geological column in Britain, and showed that everywhere he went the directions of the ancient geomagnetic field were oblique. Oblique directions were no fluke. Apparently they recorded, albeit in a very imperfect way, the motions of Britain relative to the time-averaged field, or as Creer preferred to think about it, the movement of the time-averaged field and the poles which defined it, relative to Britain. In this way Creer ushered in the concept of apparent polar wandering. Others in Runcorn's group played a part, but it was Creer who first (September 1954) calculated and plotted the path of apparent polar wandering for Europe. Such paths, which soon were constructed for other continents, provided the basis for the first physical test of Wegener's hypothesis of continental drift. They continue to provide the basis for all latitude grids on the myriad paleogeographic maps that now adorn the literature of the Earth sciences . . . That the concept of apparent polar wandering should have become embedded so swiftly in our subject is a token of its importance. We take it for granted. But when we use it we, consciously or unconsciously, salute the energy, drive, and imagination of a sandy-haired research student who was hurrying to complete his Ph.D. thesis in the two short years 1952 through 1954 . . . It is worth commenting that Creer's was not a thesis with only one idea. It contained also one of the formative discussions of partial magnetic instability in rocks, and it contained the first attempt to analyze the geomagnetic field statistically in such a manner that comparisons with paleomagnetic observations could be made.

*(Irving, 1991: 54–55)*

Creer's achievements had been unknown to many, especially those outside of paleomagnetism. Irving also mentioned some of his other accomplishments.

It is not my intention, nor is this the appropriate occasion, Mr. President, to give details of Creer's other many accomplishments: his initiation of paleomagnetic work in South America, his discovery of supermagnetism in red sediments, his prescient speculations on magnetizations produced by chemical precipitation from water circulating in the crust, his work on paleosecular variation and westerly drift of the geomagnetic field ...

*(Irving, 1991: 55)*

Runcorn also spoke about his and Irving's election to the Royal Society, and the failure to get Creer elected.

You have a proposer and a seconder. They write the citation, and then you get other people to sign to strengthen the case, and then, of course, it goes before the subject Committee, and then the Council usually takes several years before [deciding]. People are cautious, even people like Bullard and Blackett. Bullard proposed me, and Blackett seconded me. You know they didn't advocate my election to the Royal Society because of what I had done about continental drift. The citation, which, of course, came from them, says my work on the history of the Earth's magnetic field, paleomagnetism, and my study of secular variation. They obviously thought that it wouldn't do my chance any good to start talking about continental drift. Lots of people, although I did not know at the time, supported me: Holmes, O. T. Jones, Chapman, Cowling, Chandrashakar.

*(Runcorn, August 1993 interview with author; my bracketed addition)*

Turning to Irving's election, Runcorn simply noted that he proposed him, Blackett seconded, and he was elected in 1979.

Runcorn had much more to say about what happened to Creer.

But luck enters in. Ken should have been elected but unfortunately it depends on the people who were on the Committee. It is not so much who the proposer and the seconder are. It is the people who consider it on the Committee. You see he was very unfortunate because by then the Cambridge people, Vine and Matthews, and McKenzie, and so on, were preferred to him. But, I just put him up again. The first time you can go up again, if you don't get in the first time, is after seven years. Then, if you don't get in, you have a gap of three years, and then you have another chance. We have put Ken up three times. He was very unfortunate then; he nearly got in. He nearly got in last time. Well you see he got to the final list. Then someone at the Council meeting changed the order because, you see, of course, you can never avoid this sort of thing happening. There happened to be a powerful man on the Council who said there was another better than Ken. I had been on the Committee at that time, and Ken was not elected. But, I put him up again for next year. [Runcorn soon found out after the interview with me that Creer did not get elected.]

*(Runcorn, August 1993 interview with author; my bracketed addition)*

I do not know when Runcorn first put up Irving and Creer. Did Runcorn put up Creer before Irving? Did he first put them up in 1972? I do not know. Letters at Imperial College in Runcorn's papers show that Creer was put up as a candidate in 1983 and 1987. Vine and Matthews were elected in 1974; McKenzie, in 1976; Irving and Roberts were not elected until 1979. Hide, another Runcorn student, was elected in 1971.

Runcorn also thought that neither the geologists nor the astronomers had helped Creer's chances.

The geologists were always a bit of a nuisance on these committees because they are a closed group and they always support each other. The astronomers are a bit like that. It is a club.
*(Runcorn, August 1993 interview with author)*

The geophysics "club" was too small to have much clout. I think Runcorn's comment, especially about the geologists, points to another reason why Creer was not elected to the Royal Society. Just as most geologists failed to appreciate the developments in paleomagnetism, they failed to appreciate, and perhaps even viewed with antagonism, Creer's work. And, if they thought his work was important, they might have thought that it was not sufficiently important; recall the stiff reception Bullard received when he spoke in favor of mobilism at the Geological Society. Moreover, Creer never worked in a geology department. He was in the Department of Geodesy and Geophysics at Cambridge, in the School of Physics at Newcastle, and in the Department of Geophysics at Edinburgh. It is doubtful that Creer would have received much support from astronomers and physicists because his work was far from the frontiers of either science.

But despite all these extenuating circumstances, Runcorn would have served Creer better had he encouraged him to submit papers on his early work in a timely manner. Runcorn's decision to hold back publication of Creer's and Irving's work until 1957, when it was already outdated, may have helped Runcorn's own chances of being elected to the Royal Society, which he was in 1965, but it did less than timely papers would have done in the early days to convince non-paleomagnetists that mobilism was correct or at least worth reconsidering.

## 2.22 Waiters and actors: taking the paleomagnetic support for mobilism seriously

After Irving left Cambridge, he and Fisher kept in touch. Irving (2000) recalls that he wrote Fisher in early 1956 sending him reprints, and probably complaining about the resistance to continental drift by most Australian geologists. As already noted (I, §1.15), Fisher compared the situation of the reception of continental drift to the earlier reception of organic evolution.

I think there is a parallelism in the nature of scientific controversy between continental drift in the last 80 years or so, and organic evolution about 100 years earlier. Each idea as it originated was necessarily speculative, and not accompanied certainly by sufficiently cogent evidence to carry final conviction. There were, however, many suggestive pointers. In consequence of this natural situation both questions have been argued with imperfect facts, incorrect theories, and often incompetent reasoning, over a long period during which many people have committed themselves to impossible positions, and many more fearing to burn their fingers have enclosed themselves in towers not of ivory, but of solid wood. In the period about 1800–1850, although geological specimens were being collected and described, experiments in plant hybridization carried out, the classification of animals and plants greatly improved, and embryological studies at least have attracted attention, yet so unwilling are ordinary men to run the risk of contemptuous ridicule, no one of consequence attempted to revive what had been left as speculative, and almost poetical, ideas by Buffon, Erasmus Darwin, and Lamarck. Darwin worked on the problem almost secretly from 1838, and only published in 1859 because he was forced to. After that the ice came down like Niagara, but of course the new idea was still ill-understood and ill-expounded for at least the next 50 years, during which a few subordinate causes of error had been removed by special research.

*(R.A. Fisher, June 12, 1956 letter to Irving)*

Perhaps some geologists and geophysicists were impressed with the paleomagnetic support for mobilism but said or wrote nothing about it because they were afraid of what their fixist colleagues would say. Fisher, again as noted already (I, §1.15), certainly thought as much.

I think a lot of geologists must be timidly peering out of their holes on hearing the strange news that geophysicists are talking about continental drift, and I have often wondered how many scientific discoveries of importance have been left unmade for lack of the quality called moral courage.

*(R.A. Fisher, June 12, 1956 letter to Irving)*

To have the ability to dig somewhere but realize that it may be in the wrong place, and start digging elsewhere where good things are to be found, takes not only the ability to live with doubt, but the courage to change one's mind, and the energy and determination to act in a world that may not appreciate these good things. Most geologists and geophysicists, especially from North America, were either oblivious to the paleomagnetic studies in Britain, Australia, the Soviet Union, and southern Africa or were so sure of the correctness of their fixist view that they could not appreciate, or were afraid to acknowledge the consequences of these studies. As I shall soon show, Harry Hess realized it was time to dig elsewhere, and it was paleomagnetic results of a different sort that assured him that he was digging in the right place. Dietz realized it too. Moreover, both did not renounce their appeal to paleomagnetism if its results began to run counter to their views.

## Notes

1  I have summarized Gutenberg's life and career from Byerly (1960), Jeffreys (1960), Richter (1962), and Knopoff (1999).
2  Ursula Marvin (1973: 99–105) offers an excellent account of Gutenberg's 1927 version of his theory of continental spreading. This much shorter summary is derived from hers, Gutenberg's (1951b: 211–214) own remarks, and Wade's (1934, 1935) discussions.
3  It is of interest to note that Vening Meinesz does not mention Ken Creer but credits Runcorn with Britain's APW path. Probably he did not even realize that Creer was responsible for it. Runcorn, I suspect, gave the impression when presenting the ideas and work of his students that he had played a vital scientific role. He certainly played the key role in getting his students together, getting geologists Irving and Opdyke accepted as graduate students in Cambridge's Department of Geodesy and Geophysics, using his contacts to help them, but they came up with their own ideas and did the work.
4  Jeffreys had used the same argument against Holmes, contending that the mantle possessed great strength. Holmes turned this argument on its head. He argued that the lack of isostatic adjustment implied the existence of opposing forces within Earth's interior which, in turn, implied that the mantle lacked strength (I, §5.4).

5  Jeffreys is ripe for a scientific biography. Besides his work in geophysics, there are his contributions to cosmogony, meteorology, and probability and statistics.

6  This account of Jeffreys is taken from Alan Cook's (1990) excellent biographical memoir.

7  Jeffreys' work in probability is still discussed. The Department of Probability and Statistics at Carnegie Mellon University, for example, recently offered a course on Jeffreys and his ideas, and *Theory of Probability* was reissued in 1998.

8  I highly recommend McKenzie's biographical memoir on Bullard. It is informative, insightful, and entertaining. McKenzie managed to show his respect and fondness for his former supervisor without failing to give a full sense of the man.

9  Ben Browne comes off rather poorly in my account of the rise of paleomagnetism in the Department of Geodesy and Geophysics at Cambridge (II, §1.12, §1.20, §1.21, §2.10, §3.4). McKenzie, however, commends Browne highly for helping get Bullard back to Cambridge and appropriately settled.

> Runcorn's departure left a vacancy as an Assistant Director of Research, and Ben Browne, who was then Reader and Head of the Department of Geodesy and Geophysics, immediately appointed Teddy to this post. In doing so he showed both vision and generosity. He must have been aware that his position as Head of the Department would be impossible once Teddy returned, yet he worked steadily to find a University post for Teddy while he was at the N.P.L. [National Physical Laboratory], and some time after Teddy had returned to Cambridge resigned his own post in Teddy's favour.
>
> *(McKenzie, 1987: 83; my bracketed addition)*

10  Hans Pettersson led the1947–8 Swedish Deep-Sea Expedition aboard the *Albatross*. See §3.5 for further discussion.

11  Bullard and Day were not proposing seafloor spreading. The currents spread out, and move horizontally below the crust of the seafloor. Hot material rises along ridge axes, and volcanism and high heat flow are the consequences. Although the convection currents produce forces that open tensional cracks in the crust, they do not produce new seafloor that spreads out from the ridge. The currents spread, seafloor cracks at ridge axes, but seafloors do not on their model spread away from ridge axes and create new ocean basins.

12  Bullard's reference to Blackett *et al.* (1960) does not entirely exclude the possibility that Bullard spoke about their work during his talk at the Oceanographic Congress. Blackett *et al.* was received for publication on December 14, 1959. Bullard gave his talk in late August or early September 1959. But, Bullard conceivably could have received a version of Blackett's paper before giving his talk. Or Bullard may have talked to Blackett before the meeting.

13  Bullard's omission of any reference to the Cambridge work particularly hurt Creer, who was still in Runcorn's shadow. Irving, having escaped to Australia and built up his own group, was less vulnerable.

14  Bullard's claim "that paleomagnetic studies cannot detect movement in longitude" expresses his belief that relative longitude of two or more continents cannot under any circumstances be determined paleomagnetically, and it is mistaken (II, §5.16); the fact that Europe and North America have two matching APW paths alongside one another implies that formerly they were longitudinally close together as McElhinny (1973: 239) well illustrates.

15  It is worth emphasizing that Bullard was not proposing seafloor spreading, even though he invoked convection and used some of the same arguments used by Hess and Dietz. Again it is impossible to determine what, if anything, Bullard added to his talk in his final manuscript that was received for publication on July 7, 1961. Bullard, of course, never claimed that he had put forth seafloor spreading independently of Hess or Dietz. I do not know if Bullard had received a preprint of Hess's December 1960 paper, or had read Dietz's paper on seafloor spreading, which appeared in *Nature* on June 3, 1961, one month before Bullard submitted his paper. Regardless, Bullard took the lack of sediments as an argument in favor of mobilism.

16  Walter Elsasser (1966) used Cox and Doell's review in precisely the same way as Bullard; he acted like they supported mobilism, and appealed to them to claim that paleomagnetism offered "remarkable confirmation" of continental drift.

Since, however, it has seemed impossible to provide conclusive proof on a geological basis, the idea of continental drift has for a long time led a marginal existence in the limbo of geological speculation. Only in recent years has it reentered the stage, largely as a result of the remarkable confirmation of the hypothesis provide by paleomagnetic methods [see Cox and Doell, 1960, and Runcorn's paper in Runcorn, ed., 1962].

*(Elsasser, 1964: 462)*

I add that Elsasser made the comment in September 1964 during his talk at MIT at an international conference.

17 Bullard also wrote a very brief review of Runcorn's *Continental Drift* – one page with little more than a summary of included papers. He characterized the book as "not a complete or systematic account of the evidence concerning continental drift, but . . . nonetheless welcome as a review of current work on the more physical aspects of the question" (Bullard, 1963: 147).

18 Jones served as Irving's official supervisor when Runcorn was off on his extensive jaunts (Irving, November 1999 comment to author).

19 G. L. Herries Davies (2007: 240), commenting on Bailey's announcement about Lamplugh after Bullard's talk, has found support for Bailey's claim.

> Bailey now claimed that the concept had really originated with his Geological Survey colleague, George William Lamplugh, sometime between the visit of the British Association to South Africa in August – September 1905, and a date during 1910. Bailey explained that it was an encounter with the Dwyka Tillite of southern Africa which had inspired Lamplugh to reflect upon a rearrangement of the continental masses, and at Johannesburg on September 1, 1905, Lamplugh certainly lectured upon a Kimberley occurrence of the "Dwyka Conglomerate."

> Lamplugh thus is another example of a UK geologist who became a mobilist because of what he saw in South Africa, giving further support to the importance of regionalism in shaping the views of Earth scientists toward mobilism (I, §1.14, §8.13).
> Lamplugh (Anonymous, 1923: 188) also spoke in favor of drift in 1923 at a meeting on Wegener's hypothesis sponsored by the Royal Geographical Society. Acknowledging that Wegener's theory "is so vulnerable in almost every statement," he still maintained that it "is of real interest to geologists, because it has struck an idea that has been floating in the minds for a long time." The idea he had in mind was "the big overthrusts that we know of in many parts of the world . . ." He did not see how such overthrusts, "as much perhaps as 100 kilometres," could occur without "some slow irresistible creep of the continental masses." He also argued that Wegener's drift offered the only reasonable explanation of tropical glaciation. Presumably he had in mind Permo-Carboniferous glaciation, although he did not mention it specifically. Lastly, he mentioned biological disjunctions, singling out *Glossopteris*. See Marvin (1973: 85) for additional discussion of Lamplugh's remarks at the meeting.

20 Holmes apparently was always on the lookout for quotations to use in his second edition of *Principles of Physical Geology*, and he wrote them in a notebook, grouping many under subject headings. For example, the four quotations at the beginning of §2.16 were listed under the heading "Continental Drift." I want to thank Gordon Craig for telling me about the notebook, showing it to me, and making arrangements to have it photographed for me. I also want to thank the University of Edinburgh for giving me permission to quote from it.

21 Aware of the desirability of using radiometric dates, paleomagnetists had for several years been sampling rocks that could be radiometrically dated (Irving, Robertson, and Stott, 1963; Gough, Opdyke, and McElhinny, 1964 (§1.16, §1.17).

22 Irving was also asked again by Blackett to join his group and go to India, but declined because "at the time (1954) Australia beckoned with the opportunity of setting up my own show independently of Runcorn and Blackett, so I declined his offer" (Irving, March 2001 note to author; Runcorn, 1984 interview with author).

# 3

# Harry Hess develops seafloor spreading

## 3.1 Harry Hess, seafloor spreading, and revisionist history

In a 1960 preprint of his classic 1962 paper, Harry proposed that oceanic crust is created at the centers of ocean ridges, from whence it moves sideways, sweeping across broad expanses, creating new oceans. Seafloor spreading, so the hypothesis goes, is driven by mantle convection: hot mantle rises beneath ridges and flows sideways carrying along with it any less dense continental crust that happens to be present. Although some of Hess's earliest work before World War II was on deep ocean trenches, he said nothing at the time about where oceanic crust went, that is, about what we now call subduction. This came six months later, when R. L. Fisher, then the leading investigator of trenches, argued that seafloor descends into the mantle at trenches (Fisher and Hess, 1963).

For me, Hess's switch to mobilism and the manner in which he came to develop seafloor spreading are two of the most intriguing aspects of the entire mobilism debate.[1] Hess came to believe in continental drift because of its paleomagnetic support, and this change in attitude toward drift occurred before he thought of seafloor spreading. As for the evolution of ocean basins, he was forever curious about the ways in which features of the ocean floor came into being – ridges, submarine peaks (guyots), trenches, and the vast, flat, largely featureless, sub-oceanic plains. In the fifteen years after World War II, he speculated incessantly about what these processes might be. Always he readily embraced new data produced by others, and variously criticized their ideas as well as his own, rejecting or modifying some and accepting others; he made use of all new data and ideas in some fruitful way, employing them to suit his purpose. His ideas darted like quicksilver from place to place.

That his hypothesis of seafloor spreading echoed that of Arthur Holmes thirty years earlier, especially regarding the role of mantle convection, can hardly be doubted, something Hess never seemed to have fully acknowledged. As this story unfolds during this final phase of the mobilism debate, scholarly amnesia, obsessive focus on the present, lack of curiosity or disinterestedness in the past, call it what you will, was not uncommon, and may in fact be a necessary sociological feature of times of revolutionary change – it may not be a good idea for those submitting what appear to be new ideas to let their readers know that a similar idea was overwhelmingly rejected in years past. But it is one thing to make such a choice and another not to even know, e.g., the history of work on reversals! To the victors go the rewriting of history in both politics and science. Earth scientists were not mistaken to ignore a well-supported idea or result in the past (paleomagnetic support of continental drift), the support was simply no good!

## 3.2 Harry Hess, the man

Harry Hess (1906–69) was the elder of two sons born to Elizabeth Engel Hess and Julian S. Hess. Hess's father was a member of the New York Stock Exchange. Born in New York City, Hess attended Asbury Park High School in New Jersey. He entered Yale University in 1923, soon switching from electrical engineering to geology. He later recalled his education at Yale.

The story would begin at Yale in 1925 when, bored with the routine of electrical engineering, I sought something which would give a freer rein to the imagination than drawing cross sections of spark plugs. Entering the Geology Department, I was guided and encouraged by Alan Bateman for the next two years, something for which I have always been grateful. At the time there were only two undergraduates in the Department. While there were many undergraduate courses listed in the catalog, it was not worthwhile to give them for so small a class. Consequently I was enrolled in graduate courses, but undergraduates were required to take five courses and graduate students three. This nearly sank me, but in 1927 I graduated as, I believe, the first man to get a B.S. in geology at Yale. I owe my foundation for a career in geology to Bateman, Adolph Knopf, Chester Longwell, and Carl Dunbar.

*(Hess, 1968a: 85)*

He spent the next eighteen months in Rhodesia as an exploration geologist where, as he later remarked, "At 17 miles a day, I developed leg muscles, a philosophical attitude toward life, and a profound respect for fieldwork" (Hess, 1968a: 85). He also realized the need to further his education.

Starting at Harvard, I found "No Smoking" signs on every wall and decided I could not be happy there. Continuing south to Yale, I had a five-minute interview with Dean Warren, who allowed Yale had seen enough of me, and asked why I didn't go to study under Sampson at Princeton.

*(Hess, 1968a: 85–86)*

Hess obliged, and continued south to Princeton where after a brief misunderstanding he was hired as a part-time instructor and accepted into the Ph.D. program.

At Princeton I found Buddington in his office. He greeted me warmly, calling me "Harry," and immediately offered me a job as a part-time instructor. We went together to Professor Smyth's office, and Buddington began to fill out the necessary forms. Halfway through Bud discovered to his horror that I was not Harry Cannon. The situation was saved by R. M. Field, who came by at that moment. I had been on his summer school on a railroad car in Canada two years before. On the last night before crossing back into the United States he told us all liquor on the car had to be disposed of before the border was reached. It was during prohibition in the United States. I consumed far more than my share. Egged on by my fellow students, I was put on the platform and gave a lecture on the Precambrian stratigraphy of Canada. I have no recollection of this lecture but am told it was good. The crucial point is that Field, remembering this lecture, interceded for me and said I would do for the job.

*(Hess, 1968a: 86)*

Hess's major professors at Princeton were A. F. Buddington (petrology), A. H. Phillips (mineralogy), R. M. Field (oceanic structure), and Edward Sampson (mineral deposits). He became quite close to Buddington and, except for Sampson, eventually co-authored papers with all of them. Hess acknowledged their support in his acceptance of the 1966 Penrose Medal of the GSA. He also mentioned Vening Meinesz's influence (§2.5, §2.6).

Ed Sampson had been working on talc, soapstone, asbestos, and chromite deposits in various parts of the world. He shoved me into a study of the Schuyler soapstones, although I was somewhat reluctant to embark on the project. I shall always be indebted to him for this not too gentle push, because it started me on a fascinating career, a large part of which has been spent on research on ultramafic rocks. Aside from my research activities with Sampson, I am indebted for most of my formal education at Princeton to Arthur Buddington, and secondarily to Alexander Phillips, whom I succeeded at Princeton upon his retirement. Dick Field, brilliant and erratic, introduced me to research at sea and gave me the unique opportunity of going on the submarine S-48 with Vening Meinesz to measure gravity in the Caribbean island arc and Bahamas. Field opened up for me the exciting possibilities in exploration of the oceans and a parallel second career in research. Had it not been for him and this opportunity, I probably would not be standing here this evening. Finally, learning the rudiments of geophysics from Vening Meinesz in the cramped quarters of a World War I submarine was a most fortunate experience.

*(Hess, 1968a: 86)*

Hess received his Ph.D. in 1932. Encouraged by Sampson, he studied the serpentinization of a large peridotite intrusive in Schuyler, Virginia. These two pursuits, understanding serpentinization and the seafloor's major features, preoccupied him throughout his career. He often combined them by invoking serpentinization to explain seafloor features.

Hess taught at Rutgers University (1932–3) and spent several months at the Geophysical Laboratory of the Carnegie Institution of Washington (1933–4) before accepting a position in the geology department at Princeton. In 1934 he married Annette Burns, daughter of George Plumer Burns, a professor of botany at the University of Vermont. Except for visiting positions at the University of Cape Town, South Africa (1949–50), and the University of Cambridge (1965), he remained at Princeton, serving as Chair of the Geology Department from 1950 to 1966. In 1964, he was appointed Blair Professor of Geology.

Hess, a reserve officer in the US Navy at the time of the attack on Pearl Harbor, December 7, 1941, left Princeton the next morning on the 7:42 AM train for New York to report for active duty. Initially stationed in New York, where he headed an operation charged with predicting the movement of German submarines in the Atlantic, he volunteered for active sea duty and eventually took over command of the assault transport USS *Cape Johnson*. He took part in four major combat landings in the Pacific, and at the close of the war, returned to Princeton with the rank of Commander. He remained active in the Naval Reserves, and was on call for advice during crises, among them the Cuban missile crisis, the loss of the submarine *Thresher* and the *Pueblo* affair. He eventually rose to the rank of Rear Admiral.

Hess received numerous scientific honors, and devoted considerable time to various scientific organizations. He was elected to the NAS (1952), American Philosophical Society (1960), and was a Foreign Member of the Geological Society of London, the Geological Society of South Africa and the Sociedad Venezolana de Geologos. He received the Penrose Medal from the GSA (1966), the Feltrinelli prize from the Academia Nazionale dei Lincei (1966), and was posthumously awarded the Distinguished Public Service Award by the National Aeronautics and Space Administration (1969). He was President of two sections of the AGU: Geodesy (1951–3) and Tectonophysics (1955–8), President of the Mineralogical Society of America (1955) and of the GSA (1963).

Hess was a central figure in the American Miscellaneous Society, which was formed to evaluate novel ideas that might merit funding. He secured funding for one of them, "Project Mohole," to drill beneath the ocean through the Mohorovičić discontinuity into the mantle. The project was originally suggested to Hess by Walter Munk in 1957. The National Science Foundation funded it from 1958 to 1966, and he chaired the committee charged with determining where to drill. Congress terminated the project in 1966 before any cores were obtained. The project was, however, reborn, and the first core sample was obtained in 1968 under JOIDES (Joint Oceanographic Institutions Deep Earth Sampling) (Menard, 1986: 114–118; Allwardt, 1990: 207–209).

Hess was not afraid to propose solutions to fundamental problems. He repeatedly tried to explain the origin of island arcs and mountains, oceanic trenches, ridges, and crust. He typically utilized data from exploration geophysics and petrology in formulating his ideas. He thought it a mistake to collect data without trying to figure out their significance. He thought hypotheses could help direct further research, and unless based on faulty logic, often serve as stepping stones to better ones. He summarized his views about the value of well-reasoned but erroneous hypotheses and the danger of poorly reasoned ones in a lecture he prepared for his students, I believe, in the late 1950s or early 1960s.[2] Under the heading "Personal research" and sub-heading "Reasoning and Philosophy of the Science," he wrote:

1. The incorrect hypothesis derived from a logical analysis of facts at hand is an invaluable often necessary stepping stone to a better hypothesis. To advance further from an existing hypothesis one must usually be aware of the preceding reasoning even though some of it is now considered to be invalid.
2. The incorrect hypothesis derived from faulty logic may greatly retard progress towards an eventual solution. Some problems have been set back 50 years by tenacious protagonists of a fundamentally illogical point of view (e.g., coral reef problem).

Hess himself certainly used his proven incorrect hypotheses as stepping stones to new ones. Conscious of and often the author of difficulties faced by his hypotheses (RS2), he would amend them or develop new ones (RS1).

### 3.3 Hess's early career, 1932–1950: a preview

Hess's research career began aboard submarine USS *S-48* in 1932 when he joined Vening Meinesz and they made gravity and depth measurements of the Bartlett Trough in the Caribbean Sea. Hess and Ewing undertook another survey aboard submarine USS *Barracuda* in 1937. Ewing, another ambitious young scientist, eventually became director of Lamont Geological Observatory at Columbia University; he argued vehemently against mobilism and opposed seafloor spreading until 1967 (IV, §6.14). They made gravity and depth measurements in the Caribbean. They found the same relationship between negative gravity anomalies, ocean deeps, and island arcs as Vening Meinesz and his Dutch co-workers Ph. Kuenen and Umbgrove had observed in the Dutch East Indies, and Hess (1932) adopted their downbuckling or tectogene hypothesis to explain the origin of this relationship (I, §5.6, §8.14). Hess (1937, 1938a, 1938b, and 1939) combined Vening Meinesz's ideas with his own newly formed views about serpentinization. He argued that island arcs evolved into mountain belts. Hess (1946) offered a solution to the origin of guyots, flat-topped seamounts that he had observed by echo-sounder in the Pacific while commanding USS *Cape Johnson* during World War II. He explicitly adopted mantle convection, appending it to the downbuckling hypothesis in order to explain how ocean deeps could be maintained for extended periods of time. He also invoked mantle convection to explain the complicated pattern of deep-focus earthquakes typically found on the continental side of island arcs (Hess, 1951).

I shall now describe in more detail how Hess extended the tectogene hypothesis. I shall monitor his increasing support of convection and his consistently negative attitude toward continental drift during his early career. Hess's solution to the origin of guyots is particularly interesting; although he was initially very proud of it, he quickly abandoned it once he realized it faced insurmountable difficulties, showing how open he was to changing his mind.

## 3.4 Hess views island arcs as evolving into mountain belts

Meinesz's discovery of huge negative anomalies in the vicinity of island arcs is probably the most important contribution to knowledge of the nature of mountain building made in this century.

*(Hess, 1938a: 71)*

For a number of years the writer has been occupied in research in two widely different fields; one, the investigation of ultramafic rocks, their origin and alterations, and the other, the study of gravity anomalies in island arcs. It was most unexpected that these two divergent fields should come together and afford solutions for problems in each based on data obtained in the other field.

*(Hess, 1939: 263)*

Hess wholeheartedly adopted Vening Meinesz's downbuckle explanation of the negative gravity anomalies that they and Maurice Ewing had discovered in the Caribbean (§6.2). These long and narrow anomalies are located on the outer (convex) side of island arcs and are associated with oceanic trenches or deeps.

Meinesz's explanation for the anomalies of the negative strip is that the Earth's crust buckles downward in a huge vertical isoclinal fold, and thus the light material of the upper crust might extend downward to a depth of 40 to 60 km. This would be sufficient to give the observed anomalies if the lower portion of the downbuckle had a specific gravity deficiency of about 0.3. No other adequate explanation has yet been advanced (Fig. 1). Application of the theory to the West Indian region results in the coordination of many geologic facts previously merely a collection of observations with no apparent relationship to one another. So successful is the Meinesz theory in the West Indies, both in predicting relationships which are to be expected (and have since in a number of cases been found) and in joining together in a single structural entity observations which formerly fit into no definite pattern, that the writer has become completely convinced of the soundness of the theory in general, though in detail it may need some modification.

*(Hess, 1938a: 74–75; Hess's Figure 1, identical to his Figure 1 in Hess, 1938a, is reproduced in Figure 3.1)*

Hess also invoked downbuckling as the cause of several great faults in the West Indies. The largest, the Bartlett Trough fault, follows the northern scarp of the Bartlett Trough eastward from Guatemala, skirts by the southern coast of eastern Cuba, and extends to the Windward Passage, where, like other faults, it terminates in a downbuckling region.

All these faults, except the last mentioned [from Trinidad to northeastern Venezuela] run up to and probably terminate in the geotectoclinal zone. It seems to the speaker they were formed by adjustments of the crust as it moved up to the tectogene and down into it. Relative differential horizontal movements of one sector as compared to an adjacent one would be expected to take place as the crustal downfold formed.

*(Hess, 1938a: 88; my bracketed addition)*

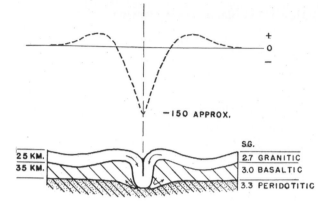

Figure 3.1 Hess's Figure 1 (1938a: 75; 1938b: 343) and Figure 2 (1939: 265). Hess (1938a: 75) captioned the figure, "Crustal buckle, specific gravity distribution and resultant anomaly curve." In Hess (1938b: 334), his caption read "Gravity anomaly curve and downbuckle of light upper crust postulated to account for the anomaly curve." According to Hess, ocean crust was made up of a top layer of sediment (not shown), a middle granitic layer, and a lower basaltic layer. The top of the mantle was peridotite. Following Kuenen, Hess often referred to Vening Meinesz's downbuckle as a tectogene.[3]

Hess estimated that the downbuckling had caused a shift of oceanic crust of approximately 50 km along the Bartlett Trough fault. He claimed that the crust south of the fault had moved NNE along the fault, causing the islands south of the fault to move eastward relative to Cuba, which lay just north of the fault.

Horizontal movements of the order of magnitude of 50 km. along such faults as the Bartlett Trough fault are not only likely, but inevitable, if the crust is to adjust itself to a curved downbuckle of this size. Thus it is not inconceivable that Haiti has moved N.N.E. 50 km or more, and once stood below and adjacent to Cuba, or that the portion of the crust now forming the Lesser Antilles geanticline once stood far to the west of its present location, so that it formed a continuous chain with the southeastward trending folds to be found in the Cretaceous rocks of Puerto Rico, which now strike into the Caribbean, where no trace of them is discernible on the sea floor.

*(Hess, 1938a: 88–89)*

Although Hess proposed horizontal displacement of oceanic crust along what was later identified as part of the boundary between the North American and Caribbean plates, and even supposed that oceanic crust south of the fault moved NNE relative to the crust on the other side of the fault, he was not invoking mobilism. The movement was small ~50 km, was caused by crustal downbuckling, and was in the opposite direction to that later proposed by mobilists. Thus, even though du Toit, for instance, was pleased that Hess argued for horizontal movement along the Bartlett Deep, he was not pleased with Hess's "orthodox" interpretation.

Hess has ably shown that the Bartlett Deep may be ascribed to a horizontal shear, but, in accordance with orthodox ideas, makes the displacement one to the N.N.E. instead of W.S.W. as under our view.

*(du Toit, 1937: 208)*

Du Toit envisioned a WSW movement of North America relative to Africa, and a WNW movement of South America relative to Africa; the two Americas had converged and what was now Central America had moved westward producing the Bartlett Deep. Ironically, Hess's proposed NNE movement of the oceanic crust south of the Bartlett Deep with the eastward movement of Central America relative to North and South America was later shown to be more or less correct.[4]

Hess (1937, 1938a, 1938b, and 1939) extended the tectogene hypothesis of Vening Meinesz and Kuenen and proposed that island arcs evolved into Alpine-type mountain belts. It was in the Caribbean region that he first described the association of serpentinized peridotite intrusions with belts of negative gravity anomalies. He noted the same association in the East Indies, and related both to the serpentine belt of the Alps:

Serpentinite intrusions and the negative strip – Serpentinized peridotite intrusions occur all along the great negative strip of the East Indies, and similarly are present in the West Indies. Apparently they come up along each side of the strip, but only in a few cases are both sides exposed and not covered by younger rocks ... These serpentinite intrusions thus become useful guides in the interpretation of any region such as the West Indies. The serpentinites clearly indicate the former extension of the negative strip from the west end of Haiti along the north coast of Cuba, and thence probably to Guatemala ... It is interesting to note that the serpentinite belt of the Alps has been long known and studied. The anomaly-field there is also a broad but relatively low negative one, so a series of events similar to that in Cuba might be postulated for it. At the southwest end of the West Indian negative strip, a double belt of serpentinite intrusions is found which follows the Cordillera Central across Colombia into Ecuador. No doubt the negative strip once existed between these two zones of intrusions but has probably now disappeared, as suggested by the high mountains indicative of isostatic uplift.

*(Hess, 1937: 75)*

Before adopting the downbuckling hypothesis, Hess was already familiar with the problem of serpentinization, from his studies of ultrabasic intrusions in the Appalachians, Stillwater Complex in Montana, and Great Dyke of Southern Rhodesia. After accepting the downbuckling hypothesis, he modified it to resolve a conflict between laboratory and field evidence (1938b: 323).

According to Hess (1938b: 321), N. L. Bowen's (1927 and 1928) laboratory work had shown that ultramafic intrusions, including serpentine intrusions, could not form directly as magmas because as they ascended they would soon become too cool to remain molten, yet the serpentine bodies he had seen looked like molten intrusions. Hess sided with the field evidence. Arguing (1938b: 328–329) that serpentine, which is a water-rich ultramafic, can form as magma, he had to find a way to transport it from the peridotite substratum to the upper crust. He appealed to the tectogene. Referring to his Figure 1 (my Figure 3.1), he proposed:

The crustal fold buckles down, as shown in Fig. 1, so that the base of the relatively strong upper crust comes in contact with the peridotitic substratum. Assuming a gradual decrease in strength downwards in the crust, the intermediate layer of supposedly basaltic composition deforms by flowing laterally rather than buckling as does the stronger upper crust. During the forcing of the bottom of the downbuckle into the upper part of the peridotitic substratum, sufficient stress is present to permit squeezing off of a product of partial fusion of the peridotite substratum (this is the hydrous peridotite magma [i.e., serpentine]). Under any other conditions this product of partial fusion would merely remain between the interstices of the grains of the peridotite substratum, because as a rule probably insufficient stress would be present to squeeze it off or, if squeezed off, it might not be able to penetrate the relatively plastic basaltic layer above it. The relatively rigid and strongly deformed downbuckle allows the hydrous peridotite magma to migrate up its vertical structures, and thus the products of this magma are formed over or near the axis of the downbuckle or in the belts one on either side of the downbuckle.

*(Hess, 1938a: 333; my bracketed addition)*

He thought he had found a process to produce, separate, and transport molten serpentine to higher levels in the crust and to intrude it beneath belts of negative gravity anomalies.

Hess then explained the formation of Alpine-type mountain belts in terms of Vening Meinesz's downbuckling hypothesis. It was at this point that he became obsessed by the relationship between serpentinites and geosynclines – between them and the down-buckling tectogene. He was to gnaw away at this relationship like a dog with a favorite bone for two decades or so. He began with Ph. Kuenen's (1936) laboratory simulation of Vening Meinesz's downbuckle and was impressed by his choice of materials that had relative strengths comparable to crust and mantle (Hess, 1938a).[5]

Kuenen's [1936] analytical examination and experimental test of Meinesz's theory offers strong support. The experiments performed by Kuenen are of particular validity because two factors have been taken into consideration which are commonly neglected in experimentation with geologic models. The first is that the strength of material used has been chosen so as to be of the right order of magnitude for the scale of the model; and second, the crust of the model has been floated on liquid paraffin to supply the zone of no strength beneath the crust.

*(Hess, 1938a: 76)*

Kuenen found that he could simulate folding of sediments.

In another set of experiments, Kuenen covered the crust with a very weak layer to represent a cover of sediments. In the first stages of deformation, the sediments are carried along on the competent crust; but when the crust downbuckles, the weak sediments are squeezed out of the core of tectogene like toothpaste out of a tube, forming alpine types of structures (Figs. 3, 4, 5).

*(Hess, 1938a: 77; Hess's Figures 3 and 5 are reproduced as my*
*Figures 3.2 and 3.3)*

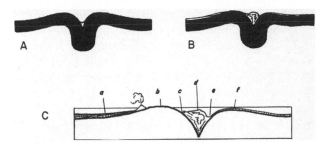

Figure 3.2 Hess's Figure 3 (1938a: 77). His 3A shows a buckling crust with little or no sediment cover. 3B shows sediments being squeezed up in geotectoclinal zones. 3C shows 3B in detail: a, geosyncline; b, emergent inner geanticline; c, ocean trough of moderate depth; d, geotectocline – upsqueezed zone; e, great ocean deep; f, submerged outer geanticline. Hess (1938a: 79) named the basin over the tectogene the "geotectocline."

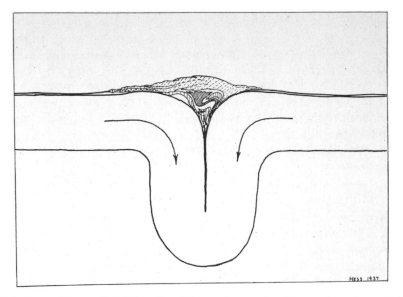

Figure 3.3 Hess's Figure 5 (1938b: 79). His caption reads, "General section of the Alps superimposed on the tectogene. Both features drawn to the same scale with no vertical exaggeration."

Folded sediments gave Hess his Alpine mountain belts centered above tectogenes and between serpentine belts (see my Figures 3.2 and 3.3).

He thought that formation of island arcs and their evolution into mountain belts had two stages. First the tectogene formed, serpentinization occurred, and the original layer of sediment was forced down with the downbuckling crust, and squeezed up out of the tectogene's core (Figure 3.2C). Then, much later, Alpine mountain belts were formed (Figure 3.3).

In both the East and West Indies, according to the writer's concept, a second great deformation has occurred a considerable time after the first one, during which the tectogene originally
was developed. In the interval between the first and second great deformations, one or both
of the geanticlines on either side of the tectogene may have emerged above sea level [see
Figure 3.2C (Hess's 3C)]. Erosion of these emergent portions, plus a great contribution of
volcanics from the concave side of the arc, may deposit great thicknesses of material in
"geosynclines" within the inner geanticline, and perhaps also outside of an outer geanticline,
as well as in the central basin over the tectogene itself ... The second deformation will deform
very intensely the material of the geotectocline [the basin over the tectogene]. Strong folding
and perhaps thrusting of the interdeformational sediments, if deposited, will occur, and
probably further upthrusting of material originally squeezed out of the tectogene, if present
will take place. This happens because the material in the geotectocline is pinched between a
sort of jaw-crusher as the main crust moves toward the tectogene and down over its rolling
hinges. Furthermore, the material which may be on the sides (or side) away from the geotectocline then impinge against the upsqueezed mass in the geotectocline. Upon coming against this
bulwark, the weak upper part may be literally scraped off the main crust as it rides forward and
down into the tectogene ... The result will be that the cover will be thrown into folds and
perhaps develop a schuppen structure [a series of high-angle reverse faults] between pairs of
thrust planes as its forward progress is topped by the bulwark and the main crust under-rides
or in reality underthrusts it.

*(Hess, 1938a: 77–80; my bracketed additions)*

By projecting forward the downbuckling hypothesis, he (1939) explained the origin
of mountain belts, citing field evidence in support. He argued (1939: 272) that in the
West Indies the negative strip of gravity anomalies and serpentine intrusives formed
during the Eocene with subsequent deformation in Late Miocene and Pliocene time.
In the East Indies, he noted that a belt of serpentines very closely follows the present
negative strip of gravity anomalies, and argued (1939: 272–274) that it formed during
Late Cretaceous or Early Eocene when the tectogene originated; subsequent Miocene
deformation of the tectogene caused intense folding of sediments above. He then
identified serpentine belts associated with older mountain systems in Australia, Asia,
southern Europe and northern Africa, northern Europe, North America, South
America, and Africa (Hess, 1939: 273–280).

   Hess's theory of mountain building was very different from the then commonly
accepted view that mountains form after geosynclines become filled with sediment.
He believed that mountain formation begins with crustal downbuckling and serpentinization. Geosynclines form afterwards.

The question now arises as to the relation of geosynclines and geosynclinal sediments to the
downbuckle. Apparently thick series of sediments, commonly of shallow-water origin, are
deposited in one or both of the shallower synclines adjacent to the downbuckle. These are often
folded during secondary deformation, though they are not nearly so intensely deformed as the
material over the downbuckle itself. It thus appears that geosynclines developed as a result of
an earlier deformation, and perhaps should not be considered as localizing the deformation in
the first place. Certainly nothing in the nature of a geosyncline seems to have been present in

the East or West Indies previous to the deformation, but subsequently basins were developed which might become geosynclines. The classic example of a geosyncline is the Appalachian. It was developed as a basin parallel to and west of the Taconic downbuckle. Sediments from the Taconic mountains, as they rose, filled this basin in the course of time. Later it was folded but in a relatively mild manner, more like the Juras than the intense Alpine type of development. Thus there probably is some relation between basins of sedimentation and deformations but it may not be the direct cause and effect relationship which is usually postulated.

*(Hess, 1939: 282)*

If his theory was correct and downbuckling was the first step in forming mountain belts, then mountain belts take longer to form than formerly thought.

He also linked his theory of mountain building to continental accretion around ancient cratons. He raised the idea while discussing serpentine belts in Australia, a continent that, if considered in isolation (I, §9.1), fit nicely in the traditional idea of peripheral accretion.

The phenomenon of successively younger [serpentine] belts one outside of the other is again well illustrated in this province [Australian region]. It is very suggestive of the growth of continents by repeated mountain-building epochs building up successive accretions around a central nucleus.[6]

*(Hess, 1939: 274; my bracketed additions)*

After surveying the continents for serpentine and associated mountain belts, Hess turned to the origin of the Mid-Atlantic Ridge, and offered his first of many solutions.

The only outcrop of basement rocks on the Mid-Atlantic Ridge is on the tiny island of St. Paul's Rock. The rock is a partially serpentinized dunite. It has the chemical composition of a typical ultramafic of the type under consideration. Evidently the Mid-Atlantic Ridge is an old folded mountain system. Meinesz has made a number of gravity traverses across it, and has found a belt of weak negative anomalies such as are often found over the remains of the downbuckle of an old mountain system. A similar belt of small negative anomalies appears in several traverses across the schists of the Piedmont zone of eastern North America near the Taconic serpentine belt and can be similarly interpreted. The age of the Mid-Atlantic Ridge system is probably old, but of course cannot be dated.

*(Hess, 1939: 281)*

These last two quotations show that, in the late 1940s, he did not believe in continental drift – the Mid-Atlantic Ridge was an old folded mountain system and the Atlantic Basin was old. Hess acknowledged that his hypothesis of mountain building was controversial, and he identified and attempted to justify its major assumptions (RS1).

The Meinesz theory applied to mountain building requires, if it is correct, somewhat drastic revision of ideas concerning mountain building. The corollary hypothesis advanced by the writer concerning serpentines makes it possible to locate the position of the former down-buckle in mountain systems and thereby distinguish the main trend line from subsidiary lines of folds often far displaced from the central axis. Conclusions advanced in succeeding paragraphs are necessarily predicated on the correctness of various assumptions. If these assumptions are true, the conclusions follow; if not, they may not. They are outlined below.

(1) That the Meinesz theory explaining gravity anomalies in present island arcs by down-buckling of the crust is correct.
(2) That it can be applied to mountain systems on land as well.
(3) That the serpentinites and related rocks are products of a hydrous peridotite magma as outlined elsewhere.
(4) That these are intruded during the first formation of a given downbuckling, but not during later deformations shoving the same downbuckle deeper.

Each of these is not merely a speculative assumption, but can be supported by considerable evidence. The writer believes the assumptions to be likely ones. However, it is well to keep them in mind and neither accept them blindly nor discard them without consideration because of the radical nature of the readjustment of long established ideas which they will necessitate.

*(Hess, 1939: 271)*

Regarding the last assumption, he suggested two possibilities:

(1) that after the first squeezing off of a product of partial fusion insufficient low melting material is present to yield a second body of magma; or (2), that fusion of the bottom of the downbuckle after the first deformation forms an impermeable cap through which the ultra-mafic magma cannot be intruded. Either or both of these possibilities may be operative.

*(Hess, 1938a: 334)*

He had no evidence that either process actually occurred and it was the weakest of the four assumptions.

His attempt to justify key assumptions even before his opponents had had the opportunity to criticize his hypothesis is precisely the preemptive move that paleomagnetists used in the 1950s when presenting their case for continental drift. He attempted to prevent difficulties becoming a plague on his hypothesis before others could raise them.

Hess did not ask what causes downbuckling of the crust. He thought it was compression, but he did not know what caused the compression. He rejected Vening Meinesz's appeal to mantle convection (I, §5.6) for two reasons. First was the close tie between gravity and topographic data.

In the basins of the Caribbean and Gulf of Mexico the anomalies are generally positive. As a rule, the deeper the water for a given basin, the more strongly positive are the anomalies. The profile made by the S-48 from Jamaica to the west end of Cuba shows the relationship most clearly. Here the anomaly profile drawn to an appropriate scale of the sounding profile results in the one becoming an almost perfect mirror image of the other. Enclosed basins in other parts of the world show similar relations. The reason for this relationship is not apparent, and no adequate explanation can be offered at present. Meinesz's suggestion of downward convection current below the crust to explain the positive anomalies in the oceans does not seem applicable here, because such a process going on at considerable depth does not seem consistent with the very close correlation of anomalies and topography.

*(Hess, 1938a: 83)*

Second, as Allwardt (1990: 110–112) has emphasized, Hess did not see how mantle convection, at least of the kind envisioned by Vening Meinesz, could last long enough to keep ocean crust downbuckled long enough (many millions of years) to form mountains.

> How such a down-buckle can be maintained for so many millions of years without disappearing remains an unsolved question. Though Vening Meinesz and other geophysicists are loath to accept this conclusion, it seems inevitable to the geologist. Judging from the relations in Cuba, it seems likely that a certain amount of compression must be maintained in the crust for the down-buckle to persist. Where this is removed, the downbuckle spreads, rises, and disappears.
>
> *(Hess, 1937: 76)*

Hess continued to support his amended version of Meinesz/Kuenen downbuckling. In 1940, he again proposed evolution of island arcs into mountain belts. Beginning with the association of serpentine and mountain belts, he stressed the idea that geosynclines form after mountain building has begun, and implicitly tied his solution to continental accretion.

> A double row of small peridotite masses, intruded during the Taconic Revolution, extends from Newfoundland to Alabama. Every mountain system of the Alpine type contains serpentinized peridotites along its axis of most intense deformation. Using the peridotites as a guide to the axis of a system and examining mountain structure of varying degrees of development and dissection, a series of horizontal planes may be visualized cutting mountain structures at progressively lower levels. For such a study the following may be suggested: (1) West Indies, mountain system not yet elevated; (2) Alps, recently elevated; (3) Appalachians, dissected to moderate depth; (4) Canadian Shield, Archean, dissected to great depth. A fifth level, the bottom of the crust may be postulated from gravity anomalies in island arcs. Re-examining the Appalachians with this broad picture of mountain structure in mind drastic rearrangement of prevailing ideas seems necessary; (1) Cambro-Ordovician stratigraphy indicates sedimentation in a sea marginal to a continent (moderately deep water shale, limestone); (2) a great Taconic Revolution produced an island arc along the present Piedmont belt (volcanism on Atlantic side, peridotite intrusions); (3) geosyncline is developed between arc and continent which received rapid, coarse, elastic, shallow water sediments (Siluro-Devonian); (4) subsequent deformations fold the geosynclinal sediments. Mountain building came first; the geosyncline developed as a result not only in the Appalachians but in every mountain system. Perhaps no greater misconception exists in geology than that geosynclines localize mountain building. The Appalachian geosyncline was folded in the late Paleozoic much as was the equivalent zone of the Jura in the Alps. This happens to be the most evident feature of the present Appalachians, but it represents not the great revolution but the last and dying phase of a series of Paleozoic deformations.
>
> *(Hess, 1940: 1996)*

Teaming up with Fredrick Betz Jr., they (Betz and Hess, 1940, 1942) gave an explanation of the Hawaiian Islands and accompanying swell based on downbuckling. If downbuckling is the cause of the Tertiary island arcs and mountain systems encircling the Pacific Basin then attendant crustal shortening would produce trans-current faults (Hess, 1938a) in the floor of the Pacific.

If we assume the Meinesz theory of the downbuckling of the crust in island arcs to be correct, it is fairly evident that the crustal segments behind a horizontally curving buckle must be adjusted to one another by transcurrent faults.

*(Betz and Hess, 1942: 114)*

Ensuing volcanism along the fault would create a thick lens of volcanic ejectamenta topped by oceanic islands.

Betz and Hess's hypothesis faced a serious difficulty because there was no evidence of any large transcurrent faults beneath the Hawaiian Islands or anywhere else in the Pacific. The great fracture zones in the eastern Pacific would not be discovered and identified as faults for another decade. Of course, Hess had postulated 50 km horizontal displacement along the Bartlett Trough fault in the Caribbean, but there was no independent evidence for a similar fault or horizontal displacement associated with the Hawaiian Islands. Although they failed to acknowledge this difficulty, they identified seven major continental transcurrent faults, including the San Andreas Fault of California, Great Glen Fault of Scotland, and the Dead Sea Fault and its extension into Syria, arguing that others would likely be discovered.

Although it may well be that some of these interpretations and estimates of displacement are incorrect [horizontal displacements of 65 miles along the Great Glen fault, and 96 miles along the Dead Sea fault], it nevertheless seems likely that such great transcurrent faults do exist and that they do represent a prevalent major phenomenon in the crystalline crust. The wide geographical distribution of the areas cited may be noted, as well as the comparatively recent date of most of the references [all but two references were after 1930]. In general, geologists have not looked for, or even suspected the presence of, large horizontal movements on more or less vertical fault planes. Movements of this type have been overlooked probably because it is relatively easier to estimate vertical displacements than horizontal ones. To judge by the number of examples discussed in the last few years, it may be assumed that many more will be found, now that their existence is becoming increasingly evident.

*(Betz and Hess, 1942: 113; my bracketed additions)*


### 3.5  Hess discovers guyots and explains their origin

The writer has given a great deal of thought to the problem of origin of guyots since first encountering them in 1944 ... It now remains to account for them. During the past two years, many hypotheses were tried and discarded. Finally the writer arrived at the hypothesis here presented. Though it explains the facts at present available, it is highly speculative and might easily be wrong. Nevertheless, it seems worth presenting as a working hypothesis, particularly since it has many interesting ramifications some of which would be worthy of investigation even if the parent hypothesis were found to be invalid.

*(Hess, 1946: 782)*

While commanding USS *Cape Johnson* during World War II, Hess took soundings of the Pacific seafloor "on random traverses incidental to wartime cruising" (1946: 773). Through these soundings he discovered twenty submerged, reefless, flat-topped sea mounts. He named them "guyots" (after Arnold Guyot, the Swiss oceanographer who founded Earth sciences at Princeton).[7] He identified 140 more through examining sounding charts of the US Navy's Hydrographic Office. They were circular or oval in plan, and their tops varied in size from two to sixty miles. They were located in an area north of the Caroline Islands and east of the Marianas and Volcano Islands approximately between 8° N to 27° N and 165° W to 146° E.

He was puzzled by their flatness and reeflessness. Their flat tops suggested they had been worn down by wave erosion, but if this were so, why were there no reefs? He hypothesized that guyots were Precambrian islands that already had become submerged far enough below sea level prior to the appearance of lime-secreting organisms. He supposed that sea level had risen relative to oceanic features projecting from the seafloor since Precambrian times because of unending deposition of continental sediment on the seafloor. Sediments would raise the ocean floor and, assuming a relative constancy of oceanic water since Precambrian times, sea level would rise. Removal of sediments from continents onto ocean floors would also cause continents to rise and oceans to sink through isostatic adjustment.

Since most of the material deposited on the ocean floor has ultimately come from the continents, isostatic adjustment of the load on the sea floor and the loss of weight from the continents has resulted in the sinking of the former and rise of the latter so that relative sea level with respect to the continents has not changed very much. One obviously cannot put a layer of several thousand feet of sediments into the oceans without causing the water to rise by an equivalent amount (less the water included in pore spaces in the sediments). Thus, quite apart from [isostatic adjustment], every centimeter of sediment put into the ocean causes sea level to rise with respect to an oceanic island by just a little less than a centimeter (less by the amount of water in pore space of the sediment). Even though the figure cited for the rate of sedimentation [1 cm per 10 000 years for red clay, and 1 cm per 5000 years for globigerina ooze] may be inaccurate it nevertheless follows that oceanic islands are and have always been slowly sinking relative to sea level.

*(Hess, 1946: 789; my bracketed additions)*

Those oceanic islands, that had been worn down by wave action and had sunk far enough below sea level before lime-secreting organisms evolved, became guyots.[8]

Hess was proud of his discovery of guyots and of his ingenious explanation of them. He had found a new geological structure, and had offered an explanation of its origin. The discovery was his own; he had not made it while working with Vening Meinesz or other more established co-workers. He kept over forty letters asking for reprints or praising his discovery. Hess submitted his paper in early July 1946 to the *American Journal of Science*, and it was published in November. He ran out of reprints by the second week of December. With permission from the Editor of the *American Journal of Science* he had the paper reprinted by the US Navy's

Hydrographic Office (Dorothy Lull letter to Hess of December 30, 1946). He asked for a hundred reprints (Hess, March 3, 1947).

Prominent Earth scientists wrote him. H. W. Murray, who had analyzed data from the US Coast and Geodetic Survey's (USCGS) survey of the Gulf of Alaska, was impressed with Hess's discovery. He told Hess that re-inspection of fathograms from the survey of the Gulf of Alaska revealed guyots in the Gulf of Alaska.

This finds me avidly digesting your article on "guyots" published in the November 1946 issue of the *American Journal of Science*. I have also just completed a rapid perusal of available RCA type fathograms in the Gulf of Alaska for the period 1945–1947. The submarine mountains discovered are all broad as to base and width of top, and either flat-topped for a considerable distance or gently rounded ... Several seamounts were discovered and all resemble your "guyots." In fact, the resemblance is so remarkable that if your fathograms profiles were to be substituted in our fathograms, it would be difficult for some persons to detect the exchange!

*(Murray, January 9, 1948 letter to Hess)*

Mary Sears at Woods Hole Oceanographic Institution (hereafter WHOI) suggested that Hess send reprints to Hans Pettersson (1880–1966) and Borjie Kullenberg.

Yesterday, I came across your article on "guyots" in the last number of the *American Journal of Science* and it occurred to me that Professor Hans Pettersson ... would be extremely interested in it. Professor Pettersson plans to start on a fifteen months circumnavigational cruise in low latitudes early in the coming year. His primary purpose is to study the deep-sea sediments with his new coring device which he tried out successfully in the Mediterranean this spring ... I wonder whether you would be willing to send a preprint both to Professor Pettersson and to his assistant, Dr. B. Kullenberg, as soon as you receive your copies.

*(Sears, November 7, 1946 letter to Hess)*

Hess replied to Sears a week later. After telling her that he would send reprints immediately to Professor Pettersson and Dr. Kullenberg, he added, "As a geologist I can sincerely say that I know of no more satisfying compliment than to have one oceanographer recommend my paper to another" (Hess, November 13, 1946 letter to Sears). The Swedish oceanographers were about to leave on the 1947–8 Swedish Deep-Sea Expedition aboard *Albatross*. Pettersson was to lead the expedition, and they planned to use Kullenberg's newly invented piston corer that could take cores of up to twenty-four meters. Hess sent reprints; Pettersson replied.

Very many thanks for your most interesting paper on the "Drowned ancient islands of the Pacific Basin. I have handed over the second copy to Dr. Kullenberg who will no doubt acknowledge it himself. It is possible that we may touch at one or more of the "guyots" described in your paper, in which case I should very much like to make attempts at getting samples from the bottom, if not sediments, possibly rock fragments. I am ignorant in how far such fragments should be identified as of pre-Cambrian age but anyhow their volcanic nature would be most important to verify.

*(Pettersson, December 17, 1946 letter to Hess)*

J. D. H. Wiseman, who had accompanied Sewell on the *John Murray* Expedition in the Indian Ocean (I, §6.13; §6.7), found Hess's paper stimulating.

It was very kind of you to send me your recent paper on the "Drowned Ancient Islands of the Pacific Basin" which I have read with great interest. You were very lucky to discover the "Guyots" whilst you were serving with the U.S.N.: they are certainly very puzzling and your tentative theoretical remarks are most stimulating.

*(Wiseman, January 10, 1947 letter to Hess)*

Voicing the paramount opinion of the day, he also thought that paleogeographers would be unhappy with Hess's postulation of a relatively permanent Pacific Basin.

Most palaeogeographers will, of course, not agree with the relative constancy of the Pacific Ocean, but I think there is great difficulty in accounting for the emergence of large areas of the ocean floor which many palaeontologists postulate.

*(Wiseman, January 10, 1947 letter to Hess)*

M. King Hubbert, who had stressed the importance of making true-scale models to test the feasibility of hypotheses about Earth's behavior (I, §5.6), congratulated Hess on his discovery.

Thank you for the reprint of your recent article "Drowned Ancient Islands of the Pacific Basin". I think this, together with your paper just delivered before the G.S.A. is a very important addition to our information of ocean basins.

*(King Hubbert, January 7, 1947 letter to Hess; the GSA)*

Hess also received requests and comments from two rising stars who later played very prominent roles. Robert Dietz asked for a reprint.

I would appreciate your sending me a reprint of your recent paper in *American Journal of Science*, regarding the deep banks in the Central Pacific Ocean. As you may know the Navy Electronics Laboratory at San Diego, California has recently organized an Oceanography Section which will be engaged in research on both background and applied phases of oceanography. I am a member of this Section and leader of a group engaged in marine geology. Therefore, I would greatly appreciate it if you will send us in the future any reprints that might be of value in our work.

*(Dietz, June 10, 1947 letter to Hess)*

J. Tuzo Wilson, who had gotten to know Hess while studying for his Ph.D. at Princeton (1933–6), sent an amusing congratulatory note.

Many thanks for your reprint on Guyots from the *American Journal of Science*. I am extremely glad to have this paper not only for its interest but also to indicate what can be done by someone with imagination as a sideline to their other work.

*(Wilson, January 6, 1947 letter to Hess)*

Some could have later said the same of Wilson.

Hess also received an interesting letter from Henry E. Crampton, research associate in mollusks at the American Museum of Natural History. Crampton had spent much of his career studying snails of the South Pacific islands. A firm believer in a pre-Pacific continent, he asked Hess about his views.

The subject [the floor of the western Pacific] is of intense interest to me because of its connection with my own researches in the Pacific on biological evolution of certain land organisms of the high islands. I have worked in seven groups of islands, from the Society Group to the Marianas. One of the general results has been to make me an adherent of the view that an extensive continental mass must have existed at least as far back as the Tertiary. By continuing subsidence and inevitable dissection, this has disappeared except for the peaks that still persist. I am aware that geologists are divided on this proposition, and the botanists also. But the zoologists who have dealt with land forms are unanimous in support of the subsidence theory, for only if this is true could the present distributional conditions come about. You may therefore realize why I am so keenly interested in your knowledge and in your views. Is it possible that you have published something on this matter of a pre-Pacific continent?

*(Crampton, March 28, 1947 letter to Hess; my bracketed addition)*

Crampton was not the only zoologist interested in Hess's work. According to Hess, zoologists had expressed more interest than geologists. Perhaps they thought, contrary to Wiseman, that Hess provided them with their needed landbridges. However, Hess was not in favor of a Tertiary-aged Pacific continent or wholesale postulation of former landbridges. He acknowledged that some landbridges (more like Bailey Willis' isthmian connections) might have once existed in the western Pacific. He also thought that more data and cooperation were needed among geologists, paleontologists, and zoologists.

I had an article in the *American Journal of Science* November 1946 and have ever since been deluged with comments from zoologists though the paper caused very little stir among fellow geologists. The zoologists saw in the paper some encouragement for land connections which all of them sought, to explain faunal distribution. In most cases they needed bridges in places where my geological information would suggest there could not possibly have been a bridge in Tertiary or later time. Geologically bridges would be easy to postulate from New Zealand to the Fijis and thence to New Guinea and the Philippines. Or from Honshu southward along the ridges to the Bonins or Volcanoes, the Marianas and on to Palau. Bridges involving the East Carolines, Marshalls and Gilberts would be virtually impossible. I doubt if bridges involving the Society Islands are possible, but never having had any experience in that area and since data on bottom topography are scarce I cannot be certain.

There should be closer and more direct joint effort on such problems by the geologist investigating the structure and time relations and the zoologist and paleontologist on the other hand who are attempting to explain faunal distribution.

I can easily see how alluring the simple and direct solution of using land bridges can be. The geologist has erred in this respect quite as often as the zoologist, the archeologist or the anthropologist. A determination of the ages of the formations underlying atolls such as those in the Marshalls by drilling some holes would probably do more than anything else to limit the field of speculation of the Pacific Basin and put it on a sound footing. Without such information I have to admit I am making some very long guesses.

*(Hess, April 7, 1947 letter to Crampton)*

This attempt to find evidence of past landbridges in the Pacific by appealing to new work in oceanography did not stop with Crampton. Other landbridgers such as van Steenis (1962) would later appeal to Menard and Hess's work (I, §3.4).

He received an interesting letter from H. Backlund, whom he had met in 1937 at the 17th International Geological Congress. Backlund taught at the Mineralogisk-Geologiska Institutionen at Uppsala University, and had worked in Greenland studying the Caledonide disjunct along its eastern margin. He raised a difficulty (RS2).

I am very much obliged to you for sending me an offprint of your interesting investigations of the "Guyot"-islands of the Western Pacific, which pamphlet I have studied with great pleasure. But if it is permitted to deliver some objections, I should like to direct your attention on the investigations of Vologdine [A. Vologdin], his address to the Geocongress of Moscow (1937) and his later investigations of the lime algae [stromatoli] and their reef-building abilities in the early pre-Cambrian of East Siberia. Already [August] Rothpletz [1853–1918] some 40 years ago has described reef-building algae in the pre-Cambrian of southern Norway, forming thick limestone beds in the sediments below the tillite formations of pre-Cambrian (Eocambrian) age. In the Caledonians of East Greenland, stratigraphy below the tillite of about the same age, there are mighty limestones, in the whole built up by algal activity and developing characteristics of reefs. And in the limestones about 1000 and more m below the tillites plenty of traces of foraminiferous activity is met with (not yet published). It seems to me that the premises of non-existence of organogenic limestone forming activity in the pre-Cambrian is not conformable with field experiences, because many of these ancient limestones are met with in a facial milieu, which excludes every possibility of volcanic-hydrothermal origin. Please excuse me these objections, but the problem of yours is still of great interest.

*(H. Backlund, January 31, 1947 letter to Hess; my bracketed additions)*

I do not know if Hess responded to Backlund; he might have tried to avoid his difficulty by supposing that guyots had already sunk well below sea level before Eocambrian times.

### 3.6 Hess adopts mantle convection and rejects mobilism during his early career

For reasons just noted, Hess was first doubtful about Vening Meinesz's appeal to convection as a cause of downbuckling and mountain building; if he were to invoke mantle convection, it would have to be of a type that would keep crust downbuckled long enough for mountains to form. Griggs provided this (Allwardt, 1990: 111–112). Hess chaired the volatile session at the 1939 AGU meeting in which Griggs' presentation of his hypothesis of mantle convection as a solution to mountain building was received with scorn by Lawson and Willis (I, §5.10). Unlike them, Hess was impressed. Allwardt uncovered an exchange of letters between Griggs and Hess that occurred soon after the AGU meeting. Griggs began the correspondence.

I surely was glad to have the opportunity to see you at the meetings and only hope that the fates will permit longer séances in the near future. I am putting together my ideas for publication of this convection current theory as a possible mountain-building mechanism. If you could join in with an article to follow on possible geological interpretations, I should be very glad, but if you decide that you want to get more data before joining me in the big swim, then I should praise your discretion.

*(Griggs, May 1939 letter to Hess; reproduced in Allwardt, 1990: 112)*

Hess declined to "join in ... the big swim," but offered Griggs advice, and explained what especially attracted him to his version of mantle convection.

[Your convection current hypothesis] may be the major factor in development of island arc structures. It is the best hypothesis to date ... The strongest point in its favor is that it explains the maintenance of the down buckle for a long period of time – 50 or 100 million years – whereas no other hypothesis yet advanced does, and the geologic evidence necessitates such a maintenance ... I think it is a valuable hypothesis, and I am therefore anxious to have you present it in such a way that it will not give an opening for someone to jump on an irrelevant point and thereby apparently discredit the whole works. This brings up one part of your talk which certainly would be jumped on; namely, the part that dealt with a sequence of events in mountain building which started with a geosyncline, preceded into buckling, and ended with isostatic uplift after the currents stopped. That, to be sure, is the sequence one would get from all the current literature, but it doesn't fit the facts in island arcs. There often is no geosyncline before buckling (in the sense of a basin with thick sediments), and geosynclines do develop after buckling on either side of the buckle it seems.

*(Hess, May 1939 letter to Griggs; reproduced in Allwardt, 1990: 112; Allwardt's bracketed addition)*

Hess thought geosynclines were a consequence of crustal downbuckling. Whether his warning was strategically wise is questionable. Fifteen years later, Hess was still complaining that geologists continued to believe, as he had formerly advocated, that mountain building begins with the formation of geosynclines.

The concept that geosynclines (long narrow troughs containing a thick section of clastic sediments, commonly of shallow-water origin) localize mountain building was challenged on the basis that such a feature is not present in island arcs before the first deformation, but normally develops later because of that deformation. This idea has met with strong resistance, but the writer maintains his original stand.

*(Hess, 1955a: 391)*

Griggs' mantle convection met with strong resistance quite regardless of what Hess said about the order in which geosynclines formed (I, §5.6, §5.10).

Hess's first published appeal to mantle convection was in his 1946 paper on guyots. He cited Griggs (1939) but did not entirely agree with him. Griggs postulated that convection currents rise under oceans and sink at the periphery of continents where island arcs form (I, §5.6). Hess agreed that convection currents sink where island arcs form; however, unlike Griggs, Hess proposed that convection currents rise under continents.

> Many authors have correlated the observation that island arcs (and hence mountain building) develop in the ocean basins along the margins of continents with the concept that the continental massifs are strong and the oceanic crust weak thereby accounting for the localization. However, if mountain building forces are related to convection currents within the Earth (Griggs 1939), the most satisfactory of the present theories, then the localization can more reasonably be explained on the basis of heat relations within the crust. Being warmer under continents and cooler under oceans the downward follow part of the convection cell would be more likely to be localized under the ocean and would be supplemented in some cases by the outward flow of warm material from beneath the continental area.
>
> *(Hess, 1946: 787)*

From what I have read, Hess wrote nothing in favor of continental drift throughout his early career. His only published remark I have seen was negative; he mentioned continental drift in his discussion of oceanic crust, which he thought had great strength whereas mobilists believed it was weak, but they had no independent evidence of its weakness other than the hypothesis itself!

> The writer believes the oceanic crust is very strong though his opinion is at variance with existing textbooks and much of the literature. However, Jeffreys (1929), Daly (1940), and Longwell (1945) all favor a strong oceanic crust. The only bases for judging its strength are its behavior and the strength of the rocks of which it is thought to be composed. Both of these indicate strength. The reason it has been generally considered to be weak, appears to be related to calling it the exposed sima or the basaltic substratum and consciously or unconsciously bringing in Daly's theory of a weak glassy basaltic substratum. But Daly postulated a strong crust and weak substratum at considerable depth. Those favoring the hypothesis of continental drift assumed a very weak basaltic crust below the oceans without, so far as the writer is aware, presenting evidence other than the hypothesis of drift to substantiate the assumption.
>
> *(Hess, 1946: 786–787)*

Hess might have believed that there were some forms of mobilism that did not require a weak oceanic crust but he wrote Vening Meinesz a letter in July 1959 claiming that he had never favored continental drift (§3.12), and I know of no reason to believe otherwise.

This did not mean he was not thinking about drift. Hess talked about continental drift in his classes, and his papers at Princeton contain lecture notes for some of them. He kept his typed lecture notes for his 1946 Advanced Geology, Geology 405. Hess discussed the permanence of ocean basins. Sounding much like the fixists Matthew or G. G. Simpson (I, §3.6), he preferred oceanic permanence over landbridges.

Lack of oceanic sediments on continents indicates that the continents have never been deep sea. It seems very probable that no large land areas have ever sunk beneath the sea. Land areas are high because they are underlain by a greater thickness of light upper crust ("granitic") than ocean areas. If thick granitic crust were once present there is no likely mechanism by which it could be removed. Hence it would always "float" relatively high. Permanence of ocean basins originally suggested by Dana. Was basis of a great geologic argument in late 19th century and early 20th century. Data which became available on radiation of reptiles and mammals at that time necessitated "land bridges" to get them from one continent to another. Some land bridges have no doubt existed but these are across shallow seas such as Bering Strait which require relatively slight uplift or lowering of the sea to make them bridges. Other bridges no longer necessary since forms thought to be identical at that time are now known on closer study to be different or considered cases of parallel evolution.

*(Hess, lecture notes for Advanced Geology 405, Hess papers)*

Hess then turned to continental drift. Although he dismissed as unnecessary the drift solution to disjunctive distributions of certain fossils, he seems purposely to have not revealed his attitude toward mobilism because Erling Dorf, a paleobotanist who was teaching the course with Hess, was to speak about it in a later class.

Continental drift also came into vogue about the same time. Though originally invented by Wegener to account for past climates it was supported by many geologists, particularly Europeans, to account for distribution of terrestrial life. As explained above, this seems no longer necessary. (I won't tell you what I think of continental drift since Prof. Dorf will lecture on it later in the term).

*(Hess, lecture notes for Advanced Geology 405, Hess papers)*

Dorf, I believe, opposed mobilism.[9] But Hess might not have wanted to tell his students what he thought of continental drift – he may have disagreed with Dorf.[10]

### 3.7 Hess's middle career, 1950–1959: a preamble

Further studies of ocean crust, ocean floor topography and sedimentation were made in the late 1940s and early 1950s. Most of the new findings were obtained at Lamont, Scripps, WHOI, the Naval Electronics Laboratory in San Diego, the Department of Geodesy and Geophysics, Cambridge, and at the Göteborg Oceanographic Institute (Kullenberg, Pettersson, and others). Much of the work was supported financially by governments or undertaken by them, knowledge of ocean floors being militarily important (see Schlee, 1973; Bullard, 1975b; Menard, 1986; Stommel, 1994; Rainger, 2000; Hamblin, 2005; and Doel *et al.*, 2006). The USCGS, the Hydrographic Office of the US Navy, its Naval Electronics Laboratory, the United Kingdom Hydro-graphic Office, and its National Physical Laboratory, for example, carried out their own oceanographic research.

In his presentation at the February 28, 1953 meeting on the floor of the Atlantic Ocean, sponsored by the Royal Society of London and moderated by Bullard, Hess (1954: 347) applauded efforts to obtain new data by M. Ewing, J. Worzel, F. Press,

I. Tolstoy, C. B. Officer, B. C. Heezen and others of Lamont, R. R. Revelle and R. W. Raitt of Scripps, and R. S. Dietz, H. W. Menard and E. L. Hamilton of the Navy Electronics Laboratory.

Hess had no new ocean data of his own. Since his discovery of guyots, he had not himself sought data from the seafloor; instead he wanted to think about what all these results of others meant.

Compared to the pre-war era there has been a great increase in the amount of geophysical work at sea, and a correspondingly vast amount of new information acquired. The energy and ingenuity of the scientists concerned has gone largely into development of new techniques and the acquisition of data, with comparatively little time spent in meditation on the broader aspects of the meaning of the results. The ideas here presented are the product of a few weeks' thoughtful consideration, whereas a year would have been a more appropriate interval to do the data justice. I hope this will be borne in mind by the reader.

*(Hess, 1954: 341)*

For the remainder of his career Hess would rely increasingly on data collected by others about the seafloor, serpentinization, and paleomagnetism including the paleomagnetism of oceanic crust. He continually reexamined his previous opinions in light of new data, and identified and offered solutions to new problems. Hess's success depended on the productivity and generosity of the data collectors, often learning of new findings from investigators before publication. Had he not had access to data prior to publication, which sometimes took years, some of his ideas about the origin of various oceanic features may well have become obsolete by the time they appeared in print.

In the mid-1950s, Hess (1954, 1955a, 1955b) abandoned his solution to the origin of guyots and proposed a new one. He kept, but altered (1955a, 1955b), his tectogene hypothesis for the formation of island arcs and mountain belts, and he developed (1954) several new solutions to the origin of oceanic ridges. Importantly, he became disposed toward mobilism because he found the paleomagnetic case for it quite compelling. This fundamental change in his attitude toward mobilism occurred, I believe, before summer 1959, quite certainly before he developed his hypothesis of seafloor spreading.

### 3.8 Hess on mantle convection, oceanic crust and upper mantle, and mid-ocean ridges, early 1950s

Hess returned to mantle convection four years later at the 1950 Colloquium on Plastic Flow and Deformation within the Earth, where he, Griggs, and Vening Meinesz spoke favorably about it (I, §5.10). Hess invoked convection to explain the persistence of crustal downbuckling beneath island arcs.

At present it is not possible to prove that convection currents exist in the mantle nor is it possible to show that they cannot exist. The difficulty of explaining persistence of the root in island arcs without appealing to such currents leads the writer to be strongly disposed to favor them.

*(Hess, 1951: 530)*

He claimed that his version of mantle convection explained the downward-sloping pattern of deep-focus earthquakes on the concave side of island arcs (RS1). Illustrating his idea with the following diagram, he supposed tandem convection cells.

A somewhat different system of tandem convective is suggested by the writer to account for the pattern of deformation on the Earth's surface and a correlation with deep-focus earthquake data (see Fig. 16). Note shear directions at A and B are opposite to those at C ... The "mega cell" is a convection cell extending from 500 km to base of mantle. Outflow from under warmer continent and direction of flow of mega cell have to be such that both are favorable to the direction of motion of the "roller cell"; thus not all coasts of all continents need have associated island arcs. Gap in deep foci near 475 km is at level between roller and mega cell.

*(Hess, 1951: 530; his Figure 16 is my Figure 3.4)*

Convective currents arise under continents and sink beneath ocean floors. They explained the origin and maintenance of tectogenes, and the distribution of earthquakes above 475 km. The "mega-cell" was a new idea; Hess had already proposed the "roller cell" in 1946. The mega-cell extended vertically from 500 km to the base of the mantle, he left unspecified its horizontal dimension.

By insisting that crustal downbuckling occurs only if convective flow along adjacent horizontal elements of stacked cells is in the same downward direction, he avoided the potential difficulty of why island arcs fail to form along every continental margin (RS2). Location and convective flow of mega-cells were controlled by thermal conditions at the core–mantle boundary. Hess realized that mantle convection was a questionable concept and attempted to justify it by appealing to convection in the core (RS1).

Recently Elsasser and Bullard independently arrived at theories of the Earth's magnetic field based on convection in the core. To have convection in the core necessitates convection in the mantle; otherwise, there would be no effective way of removing heat from the core–mantle interface and convection in the core would stop.

*(Hess, 1951: 530)*

He also answered Birch's objection (I, §5.10) to single-tier convection (RS2).

Birch's objection to convection on the basis of a difference in character of the materials above and below a level near 900 km is not a serious obstacle. If this barrier is a pressure-induced phase change, a convection current could pass through it, the reaction taking place in opposite directions on the upward and downward limbs of the cell. If it were some other type of barrier, tandem convection cells might be set up one above the other as postulated above the core–mantle interface.

*(Hess, 1951: 530)*

At the Royal Society of London's 1953 meeting on the floor of the Atlantic, Hess took note of two significant discoveries.

The most momentous discovery since the war is that the Mohorovičić discontinuity rises from its level of about 35 km under the continents to about 5 km below the sea floor ... while ... the second most important discovery is the recognition that we have at the Earth's surface fragments of what are almost certainly the material from below the Mohorovičić discontinuity.

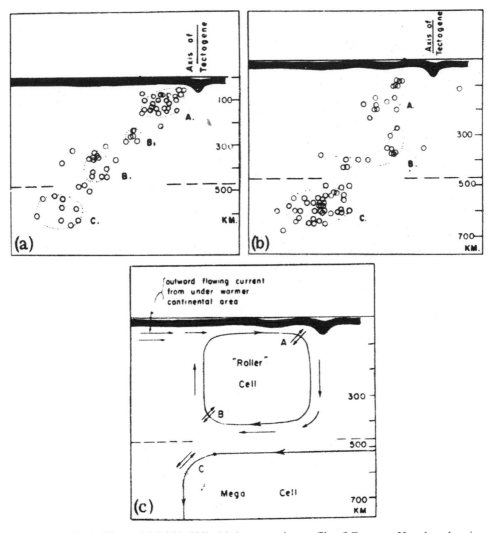

Figure 3.4 Hess's Figure 16 (1951: 530). (a) A composite profile of Guam to Honshu, showing the spatial relation between a tectogene underneath an island arc or trench and deep-focus earthquakes; (b) a composite model of Tonga-Kermadec Islands showing the same relationships; (c) Hess's hypothetical convection cells in relation to deep-focus earthquakes. The horizontal dotted line represents what Hess thought was a gap in the distribution of deep foci near 475 km.

Samples of such nodules [i.e., peridotites brought to the surface in basaltic volcanoes] in basalts from three continents and from oceanic islands were found to be virtually identical ... This substantiates the view that the material under the Mohorovičić is peridotite, as had been deduced from seismic velocities. Furthermore, we may consider this the material of the upper part of the mantle and note that it is the same below continents as below the oceans.

*(Hess, 1954: 341–342; my bracketed addition)*

He now claimed that the oceanic crust is 5 km thick, lacks a granitic layer, and is blanketed with an average of only 0.7 km of unconsolidated sediments, which was three to five times less than former estimates, which had been arrived at by extrapolation of present rates of deposition back to the Precambrian.

At the same meeting, he argued that enough was "known of the character of oceanic crust" to propose "working hypotheses on the origin and development of oceanic ridges," and noted, "Without hypotheses to test and prove or disprove, exploration tends to be haphazard and ill-directed. Even completely incorrect hypotheses may be very useful in directing investigation toward critical details" (Hess, 1954: 344). He (1954) defined oceanic ridges very broadly as topographically elevated regions, and counted among them the Mid-Atlantic Ridge, Mid-Pacific Mountain Range, Hawaiian Ridge, Caroline Ridge and the Mid-Indian Ocean Ridge. He (1954: 346) proposed explanations not only of the Hawaiian and Mid-Atlantic ridges, but also of island arcs, "merely as working hypotheses to suggest critical points to attack in further exploration at sea."

Although his explanations of the Hawaiian Ridge and island arcs remained unchanged from previously, his explanation of the Mid-Atlantic Ridge was a major departure. In 1939, he had argued that the Mid-Atlantic Ridge was an ancient folded mountain belt (§3.4). He now raised two objections to this based on new topographical and seismic studies of oceanic ridges (RS2).

Almost all of them [i.e., oceanic ridges] are devoid of linear small-scale ridges parallel to the main axes as might be expected if they were composed of folded rocks. Fault scarps commonly at high angles to the ridge axis are found on some and some show relatively high seismicity.

*(Hess, 1954: 344; my bracketed addition)*

The Mid-Atlantic Ridge was not an old, folded mountain belt that had become inactive. He now proposed that the Mid-Atlantic Ridge

involves brecciation of the peridotite substratum by great masses of basaltic magma perhaps over an upward convection current in the mantle. Some blocks of peridotite engulfed in basalt may be present at the surface as perhaps is the case at St. Paul's Rock. The somewhat lower density of this column as compared to the columns either side of it permits its surface to rise well above the ocean floor. During the time when some molten basalt was present and the temperature of the column as a whole was higher and hence less dense, the ridge might have stood high above sea level. An upward convection current beneath it would also tend to lift the ridge and cessation of the current to let it subside.

*(Hess, 1954: 346)*

Finally, he had linked mid-ocean ridges with rising mantle convection currents. The higher temperatures, brought about by molten basalt, caused isostatic uplift of the Mid-Atlantic Ridge. When convection stops, the ridge subsides (Figure 3.5).

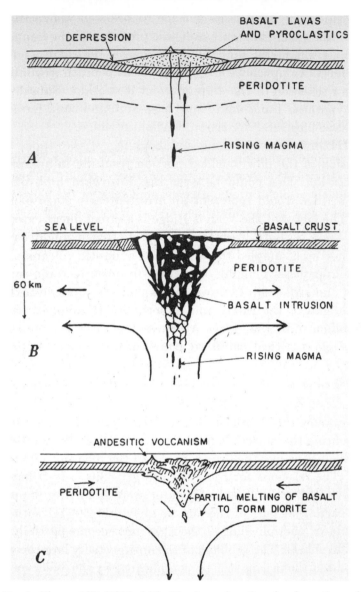

Figure 3.5 Hess's Figure 10B (1954: 345). Hess's explanation for formation of mid-ocean ridges. Upward mantle convection causes elevation of ridge; stopping of convection causes ridge to sink. Convection currents remain 60 km below, well within the upper mantle.

### 3.9 In the early 1950s Hess applies the olivine–serpentine transformation to formation of guyots and mid-ocean ridges

The two petrologists N. L. Bowen and O. F. Tuttle (1949) found that olivine + water ↔ serpentine + heat. At temperatures below 500 °C if water is present, the reaction proceeds to the right, above 500 °C it proceeds to the left as serpentine is transformed into olivine and water is released. Olivine and serpentine are both characteristic constituents of peridotite, serpentine being less dense than olivine. Serpentinization causes a column of peridotite to increase in height; deserpentinization causes a decrease. The reaction offered a reversible mechanism for vertical movements of oceanic crust. At the Royal Society of London's 1953 meeting Hess (1954: 346–347) used this to explain the origin of guyots, oceanic ridges, and oceanic crust.

He realized in 1953 that his 1946 solution to the guyot problem faced two obstacles whose combined effect was fatal to his earlier ideas. E. L. Hamilton, then a graduate student at Stanford, discovered Late Cretaceous shallow-water calcareous reef fossils atop guyots during the 1950 MIDPAC Expedition, which was jointly carried out by Scripps and the US Navy Electronics Laboratory (Menard, 1986: 54). This, plus the fact that previous estimates of sedimentation were now known to be three to five times too high, led Hess in 1953 to reject his 1946 solution. In 1955, he succinctly summarized the situation when he proposed a more detailed version of his new idea.

Originally the writer (1946) postulated that guyots were truncated in Precambrian time before lime-secreting organisms were available to protect them, and that their submergence was due to relative rise of sea level largely by sedimentation in the oceans. Hamilton's (1953) discovery of Late Cretaceous shallow-water fossils on them cuts the time to one-fifth, and recent investigation of the total amount of sediment on the sea floor cuts the rate of sedimentation by about one-fifth so that submergence as called for in the original hypothesis is 25 times too slow.

*(Hess, 1955a: 405)*

The problem of guyot development had become one of finding a mechanism for sinking guyots slowly enough to allow them to be truncated to sea level, but fast enough to sink them sufficiently far below sea level to prohibit reef formation. He realized that the reversible reaction between olivine and serpentine might provide this, and argued that such a possibility was supported by recent refraction profiles. Seismic wave velocities increase with the density of the transmitting material. Hess noticed that the deeper the water (and lower the ocean crust), the greater the velocity of the transmitted waves; the lower the ocean floor, the greater the density of the underlying mantle. Because olivine is denser than serpentine, he surmised that deeper seafloor contained less serpentinized peridotite than the shallower.

Now that we know the depth of the Mohorovičić under the oceans and have samples representing the material of the upper part of the mantle, some rather far reaching conclusions may be drawn based on the relation: olivine + $H_2O$ = serpentine + heat ... If water were gradually supplied from the interior of the earth that part of the mantle above the 500° isotherm should be subject to the above reaction going to the right and become partially serpentinized with an evolution of heat … The percentage of serpentinization might well be indicated by the small velocity differences which have been obtained below the Mohorovičić on various recent refraction profiles at sea. These velocities range from 7.9 to 8.25 km/s. and in general it seems, the deeper the water, the higher the velocity. Not enough profiles are as yet available, and the accuracy of the velocity measurements is perhaps not great enough to warrant such a conclusion, but if the above generalization is true another interesting deduction might be drawn. Serpentinization of olivine involves a density change from 3.25 to 2.60. Partial serpentinization of the material below the Mohorovičić would thus increase the height of the column. The fact that broad relatively flat areas of the ocean now stand at varying depths below sea level might be correlated with the degree of serpentinization of the underlying mantle.

*(Hess, 1954: 346–347)*

Hess had a new mechanism to sink guyots. But he also needed a way to change the depth of the 500 °C isotherm, and suggested rising convection currents or extensive basaltic volcanism.

Furthermore, were the isotherms to rise in a serpentinized area as might happen over the upward limb of a convection current or as a result of extensive basaltic volcanism, the serpentinization reaction would be reversed, water would be released to the ocean and the floor of the ocean would settle. In this mechanism we might find a ready explanation for the subsidence of the Pacific guyots in the Cretaceous as well as for atoll formation. The subsidence would of course be local and limited to the areas heated whereas any change in sea-level resulting from the water released would be very small inasmuch as it would contribute to rise of sea-level all over the Earth.

*(Hess, 1954: 347)*

Rising convection raises the 500 °C isotherm, deserpentinization follows, and ocean floor sinks; cessation of convection lowers the 500 °C isotherm, increases serpentinization, and ocean floor rises. Hess had *reversed* the role of mantle convection. Formerly he had supposed that rising convection caused crustal elevation of the Mid-Atlantic Ridge. So it is not surprising that he did not, at first, relate the olivine–serpentine transformation to formation of mid-ocean ridges. This would soon change.

Hess reintroduced the transformation in 1955.

If it be assumed that water and other volatile constituents are slowly leaking from the mantle of the earth, then such water upon crossing the appropriate isotherm, say 500 °C, would react with the olivine present to form serpentine and possibly some other hydrated phases in small amounts. A considerable volume increase would result, and the surface of the earth over this area would rise. As shown in Fig. 6 [reproduced as Figure 3.6] it might be expected that serpentinization would start at the appropriate isotherm below the Mohorovičić and gradually work up toward it as water was added. If as a result of convective overturn in the mantle or perhaps intrusion of basalt the isotherms were caused to rise, deserpentinization would occur at the base of the zone and serpentinization higher up without any necessary effect on the

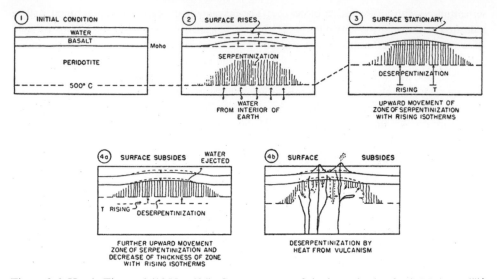

Figure 3.6 Hess's Figure 6 (1955a: 404). Consequences of the hypothesis of sub-Mohorovičić serpentinization by rising water from the mantle, deserpentinization by rising isotherms, and consequent effects on the submarine topography. Hess used the same illustration in his 1955 paper on oceanic crust (Hess, 1955b: 430).

surface topography. But once the top of the zone of serpentinization reached the Mohorovičić, then any deserpentinization at the base would cause the overlying topography to subside.

*(Hess, 1955a: 403–404; my bracketed addition)*

Applying the idea to explain the origin of the Mid-Atlantic Ridge, Hess rejected his earlier hypotheses.

Until recently most geologists would have supposed that the Mid-Atlantic Ridge was either a folded mountain system with consequent thickening of the crust above the Mohorovičić discontinuity or alternatively that it was a thick section of volcanic material lying on normal oceanic crust or intruded into it. Now it seems more likely that it represents a welt of serpentine. Serpentinized peridotite was dredged from large fault scarps on the Ridge by Ewing *et al.* (Shand, 1949). The flanks of the Ridge once stood much higher than today as indicated by paired terraces along its east and west slopes. Presumably some deserpentinization has occurred since maximum serpentinization. The problem of why serpentinization was concentrated in the Atlantic along a median line can perhaps be explained in several ways. One hypothesis could be that convective circulation in the mantle occurs, and the Ridge represents the trace of an upward limb of a cell. In this case water ejected from the top of the column might cause the later deserpentinization. Whatever hypothesis may be suggested to account for the localization of serpentinization along the Ridge is at present pure speculation. The idea that the topographic elevation of the Ridge may be due to serpentinization should be considered on its own merits apart from the above speculation.

*(Hess, 1955a: 404–405)*

Hess had reversed the direction of mantle convection. In 1953 he had argued that the Mid-Atlantic Ridge was caused by rising mantle convection; he now argued that rising mantle convection caused it to subside.

He proposed that the origin of other oceanic ridges might be explained in the same way. He also linked his solutions to the origin of guyots and oceanic ridges; some guyots, such as those among the Mid-Pacific Mountain Range, might have formed atop an ancient ridge that subsequently sank.

Such features as the mid-Atlantic Ridge, mid-Indian Ocean Ridge and several oval-shaped plateaus in the North Pacific and Indian Oceans might be attributed to water leakage from the interior of the earth under these areas and to consequent serpentinization with volume increase above the 500° isotherm. Rise of the 500° isotherm as a result of injection of basaltic magma, or perhaps convective overturn of the mantle, would cause deserpentinization and subsidence of the topography. In other words, the reaction is reversible. The submergence of the guyots of the mid-Pacific Mountain Range might be explained as a result of deserpentinization under a mid-Pacific ridge which once stood thousands of feet higher than it does today.

*(Hess, 1955b: 431)*

Although he did not explicitly claim that ridges evolve and are ephemeral, both ideas are implicit. Three years later, Menard, drawing on Hess's ideas, would argue that ridges evolve and are impermanent (§5.7). Hess did not first explicitly state that ridges are ephemeral until 1959 (§3.13).

## 3.10  Hess (1955) revises his theory of mountain formation from island arcs

The discovery that oceanic crust lacked a granitic layer and was only 5 km thick raised a serious difficulty for the tectogene hypothesis. In 1954, M. Ewing and Worzel argued that a 5 km downbuckled crust failed to produce enough mass deficiency to explain the observed negative gravity anomalies (RS2). Hess summarized their objection a year later.

The belt of huge negative gravity anomalies found in island arcs, as explained by Vening Meinesz as well as by Umbgrove, Kuenen, Griggs, Hess, and others, resulted from downbuckling or downbulging of the upper granitic crust which was then considered to be about 25 km thick. The mass deficiency of the downbulged granitic material as compared to the supposed basaltic material on either side of it was sufficient to account for the size of the anomalies and for the shape of the anomaly curve. In island arcs the mass distribution proposed in the above hypothesis is now known to be incorrect, as pointed out by Ewing and Worzel (1954) and Worzel and Ewing (1954). Substitution of new values for densities and thickness of layers, as required by recent seismic data, results in a new picture of the cross-section through the negative anomaly belt over trenches. The dimensions are changed, but qualitatively the structure remains the same.

*(Hess, 1955b: 432; Ewing and Worzel references are the same as mine)*

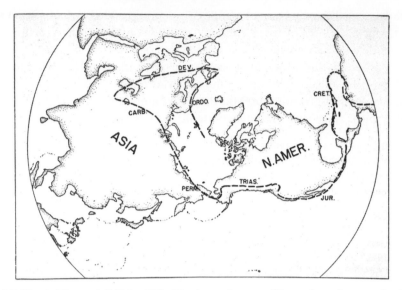

Figure 3.7 Hess's Figure 4 (1955a: 399). Hess's caption was "Extension of serpentine belts by segments along the strike of successive episodes of deformation." Serpentine belts marked former and existing mountain belts.

He offered two rebuttals (RS2). Appealing to a topographic and seismic study of the Tonga Trench by R. Raitt, R. L. Fisher and R. G. Mason at Scripps, Hess proposed that oceanic crust beneath trenches might still be thick enough to cause negative gravity anomalies. Their profiles indicated a thickened crust of 8 km, almost twice the thickness of normal ocean crust. (See §3.16–§3.20 for further discussion of Fisher's and Ewing and Worzel's analyses of trenches.) Encouraged by this doubly thickened crust but realizing that an even thicker crust was needed to explain the gravity deficiency, he proposed that sideways deflection of the seismic waves that give rise to the profile had failed to reveal the thickness of oceanic crust directly beneath the trench.

An excellent topographic and seismic survey was made over the Tonga Trench on the Capricorn Expedition of Scripps Oceanographic Institution. The seismic profile was reported on by Raitt *et al.* (1954). It shows a normal basaltic layer about 5 km thick away from the bottom of the Trench. The depth of the basaltic root under the Trench, however, may actually be considerably greater, inasmuch as the measurement may represent the slant distance to the Mohorovičić sideways at an angle of perhaps 45°. The vertical depth to the Mohorovičić could be much greater as shown in Fig. 4 and as suggested by the presence of andesitic volcanism in the island arc to the west. The critical point is that the basaltic layer does thicken and hence the structure is to be explained by lateral compression.

*(Hess, 1955b: 432)*

Hess had no independent evidence for his suggestion. He and Fisher would later write a joint paper on trenches (§3.20).

Worzel and Ewing (1954) unsuccessfully attempted to measure the thickness of oceanic crust at the bottom of the Puerto Rico Trench, but found only that its sedimentary layer was thicker than found elsewhere in the Caribbean. They also observed the usual negative gravity anomaly, attributing its presence to sediments in the trench. Hess disagreed. He argued that the anomaly would still be there even if there were no sediments, noting that the Tonga Trench had a negative gravity anomaly but no sediments.

Their statement that the anomalies can be explained largely by the mass deficiency of the great thickness of sediment, while true in a sense, is misleading, for if the sediments were not present then the anomalies would be still larger. At least half of the free air anomaly results from the mass deficiency of the water at the top of the column. In the case of the Tonga Trench one can estimate that the free air anomaly will be as large as or larger than that for the Puerto Rico Trench, though no sediments are present in the former.

*(Hess, 1955b: 434)*

He agreed, however, that a downbuckled crust did not completely compensate for the typically large negative gravity anomalies above trenches. Crust beneath trenches was not in isostatic equilibrium; some force prevented it popping up.

Considering the large horizontal area of the trenches, it is quite impossible that the crust could be this far out of isostatic equilibrium *unless a force were acting on it to keep it out of balance* ... This could mean a compressive force acting tangentially ... If this force were not acting then the trenches would rise to the equilibrium position.

*(Hess, 1955b: 435)*

Hess (1955b) left the cause of the force unspecified, although the year before he had proposed converging downward mantle convection to explain the formation of island arcs (Hess, 1954: 345).

He (1955b) continued to maintain that island arcs evolve into Alpine-type moun-tain structures. Admitting that their evolution extends over a much longer interval and is more complex than he had formerly believed, he argued that Alpine-type mountain belts form only from island arcs that impinge on continental margins. This second restriction allowed him to accept the new finding about oceanic crust, and still maintain that Alpine-type mountain belts arise from island arcs (RS1).

He began by discussing island arcs that had not evolved into Alpine-type moun-tain belts, island arcs that do not impinge on continents. The Tonga Trench, which lacked sediments, he claimed, represents the first stage in the evolution of an island arc. If deforming forces cease, trenches like the Tonga Trench become gentle ridges.

The Tonga Trench may be pointed to as the first stage of development. Here, if the deforming forces were to die out, the Trench would disappear and a gentle ridge 500 to 1000 m high on the sea floor would take its place. The thickened basaltic layer would then be floating in isostatic equilibrium.

*(Hess, 1955b: 436)*

The Puerto Rico Trench represented the next developmental stage; it became filled with a huge volume of undeformed sediments. The Barbados Ridge, whose sediments had been deformed, represents the third stage in the evolution of island arcs.

The next more advanced stage may be represented by the Barbados Ridge. It is still far out of isostatic equilibrium. A downward bulge, probably of sediments underlain by metamorphosed sediments and basaltic material, altogether perhaps 30 km thick, has developed, judging from isostatic anomalies. Active volcanism in the Lesser Antilles indicates a deep root. If the Barbados Ridge were relieved from compression, it would rise to form a low island chain with the rudiments of alpine structure. Probably, it would not, if arrested at this stage, form a high mountain range. The islands of Hispaniola and Cuba probably represent just the situation speculatively proposed above for the Barbados Ridge development.

*(Hess, 1955b: 436–437)*

Hess (1955b: 437) concluded that island arcs, which form in true ocean basins and which do not impinge on continental margins, are not destined to become Alpine mountain belts.

With the above reasoning, one may ask how one would ever develop a mountain system such as the Himalayas. The clue to this seems to be given by following the Antillean chain southward and westward into the Cordillera along the northern coast of Venezuela. Where island arc-type of deformation impinges on the edge of the continent, the thin sialic crust may be downbulged in exactly the manner postulated formerly by Meinesz for the negative strip of island arcs. The great Alpine-Himalaya chain probably began along a Mediterranean of oceanic character and transgressed in places on the border of a continent.

*(Hess, 1955b: 437)*

If deforming island arcs impinge on continental edges, their associated downbuckle is partly sialic, and so they evolve into Alpine-type mountain belts once downward convection ceases and isostatic equilibrium is achieved. Hess had found a way to retain Vening Meinesz, Kuenen's and his own downbuckled crust, turn some but not all island arcs into Alpine-type mountains, and yet accommodate the new findings about oceanic crust. Moreover, he had found a way to do it without invoking the collision tectonics of mobilists such as Argand, Staub, Holtedahl, Bailey, Collet, Wegmann, and F. E. Suess (I, §8.7–§8.13).

### 3.11  Hess (1955) continues to oppose mobilism

Although Hess wrote nothing specifically about mobilism in his published papers during the middle-1950s, I believe that he opposed it because he sought fixist interpretations of some major features. However, his was a gentle opposition, which worked in his favor. Unlike other major figures in North American geology, he never indulged in polemics against mobilism, and when he became a mobilist in 1959, he had little baggage to jettison.

As just shown, Hess explained the origin of the Alps and Himalayas without continental collision of Africa with Europe, and India and Asia. He also offered fixists an alternative solution to why early Paleozoic mountain ranges on both sides of the North Atlantic can be matched, a solution that grew out of his revised explanation of how island arcs evolve into mountains. He proposed that a crustal downbuckle formerly extended across the Atlantic and evolved into mountain belts only where it impinged on continental margins.

The apparent termination of some continental mountain systems abruptly at the sea may mean that their extensions on the ocean floor are inconspicuous and are perhaps covered at present by sediment, as indicated in the preceding paragraph [in which Hess explained how Cuba and Hispaniola formed]. This might explain why the early Paleozoic folded mountains seem to disappear at the coast of Newfoundland and reappear in northern Ireland and Scotland. Traces of this might be discovered by an increase in depth to the Mohorovičić along the line joining the two ends.

*(Hess, 1955b; my bracketed addition)*

Moreover, like most geologists who worked in North America and Australia, he believed in continental accretion, the formation of mountain belts from island arcs marginal to continental nuclei.

Looking at the continent of North America it can be seen at a glance that each successive serpentine belt is, at least roughly, concentrically outside the older belts of deformation. The development is not symmetrical but may occur first on one side of a continent and then on another. A concentric growth outward in the East Indian region is also evident. Crescentic outward growth from western Australia eastward to the youthful arcs from New Zealand to New Guinea is clear ... These cases were pointed out by the writer in 1937. In recent years J. T. Wilson (1949c) forcefully advanced the concept of continental growth by accretion along the margins as result of successive mountain-building epochs. Certainly the idea has merit. It should be pointed out that, although the accretion hypothesis may be valid, it does not necessarily mean that the continents have increased in horizontal dimensions. Much of the growth may have gone into thickening of their crust (Hess, 1954). Besides concentric growth, a growth by extension in successive segments along the strike is also commonly observed. In Fig. 4 [my Figure 3.7] is shown a series of such segments starting in Spitzbergen, crossing Europe and Asia, and following the western margin of North America and South America to Antarctica. The sequence starts in the Ordovician and ends in the Cretaceous. The development concentrically outward from the cores of continents takes place simultaneously.

*(Hess, 1955a: 398–399; my bracketed addition; Wilson (1949c)*
*is my Wilson (1949))*

### 3.12 Hess (1959) switches to mobilism because of its paleomagnetic support

Hess changed his mind about continental drift approximately a year *before* he came up with seafloor spreading. Precisely when he changed his mind is unclear, but it definitely occurred before July 6, 1959, when he wrote his old mentor, Vening

Meinesz, and told him that he had become a mobilist. Hess was impressed with mobilism's paleomagnetic support, and it was the main reason he became a mobilist. Moreover, he made it quite clear that earlier he had never favored mobilism. He also told Vening Meinesz that he had become interested in his work on spherical harmonics of crustal elevations (§2.5), and that he had developed a new explanation for oceanic ridges, which, I hasten to add, was not seafloor spreading.

Am much interested in your spherical harmonics. It seems to me that the lower orders relate to a *recent* dynamic situation within the Earth. Irving in the *Geophysical Journal*, March 1959, summarizes the paleomagnetic data. It is becoming almost compelling to accept continental drift, something I never favored. My own work on the nature and origin of great oceanic ridges such as the Mid-Atlantic suggests that they are ephemeral features with a life of something like $10^8$ years. Menard has shown their median position with regard to the 1000 metre line. If continents have moved the ridges are adjusted to present positions of the continents. High heat flow (6 times normal) from the crests of rising ridges such as the Mid Atlantic and Easter Island suggest present or recent past upward convection in the mantle under them. Some past ridges are marked by belts of guyots and atolls such as the so-called Mid Pacific mountain range. Here the subsidence has been almost 2 km on the average, and it started in the Cretaceous. This is where I derive the $10^8$ years life cycle. To sum up, this is why I believe the harmonics are related to a presently active dynamic system and probably related to recent convection in the mantle.

(Hess, July 6, 1959 letter to Vening Meinesz; Hess papers)

Irving's paper, "Paleomagnetic pole positions: a survey and analysis," appeared in March 1959. Although Hess might have been convinced of paleomagnetism's "almost compelling" support before seeing Irving's paper and merely identified it as a good paper for Vening Meinesz to read, I suspect that Irving's paper, a thorough review of mobilism's paleomagnetic support, which, I have argued, had become essentially difficulty-free by 1959, convinced Hess that continental drift was likely correct even though he had never before favored it. Regardless of precisely when Hess changed his mind, he did so, as I shall show, because of mobilism's paleomagnetic support and well before he developed his ideas about seafloor spreading. Hess was one of those superior scientists who could change his mind because of new information based on work in a field in which he had never worked. He also had the courage to adopt an unpopular position. Hess had met Irving at the 1957 IUGG meeting in Toronto. Although they did little more than introduce themselves to each other and did not discuss paleomagnetism, Hess might have heard Irving's talk at the meeting about Australia's APW path. Bullard, who chaired the session during which Irving presented his paper, was impressed with Irving's talk (II, §8.11). The room was packed, Irving argued with Rutten and Graham, and Irving cannot remember if Hess was there (Irving, May 2004 email to author). Hess was also friends with Runcorn. Perhaps his personal contacts with both convinced him that he should take their work seriously. But contacts aside, he was one of only a very few North American fixists who became a mobilist before 1960 because of its paleomagnetic support.

My explanation that Hess changed his mind about mobilism because of its paleomagnetic support is at odds with Carey's claim that he converted Hess to continental drift and seafloor spreading while visiting Princeton in 1959/1960. As already noted, Carey visited Princeton in 1959 and 1960 and claimed that when there he presented his views about seafloor generation and Earth expansion "*and had long discussions with Harry while a guest at his home on several occasions, and during these converted Harry to continental drift and ocean-floor spreading* (Carey, 1981 letter to author; emphasis is Carey's). Carey most certainly visited Princeton, was a guest at Hess's home, and talked Hess's ear off about continental drift and Earth expansion. Hess later wrote Carey in 1963 after Carey asked him to participate in a symposium at the University of Tasmania that was held in May 1963. He declined, and with his customary dry wit invited Carey to Princeton, if he decided to come again to the United States, and added, "I haven't quite recovered from your last visit even though brief. I have had to chuck out many long cherished ideas and replace them with new ones. Haven't yet found new ones for some of the gaps. Perhaps you could give me a hand" (Hess, March 15 letter to Carey). Even if it is assumed that Hess's reference to Carey's last visit was to the one in 1959, and that Hess's remarks were genuine and not merely a way to soften his rejection of Carey's invitation, I believe that Carey did not convince Hess that mobilism was correct. I think that Hess already thought mobilism was probably correct well before Carey visited Princeton. Carey spent the 1959–60 academic year at Yale University. Longwell had invited him at the close of Carey's Hobart symposium in 1956, but Carey was unable to go until the academic year 1959–60. Recalling his year at Yale and in the United States, Carey wrote:

In Yale I delivered complete courses in structural geology and global tectonics. But I also lectured in many other American universities, mostly under the American Geological Institute Visiting International Scientists Program: Brown, Columbia, Harvard, Wesleyan University, Lehigh, Princeton, Duke, North Carolina, Louisiana State, St. Louis, University of Cincinnati, and Ohio State, as well as Toronto, Western Ontario, McGill, Calgary, and British Columbia in Canada. As with Matthew's sower, some seeds did fall on fertile soil and took root, only to be choked off later when subduction weeds grew rank . . . Apart from Yale, my deepest involvement was with Princeton where I lectured several times in late 1959 and early 1960, including discussion of oroclines, the paleomagnetic evidence of large intercontinental movements, and ocean-floor growth by repeated insertion of paired slices at the mid-oceanic ridges as detailed in the Hobart Symposium (Fig. 19). Harry Hess, chairman of the Princeton geology school, and I cemented a warm friendship that deepened until his premature death.

*(Carey, 1988: 118–119; Carey's Fig. 19 is my Figure 6.18, Volume II)*

Hess wrote Vening Meinesz of his turn toward mobilism in early July 1959, two months before the 1959–60 academic year began, and Carey did not lecture at Princeton until late 1959. It was not Carey who converted Hess to continental drift; it was the favorable paleomagnetic results that moved him to think it likely.

Carey may, however, have reinforced Hess's belief in the strength of paleomagnetism's support of mobilism, and possibly helped him to see his way to seafloor spreading. Eldridge Moores certainly thinks so. Moores, then a graduate student at Princeton, was enrolled in Advanced General Geology, a required course for first year graduates. He also attended Carey's lecture, and witnessed its outward effect on Hess. Putting together what Hess had been saying about paleomagnetism before and after Carey's visit, Moores thinks that Carey "must be given the credit for 'pushing Hess over the edge.'"

When I enrolled as a graduate student at Princeton University in fall, 1959, one required course was called "Advanced General Geology." In it, several professors lectured on their current research. In his many comments, Hess and his colleague, A. F. Buddington, discussed polar wander paths, paleomagnetism, and rock magnetism. They emphasized the lack of understanding of the origin of rock magnetism, and argued that some of the apparent polar wander data could be the result of complications of the magnetization process. Hess also mentioned the seismic results for a thin oceanic crust, the evidence for only thin sediments in the ocean basins, and the recently discovered heat flow results ... Late in 1959 ... Carey came through to deliver a lecture on continental drift and earth expansion ... Carey gave a three-hour spell-binding lecture, ending completely spent, covered with sweat and chalk dust. At the end, we all filed numbly out of the room. Halfway through the talk, however, Hess bolted out of his seat and started pacing up and down the aisle. Thereafter in Advanced General Geology, there was no more talk of problems of paleomagnetism and polar wander paths. Within two months, Hess was circulating a manuscript, entitled "Evolution Ocean Basins" [sic], which was eventually published as "History of Ocean Basins" ... I believe that S. W. Carey must be given credit "for pushing Hess over the edge."

*(Moores, 2003: 21–22)*

Is there a way to reconcile Carey's and Moores' recollections with Hess's letter to Vening Meinesz? After showing Moores what Hess had written to Vening Meinesz, he was surprised, still thought Hess had expressed reservations about paleomagnetism's support of drift, but suggested a reasonable way out of the dilemma: "in the classroom" Hess "was more conservative than in his letter to Vening Meinesz."

The course was during the fall semester 1959. Several faculty members came in and talked about what was on their minds at the time ... I remember Hess principally ... Hess talked about the peridotites, the tectogene hypothesis, which he had had to modify because he had accepted the thin nature of oceanic crust. So his diagram of a tectogene was different from that he published in 1939 – rather a U-shaped furrow of oceanic crust lining oceanic trenches, as I remember. We did discuss Runcorn's and Irving's results, but until Carey's visit, Hess expressed reservations about polar wander paths and talked about the complications of paleomagnetism. So in the classroom he was more conservative than in his letter to Vening Meinesz.

*(Moores, May 24, 2010 email to author)*

So what moved Hess so much about Carey's lecture? Remember (II, §6.12), Carey maintained that he had presented what essentially became seafloor spreading in a paper he wrote in 1954 that had been rejected and never published and of which there is no extant copy. Moreover, Carey certainly discussed his account of seafloor generation as incorporated in his Earth expansion. Perhaps Carey said something that got Hess thinking about how to alter his 1955 explanation of the origin of the Mid-Atlantic Ridge, especially given that he was already reevaluating it. If so, I do not know what it was. It certainly was not the need to appeal to mantle convection because Hess, as witnessed by his letter to Vening Meinesz, had already concluded that convection was needed to explain high heat flow at ridged crests.

### 3.13 Hess (1959, 1960) reevaluates his views about ocean basins

Having become a mobilist in 1959, Hess was forced to reexamine his former views about the origin of various features of ocean basins: were they consistent with mobilism?[11]

Three abstracts and a paper of Hess's were published in 1959 in which he reevaluated his former opinions about oceanic crust, mid-ocean ridges and guyots in light of new findings. He predicted what he now thought coring through ocean crust into the upper mantle would reveal about their structure. It is, however, difficult to date precisely when he wrote them, and whether he did so before he told Vening Meinesz that he found mobilism "almost compelling" because of its paleo-magnetic support. The paper (Hess, 1959b), entitled "The AMSOC hole to the Earth's mantle" and quickly reprinted (Hess, 1960c), was, as Allwardt (1990: 166) noted, half science and half promotion. Hess briefly described the American Miscel-laneous Society (AMSOC) and outlined objectives of its proposed project to drill a hole through the Mohorovičić discontinuity. He further developed his 1955 hypoth-esis for the origin and development of mid-ocean ridges, introduced a new model for oceanic crust, offered a solution to the origin of its major layer and continued to speculate about why, as was becoming increasingly evident, the ocean floor was covered with surprisingly little sediment. Allwardt (1990: 209) argued that the paper (1959b) was written at least five months before it was published in December 1959, which suggests that it probably was written at or before early July when he confessed to Vening Meinesz that he had become a mobilist. Hess gave a talk at the first International Oceanographic Congress on either August 31 or September 1, 1959. Preprints of many abstracts were published prior to the Congress. His abstract (Hess, 1959a), I believe, was written before his letter to Vening Meinesz.[12] Hess also presented papers at the 45th Annual Meeting of the AAPG during April 1960 in Atlantic City, and at the 127th Meeting of the American Association for the Advancement of Science during December 1960 in New York – these had identical abstracts which grew out of his AMSOC 1959b paper (Allwardt, 1990: 209). Thus I will adopt the following chronology: his abstract (1959a) for the International Congress, his AMSOC paper (1959b), his letter to Vening Meinesz, and the two 1960 identical abstracts (Hess, 1960a, b). At the International Oceanographic Congress, Hess proposed a slightly amended version of his 1955 solution to the problem of mid-ocean ridges. He began by summarizing the mostly new data.

Recently Menard has shown that median ridges exist in all oceanic areas. Heezen has pointed out that the Mid-Atlantic ridge has a narrow longitudinal graben along its crest and that the shallow earthquake activities are concentrated below the graben. Investigators from Scripps (Maxwell and von Herzen) have found that the heat flow on the crest of some Pacific ridges is several times higher than normal. Gravity data obtained by Vening Meinesz and investigators from Lamont (Worzel, Ewing *et al.*) show that no conspicuous anomalies are found on ridges. Seismic data presented by Raitt, M. Ewing, J. Ewing, and others generally show that the M discontinuity cannot be found on the crests of ridges. Certain details of the seismic profiles on the flanks of the ridges gives rather clear insight into the nature of the ridges and their probable origin.

*(Hess, 1959a: 33–34)*

In 1955, Hess was undecided whether the rising of the 500° isotherm with the accompanying transformation of serpentine to olivine and subsequent subsidence of the Mid-Atlantic Ridge was caused by rising convection or upwelling basalt. He decided that the recently observed central graben and associated earthquakes, with its even higher temperatures, favored convection. Convection causes the 500° isotherm to rise, which causes deserpentinization and in turn subsidence of the ridge. Residual heat produced by convection, however, is conducted to the surface, and after convection ceases the 500° isotherm descends.

Assuming that the process forming the ridges is serpentinization which involves about 100 cal/g in heat evolved, the possibility was investigated that this might account for the high heat flow. It fails to do so by about two orders of magnitude. Somewhat more heat could be obtained by supposing it resulted from basalt intrusions into the ridge but this too fails by more than an order of magnitude.

The hypothesis is advanced that the ridges represent the trace on the Earth's surface of upward flowing limbs of mantle convection cells. Water released at the top of the column produces the serpentinization, subcrustal drag of the horizontal flow produces extension and the graben on the crust. Heat moving slowly upward by conduction accounts for the high flow and ultimately for deserpentinization and subsidence of the ridge.

*(Hess, 1959a: 34)*

Although he already had proposed that ridges subside, he now began to describe them as ephemeral. This was based in part, I believe, on his identification of the region of guyots in the southwestern Pacific as a dying and short-lived ridge, where its elevation and heat flow were lower than recorded on "young" ridges like the Mid-Atlantic. Hess's interpretation of regions of high guyot concentration as "old" ridges made perfect sense in light of his explanation of guyots.

There is some evidence that the great oceanic ridges may be ephemeral in nature. "Old" ridges of rather subdued topography such as the Mid-Pacific mountain range have abundant guyots and atolls indicating subsidence of 3.000 to 6.000 feet. On "young" ridges such evidence of subsidence is absent and high heat flow seems characteristic. The seismic profiles on "old" and "young" ridges seem to differ significantly.

*(Hess, 1959a: 34)*

He said nothing about mobilism or the paucity of seafloor sediments. Nonetheless he imagined ridges as evolving entities.

In his AMSOC paper Hess (1959b) presented a new model of oceanic crust. Arguing in favor of a three-layered oceanic crust, he identified the main layer (Layer 3), often called the ocean layer, the thickest and most uniform and most widespread, as serpentinized peridotite. In 1953/1955 he had suggested that it was basalt or serpentinized peridotite, but then favored basalt. In 1959 he changed his mind again, noting that seismic velocities allowed it to be either basalt or serpentinized peridotite.

Layer 3 is commonly referred to as "the crust" and is generally considered to be basalt. The reasons for calling it basalt are in part legendary, and in part based on its seismic velocity, which commonly ranges from 6.4 to 8.6 km/sec. Other than its seismic velocity there is no compelling reason for concluding it is basalt. It may be basalt or it may be serpentinized peridotite.

*(Hess, 1959b: 343–344)*

He opted for serpentinized peridotite.

Serpentinized peridotite is favored by the writer because such material has been dredged from fault scarps on the Mid-Atlantic Ridge in three places by investigators from Lamont Geological Observatory [Shand, 1949]. Basalts have also been dredged from this Ridge but could be attributed to debris from nearby seamounts or volcanic islands.

*(Hess, 1959b: 344)*

But in 1955 Hess knew about the abundance of serpentinized peridotite in dredge-hauls (§3.10). So there was a further reason for his choice: Layer 3 was of uniform thickness, unknown in 1955. He first discussed the new data.

The unique thing about Layer 3 is its comparatively great uniformity in thickness. Fig. 4 [which is not reproduced] is a plot of frequency of occurrences of thicknesses for this layer. This is based on all of the published seismic profiles in the deep sea plus some unpublished data of Raitt but omits those profiles which appear to be complicated by seamounts or islands in the immediate vicinity. Note that 81% of the cases in the sample examined range in thickness from 4.0 to 5.5 km. This range necessarily includes observational error which reasonably could be considered to be ±0.5 km. Raitt (personal communication) finds that the average thickness of Layer 3 for all of his profiles, selected in much the same way as the data for Fig. 4 were selected, comes to 4¾ km.

*(Hess, 1959b: 344–345; my bracketed addition)*

He then suggested that its uniform thickness not only ruled out magmatic generation of basalt but also placed an important constraint on the way in which Layer 3 had formed.

The surprising uniformity in thickness of Layer 3 requires that the bottom of the layer represent the position of an isotherm or past isotherm, and that this is a level at which a reaction or phase transition has taken place. If the layer were basalt flows, one would expect great variability in the thickness. Flows would be many times thicker near a vent or fissure from which they issued than at greater distances from their source.

*(Hess, 1959b: 345)*

The uniform thickness and wide extent of Layer 3 (sometimes called the ocean layer) indicated to Hess that it had to have been formed *in situ* – it was just too uniform to have been formed piecemeal and then transported to its present location.[13] With this explanation for Layer 3, Hess thought he had discounted the possibility that Layer 3 was basalt.

This leaves two alternatives: (1) that Layer 3 is basalt but that rocks of this chemical composition extend down into the mantle and are converted to eclogite ... or (2) that the "crust" and material below are peridotitic in composition and an abrupt change from partially serpentinized peridotite to unserpentinized peridotite occurs at the Moho.

*(Hess, 1959b: 345)*

Needless to say, Hess preferred serpentinization. The upper mantle, which he still believed was olivine peridotite, was transformed into serpentinized peridotite creating crustal Layer 3. He suggested that serpentinization of the upper mantle occurred at a much earlier time when Earth was much hotter.

In this case the Moho under the oceans would represent some ancient time when the 500 °C isotherm stood at this datum plane below sea level. A glance at Fig. 1 [which illustrated relationships between heat floor, conductivity and temperature at depths of 5 to 6 km below the ocean floor] shows that it must be much deeper than this today.

*(Hess, 1959b: 345; my bracketed addition)*

When Earth was much hotter, the 500 °C isotherm would have been considerably closer to the surface than it is now, and above it, water escaping from Earth's interior would have reacted with the olivine peridotite and formed a primordial crust of serpentinized peridotite. Allwardt discovered a report issued by the AMSOC Committee in September 1959 that further explains Hess's view about the formation of the primordial crust.

Some scientists believe that the Moho is an abrupt change, perhaps representing the original surface of the earth and that the materials immediately above it are later volcanic outpourings. Others believe that the Moho is a transitional zone perhaps representing a phase change or a *"frozen isotherm" that developed as the surface of the earth cooled. If the latter is true, then the top of the deep crustal [sic] rather than the top of the mantle may be the primordial surface of the earth.*

*(Allwardt, 1990: 167; emphasis is Allwardt's)*

The report is unsigned, but Allwardt (1990: 209–210) convincingly argues that even if Hess did not write the report, he would have to have approved it before its release. Surely, Hess's view is the latter one. It should also be noted that this hypothesized "deep crust" would have covered the entire earth: continents would have formed later.

   He also modified his view about seafloor sediments. Hess (1955b: 428) had previously claimed that oceanic crust contained only one layer of sediments, although he thought that some might have become consolidated into the main layer of oceanic crust. He now argued that there was an upper layer, Layer 1, consisting of

unconsolidated sediment and a lower one, Layer 2, comprising "consolidated sedimentary rocks or volcanic rocks or both" (Hess, 1959b: 260).

Hess, nevertheless, remained puzzled about the paucity of sediments on ocean
basins.

> The total thickness of Layers 1 and 2 is of the order of 1.3 km and if considered to be all
> sedimentary rock it is surprisingly small in amount considering present-day rates of sedimen-
> tation. Measurements made on cores of this commonly give a rate of about 1 cm/1000 yr. If the
> oceans are postulated to be three billion years old, this would mean 30 km of sediment at the
> faster rate or 3 km at the slowest. The quaternary may be abnormal in its contribution of
> sediment to the sea because of Pleistocene glaciation, but this argument does not seem
> particularly convincing to account for a rate of perhaps 50 times normal.
>
> *(Hess, 1959b: 343)*

Hess suggested three possibilities.

> The most obvious alternatives are: (1) The oceans are relatively young. At 1 cm/100 yrs the
> sediment could be accounted for if sedimentation only started in the Cretaceous. (2) The pre-
> Cretaceous sediments have in some manner been removed; for example, by incorporation into
> the continents by continental drift. (3) Nondeposition of any sediment over much of the ocean
> floor was a common attribute of the past. In any case those who expect a complete record far
> back to billions of years ago are doomed to disappointment. It will be extremely interesting
> when the well is drilled to find out which of these alternatives (if any) prove to be correct. In
> any case I would predict (though I am rooting against this prediction) that a very incomplete
> section will be found.
>
> *(Hess, 1959b: 343)*

This contains, I believe, Hess's first published mention of continental drift since his
passing reference to it in his 1946 guyot paper (§3.5). In the year to follow, Hess would
choose an option that he had not yet thought of, that it was ocean floors not oceans
themselves that are very young, being continually created through seafloor spreading.
His suggestion that continental drift had somehow removed pre-Cretaceous sediments
was not seafloor spreading; he was referring to the idea that continents lost sediments
as they plowed through the ocean floor. His comment that he was "rooting against"
his own prediction that a very incomplete section would be found may seem puzzling.
Like many geologists, he would have loved to find a complete and uninterrupted
sedimentary column, but by early 1959 he seems not yet to have quite freed himself
from a nagging prejudice against continental drift. By mid-1959, however, as shown
by his July 6, 1959 letter to Vening Meinesz (§3.12), he had accepted mobilism as a
possibility but had not yet come up with seafloor spreading. His view of mid-ocean
ridges was essentially the same as in the abstract for his talk at the First International
Oceanographic Congress in which he had accented the ephemeral nature of ridges,
their high heat flow, and had settled on $10^8$ years for their lifecycle. Heat flow
measured at the crest of ridges was six times greater than that from other parts of
the seafloor, which suggested "present or past upward convection in the mantle under

them" (Hess, July 6, 1959 letter to Vening Meinesz). Having adopted continental drift, Hess had something new to add about oceanic ridges, which was: "Menard has shown their median position with regard to the 1000 metre line. If continents have moved the ridges are adjusted to present positions of continents" (Hess, July 6, 1959 letter to Vening Meinesz). This is not seafloor spreading. There is no suggestion that what occurs at mid-ocean ridges is connected to continental drift. There is no suggestion that Layer 3 is created at mid-ocean ridges, and that continents split apart and drift away from each other on the backs of convection currents as he was later to propose. At this juncture, Hess was thinking that ridges somehow adjust their position in response to continental drift.

Hess's letter to Vening Meinesz also reveals that he had begun to think about the distribution of crustal surfaces. During the early 1950s, Vening Meinesz (§1.5) had combined his idea of mantle convection with A. Prey's analysis of surface elevations expressed as spherical harmonics, and to explain them he proposed that there had been episodes of intensive convection separated by long quiet periods. According to Vening Meinesz, during the first episode a single huge convective cell brought about the separation of core and mantle, and the formation of a sialic proto-continent. Subsequent, but still very ancient, and increasingly higher order convective cells then split apart the protocontinent into the present-day continents and caused polar wandering. These episodes all occurred well before Wegener's continental drift supposedly happened. Later still, according to Vening Meinesz, higher ordered convection caused and continues to cause small-scale, smaller features such as island arcs and mountain belts. Hess, "much interested" in Vening Meinesz's "spherical harmonics," wanted to apply lower order convective cycles to "a *recent* dynamic situation within the Earth." He wanted to introduce the idea that lower order "harmonics are related to a presently active dynamic system and probably related to recent convection in the mantle" (Hess, July 6, 1959 letter to Vening Meinesz). Vening Meinesz thought that continents had split apart when Earth was very young. Unlike Vening Meinesz, Hess had become a mobilist and thought that major continental drift was still occurring. Although Hess did not know *how* convection and mid-ocean ridges related to continental drift, he thought they were related.

Hess's other two abstracts, the identical ones, show that at the time of writing he still had not thought of seafloor spreading.

The major scientific objective of the AMSOC project is to obtain samples of the upper part of the earth's mantle and to determine the nature of M discontinuity ... Above the [M] discontinuity is a layer commonly called the "crust" and generally considered to be basalt. It is extraordinarily uniform in thickness, suggesting that its base represents a phase transition and that the discontinuity under the oceans might represent the level of some isotherm, or past isotherm, at which a reaction took place. Above the "crust" is a layer of variable thickness and seismic velocity thought to be consolidated sedimentary or volcanic rocks. Finally one comes to the unconsolidated sediments of the ocean floor, which are a few hundred meters thick and no doubt resemble the material obtained from shallow cores. In these last two layers one might hope to find fossils going far back in the history of the oceans and to derive from this record information of extraordinary scientific importance.

The following predictions are made:

(1) The mantle will be peridotite resembling the olivine nodules found in basaltic volcanoes and St. Paul's Rock.
(2) The "basalt crust" will be serpentinized peridotite, hydrated mantle material.
(3) The M discontinuity represents a past isotherm above which serpentine was a stable phase.
(4) The sedimentary column will be very incomplete and have many great hiatuses.

*(Hess, 1960a: 1250; 1960b: 2097; my bracketed addition)*

Hess still believed that Layer 3, the dominant layer of oceanic crust, had been formed *in situ* when the Earth was young, and that its formation was unrelated to the origin and development of mid-ocean ridges.

### 3.14 Hess (1960) comes up with seafloor spreading

Hess (1960d) presented his seafloor spreading hypothesis in December 1960 in a preprint entitled "Evolution of ocean basins." There is not enough information to determine with assurance how he developed key aspects of his new solutions to ridge elevation and the formation of the oceanic crust (Layer 3). Some things, however, are definitely known. First, he did not change his mind because he learned something new about the seafloor; he had all the oceanic information he needed in 1959. Second, he rejected his 1955/1959 serpentinization solution for the rise and fall of mid-ocean ridges because he recognized that it faced a severe difficulty, as will be explained in a moment. Third, he reversed yet again the role convection had in changing ridge-elevation; with his serpentinization solution, rising convection currents explained the sinking of ridges; with his new solution (as in his 1953 solution), rising convection caused ridge-elevation, and the cessation of convection caused ridge-sinking. Fourth, as in his 1955 explanation, he thought the ocean layer (Layer 3) is serpentinized peridotite and was created above the 500° isotherm by hydration of olivine peridotite, and it is this process that explains its uniform thickness. The latter is the only feature his 1955 solution and seafloor spreading have in common. By seafloor spreading, new oceanic crust is created at ridges where water is released from the mantle over the rising limbs of convection currents, and it is these currents that raise the 500° isotherm to within 5 km of the surface. Newly created oceanic crust is carried away from ridge crests by the horizontal limbs of convection currents, thereby extending new seafloor and making room for further hydration of peridotite at ridge crests. Thus convection plays a huge role in the creation and lateral movement of new oceanic crust. Fifth, Hess had already become a mobilist and had begun thinking in terms of huge mantle convection cells before he developed his new explanation of ridge elevation and the formation of oceanic crust.

Hess (1960d) began his discussion of seafloor spreading by summarizing what was known about the upper mantle, oceanic crust, and mid-ocean ridges. Although his discussion was more detailed than it was in 1959, he reported no new data. What led to his change of mind was not new information, but a serious difficulty with his 1955/1959 solution, a difficulty he had not recognized when he wrote his abstract for the first International Oceanographic Congress.

The most significant information on the structural and petrologic character of the ridges comes from refraction seismic information of Ewing and Ewing (1959) on the Mid-Atlantic Ridge and Raitt's (1956) refraction profiles on the East Pacific Rise. The sediment cover on the Mid-Atlantic Ridge appears to be thin and perhaps restricted to material ponded in depressions of the topography. On the ridge crest layer 3 has lower than normal seismic velocity ranging commonly from 4 to 5.5 km/sec instead of 6 to 6.9 km/sec. The M discontinuity is not found or is represented by a transition from layer 3 to velocities near 7.4 km/sec. Normal velocities and layer thicknesses, however, appear on the flanks of ridges.

Formerly the writer (1955, 1959a) attributed the lower velocities (ca. 7.4 km/sec) in what should be mantle material to serpentinization, olivine reacting with water released from below. The elevation of the ridge itself was thought to result from the change in density (olivine 3.3 g/cc to serpentine 2.6 g/cc). A 2 km rise of the ridge would require 8 km of complete serpentinization below, however a velocity of 7.4 km/sec is equivalent to only 40% of the rock serpentinized. Thus serpentinization would have to extend to 20 km depth to produce the required elevation of the ridge. But this reaction cannot take place at a temperature much above 500 °C which considering the heat flow probably lies at the bottom of layer 3, about 5 km below the sea floor, and cannot reasonably be 20 km deep.

> *(Hess, 1960d: 10–13; Hess (1955) is my Hess (1955b),*
> *other references are the same as mine)*

He no longer could explain the elevation of ridges by appealing to serpentinization of the upper mantle. Serpentinization occurs only at temperatures below 500 °C. Heat-flow data at ridges indicate that the 500 °C isotherm is approximately at the bottom of Layer 3, 5 km below the surface, and at this depth seismic velocities are 7.4 km/sec indicating only 40% serpentinization; for the ridge to be elevated the needed 2 km a block of 40% serpentinized peridotite would have to extend 20 km below the surface. A 5 km thick block of 40% serpentinized peridotite could account for only one quarter of the observed elevation of the ridge.

Hess had already thought about the high heat flow observed at ridge crests when he wrote the abstract for his talk to the 1959 Oceanographic Congress. When in July 1959 he told Vening Meinesz (§3.12) that he had become a mobilist, he also told him that convection was needed to explain heat flow at ridge crests: "High heat flow (six times normal) from the crests of *rising* ridges such as the Mid-Atlantic and Easter Island suggests present or recent past upward convection in the mantle under them" (Hess, July 6, 1959 letter to Vening Meinesz; my emphasis). Even after invoking convection to explain the heat flow at the crests of rising ridges, he did not immediately realize the conflict between this and serpentinization as an explanation of ridge-elevation. Allwardt (1990: 171) makes this point about Hess's abstract for the 1959 Oceanographic Congress: although in this abstract Hess still invoked serpentinization to explain ridge elevation (that is, convection causes the *sinking* of *old* ridges), he also invoked convection to explain the heat flow found at crests of *young* ridges, which, of course stand high.

Allwardt (1987) also had the good sense to ask Carl Bowin if he remembered what Hess was thinking about in late 1959. Bowin is the only person Hess thanked "for critical evaluation of a number of the ideas discussed" in his preprint. Here is what Bowin recalled.

Harry and I had many conversations in those months preceding December 1960. I had completed my dissertation in September 1960 (Bowin, 1960, 1966), and at Harry's request, I remained on at Princeton as an instructor for a year. In Harry's initial preprint, I was the only person named in the acknowledgments, and from time to time I have been asked for insight into Harry's development of his ocean spreading hypothesis. From my review of letters and preprint and reprint materials, I can now, I believe, see a link between Harry's and my conversations and the development of his hypothesis ... Using a CDC 1604, I was attempting in 1960 to calculate the time it took for the 1½-inch long chiastolite crystals to grow during contract metamorphism in the wall rock of a gabbro intrusion in northern Maine that I had mapped for a Master's thesis for Northwestern University. The heat-flow conduction, temperature, and time relations being determined also had impact upon our discussions of Hess's serpentinization hypothesis for ridge elevation. The increase in temperature gradient and a rise in the 500° isotherm above an upwelling convection cell to a very shallow level, i.e., above a 20-km depth required for uplift of the ridge by serpentinization. Hence, I now infer that prior to our heat-flow discussions, Harry's primary concern had been to raise the ridges and later have them go back down, and that serpentinization had been the consensus mechanism.

*(Bowin, 1987: 475; Bowin references are the same as mine)*

Bowin probably helped Hess realize that his 1955/1959 serpentinization solution was no longer tenable; he certainly seems to have forced Hess to think about whether the new heat-flow data created difficulties for his old solution.

So why did Hess have trouble grasping this difficulty? Allwardt (1990) provides a plausible answer. With his 1955/1959 solution, rising convection causes old ridges to sink; with seafloor spreading it causes young ridges to rise. Hess reversed convection's key role in explaining ridge elevation. In his 1959 solution, watering above the 500 °C isotherm indicated rising convection currents below, currents that did not reach the 500 °C isotherm until the area below the Moho had already been hydrated. With his new solution, upward convection caused young ridges to rise; upward convection was now associated with young ridges and water reached the top of the mantle as ridges elevate.

Although I believe Allwardt is correct, there is an additional point. Hess had once before reversed the role of convection in the formation of mid-ocean ridges, for he had appealed to convection in 1953 to raise ridges (§3.8) when he proposed that upward convection would raise a ridge and cessation of the current would cause it to sink. So it might seem that the prospect of once again reversing the role of mantle convection would not have troubled him. Hess's view of convection in 1960, however, was fundamentally different from that in 1953, when he envisioned a tiny convective cell that remained submerged 60 km in the upper mantle and was linked with upwelling basalt (see Figure 3.5).

Having explained the difficulty with serpentinization, Hess (1960d) introduced some elements of his solution to the formation of Layer 3 and gave his new solution to the rise and fall of mid-ocean ridges.

Layer 3 is thought to be peridotite 70% serpentinized. It would appear that the highest elevation that the 500 °C isotherm can reach is approximately 5 km below the sea floor and this supplies the reason for the very uniform thickness of layer 3. The cause of the actual elevation of the ridge will be considered in the succeeding section.

*(Hess, 1960d: 13)*

Hess learned from his mistakes. Once he understood that the critical 500 °C isotherm is at a depth of only 5 km, he realized that his idea of serpentinization, as an explanation of the rise and fall of ridges, was doomed. However, he also realized that serpentinization at ridges offered a solution to the problem of the origin of Layer 3. The very fact that the 500 °C isotherm critical to serpentinization of peridotite is only about 5 km below the surface now helped Hess see that serpentinization at ridges might explain the formation of Layer 3.

There was more to Hess's solution to the formation of Layer 3 than its creation; for he had to find a way to move newly created 5 km thick slices away from ridge axes without altering their thickness. Mantle convection again was his answer, and once again Hess had to reconfigure its role. As just pointed out, when Hess had invoked convection in 1953 to explain why ridges are high, he had been thinking in terms of small convective cells that remained 60 km below the surface. Hess, as his July 6, 1959 letter to Vening Meinesz indicates, had already begun thinking about huge convective cells which he linked to Vening Meinesz's ideas about topographic spherical harmonics, and he now envisioned convection as extending up to the Moho where he placed the critical 500 °C isotherm. Moreover, Hess in this letter to his old mentor shows that he had become a mobilist, had linked mantle convection with mobilism and the formation of oceanic ridges, and also reveals that he did not know how the linkage works: if anything, he had it backwards, thinking that mid-ocean ridges owed their position to continental drift. He wrote, "If continents have moved the ridges are adjusted to present positions of the continents." Hess thought the answer to the median position of mid-ocean ridges was embedded within his idea that newly created seafloor (Layer 3) is continually moved off ridge crests. Moreover, his realization that convection reached the Moho, I believe, was instrumental in positioning him to come up with the idea that convection moves newly formed 5 km thick blocks of oceanic crust away from ridge crests, thereby creating a new Layer 3 and making way for the serpentinization of more rising peridotite along ridge axes.

Once he had his solution to the horizontal movement of newly formed Layer 3, I suspect that he soon realized he had an explanation for the median position of ridges and, perhaps, a solution to mobilism's mechanism problem.

## 3.15 Hess (1960) explains how seafloor spreading solves many problems

We have a great deal to talk about as is usual after a year, in particular the ideas on spherical harmonics of the Earth's topography. I have been thinking about this a lot and developed a number of tangent hypotheses. O'Keefe is talking here next week on harmonics of the gravity field as determined by satellite observations.

*(Hess, February 19, 1960 letter to Vening Meinesz)*

Considering the present keen interest in your work on spherical harmonics of the Earth's topography and your recent extension of the ideas presented here in December 1958, would you be willing to add to the manuscript such new ideas and data as have accumulated in the intervening time?

*(Hess, May 25, 1960 letter to Vening Meinesz)*

I believe the spherical harmonics of Earth's topography is an extremely important new development perhaps equaling gravity in island arcs.

*(Hess, July 11, 1960 letter to Vening Meinesz)*

After admitting that his hypothesis was controversial, his first line of defense was to note that it was in good company. After all, other hypotheses about the formation of continents and oceans were equally controversial. He characterized his essay as geopoetry, borrowing from Umbgrove (I, §8.14). By so categorizing his essay, Hess may have hoped that readers would be more likely to forgive him for proposing solutions based on dubious assumptions by making plain that he was not trying to hide the controversial nature of his hypothesis. He also noted that his own speculations were not entirely undisciplined because he would hold as best he could to uniformitarianism.

The birth of the oceans is a matter of conjecture, the subsequent history is obscure and the present situation as to structure is just beginning to be understood. Fascinating speculation on these subjects has been plentiful but not much of it predating the last decade holds water. Little of Umbgrove's (1947) brilliant summary remains pertinent when confronted by the relatively small but crucial amount of factual information collected in the intervening years. As did Umbgrove in his chapter, I shall consider this chapter an essay in geopoetry. In order not to travel any farther into the realm of fantasy than is absolutely necessary I shall stick as closely as possible to a uniformitarian approach; even so, at least one great catastrophe will be required early in the Earth's history.

*(Hess, 1960c: 1)*

Hess began at the beginning, adopting Urey's views that primordial Earth rapidly condensed to a solid planet losing most of its original mass. Because condensation was so rapid, the newly solid planet had little atmosphere, no oceans, and a low

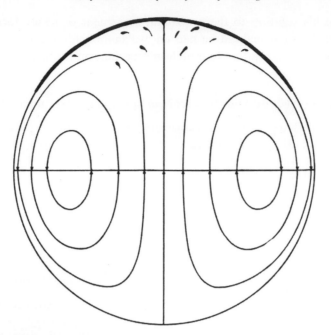

Figure 3.8 Hess's Figure 1 (1960d: 6). Two-cell convective overturn of Earth's interior after Vening Meinesz. Continental material extruded over rising limb but would divide and move to descending limb if convection continued beyond a half cycle. Possible segregation of liquid iron core might prevent more than a half cycle.

surface temperature. Once formed, trapped radioactive material caused Earth's temperature to rise, which decreased its strength and led to what Hess called "The Great Catastrophe." Hess, like Runcorn (II, §9.5), appealed to Vening Meinesz's ideas about spherical harmonics of topography and Chandrasekhar's thoughts about two-cell convection; he suggested a one-time two-cell convective overturn that separated core from mantle and formed a single huge primordial continent.

The stage is now set for the great catastrophe which it is assumed happened forthwith. A two-cell convective overturn took place (Vening Meinesz, 1952) resulting in the formation of a nickel-iron core and at the same time extruded over the rising limbs of the current, the low melting silicates to form the primordial single continent (fig. 1 [reproduced as my Figure 3.8]) ... The proposed two-cell overturn brought about the bilateral asymmetry of the Earth, now possibly much modified but still evident in its land and water hemispheres. After this event, segregating the core from the mantle, two-cell convection was no longer possible in the Earth as a whole (Chandrasekhar, 1953).

*(Hess, 1960d: 3; my bracketed addition)*

Hess then estimated the percentage of continental crust and ocean water that would have been forced to the surface by the original convective overturn. Assuming uniformitarianism, and, using estimates of how much continental material had surfaced through volcanic activity over the past 400 years, he (1960d: 4) proposed, "approximately 50% of the continents had been extruded during the catastrophe." He concluded further that the "best guess that the author can make is that something between zero and 1/3 of the oceans appeared on the surface during the catastrophe."

He ended his account of primordial Earth by remarking that his geopoetry had been tempered by explicit and reasonable assumptions.

We have now set the stage to proceed with the subject at hand. Dozens of assumptions and hypotheses have been introduced in the paragraphs above to establish a framework for consideration of the problem. The writer has attempted to choose reasonably between a myriad of possible alternatives. No competent reader with an ounce of imagination is likely to be willing to accept all of the choices made. But unless some such set of confining assumption is made, speculation spreads out into limitless variations and the resulting geopoetry has neither rhyme nor reason.

*(Hess, 1960d: 5)*

Looking ahead, those who later came to accept seafloor spreading divorced it from Hess's musing about Earth's early history, but he felt it necessary to try to set the opening stage. His 1959 and 1960 letters to Vening Meinesz show that he considered spherical harmonics of present topography important, which he linked to mantle convection, which itself was central to his account of the evolution of ocean basins. Mantle convection as a cause of seafloor spreading was speculative but not unreasonable, and Hess defended it by showing that it could explain the evolution of ocean basins.

Convection currents in the mantle were long ago suggested by Holmes to account for deformation of the Earth's crust (Vening Meinesz, 1952; Griggs, 1939, 1954; Verhoogen, 1954 and many others). Nevertheless mantle convection is considered a radical hypothesis not widely accepted by geologists and geophysicists. If it were accepted a rather reasonable story can be constructed which describes the evolution of ocean basins and the waters within them. Whole realms of previously unrelated facts fall into a regular pattern which is highly suggestive that close approach to satisfactory theory is being attained.

*(Hess, 1960d: 13–14)*

To begin, he argued that seafloor spreading explains the formation of Layer 3 and the elevation of ocean ridges. Seafloor spreading explained the main features of the seafloor, and is a key element of mobilism, offering a possible mechanism for drift. Unlike almost every North American Earth scientist raised on fixism, Hess recognized the strength of the paleomagnetic case for mobilism. Characterizing this support as "compelling," the same word he had used in his July 1959 letter to Vening Meinesz (§3.12), Hess explained why mobilism was acceptable.

Paleomagnetic data presented by Runcorn (1959), Irving (1959) and others strongly suggest that the continents have moved by large amounts in geologically comparatively recent times. One may quibble over the details but the general picture on paleomagnetism is sufficiently compelling that it is much more reasonable to accept it than to disregard it. The reasoning is that the Earth has always had a dipole magnetic field and that the magnetic poles have always been close to the axis of the Earth's rotation which necessarily must remain fixed in space. Remanent magnetism of old rocks shows that position of the magnetic poles has changed in a rather regular manner with time but this migration of the poles as measured in Europe, North America, Australia, India, etc. has not been the same for each of these land masses. This strongly indicates independent movement in direction and amount of large portions of the Earth's surface with respect to the rotational axis. This could be most easily accomplished by a convecting mantle system which involves actual movement of the Earth's surface passively riding on the upper part of the convecting cell. In this case at any given time continents over one cell would not move in the same direction as continents on another cell.

> *(Hess, 1960d: 15; Hess's references are equivalent to my Runcorn,*
> *1959c and Irving, 1959)*

By combining seafloor spreading with continental drift, he could provide answers for many problems. Viewing the evidence for separation of Africa and South America as the strongest argument for drift, he proposed an average spreading rate on either side of the mid-Atlantic ridge of 1 cm/yr based on separation of South America from Africa since the end of the Paleozoic at a rate of 2 cm/yr. This rate falls within those based on paleomagnetism as well as those based on rates of extension of the Mid-Atlantic rift zone in Iceland.

The rate of motion suggested by paleomagnetic measurements lies between a fraction of a cm/yr to as much as 10 cm/yr. If one were to accept the old evidence which was the strongest argument of the continental drifters, namely the separation of South America from Africa since the end of the Paleozoic and apply uniformitarianism, a rate of 1 cm/yr results. Heezen (1960) mentions a fracture zone crossing Iceland on the extension of the Mid-Atlantic rift zone which has been widening at a rate of 3.5 m/1000 yrs.

> *(Hess, 1960d: 15; Heezen reference is my 1960b)*

Hess's integration of continental drift and seafloor spreading explained the median position of most mid-ocean ridges.

Menard's theorem that mid-ocean ridge crests correspond to median lines, now takes on new meaning. The mid-ocean ridges could represent the traces of the rising limbs of convection cells ... The Mid-Atlantic Ridge is median because the continental areas on each side of it have moved away from it at the same rate – a centimeter a year.

> *(Hess, 1960d: 16)*

Hess then offered a new solution to mobilism's mechanism problem, quite different from Wegener's.

This is not exactly the same as continental drift. The continents do not plow through oceanic crust impelled by unknown forces, rather they ride passively on mantle material as it comes to the surface at the crest of the ridge and then moves laterally away from it.

> *(Hess, 1960d: 16)*

Of course, over thirty years before, Holmes had already provided mobilists with a very similar solution to the mechanism problem of continental drift. But, Hess said of Holmes only that he, like Vening Meinesz, Griggs, Verhoogen, and others, had proposed mantle convection to account for deformation of Earth's crust, without even mentioning that Holmes, unlike the others who were not then mobilists, specifically had invoked mantle convection as the cause of drift. Once again, Holmes' work was ignored. Even though there were differences between the two hypotheses, Holmes' 1944 version (I, §5.8), which appeared in his *Principles of Physical Geology*, was sufficiently similar to Hess's seafloor spreading that Hess surely should have given him more credit. Hess also did not mention Carey. Although Carey did not convert Hess to mobilism, he might have influenced his thinking in important ways as Moores claimed he did (§3.12).

Seafloor spreading explained other features of mid-ocean ridges. Hess noted that ridges "have unusually high heat flow along their crests," that "a median graben exists along the crests of the Atlantic, Arctic and Indian ocean ridges," and that "shallow depth earthquake foci are concentrated under the graben" (1960d: 9–10). Rising convection currents beneath ridges explained their high heat flow, and according to Hess, the central graben and its shallow earthquakes were caused by the movement of newly created seafloor away from ridge axes carried on the backs of diverging horizontal limbs of convection currents. "Mid-ocean ridges are ephemeral features having a life of 200 to 300 million years (the life of the convecting cell)" (Hess, 1960d: 32). Thus seafloor spreading driven by convection explained the elevation, central graben, high heat flow, and ephemeral nature of oceanic ridges, and when combined with continental drift, explained their median position.

Hess (1960d: 32) then sought to explain four general features of the seafloor: "the relative thin veneer of sediments of the seafloor, the relative small number of volcanic sea-mounts," and the absence "of rocks older than Cretaceous in the oceans," and of ancient oceanic ridges. These, he argued, arose because they were relatively young – 300 to 400 Ma. Sediments could be thin because deposition rates were much lower in the past, and perhaps oceans did not appear until the Paleozoic. But these clashed with uniformitarianism, and he turned the reasoning inside-out. Seafloors, he said, were short-lived; oceans, ancient.

One gets the impression from reported values of the thickness of deep sea oceanic sediments that the rate of sedimentation has been a few millimeters per thousand years ([E. L.] Hamilton, 1960). At a rate of 3 mm/1000 years in 4 aeons the thickness should be 12 km (perhaps 9 km after compaction) instead of a thickness of about 1.3 km found by seismic refraction. This one order of magnitude discrepancy has led some to suggest that the water of the oceans may be very young, that oceans came into existence largely since the Paleozoic. This violates uniformitarianism to which the writer is dedicated.

*(Hess, 1960d: 16–17; my bracketed addition)*

Alternatively, during the Pleistocene glaciations the rate of sedimentation could have greatly increased, and the average rate generally could have been much less; Hess was skeptical of this, but did not explain why. He then explained why the seafloor had so few volcanoes.

Another discrepancy of the same type [as the lack of sediments], the small number of volcanoes on the sea floor, also indicates the apparent youth of the floor. Menard estimates there are in all 10 000 volcanic sea mounts in the oceans. If this represented 4 aeons of volcanism and volcanoes appeared at a uniform rate, this would mean only one new volcano on the sea floor per 400 000 years. One new volcano in 10 000 years or less would seem like a better figure. This would suggest an average age of the floor of the ocean of perhaps 100 to 200 million years.

*(Hess, 1960d: 17; my bracketed addition)*

The comparative rarity of volcanic sea mounts was most readily explained by supposing short-lived seafloors, which "would account also for the fact that nothing older than late Cretaceous has been obtained from the deep sea or from oceanic islands" (Hess, 1960d: 17–18). He then wondered why there were no remaining Paleozoic or Precambrian ridges. After all, there should be some, if existing ocean floors were permanent.

Still another line of evidence pointing to the same conclusion relates to the ephemeral character of mid-ocean ridges and the fact that evidence of only one major ridge still remains on the ocean floor [Hess' hypothesized Mesozoic Mid-Pacific Ridge]. The crest of this one began to subside about 100 million years ago. It will be described in more detail below. The question can be raised: where are the Paleozoic and Pre Cambrian mid-ocean ridges, or did the development of such features begin rather recently in the Earth's history?

*(Hess, 1960d: 18; my bracketed addition)*

In this way, seafloor spreading and its corollary that ocean floors were young, explained these four outstanding features of ocean basins.

Hess then sought to explain guyots and atolls in terms of seafloor spreading. His new solution, like that of his 1953 hypothesis, assumed they had formed at ancient ridge crests, in this case an old Mesozoic Mid-Pacific Ridge (Figure 3.9)

Noting that Cretaceous shallow water fossils had been found on guyots by Edwin Hamilton (1952, 1956), he rejected his 1946 solution.

Hess (1946) had difficulty in explaining why the guyots of the mid-Pacific mountain area did not become atolls as they subsided. He postulated a PreCambrian age for their upper flat surfaces moving the time back to an era before lime secreting organisms appeared in the oceans. This became untenable after Hamilton found shallow water Cretaceous fossils on them.

*(Hess, 1960d: 20)*

He proposed that guyots and atolls evolve from volcanic peaks that form at ridge crests, which then sink as they move down ridge flanks and their tops are flattened by wave erosion. He noted that atolls dominate on the southern flank and guyots on the northern flank of his Pacific Mesozoic Ridge. He proposed that the flanks were at different latitudes, and suggested that guyots on the northern flank had formed in water that was too cold to support reef-forming organisms. He appealed to paleomagnetic results.

Looking at the same problem today and considering that the North Pole in early Mesozoic time as determined from paleomagnetic data from North America and Europe, was situated in

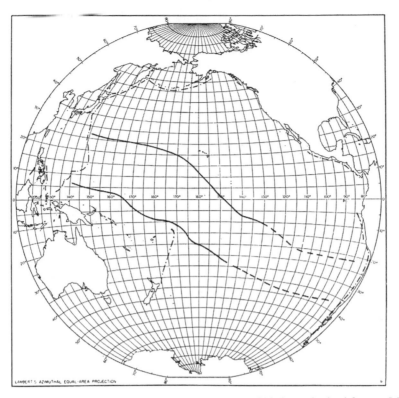

Figure 3.9 Hess's Figure 4 (1960d: 9) shows the location of his hypothesized former Mesozoic Mid-Pacific Ridge. The area of subsidence, where the ridge once stood, bounds an area of guyots and atolls.

southeastern Siberia, it seems likely that the Mid-Pacific mountain area was too far north for reef growth when it was subsiding. The boundary between reef growth and non-reef growth in late Mesozoic time is perhaps represented by the northern margins of the Marshall and Caroline islands, now a little above 10° N, then perhaps 35° N. Paleomagnetic measurements from Mesozoic rocks, if they could be found within or close to this area are needed to substantiate such a hypothesis.

*(Hess, 1960d: 21)*

Hess (1960d: 33) then briefly considered what happens at descending limbs of convection cells. He hinted at the formation of island arcs, remained silent about trenches, and suggested that oceanic sediments and seamounts are metamorphosed and welded on to continents as they "ride down into the jaw-crusher of the descending limb." Leading edges of drifting continents are not forced down into the mantle and they may sometimes override the downward flowing current as has happened in the Pacific Ocean. When this happens, island arcs often form.

A continent will ride on convecting mantle until it reaches the downward plunging limb of the cell. Because of its much lower density it cannot be forced down so that its leading edge is strongly deformed and thickened when this occurs. It might override the downward flowing mantle current for a short distance but thickening would be the result as before. The Atlantic, Indian and Arctic oceans are surrounded by the trailing edges of continents moving away from them. Whereas the Pacific Ocean is faced by the leading edges of continents moving toward the island arcs representing downward flowing limbs of mantle convection cells or, as in the case of the eastern Pacific margin, have plunged into and in part overridden the zone of strong deformation over the downward flowing limbs.

*(Hess, 1960d: 31)*

He and Robert L. (Bob) Fisher provided a more extensive discussion of the origin of trenches and island arcs in terms of seafloor (see §3.16).

Hess also considered Earth expansion, which was currently in vogue in some quarters (II, §6.2; II, §6.14; §6.9). He raised two serious difficulties against it (RS2), and argued that his overall hypothesis was superior to Earth expansion (RS3). He (1960d: 18–19) admitted that Earth expansion explained much of what seafloor spreading did, especially the paleomagnetic support for continental drift, and paucity of oceanic sediments, volcanoes, and old mid-ocean ridges. Nevertheless, he hesitated "to accept" what he described as "this easy way out" because:

First of all, it is philosophically rather unsatisfying, in much the same way as were the older hypotheses of continental drift, in that there is no apparent mechanism within the Earth to cause a sudden (and exponential according to Carey) increase in radius of the Earth. Second, it requires the addition of an enormous amount of water to the sea in just the right amount to maintain the axiomatic relationship between sea level-land surface and depth to the M discontinuity under continents which is discussed later in this chapter.

*(Hess, 1960d: 19)*

Hess's "philosophical" objection was simply to raise a mechanism difficulty against Earth expansion. He thought that mantle convection and seafloor spreading removed the mechanism difficulty that been raised against older versions of mobilism. Turning to the second objection, he argued that with Earth expansion the volume of ocean basins increases, the amount of oceanic water remains constant, and therefore, ocean basins become shallower and shallower. This objection already had been raised by Longwell in his 1958 summary comments of Carey's symposium (II, §6.16).

### 3.16 Fisher's work on trenches; he teams up with Hess

Fisher of Scripps, whose work deserves special attention, and Hess agreed to write a paper on trenches in December 1959 before Hess had thought of seafloor spreading. They submitted the final manuscript in May 1961, approximately six months after Hess had finished his preprint of seafloor spreading. Fisher was the acknowledged expert on trenches. Hess (1955) already had appealed to Fisher and his colleagues' work at Scripps in defending Vening Meinesz's idea that trenches were caused by a downbuckling crust from the attack against it launched by M. Ewing, Heezen,

Worzel, and others at Lamont. Fisher and Hess rejected the "Lamont" view that trenches are tensional features. Fisher turned out to be an excellent co-author for Hess.[14]

In their paper (Fisher and Hess, 1963), Hess wrote only the brief section summarizing the origin of trenches and island arcs. He offered an explanation in terms of seafloor descending into Earth's interior at trenches and deserpentinization of Layer 3 above the 500° isotherm. However, he said nothing about the generation of seafloor at ridges, its migration away from them, the youthfulness of ocean basins, or mobilism generally. He presupposed seafloor spreading, but said nothing directly about it. Their paper could be read and understood without knowledge of seafloor spreading. He did not even tell Fisher about seafloor spreading; Fisher did not know about it when they submitted their paper. Ironically, Fisher was not at all opposed to seafloor spreading once he found out about it (Fisher, January 7, 2006 interview with author). Hess could have told him, and not risked disagreement. Indeed, they agreed about much. They agreed that trenches are compressional. Both thought oceanic crust sank into the mantle beneath trenches. They agreed that the "Lamont" view was mistaken; trenches are not caused by extension.

They disagreed, however, about the angle of descent of ocean floor into the mantle. Hess argued that oceanic crust descended vertically directly beneath trenches. In a sense, he simply broke Vening Meinesz's downbuckle at its base, and proposed that oceanic crust continued to descend into the mantle directly below trenches carried on the backs of convection currents. Fisher's 1950s work on trenches indicated that oceanic crust descends diagonally beneath trenches and continental margins before reaching the mantle. Fisher and Hess's agreements and disagreements about trenches cannot be fully appreciated without understanding the disagreements that arose across the community during the second half of the 1950s about the interpretation of new information gained primarily through seismic refraction studies by workers mainly at Scripps, Lamont, and WHOI. I shall explain these disagreements below.

I begin with Fisher's work on trenches, through the mid-1950s.

Fisher's main interest is trenches. He studies them whenever he can find them – anywhere in the Pacific Ocean. They are as exciting to him as bull-fighting, his other great passion.

*(Raitt, 1956: 67)*

Bob Fisher entered Caltech in mid-1943. After completing his freshman year, he joined the Navy until mid-1946 when he returned to Caltech. He graduated in the class of 1949, "the biggest class Caltech has ever had because it swept up people from the different classes that had gone into service" (Fisher, November 28, 2005 interview with author). He took a class there from Benioff, when he was "putting together his dipping fault plane ideas" (Fisher, November 28, 2005 interview with author). During his junior year he also took a special single-term course from Gutenberg and Richter. "Gutenberg didn't have any problem with continental drift" (Fisher, January 7, 2006 interview with author). He read Wegener and du Toit in the class, and Fisher "realized that you cannot look at the Southern Hemisphere and not be a continental drifter" (Fisher, January 7, 2006 interview with author). Although Fisher did not argue for mobilism before the development of plate tectonics, he was not adverse, perhaps was even sympathetic, toward it. Gutenberg's favorable introduction to mobilism helped. He later read the proceedings to Carey's Hobart symposium soon after their publication in 1958. He found them interesting, although he rejected Carey's Earth expansion (Fisher, January 7, 2006 interview with author). Unlike most North American Earth scientists who were educated during the 1940s and 1950s, Fisher escaped the influence of vehement fixism.

After a long summer as a USGS field geologist on Saint Lawrence Island, Alaska, Fisher began graduate work at Northwestern University in 1949, planning to study under W. C. Krumbein (1902–79), a leading expert in statistical sedimentology. Within a year, however, he changed his plans and returned to California to study at Scripps.

I was there [at Northwestern] for about a year and one evening in spring 1950 in the library at Northwestern I read a paper by Munk and Traylor and there was a picture of wave refraction in La Jolla Bay. I read that thing and I was very interested; I was fascinated. I went in the next day to talk to Krumbein who had been at Scripps during the war in one of those programs on beach studies. He said, "Oh yes, I know a man out there named Francis Shepard and also Roger Revelle, and when you finish your degree here maybe you would like to write them to see about getting a job." I went home that evening and I talked to my beautiful young wife who also was a Californian, and she said, "Why don't you just write to one of the names you have?" I wrote to Fran Shepard, and that was at the end of my first year at Northwestern, and he said, "Why don't you come out here? We need you." So I went back with that letter and showed my wife, and we said good bye to Northwestern. I came out to Scripps during the last week of June 1950, the same week the Korean War started.

*(Fisher, November 28, 2006 interview with author; my bracketed addition)*

Fisher was one of several Caltech students who spent most of his career at Scripps. Like Munk, whose paper with its picture of La Jolla Bay inspired Fisher to return to California, Fisher received his undergraduate degree from Caltech, a Masters from UCLA, and his Ph.D. from UCLA-Scripps based on his work wholly at Scripps. At Scripps Fisher customarily worked with R. W. Raitt, G. G. Shor, and H. W. Menard, all of whom earned degrees of one sort or another from Caltech. Fisher and Menard were trained there in geology with introductory courses in seismology and exploration geophysics. Raitt and Shor received their undergraduate and graduate degrees from Caltech. Shor was trained in geophysics, Raitt in physics; both became exploration geophysicists. Raitt already was at Scripps before Fisher arrived; Shor came three years later. Raitt and Shor were the key figures at Scripps responsible for seismic refraction studies that revealed the structure of the Pacific oceanic crust.

Menard, who became a major figure at Scripps was still at the nearby US Navy Electronics Laboratory when Fisher arrived; he and Fisher soon became collaborating friends. Within a few years they tacitly agreed that Fisher would concentrate on trenches, while Menard would continue to work on fracture zones and ocean ridges (Fisher, November 28, 2005 interview with author).

When Fisher arrived at Scripps in June 1950, he met Roger Revelle, who was then Acting Director, soon to become Director. Revelle became Fisher's mentor, and introduced him to Hess.

Before long Roger Revelle and I became friends. Revelle and I became like father and son or more so like older and younger brother. He was my mentor in terms of humanism. I would do the field observations, and we would publish together. Revelle was a great admirer of Harry Hess, and that is the background for my connection with Hess.

*(Fisher, November 28, 2006 interview with author)*

Fisher logged many miles at sea studying trenches. When asked to co-author the paper with Hess he already had planned and led five deep-sea expeditions each of more than one-month duration, had participated in four more, and had become an acknowledged expert on trenches.

Using serial half-pound TNT charges and echo-sounder, Fisher took soundings of the Tonga Trench in December 1952 and January 1953 during Scripps' Capricorn Expedition. Revelle and he concluded that the trench was at least $10\,633\,\text{m} \pm 27\,\text{m}$ deep, which appeared to be the deepest place in the Southern Hemisphere (Fisher, 1954; Fisher and Revelle, 1955).

Fisher and Revelle (1955) speculated about the origin of trenches, presenting a paper at Columbia University's symposium "The Crust of the Earth." They noted that the largest earthquakes occur beneath trenches, that they are underlain by oceanic crust, and are associated with huge negative gravity anomalies, which implied, they argued, that some force acts against gravity "to pull the crust under the trenches downward" (1955: 38). They proposed mantle convection as the driving force, although they neither mentioned it by name nor stated whether convective currents rise under continents or oceans. They appealed to recent heat flow measurements that Maxwell had taken in the region of the Acapulco (Middle America) Trench, which were published the following year in Bullard, Maxwell, and Revelle (1956: 167) (II, §2.12).

What may these forces be? Here studies of heat flow in the crust suggest a possible answer ... Near the surface of the earth the heat is transported outward principally by conduction, but at greater depths there may be a slow upward movement of the hot rock itself, carrying heat toward the surface. If rock at these depths moves upward in some regions of the earth, there must be other regions where cold rock moves downward. This movement would reduce the outward flow of heat. Now measurements near the floor of the Acapulco Trench show that the flow of heat there is less than half the average for the earth's surface ... So it may be that relatively cool rocks are slowly moving downward under the trench. Such a downward flow would tend to drag the crust down with it and may well account for the formation of the trench.

*(Fisher and Revelle, 1955: 38)*

They further supported their idea that crustal material is dragged into Earth's interior by appealing to the unexpected thinness of sedimentary deposits in the deepest trenches.

Speculating from what we know, we may imagine that a trench has the following life history. Forces deep within the earth cause a foundering of the sea floor, forming a V-shaped trench. The depth stabilizes at about 35 000 feet, but crustal material, including sediments, may continue to be dragged downward into the earth. This is suggested by the fact that the deepest trenches contain virtually no sediments, although they are natural sediments traps.

*(Fisher and Revelle, 1955: 39–40)*

When convection ceases, trenches fill with sediment, rise up isostatically, eventually becoming island arcs.

Fisher next co-authored a paper on the topography and structure of the Tonga Trench with Raitt and Mason (Raitt *et al.*, 1955) – this is the paper Hess quoted in his defense of Vening Meinesz's tectogene against the Lamont attack (§3.17). Fisher and Raitt were the main contributors; Fisher primarily wrote the paper, and was responsible for the geological interpretation; Raitt did the seismic refraction (Fisher, January 7, 2006 interview with author). The fieldwork was done during the same Capricorn Expedition (1952–3) on which Fisher did the echo sounding of the Tonga Trench. Presenting their results at the "Crust of the Earth" symposium in October 1954, they noted that the Tonga region possesses features common to other oceanic trenches, a trench associated with shallow earthquakes, volcanoes, and "deep earthquakes concentrated in a fault zone dipping steeply away from the oceanic side of the trench" (Raitt *et al.*, 1955: 238). They found that the Moho is 20 km below sea level under the trench, but only 12 km below sea level just to the east of the trench. They discovered that the main oceanic crustal layer (Layer 3) thickens to 8 km beneath the trench, approximately 4 km thicker than east of the trench. They also found "no evidence for a substantial thickness of continental sialic material" (Raitt *et al.*, 1955: 253). They proposed that the trench formed by "down-bowing" of ocean crust, and supported this by describing how the flattened top of the Capricorn seamount (guyot), which they found just east of the trench, tilted down toward the trench.

If trench formation is dominantly a continuing process of down-bowing, the lesser tilt of the seamount surface suggests relatively late formation, or at least truncation. However, it is entirely possible that the tilt of the seamount surface is due purely to local adjustments and bears no relation to broader trench-forming processes.

*(Raitt* et al.*, 1955: 253)*

They remained silent about whether mantle convection was actively dragging the Earth's crust downward under the trench and was creating the trench, as Fisher and

Revelle had proposed a year before, or whether the crust was simply bent down and was not actively disappearing into the Earth.

### 3.17 Lamont's view of trenches, 1954–1959

M. Ewing, Worzel, L. Shurbet, and Heezen co-authored several papers in 1954 and 1955 on the Puerto Rico Trench (Ewing and Worzel, 1954; Ewing and Heezen, 1955; Worzel and Shurbet, 1955). Ewing and Worzel leveled the first attack on Vening Meinesz's tectogene hypothesis as a solution to the large negative gravity associated with trenches. They claimed that a thick layer of sediments explained the anomaly.

Seismic-refraction results and gravity data have been used to deduce the crustal structure from the ocean basin north of the Puerto Rico trench to the Caribbean Sea. It is concluded that the Mohorovičić discontinuity (characterized by compressional-wave velocities of about 8 km/sec) lies under the trench, and at slightly shallower depth under Puerto Rico. The large negative gravity anomaly is attributed to a great thickness in sediments in the trench rather than to a "sialic root" due to a down-buckle of the crust under the trench as formerly thought.

*(Ewing and Worzel, 1954: 165)*

Previous analyses of "typical" oceanic crust in the Caribbean, which showed only a 3 km thick oceanic crust, a 1 km thick layer of sediments, and no sialic crust, led them to question Vening Meinesz's solution and to expect that they would find a thick layer of sediments in the Puerto Rico Trench, which would explain the gravity anomaly (Ewing and Worzel, 1954: 168). As they expected, they did find a thick layer of sediments in the trench. They admitted, however, that their seismic-refraction measurements were incomplete; they failed to find the bottom of the sedimentary layer and did not penetrate the main layer of oceanic crust.

At all the stations located within the trench ... the seismic measurements gave positive information only for the unconsolidated sediments, i.e., neither basement rocks nor consolidated or semi-consolidated sediments were detected. There is no question about the proper operations of the equipment since satisfactory results were obtained with it on days before, after and between the several observations in the trench. In more than 100 sea seismic stations, layers below the unconsolidated sediments have always been observed. On any interpretation of these results it must be concluded that the unconsolidated sediments are many kilometers thick.

*(Ewing and Worzel, 1954: 168)*

They explained the negative gravity anomaly in the Puerto Rico Trench by proposing a very thick layer of sediment underlain by normal oceanic crust (Ewing and Worzel, 1954: 171).

The next two Lamont papers on trenches were presented at Columbia University's symposium "The Crust of the Earth," the same meeting at which Hess discussed the formation of island arcs and Alpine mountain belts, and Fisher and Revelle discussed their view of trenches. Worzel and Shurbet described their work on the Puerto Rico Trench. Their seismic data were still incomplete. They explained the negative gravity anomalies associated with the Puerto Rico Trench by appealing to its thick layer of sediments and hypothesized thin layer of oceanic crust beneath the trench. Turning to the Mindanao Deep for which there also were no available seismic data, they assumed that it contained little sediment and an especially thin underlying ocean crust. They proposed that tensional forces had caused crustal thinning.

The gravity anomalies in the Puerto Rico region in conjunction with the seismic data indicate a great thickness (about 6 km) of sediments over a thick layer of crustal rocks (basalt?) in the trench ... The great deeps such as the Mindanao [Philippine] Deep cannot have a very large accumulation of sediments in them, and their crustal thickness is reduced. This reduction of crustal thickness points to a tensional cause for the trenches, as it is believed no other process could cause a thinning of the crustal layers.

*(Worzel and Shurbet, 1955: 99; my bracketed addition)*

Ewing and Heezen (1955) also presented a paper on the Puerto Rico Trench. Undeterred by the lack of any new seismic refraction data, they claimed that trenches are tensional features, and claimed that their explanation of the formation of them was better than Vening Meinesz's tectogene hypothesis (RS3).

The work of Worzel and Ewing (1954) and Worzel and Shurbet [presented at his symposium] ... has demonstrated that a downbuckle ("tectogene") is not required to explain the gravity anomalies observed over the deep oceanic trenches. They find that the gravity anomalies can be more satisfactorily interpreted as the effects of a narrow band of exceptionally thin crust overlain by sediment of negligible or greater thickness. The fact that a local thinning of the crust can better explain the gravity results leads one to the conclusion that tension rather than compression may be the dominant force involved in the formation of the trenches.

*(Ewing and Heezen, 1955: 206; my bracketed addition)*

Concentrating on the Puerto Rico and Cayman trenches, they speculated about the evolution of trenches into island arcs (RS1).

Under the explanation offered here both features [i.e., the Puerto Rico and Cayman trenches] are the result of tension. It is supposed that along the axis of the present trenches the crust has either fractured or thinned owing to tension. In order that equilibrium be reached, the thin crust will eventually be forced up, or the fissures partly filled by upward flow of subcrustal material under the trench. It can be supposed that the viscosity of the subcrustal material is sufficiently great that a considerable time lag will occur between the time of the fracture and the time the subcrustal material has finished its upward flow. The sediments deposited in the trench during this interval will be uplifted and deformed. This provides a mechanism for the formation of the islands of the arcs and for the periodic extension of the continents.

*(Ewing and Heezen, 1955: 266–267; my bracketed addition)*

They offered no cause for the tension. They also needed seismic profiles that would penetrate to greater depths and give the thickness of the sediment and underlying crustal layers, especially Layer 3, the main layer of oceanic crust. They needed what Fisher and his colleagues at Scripps had already begun and continued to provide, and what Charles Officer and Maurice Ewing's brother, John Ewing of Lamont, would soon begin to provide from the Puerto Rico Trench (§3.17).

Talwani, George H. Sutton, and Worzel wrote another Lamont paper on the Puerto Rico Trench. With more seismic data than before, they proposed a slightly thickened crust underneath the trench, but still declared that their findings generally agreed with those of M. Ewing, Shurbet, and Worzel (Talwani *et al.*, 1959: 1554). Trenches were tensional features.

### 3.18 Fisher's Ph.D. dissertation (1952–1956) and his continued work on trenches

In those years when I was writing my thesis [Fisher, 1957] I read Ross Gunn and others. I had no problem later with what came to be called subduction. Now that is because I had been in Benioff's classes at Caltech, and at that time he was showing the dipping faults. So that was no problem. At that time I had George Shor, and that was George Shor's first real expedition. While George is a little older than me and had gone to Caltech earlier, he had just then come to sea. I took him to sea and we did the work. George didn't know how difficult trench work was. He went along and he and I shot the lines there off Champerico and in the Middle America Trench, which clearly shows the oceanic layer going right down under the shelf, dipping down, not being thick. And in the illustration in that paper [reproduced below as Figure 3.10] which is from my thesis, you'll see the perfect picture of subduction.

*(Fisher, November 28, 2006 interview with author; my bracketed addition)*

In 1952 Fisher began working on the Middle America Trench; it is a continuous depression between 30 km and 160 km offshore, which stretches from the Islas Tres Marías along the west coast of Mexico to the west coast of Panama, and includes what were formerly called the Acapulco and Guatemala Trenches. Fisher led most geological–geophysical explorations of the trench, including the Toro (1953), Chubasco (1954), and Acapulco Geological (1956). He used the results as the basis

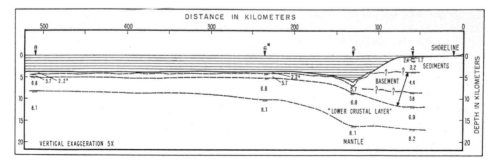

Figure 3.10  Fisher's Figure 5 (1961: 715) displays his analysis of oceanic crust and its diagonal descent beneath the Middle America Trench and continental margin. The "Lower Crustal Layer" is Layer 3.

of his Ph.D. thesis "Geomorphic and seismic-refraction studies of the Middle America Trench, 1952–56," which he completed in 1956 (Fisher, 1957). He summarized key findings in two papers which he submitted in February 1959 but were not published until two years later (Fisher, 1961; Shor and Fisher, 1961).

Fisher's structural interpretation of the Middle America Trench differed little from Raitt's and his previous interpretation of the Tonga Trench (Raitt *et al.*, 1955). He claimed that the M-discontinuity beneath the trench is approximately 16 km below sea level, about twice as deep as in the Pacific Basin west of the trench, and that it continues to descend diagonally beneath the continental shelf. Layer 3 of oceanic crust beneath the trench is about 7.5 km thick, approximately one-and-half times thicker under the trench than in the Pacific Basin, and it too continues to descend diagonally beneath the continental shelf. He identified two layers overlying Layer 3, an upper 1 to 2 km-thick layer (Layer 1) of unconsolidated sediments, and a slightly thicker underlying layer (Layer 2) of volcanic rocks or consolidated sedimentary rock. Layer 2 thickened considerably under the continental shelf (Fisher, 1961: 715–716). He illustrated his findings as shown in Figure 3.10.

Fisher speculated about the origin of trenches and the negative gravity anomalies associated with them. He argued against the early Lamont interpretation, accepted Benioff's idea that earthquake foci lie along a fault plane beginning underneath the Middle America Trench and extending eastward and diagonally downward underneath Central America; he adopted Benioff's characterization of the region as a "marginal reverse fault complex" with underthrusting of oceanic crust beneath the continental margin. He appealed to Ross Gunn's view that continental overthrusting could lead to trench formation and a parallel line of volcanoes (Fisher, 1961: 704). Gunn wrote (1936, 1937, 1947, and 1949) several papers during the late 1930s and 1940s, often in obscure journals, proposing that island arcs and trenches form through bending and subsequent fracturing of the crust at continental–oceanic boundaries with subsequent over-riding of lighter continental material over oceanic crust.

Fisher adamantly opposed the contemporary Lamont view, retrospectively remarking:

We knew it [i.e., the Lamont view] was garbage. I mean the idea that if you have tension you wouldn't have that sort of thing. The gravity picture doesn't fit it and so on. As Russ Raitt once said, "The reason they never got anything under it [i.e., the Puerto Rico Trench] was maybe because the electricity failed." In other words, their interpretation under the Puerto Rico Trench was just faulty. They didn't get the mantle underneath the trench. They don't get it under the trench whereas we got the mantle – Raitt and I, and Raitt, Fisher and Mason, and when I worked with Shor later on. Then Shor worked later by himself up in the Aleutian Trench and found not as complete but similar pictures, in other words, nothing to argue against our view. So from 1953 on I saw no reason to think that trenches were due to anything but a basically compressional force.

*(Fisher, November 28, 2005 interview with author;*
*my bracketed additions)*

He originally stated his opposition more diplomatically, raising the following difficulty (RS2).

> According to Ewing and Heezen (1955, p. 266), "The fact that a local thinning of the crust can better explain the gravity results leads one to the conclusion that tension rather than compression may be the dominant force involved in the formation of trenches." Without going more deeply into the question of forces forming trenches, the present writer wishes to point out that the interpreted results for the Middle America Trench off western Guatemala as here reported, and those from the Tonga and Peru-Chile trenches, apparently do not support this conclusion.
>
> *(Fisher, 1961: 716)*

Fisher's finding was that oceanic crust was not exceptionally thin beneath trenches; it was actually one-and-a-half times thicker than typical oceanic crust. Moreover, Fisher thought something was stopping the oceanic crust from rising isostatically. Fisher also rejected Vening Meinesz's downbuckling crust. "There is no indication of a tectogene-like bulge of crustal material into the mantle beneath the trench" (Fisher, 1961: 716). The crust descended diagonally below the continental margin; it did not double upon itself (RS2). Fisher's was a remarkable piece of work.

Fisher remained silent about possible forces. Benioff and Gunn, however, provided him with a viable process. The diagonally downward plunging mantle resembled Benioff's marginal reverse fault complex.

> Gunn (1947) attempted a quantitative treatment of the Middle America Trench–Coastal Mountain–volcanic Chain relations. He assumes a strong elastic lithosphere supported on a weak magma and examines the mechanics of a "compressed shear thrust fault" formed in such a crust. From his calculations, Gunn states that shear faults resulting from horizontal compression should be localized at continental margins and that with continuing compression the higher continental mass would overthrust the ocean basin and simultaneously form a linear deep at the toe of the overriding block and a line of volcanoes, parallel to the deep, 50–90 km inland from the continental margin.
>
> *(Fisher, 1961: 704; Gunn reference is the same as mine)*

Fisher and Revelle (1955) had invoked a convection-like process without using the name. Fisher did not mention convection in his 1961 paper; he recalled that he was not opposed to mantle convection; he just said nothing about it in the paper (Fisher, November 28, 2005 interview with author). Gunn, however, explicitly rejected convection (Gunn, 1947: 240–241). He argued that there was no identified place where convection currents rose (RS2). If this difficulty bothered Fisher in February 1959 when he submitted his paper on the Middle America Trench, it certainly did not bother him once he learned about seafloor spreading.

### 3.19 C.B. Officer and company's solution to the origin of trenches

Charles (Chuck) Officer was born in 1926. His father was an engineer, and he figured that he should become one too. He spent his last three years of high school at Philips

Exeter Academy and showed an aptitude for mathematics and science. He joined the US Navy at age seventeen in 1944 immediately after graduation. The Navy sent him to Brown University where he remained until discharged in 1946 and returned to complete his B.S. in physics in 1947. He spent the next year at Wesleyan University where he obtained an M.A. in physics. With an offer of a graduate assistantship from Yale University, he began a Ph.D. program in nuclear physics. Not interested enough to continue, Officer looked into the possibility of switching to economics. However, a summer job in 1949 at WHOI introduced him to geophysics, and he found himself in fall 1949 pursing a Ph.D. at Columbia University with Maurice Ewing as his supervisor.

My friend Arthur Voorhis saw the advertisement. We were both looking for a summer job, so I said, "Let's call." So we did. I had a car and we both drove up there and saw Brackett Hersey. He hired us on the spot for the summer ... The physics I had known was nuclear physics. I had never even heard the word "geophysics" before I went up to Woods Hole. I worked with Hersey. I found it very interesting, and the man to go to to get a degree was Maurice Ewing ... He was at Columbia, in fact, had just started there with a graduate program. It turned out that he was out at sea on the Woods Hole ships. So Brackett sent him a cable and he sent a cable back saying, "Sure, you're accepted as a graduate student." No exams, none of that kind of stuff.

*(Officer, February 20, 2006 interview with author)*

Officer divided his time between Columbia University's main campus where he took his first geology classes and Lamont where he worked on research and learned geophysics. He took classes from Bucher and Kay; neither, as far as he remembers, discussed mobilism. He enjoyed Bucher's enthusiasm and openness. Officer took seismic refraction measurements in the Atlantic aboard WHOI ships with Ewing in summer 1950. The work led to his Ph.D. thesis; he submitted his findings for publication in October 1951, and he received his Ph.D. the same year (Officer *et al.*, 1952). He then took a position as geophysicist at WHOI, where he remained, except for a Fulbright Research Fellowship to New Zealand in 1953–4, until 1955 when he accepted a position in the Geology Department at Rice Institute, which became Rice University in 1960. He also arranged to continue working at WHOI during the summers, using their ships.

Before leaving WHOI for Rice Institute, Officer teamed up with John Ewing of Lamont, and R. S. Edwards and H. R. Johnson of WHOI. They took almost fifty seismic-refraction profiles in the Eastern Caribbean, including three within and two bordering the Puerto Rico Trench, during a WHOI expedition from January to March 1955 (Officer *et al.*, 1957: 364). They succeeded in obtaining the first seismic profile beneath the Puerto Rico Trench that extended below the Moho. Submitting a paper in April 1956, they reported that the main layer of oceanic crust thickened beneath the Puerto Rico Trench from 4 km to 7 km, that the sedimentary layer also thickened in the central portion of the trench, and noted that their results were "remarkably similar to" those obtained by Raitt, Fisher, and Mason in 1952–3 from the Tonga Trench (Officer *et al.*, 1957: 371).

Officer also proposed an explanation of the Puerto Rico Trench and associated island arc that was remarkably similar to that which Fisher was developing and would later propose for the origin of the Middle America Trench.[15] Officer, like Fisher and Raitt, appealed to Ross Gunn's work, noted that a zone of earthquakes descend diagonally at about 45° from trenches under their associated island arcs, and argued that trenches are caused by compression.

Officer placed his solution within the context of the overall Caribbean Basin. He argued that "island arcs and deep-sea trenches have been formed by horizontal compression along the border of the altered Caribbean and normal Atlantic basin" (Officer *et al.*, 1957: 359). The seismic profiles indicated that the crust of the Caribbean Basin is less dense than that of the Atlantic. He proposed that the Caribbean Basin is neither continental nor oceanic, but is more oceanic than continental, and with the intrusion of material from deep within the mantle is becoming more continental. This influx causes outward horizontal compression on the boundary of the Caribbean Basin, which led to the formation of the Puerto Rico Trench and associated island arcs. The outward horizontal compression, he argued, caused the crust to fracture at the boundary with the lighter Caribbean crust overriding the heavier Atlantic crust. The resultant configuration led to the formation of the trench within the Atlantic Basin and island arcs within the Caribbean.

Officer also agreed with M. Ewing, Heezen, Worzel, and others at Lamont that Vening Meinesz's tectogene hypothesis was incorrect. Seismic results showed that although the crust beneath the Puerto Rico Trench is thicker than normal oceanic crust, there is no large downbuckle beneath trenches. He added, however, that their seismic profiles, which extended only slightly below the Moho, did not rule out "the Vening Meinesz hypothesis of convection cells at greater depths" (Officer *et al.*, 1957: 369). They said nothing about their explanation for the formation of oceanic trenches through compression being at odds with the tensional explanation offered by M. Ewing, Heezen, Worzel, and others at Lamont. Officer and company ended their paper by suggesting that their explanation of trench and island arc formation could be incorporated into Marshall Kay's idea of continental growth by accretion of mountain belts that were evolved from island arcs (Officer *et al.*, 1957: 377). Officer knew of Kay's views from the class he took from him.[16]

### 3.20 Fisher and Hess's joint paper

Fisher and Hess were approached by the publishers John Wiley & Sons at the 1959 First International Oceanographic Congress to write a paper on trenches for their forthcoming series "The Sea." Fisher and Hess agreed. On finding out about it, M. Ewing wanted to get involved. Fisher recalled:

In 1959 there was an International Oceanographic Congress in New York. We were all there … John Wiley and Sons had a very wet cocktail party in one of the rooms of the Commodore Hotel. What they were doing was getting people sloshed enough so that they would agree to write the series they were going to have called *The Sea* … There was Harry. I had known Harry from my telephone calls before that but I had never met him. So at that meeting he was there and Ewing was there and just about everybody who was anybody was there in '59 – there was a picture that was taken out at Lamont and it is like the Athenian Senate and so on. OK. There they all are from the fifties and earlier … But anyway, so there we all were getting sloshed and then their people came around and said, "Will you write … would you be willing to?" So Harry and I knew each other and had a lot of respect for each other and we will do it, and Maurice Ewing said "Well, we have worked there, I'll be on it too." Now, of course, Hess had a history with Ewing; I hadn't, I had not met Ewing. In other words, to him I was still a graduate student although by that time I had my degree – not quite a graduate student. He wanted to be the third one on the paper. In fact he insisted on it. And Harry didn't really want him, but [he said] "Ok, we'll do it." As time went on, we didn't do anything and Maurice Hill [editor of the series] was asking if we would get started. Actually we did it fairly fast. That was in '59, and we got the thing written by '61. But at no time did Maurice Ewing contribute squat. All he did was want his name on the paper, and it would have been Fisher, Hess and Ewing, and it might have been, if he did like he did at Lamont, Ewing, Fisher and Hess, or whatever. OK. Harry had no problem with me being senior author because four-fifths of it was my stuff in terms of the fieldwork. After we all got sloshed and agreed to do this, we separated. Hess didn't have much interest in involving Ewing at all. But that is the background to how we came to write that paper because by that time Hess knew I had done all that work all the way up to the Peru–Chile Trench. And I had known that he was very interested in trenches and had worked all through the Caribbean area.

*(Fisher, November 28, 2005 interview with author)*

Hess and Fisher were not displeased that Ewing contributed nothing (Fisher, November 28, 2005 interview with author).

   Although Fisher had spoken a few times to Hess on the phone, he met him first at the meeting. Hess, of course, already knew about Fisher's work on the Tonga Trench and he had used it to argue against Lamont's extensional account of trenches. Hess was friends with Raitt and Revelle, and knew about Fisher and Raitt's IGY Downwind work on the Peru–Chile Trench. Hess was pleased to work with Fisher (Fisher, interview with author, January 23, 2006).

   While both were in Washington, DC, attending the AGU meeting (December 27 to 30, 1960), Fisher and Hess met for about two hours to discuss what they would put in the paper. Fisher was quite sure that they met at the Cosmos Club where he was staying.

We were sitting in a car outside the Cosmos Club on a rainy night just talking about these things and then we were in my room with all of this stuff spread out on the bed getting ideas, and Harry said, "You and I rather agree. We don't need to worry about this." And then he said, "I'll write up that stuff."

*(Fisher, November 28, 2005 interview with author)*

Fisher recalled that they began talking about whether trenches were compressional or tensional.

We just spread out some of our notes and a couple pieces of paper on my double bed. The way it went was Harry said, "What do you think about this? What is your feeling?" I said my general feeling is that trenches have to be compressional rather than tensional. He said, "Well we can agree on that." From then on we just started building.

*(Fisher, January 7, 2006 interview with author)*

They parted. Fisher sent Hess his part of the paper before he left for sea in August 1960 as expedition leader of Scripps' Monsoon Expedition to the Indian Ocean. He returned in April 1961, and it appears that Hess had worked on the paper while Fisher was gone and had returned the manuscript. Fisher commented on the manuscript and returned it to Hess, who then worked on it and returned it to Fisher along with this note dated Sunday, May 7, 1961, which Fisher kept.[17]

Dear Bob

Having had first the Beloussoves and later Meinesz on my hands plus a symposium I have not had ten minutes to work on our manuscript for the past three weeks. Finally holed up in the Cosmos Club today and wrote five more pages. I have also changed two sentences in older part of the manuscript as a result of criticism by Meinesz.

Had trouble reading your notes on Fig. 10 [reproduced here as Figure 3.11] and the other sheets you sent back. The notes probably are fading. Intended to have a long caption to go with Fig. 10 and not include this in the text. I have changed page 9 somewhat to take care of this. Fig. 10 represents a layer of arbitrary thickness not necessarily the "crust." Add this to the caption if you wish. Make any other changes in the manuscript that seem necessary.

Have to dig up a few more of the references when I get back to Princeton tomorrow.

Sorry about the delay

Best regards

Harry (my bracketed addition)

Fisher made several minor corrections and sent the manuscript to Maurice Hill on May 15, 1961. He also sent Hess a copy of the paper along with a note dated May 16, 1961.

Harry:

Here's your copy of our paper as I sent it to Hill yesterday. Let's hope he's in tolerant mood, or is too busy to be critical.

It has been a pleasure to work on this paper with you.

My best wishes to you.

Bob

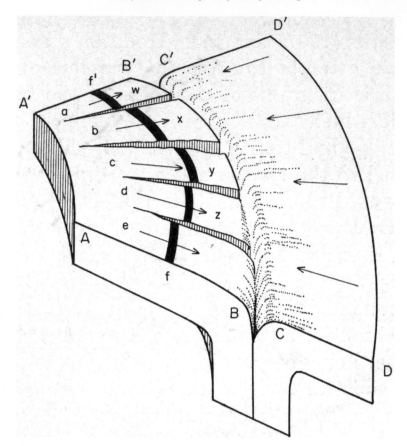

Figure 3.11 Hess's Figure 10 in Fisher and Hess (1963: 431) illustrating his solution to the origin of trenches and parallel line of volcanoes. Hess postulated movement of crust from A′A to B′B creates extension of crust causing formation of fissures through which upwelling magma forms volcanoes at w, x, y, and z. If individual sectors a, b, c, d, and e move at different rates, as indicated by displaced bed f′f, strike-slip faulting would occur along vertical planes perpendicular to the trench.

The question of Ewing's co-authorship had disappeared.

They divided their paper into three sections. Fisher was entirely responsible for the first two sections, "Previous work" and "Topography of trenches." In the first section, he favorably mentioned Gunn's ideas, but noted, "more data on the physical properties of crust and mantle are needed to test his hypothesis" (Fisher and Hess, 1963: 415). In the second section he provided this information by summarizing topographical studies of twenty trenches, and reviewed his, Raitt's, and Shor's seismic-refraction studies of three trenches, Tonga, Peru–Chile, and Middle America. Fisher accented the lack of sediment within them, thickening of oceanic crust beneath them, and their diagonal descent under trenches and island arcs or continental margins. The third section, "Structure of trenches," contains a comparison of the crustal structure beneath these three, and the Puerto Rico Trench, a critique of the Lamont interpretation of trenches, and Hess's solution to the origin of trenches and island arcs. Fisher wrote the first part, although Hess added the sentence, "This [change in velocity of P waves passing through the crust on the island side beneath the Tonga Trench] could be the result of serpentinization" (Fisher and Hess, 1963: 427; my bracketed addition). Hess wrote the remaining part; his contribution begins with the first full paragraph on page 427 (Fisher, January 23, 2006 interview with author).

Fisher constructed a figure, Fig. 8 of their paper, comparing the structure of all four trenches that he and Hess used in their attack on Lamont's extensional interpretation. Each cross section, including the one of the Puerto Rico Trench which he redrew from data in Talwani *et al.* (1959), showed a thickened ocean crust diagonally descending under the trench and continuing either under an island arc or continental margin. Fisher described the tilted top of the Capricorn guyot, which inclined downward toward the Tonga Trench, and a tilted guyot on the lip of the Aleutian Trench that Menard and Dietz had noted in 1951 (§4.8, §4.9, §5.4). The tilted tops formed with the downward bending of the crust.

Hess raised two difficulties with Lamont's view (RS2). He argued that if trenches had formed by extension, the crust would quickly rise isostatically. The second "even more serious" difficulty was "that there is no reason why so large a departure from equilibrium should ever have come about" in the first place (Fisher and Hess, 1963: 429).

Hess was upbeat, as he was wont to be, admitting that his explanation was speculative, but claiming that it was at least reasonable, which made it preferable to others (RS3).

It now seems necessary to construct a hypothesis which would adequately account for the mass distribution, the seismic velocities, and the forces acting to keep trenches far out of isostatic equilibrium. Such a hypothesis is necessarily highly speculative but it is at least better than no reasonable hypothesis – which is the present status with respect to this problem.

*(Fisher and Hess, 1963: 429–430)*

Implicitly linking his hypothesis with seafloor spreading, he argued that downward convection better explains the origin and maintenance of trenches than Earth contraction. Layer 3 of oceanic crust with its overlying volcanics and sediments descend into the mantle. Water is released with deserpentinization of Layer 3 as it descends below the 500 °C isotherm, the serpentinized Layer 3 reverts to peridotite, water is released, the accompanying sediments and volcanics melt, and the water, and the resulting fluid magma migrate upwards on the convex side of island arcs.

The forces necessary to produce the structure and departure from isostatic equilibrium will be attributed to mantle convection currents. The only other source which has been suggested, thermal contraction of the earth, seems the less likely of the two at this time ... The crustal layer goes down with the descending limb of the convection cell until it reaches a temperature in the neighborhood of 500 °C where a deserpentinizing reaction takes place, releasing water.

Similarly, the original supracrustal volcanics and sediments descend until melting or partial melting occurs. Fluids, magma or water, rise, migrating toward the island arc of concave side of the structure.

*(Fisher and Hess, 1963: 430)*

Remarkably Hess then went in a direction contrary to Fisher's findings, and to what Fisher had written in earlier parts of the paper. Hess rejected Benioff's idea of motion along diagonally descending fault planes. He first introduced Benioff's position.

The above discussion treats largely with the curved trenches flanking island arcs. The question might be raised whether the linear trenches, such as those off Chile or middle America, have the same structure. One might reasonably propose that these trenches are formed by the continental block overriding the oceanic area or the downward flowing limb of a convection cell. Benioff (1949, 1955) postulated an overthrust mechanism and showed that the deep earthquake foci lie along a plane dipping under Chile at approximately 40° ... Before proceeding to this subject, it should be noted that the Tonga-Kermadec Trench is also a linear feature but in other respects it is related to a province having all the characteristics of an island arc. Coulomb (1945) showed that the deep earthquake foci for island arcs also lie on a plane dipping under the concave side at about 45°.

*(Fisher and Hess, 1963: 431–432)*

Hess then argued that recent first-motion studies created a difficulty with descending Benioff zones (RS2).

Hodgson (1957) studied the first motion of large deep-focus earthquakes of the Tonga-Kermadec and South American regions. From these he deduced that the movement causing the earthquakes occurred as nearly horizontal displacement on nearly vertical planes. This type of analysis results in defining two planes at right angles to each other. Which is the real fault plane is indeterminate. In either case Benioff's postulated 40–45° thrust motion is ruled out.

*(Fisher and Hess, 1963: 432)*

Hodgson pioneered first motion studies at Dominion Observatory in Ottawa, Canada, arguing that the earthquakes in both regions were caused by horizontal movements across near vertical strike-slip faults (see IV, §6.11 for the later importance of first motion studies). If Benioff's view was correct, earthquakes should have occurred along fault planes descending diagonally at 40–45°. Relying on Hodgson's work (Hodgson and Stevens, 1964), which was later shown to be based on limited data that were collected before the World-Wide Standardized Seismograph Network of the USCGS was in operation (Isacks *et al.*, 1968: 5872), Hess claimed that the seafloor and uppermost mantle plunged vertically beneath trenches: horizontally converging convection currents met, turned 90°, and dove directly into Earth's interior. Illustrating as shown in Figure 3.11 (Fig. 10 mentioned in Hess's May 7, 1961 letter to Fisher), Hess argued that his solution explained how a parallel line of volcanoes formed on the concave side of a trench.

However, Fisher's findings had supported Benioff's idea of thrust-faulting along diagonally descending planes that extended underneath trenches and continued underneath continental margins. Occurring as they did in the same paper, the contrast between Fisher and Hess's figures (Figures 3.10 and 3.11) is striking. Fisher had no problem with seafloor spreading and was willing to defer to Hess's idea of serpentinization. Fisher rejected Hess's appeal to Hodgson's work, wisely believing his own data were more reliable and complete than Hodgson's. The seismic-refraction data that Fisher, Raitt, and Shor had collected by 1960 from the Tonga, Chile, and Middle America and Sunda Trenches showed that oceanic crust descended diagonally beneath each trench and continued its descent beneath the island arc or continental margin. Fisher (1961) maintained that movement was neither horizontal nor along a vertical plane; movement was along a <45° diagonally descending fault plane. Despite his disagreement with Hess, Fisher, as indicated by the manuscript he sent to Maurice Hill (an editor of the volume) and his May 16, 1961 letter to Hess, did not explicitly argue his case with Hess or attempt reconciliation. And so the reader is left with this interesting record of what, in retrospect, can be seen as the beginning of the demise of the Vening Meinesz/Kuenen tectogene and Hess's ideas on the role of serpentinization in the evolution of oceanic crust.

However, Hess's instincts regarding the interpretation of first motion studies were sound: first motion studies were crucially important, but until the early 1960s when the World-Wide Standardized Seismograph Network was in operation, they were based on unreliable and incomplete data. Indeed, as I shall show later (IV, §7.14), McKenzie relied heavily on first motion studies of earthquakes around the perimeter of the Pacific Basin when he first proposed plate tectonics (McKenzie and Parker, 1967). They became a cornerstone of plate tectonics. Later data showed that the dip-slip faulting is responsible for most earthquakes beneath island arcs. Some shallow earthquakes are caused by strike-slip faulting where lithosphere begins to bend downward, but once the lithosphere begins to descend, dip-slip faulting predominates (Isacks *et al.*, 1968: 5871–5872).

### 3.21 Hess, the scientist

In closing this discussion of Hess and in particular his development and defense of seafloor spreading and mantle convection, I want to emphasize his longtime interest in several problems, his use of the standard research strategies in defending and criticizing solutions both his own and those of others, his global outlook and penchant for working outside his specializations, and above all his willingness to change his mind because of new research outside his own area of expertise, notably paleomagnetism.

Hess's seafloor spreading hypothesis arose out of his long-time interests in solving a nest of problems in oceanography, geotectonics, and geophysics; for some of which he had earlier offered and rejected solutions. He brought to bear his long-time concerns with the origin of serpentine from peridotite, of ocean ridges, trenches, island arcs and mountain belts, the formation of the oceanic layer (Layer 3), the formation of guyots, the absence of thick oceanic deposits, and mantle convection. He repeatedly returned to these problems (1932, 1937, 1938a, 1938b, 1939, 1940, 1951) and had thought about some of them since the beginning of his career. He sought a unified solution to them in terms of the Vening Meinesz/ Kuenen idea of a tectogene. He discovered guyots in the mid-forties and proposed (1946) that they were ancient drowned oceanic islands; in all, proposing successively three solutions of their origin and development. As new information on oceanic features appeared, he revised (1954, 1955b) his model of oceanic crust, adopted (1955a) the reversible serpentine to peridotite process as a means of elevating and depressing the seafloor, and used it in his second explanation of guyot formation (1955a). He later proposed that seafloor spreading explained the origin of guyots; they formed atop ridges and became flat-topped by wave action as they moved away from ridges and sank (1960d, 1962).

He also had worked on the problem of the origin of oceanic ridges for much of his career. In (1939) he suggested that the Mid-Atlantic Ridge was an ancient-fold mountain belt. Later he proposed, successively, three new solutions, swiftly changing his ideas as new data appeared. He (1955a) used his newly developed ideas about serpentinization to propose a new solution to the origin of mid-ocean ridges, which also accounted for the common presence of serpentinized peridotite in dredge-hauls from the Mid-Atlantic Ridge. He (1959a) proposed that upward convection currents cause the 500 °C isotherm to rise as evidenced by the higher than normal heat flows on ridges. He also became more concerned with the ephemeral nature and median position of oceanic ridges. He continued to be puzzled by the thinness of the sedimentary deposits on the seafloor and in 1959 suggested that they indeed were incomplete, although he was unsure why this was so.

Hess (1937) first rejected mantle convection but then became receptive to it after he heard Griggs in 1939 speak favorably about it. He (1946) appealed to it in his first explanation of guyots, and (1951) invoked it to explain the persistence of crustal downbuckling beneath island arcs. Three years later, he (1954) first invoked convection currents for the formation of ocean ridges, and, up through his presentation and defense of seafloor spreading, continued to do so, albeit changing its role in explaining their birth, evolution, and death. In his initial development of seafloor spreading, Hess (1960d) brought every one of these concerns together. He discarded his 1955/ 1959 solution of the problem of ridge origin; serpentinization of peridotite could be expected to occur in the mantle below ridges because heat flow data required that the critical isotherm be only 5 km below the seafloor and serpentinization could not therefore occur at and above this level precisely where the major layer (Layer 3) of oceanic crust should be being created. Convection currents began to play a more significant role; upward currents caused the elevation of ridges, horizontal currents passively carried oceanic crust away from ridges, downward currents pulled oceanic crust down into the mantle providing the deep trenches, cessation of mantle convection caused ridges to sink, and the relatively short periods of convection cycles explained the ephemeral nature of ridges. Guyots formed at ridge sites and drowned themselves as, under the influence of convection, they moved off the ridge carried by oceanic crust. Once he realized that the seafloor was short-lived, relative to continents, he had an answer to the incompleteness of the sedimentary layer (Layer 1); it was no longer puzzling because seafloor itself, in the geological scale of events, is youthful.

Hess defended his various hypotheses in terms of their problem-solving effectiveness, and employed the standard research strategies in attempts to improve their effectiveness and decrease that of competing solutions. Hess (1932) first employed the Vening Meinesz/Kuenen downbuckling hypothesis to explain the huge negative gravity anomalies that were associated with trenches of the Caribbean region. Later (1937, 1938a, 1938b, 1939) he used the downbuckling hypothesis to account for formation of Alpine-type mountain belts and their serpentine intrusions (RS1). He also argued that downbuckling caused horizontal displacements of the seafloor in the Caribbean (1938a), and Betz and he (1942) explained the formation of the Hawaiian Islands and accompanying swell in a somewhat similar way (RS1). He (1951) combined downbuckling with convection currents to account for the observed pattern of deep-focus earthquakes on the continental side of island arcs; convection currents ensured the longevity of downbuckling (RS1), which he maintained was not solved by the contractionist theory of mountain building (RS2); and he (1955a) extended downbuckling to encompass the old idea of continental accretion (RS1). Hess thought guyots and ridges were intriguing problems and presented a succession of improved solutions, rejecting earlier ones. He (1954, 1955a) recognized that the discovery of Cretaceous fossils atop guyots raised serious difficulties with his 1946 hypothesis (RS2), and proposed a new solution that depended on serpentinization of peridotite, which both avoided the difficulties faced by his former solution and incorporated his most recent ideas about oceanic crust (RS1). He (1939) first proposed that the Mid-Atlantic Ridge was an ancient fold mountain belt. Then, he (1954) later raised two difficulties made apparent by new topographical and seismic studies (RS2); the Mid-Atlantic Ridge did not have small-scale ridges running parallel to its main axis, and it was seismically active. He (1954) proposed a new hypothesis, that the Mid-Atlantic Ridge is a mass of hot basalt atop an upward mantle convection current, which explains its elevation; cessation of the mantle current could explain its eventual subsidence. A year later, Hess (1955a) raised difficulties against his former views (RS2), and proposed that the Mid-Atlantic Ridge was a welt of serpentinized peridotite. His new proposal enabled him to explain the discovery of serpentinized peridotite found in ridge dredge-hauls, which was left unexplained by his two former solutions (RS3); he noted that his new solution, like his previous ones, could also explain elevation and subsequent subsidence of the ridge. Later he (1960d) rejected his former solution of ridge elevation, replacing it

with his seafloor spreading hypothesis. Again, he raised a difficulty with serpentinization as a cause of ridge elevation; serpentinization could cause only a quarter of the
needed elevation (RS2).

Hess's defense of seafloor spreading again shows his penchant for using the
standard research strategies to argue in favor of his hypothesis and against alternatives. He (1960d) detailed all the problems that his hypothesis solved (RS1). It
explained the origin of oceanic crust, elevation, subsidence, ephemerality, central
graben, high heat flow, and seismicity of ocean ridges, and, when coupled with
mobilism, their common mid-ocean position. He also showed that seafloor spreading
explained the thinness of sediments, the absence of rocks older than Cretaceous, the
small number of volcanic seamounts, and paucity of ancient ridges on the seafloor
(RS1). Fisher and he (1963) extended seafloor spreading to account for the origin of
oceanic trenches (RS1). Impressed with the paleomagnetic case, he adopted mobilism, and argued that seafloor spreading provided mobilism with a mechanism (RS1).
He also invoked seafloor spreading to explain the origin of seamounts, and explained
the difference between atolls and guyots by appealing to latitude differences (RS1).
Besides showing that his former explanations of ridge-elevation and guyot-formation
faced difficulties (RS2), he argued that alternative explanations of the lack of
sediments and volcanoes on ocean floors, which depended on sharp decreases in
rates of sedimentation and volcanism, clashed with uniformitarianism. He also raised
difficulties with Earth expansion (RS2).

Hess, like Wegener, Daly, and Holmes, was a generalist and a globalist. Like Daly
and Holmes, he was trained as a geologist but was comfortable working in some
areas of geophysics. Hess received most of his formal training in petrology and
mineralogy. A. F. Buddington supervised his dissertation, which was on peridotites
and serpentinization, an interest that captivated him throughout his career. He
engaged in detailed mineralogical studies (Buddington, 1970: 4–5). R. M. Field
inspired him to take an interest in oceanography, and Vening Meinesz introduced
him to exploratory geophysics at sea. Of course, his most important work, work that
led to his tackling major problems, concerned the geological structure and composition of ocean floor. So Hess gained expertise in several areas, and used it to tackle
major problems. Hess, again like Wegener, Daly, and Holmes, relied primarily on the
data of others. Although he had sailed with Vening Meinesz, and, as a result of
wartime service at sea, had discovered guyots, he had not been to sea for almost
fifteen years when he came up with seafloor spreading. But, as a member of the
oceanographic community with excellent connections, he had access to a steady
stream of new oceanographic data. He obtained data from researchers at Lamont,
Scripps and the US Navy Electronics Laboratory (NEL) in San Diego, the Department of Geodesy and Geophysics, Cambridge, and WHOI. He often got to see data
directly from researchers that sometimes took several years before it was published
(Menard, 1986: 111). Menard makes this general point when discussing his work and
the importance of an "invisible college" during a scientific revolution.

Here we see an enormous advantage of the members of an invisible college during a scientific revolution. A few insiders had about five years to digest the implications of a flood of observations before the outsiders ever saw the bare data. Judging solely by the results, it also appears that the most favorable position is to be an insider who is not distracted by actual collection of data.

*(Menard, 1986: 111)*

Hess was in a most fortunate position; he had access to much data without the bother of collecting it. This was doubly beneficial; he was not open to the temptation of believing his own data more important than that of others. As Menard (1986: 24) put it when reminiscing about Wegener's work but could have just as easily been talking about his own work and that of other marine geologists, "Some earth scientists believe in God and some in Country, but all believe that their own field observations are without equal and they adjust other data to fit them." There was the practical benefit. Hess did not have to spend time getting data; he could concentrate on explaining it.

Hess was also a globalist when it came to the oceans. Guyots were a problem of the Pacific Ocean, the origin of the Mid-Atlantic Ridge, of the Atlantic Ocean. He worked on the origin of the Hawaiian Islands and Caribbean problems. He was interested in trenches and ridges in the Pacific and Caribbean, and especially the Mid-Atlantic Ridge. As I shall later show, two of Hess's important fellow oceanographers, Menard and Heezen, tended to concentrate on the oceans where they collected data. Having little or no oceanographic data of his own to feel wedded to, Hess had no favorite ocean or body of data.

Hess was an exceptional scientist who had well-developed ideas of his own but was willing to change them, sometimes repeatedly if he thought others were finding out contrary things. Preoccupied with ocean floors, he changed his mind about the major problems that he wrestled with throughout most of his career, the origin of oceanic ridges, guyots, trenches, and the formation of oceanic crust. As he told Vening Meinesz, Hess, then a mature scientist, changed his mind about mobilism before 1960 because of the paleomagnetic evidence, and in this respect he seems to have been unique among North American Earth scientists. Heezen and Menard, as I shall show, also welcomed the paleomagnetic evidence, but only insofar as it supported their other beliefs. Neil Opdyke, a fixist when an undergraduate at Columbia University, had become a mobilist a few years earlier than Hess because of the paleomagnetic evidence, but Opdyke was then just starting his research career. The paleomagnetic support of mobilism forced Hess to reconsider his views about mobilism, and he developed solutions to the origin and development of ocean floors that were consistent with it. He went even further; he also provided a solution, or so he and others thought, to mobilism's mechanism problem.

## Notes

1 See Frankel (1980) for an early attempt to trace the development of Hess's ideas. I have since greatly benefited from Allwardt's excellent 1990 Ph.D. dissertation, "The roles of Arthur Holmes and Harry Hess in the development of modern global tectonics." Although I made

some use of Hess's papers in discussing the reception of Hess's seafloor spreading (Frankel, 1982), Allwardt put them to much better use than I did. Oreskes (1999) also made extensive use of Hess's papers, and presents her own account of the development of Hess's ideas. The account offered here is much closer to Frankel (1980) and Allwardt (1990) than it is to Oreskes (1999). Allwardt (1990) provides an excellent discussion of Hess's adoption, modification, and extension of the downbuckling or tectogene hypothesis.

2 Hess's papers contain many hand-written lecture outlines. Those whose ages are identified range from the late 1950s through the middle 1960s. These handwritten notes are easily distinguishable from the typed and more complete notes of 1946.

3 Although Hess and many English-speaking geologists thought that Kuenen coined the term "tectogene," it turns out that the term was actually coined by E. Haarmann. Finding a letter by Kuenen addressed to Robert Dietz in Hess's papers, Allwardt (1990, 109–110, footnote 70), straightened out the matter.

Most English-speaking geologists had never heard of the term "tectogene" before Kuenen (1937) used it to describe Vening Meinesz's downbuckles. Those unfamiliar with the German origins of the term have generally, but mistakenly, credited Kuenen with its invention – much to his annoyance (Philip H. Kuenen, written communication to Robert Dietz. A copy of this letter is part of the Harry H. Hess Collection at Princeton University). In his letter to Dietz, Kuenen gave the original reference as follows: Haarmann, E., 1926, "Tektogenese" oder Gefügebildung statt Orogenese oder Gebirgsbildung, *Zeitschr. Deutsche Geol. Gesellsch., Bd. 78*, Monatsb. 3–5, p. 105–107.

Oreskes (1999: 355, footnote 89) could be taken as implying that Allwardt himself was confused, claiming that "Allwardt (1990, p. 69, footnote 70) credits Kuenen with supplying the term *tectogene*, but Kuenen says he borrowed it from the German geologist Erich Haarmann."

4 In 1939 Du Toit also wrote Hess about his work in the West Indies, and suggestion of horizontal movements. Oreskes found the letter among du Toit's papers. She reproduced part of the letter, and I have included it in Note 9 below. She appears to think that du Toit's letter is evidence that Hess linked convection to continental drift. I disagree (see also Note 10 below).

5 See Oreskes (1999: 249–251) for a summary of Kuenen's experiments.

6 It is not surprising that Kay, champion of continental accretion, later referred to this paper of Hess's as offering support for his own account of mountain building and continental accretion (II, §7.3), because he also believed that mountain belts evolved from island arcs.

7 Hess received many reprint requests for and laudatory letters about his paper on guyots, but none were as praiseworthy as this one.

February tenth, 1947

My dear Professor Hess:

There arrives in morning mail, kindly sent to me by a friend, the article by Doctor Roy K. Marshall, in a recent THE BULLETIN, Philadelphia, discussing Your wonderful discovery in connection with Peaks of the Pacific Ocean.

Congratulations not merely at great discovery but that, at one bound, You have achieved surpassing Immortality in Scientific Annals. And further felicitations that You have so very happily and strikingly linked Your disclosures with a Name that will add, with its own historic importance in Geology, to Your status as a leader in solutions of the problems of the Universe.

You will gauge the gladness with which I hail your association of the Name: Guyot with this remarkable work of Your own. Arnold Guyot was my Great-uncle and my Godfather, as he was the Godfather of my Mother – this means more European-wise than here.

It was of him that a great Judge said: "You will go from Connecticut to California and you will not see his equal." Princeton University may well be proud of the new luster added to its name by Yourself and by tribute to Arnold Guyot. In hopes of soon meeting You, believe me Gratefully and Joyfully Most Faithfully Yours

A. Guyot Cameron

Hess responded one week later. Directing attention away from himself and toward the Guyot family he wrote:

My dear Mr. Cameron:

Thank you for your kind letter concerning "guyots". During the months at sea when these features were being discovered I thought for a long time about a suitable name for them. My final choice – guyot – I think was most appropriate, both because Arnold Guyot was the founder of the Earth Sciences at Princeton and because of his great interest in mountains as land forms, a field of investigation in which he was a leading authority in his day. I am most happy that a member of Professor Arnold Guyot's family may have the pleasure of knowing that his name still received the recognition that is due him.

Very sincerely yours
H. H. Hess
*(Hess, February 17, 1947 letter to Cameron)*

I should like to thank Princeton University and George Hess for allowing me to reproduce these letters and other documents from Hess's papers at Princeton University Library.

8   Hess's explanation of guyots as drowned very ancient islands recalls Darwin's explanation of coral reefs. But Hess did not cite, as Menard notes, Darwin's 1842 *Structure and Distribution of Coral Reefs* (Menard, 1986: 6). As Menard (1986: 6) further notes, "it is hardly conceivable that he [Hess] had not read" Darwin's work, and adds that Darwin even had as a corollary that islands that sink in high latitudes become drowned rocky platforms. Indeed, Hess did read Darwin, and knew of the similarities between his own theory and Darwin's. In the lectures notes, which he prepared on November 17, 1946, Hess discussed Darwin's theory, the history of the coral reef problem, and his own theory, saying of it that it is the same "as a subsidence theory but relatively the water rose instead of the island sinking . . ." (Hess, "The Coral Reef Problem," which was part of his lecture notes for 1946). In defense of Hess, however, it is worth noting that Hess could not appeal to Darwin's explanation of the formation of reefless drowned islands because Hess had found guyots in low latitudes. But Menard was not criticizing Hess. Indeed, Menard also noted that he and others such as E. L. Hamilton also failed to note that Darwin had already proposed a solution to the formation of reefless seamounts. Moreover, Menard (1986: 300) concluded from his personal account of the revolution in the Earth sciences that the quality of scholarship in a scientific revolution is "regrettable."

9   Erling Dorf (1905–84) received his B.S. (1925) and Ph.D. (1930) from the University of Chicago. Dorf joined Princeton's faculty in 1928, where he remained until 1975. Dorf was a member of the geology department and curator of paleobotany. He studied (1933, 1938, and 1942) North America flora, concentrating on its Tertiary and Upper Cretaceous western flora. Dorf (1959, 1964) argued (correctly) that ancient plant remains in the western United States indicated that North America had enjoyed a warmer climate during the Cretaceous, and much of the Tertiary. He (1959) made similar (correct) claims about Western Europe's Tertiary climate. Dorf said nothing about mobilism in any of his papers cited above, including those published after the rise of paleomagnetism and publication of Hess's 1962 paper on seafloor spreading. Dorf appealed to worldwide cooling and topographical changes in elevation to explain climatic changes during the Tertiary. Apparently, he must have thought mobilism was worth mentioning. Here is what he said in 1964:

Thus we conclude that the Eocene climate in the vicinity of what is now Yellowstone Park was essentially the same as that which now prevails in the Gulf Coast region of southeastern North America. The change from a humid, nearly subtropical climate to the present cool-temperate to subarctic conditions was probably the result of a general world-wide cooling accompanied by a gradual uplifting of the entire Rocky Mountain area by as much as 7000 feet over the past million years.

*(Dorf, 1964: 111–112)*

10  Oreskes seemed to think that Hess and Griggs supported mobilism before the 1960s. She is not too sure about Vening Meinesz. She wrote in an endnote:

Did Vening Meinesz and Hess link convection to continental drift? Vening Meinesz's discussion of Holmes and Hess's discussion of Griggs strongly suggests that they did: the whole point of convection, for Holmes and Griggs, was to explain the cause of drift. Alexander du Toit also made the connection between gravity anomalies, convection, and drift, in a letter to Hess in 1939. "Thank you for your paper on "Gravity Anomalies in the West Indies," he wrote, "which sheds such a great deal of light on the tectonics of that region … I have been much struck by the evidence therein submitted of big horizontal movements, such as would support the hypothesis of continental drift" (A. du Toit to H. H. Hess, March 16, 1939, HHP 7: WI).

Vening Meinesz's later work is more ambiguous. His papers from the 1940s and 1950s clearly advocate a close tie between subcrustal drag and crustal compression, but they sidestep the question of large-scale continental migrations (Vening Meinesz, 1941, 1952). In Daly's (1940) *Strength and Structure of the Earth*, he quotes a passage in which Vening Meinesz seems to suggest convection currents as the result rather than the cause of crustal motions (Daly, 1940, pp. 267–268). As for Hess, his course notes from the 1940s and 1950s show that, at minimum, he assigned readings from Daly, Holmes, du Toit, Vening Meinesz, and Griggs in his classes; in one course handout, he lists Wegener among famous geological luminaries!

The most plausible interpretation is that both men did initially link convection to large-scale crustal motions of some kind, but they lost the momentum of this work during the early years and took more circumspect positions. Then, in the 1950s, with the emergence of the paleomagnetic evidence Hess returned to these earlier ideas.

*(Oreskes, 1999: 358–359)*

Griggs did not link convection with continental drift; he linked it with mountain building (I, §5.10). Vening Meinesz did not support mobilism until the early 1960s (§2.6); he did support polar wandering. Although Alex du Toit linked substantial horizontal movements along with continental drift, Hess did not. In Betz and Hess (1942) there is no mention of continental drift in those places where they discuss horizontal movements along the San Andreas, Great Glen, and the Dead Sea faults. Of course, du Toit saw a connection. Hess did argue that such movements were caused by downbuckling oceanic crust. Indeed, the fact that Hess did not mention mobilism when discussing horizontal displacements along transcurrent faults suggests to me that he did not favor mobilism, as he also told Vening Meinesz, until soon before writing him in July 1959.

When I examined Hess's papers at Princeton, I missed the class handout in which he listed Wegener among famous geological luminaries. I did find a handout for Hess's History of Geology Class, Geology 406. Hess included Wegener among the "Pioneers in Geology," and assigned a student to read some of Wegener's works. He assigned other students to read Hutton, Werner, Playfair, Lyell, Sorby, Brogniart, Desmarest, James Hall, Agassiz, W. M. Davis, J. D. Dana, Rosenbush, William Smith, V. M. Goldschmidt, T. C. Chamberlain, Barrell, Schuchert, Sedgwick, Murchison, Gilbert, Wegener, and du Toit. Thanks to Bill Bonini, Professor Emeritus, Department of Geosciences at Princeton, Hess's class can be dated within a few years. One of the students in the class was named Lowe. Looking through the Geoscience archives at Princeton, Bonini found a Robert Arthur Lowe, Class of 1951. Bonini speculates that Lowe "would have taken the 406 course in his senior year, 1950–1951" (Bonini, email to author, April 16, 2004).

Even though Hess included Wegener among the "Pioneers in Geology" and assigned students to learn about him, I doubt if he thought very highly of Wegener. Hess was asked in 1968 by *Science* to review a manuscript by Tuzo Wilson entitled "A revolution in Earth sciences; life cycle of ocean basins" which contained an historical section on the development of continental drift. Hess had many objections to the paper, and recommended to the editors of *Science* that it should not be published. He wrote Wilson of his decision, and sent him a copy of his review. Wilson had praised Wegener; Hess told him what he thought of Wegener.

I also object to making Wegener as much of a hero as you do. I regard him as rather a wind bag. Taylor presented continental drift in a lecture in 1908 (published in 1910, *Bull. GSA*) and Baker did it in 1911. Others back to 1850 suggested the basic idea but without much documentation.

*(Hess, March 25, 1968 letter to Wilson)*

Of course, it does not follow that Hess opposed mobilism because he thought Wegener was a windbag. Nor does it follow that he favored mobilism in the 1940s and 1950s because he assigned Wegener in the reading list. He also assigned Dana, T. C. Chamberlain, and Schuchert.

11 I cannot confidently unravel the details of Hess's path from his 1955 views about the seafloor through his attempts to reconcile them with his adoption of mobilism and the influx of new data. I shall only attempt to present a rough guide, relying on his publications, unpublished items from his papers, and Allwardt's (1990) account.

12 Mary Sears wrote a preface to the collection of abstracts, which she dated as July 10, 1959. Hess therefore probably submitted his abstract before he wrote his July 6 letter to Vening Meinesz. I do not know if Hess submitted the abstract before he became a mobilist.

13 Hess at this juncture failed to consider a slightly different alternative, namely, transport of newly created seafloor by a passive process that would not bring about an uneven layer. Indeed, movement of newly created seafloor from ridge sites by a passive process is central to his seafloor spreading concept. This reflects the fact that in 1959 he had not developed his concept of seafloor spreading.

14 Hess continued to have enormous respect for Fisher's work on trenches. In giving one of Fisher's NSF proposals the highest possible rating, he noted on November 2, 1966, "Fisher is probably the most experienced man in the country in research on trenches and in doing the type of work at sea described in this proposal." I should like to thank Princeton University and George Hess for allowing me to quote from this document from Hess's papers at Princeton University Library.

15 The authors noted in their paper that the "final analysis, interpretation and conclusion are those of the first two authors" who were Officer and J. Ewing (Officer *et al.*, 1957: 365). Officer, however, when asked if he and Ewing were both responsible for the solution offered for the origin of the trench and island arcs, recalled that the solution was his alone (Officer, February 8, 2006 phone conversation with author). Because the proffered solution differs substantially from M. Ewing's solution, I suspect that Officer's recollection is correct.

16 Officer left academe soon after publication of this paper, giving up a tenured position at Rice University in 1958 to begin Marine Geophysical Company, which morphed into Marine Geophysical Services in 1958, and Marine Geophysical International a year later. His company eventually merged with Alpine Geophysical Associates in 1961, just before Alpine became publicly owned. Officer resigned as Chairman of Alpine in 1968, and sold his share holdings. With no financial worries, he twice ran as the Democratic candidate for the US House of Representatives (1964 and 1972). He lost both elections, but came very close to winning in 1964, when he lost by only 298 votes. He returned to academe in 1975 when he became an adjunct professor, Earth Sciences Department, Dartmouth College. During his second academic career, he wrote on a variety of topics including Cretaceous/Tertiary extinctions. He argued, often with Charles Drake, against Lois Alvarez's view that extinctions were caused by asteroid impact (see, for example, Officer and Drake, 1986 and 1989).

17 I want to thank Bob Fisher for taking the time to find the note, and giving me permission to include it.

# 4

# Another version of seafloor spreading: Robert Dietz

## 4.1 Introduction

Robert Dietz (1961a) first presented his version of seafloor spreading in *Nature* on June 3, 1961, a year before Hess's paper on seafloor spreading finally appeared. Notwithstanding, as Dietz (1962c and 1968) acknowledged, Hess deserves priority for seafloor spreading. It is not clear to me whether Dietz came up with the idea independently, took it from Hess, or forgot that Hess had told him about seafloor spreading before he presented his own version. I will weigh the alternatives at the end of this chapter. Regardless of its origin, Dietz's version merits close attention. He presented a slightly different defense of it than Hess.

## 4.2 Robert Dietz, the man

Robert Sinclair Dietz (1914–95) was born in Westfield, New Jersey, approximately 40 km from New York City, the second youngest of seven children: six boys and one girl.[1] Dietz's mother died before he finished high school, and his father died several years later. Dietz characterized his older sister as his surrogate mother. She was an assistant librarian and introduced Dietz to the world of books. Once in high school, he read James Jeans, Arthur Eddington, and H. G. Wells, decided to become a scientist, and rejected religion. He and a school friend became ardent rock hounds, often visiting the American Museum of Natural History, where they were befriended by a retired Yale professor who encouraged their interest in geology.

Dietz decided on geology before attending college. He (1994: 4) "hitchhiked west to the University of Illinois choosing that school mainly because the Chicago World's Fair (1933) was in progress," and out-of-state tuition was only $65 per semester. There he obtained his B.S. (1936), M.S (1939), and Ph.D. (1941) in geology, although most of his graduate work was done while at Scripps. Dietz's mentor, Francis Shepard, one of the pioneers in marine geology in the United States, taught at the University of Illinois but did his research during summers at Scripps. Dietz and his fellow student, K. O. Emery, became Shepard's first graduate students in marine geology. Before deciding to join Shepard, Dietz wanted to write his Ph.D.

thesis on the Moon's surface features, but that was vetoed. "The idea was chided as totally bizarre and besides 'there was no one to check my field work.' I was told that speculation does not become a young student and it would be better to do some real geology, like mapping a quadrangle in Vermont" (Dietz, 1994: 21). Dietz turned to Shepard and with Emery helped map submarine canyons in the seafloor off California. He worked with Shepard from 1936 until 1941, and later characterized their association as "the defining event" of his career (Dietz, 1994: 6).

When Dietz received his Ph.D. in June 1941, he had little prospect of getting an academic position. Marine geology was not considered important, and positions in geology were few. A graduate of the Reserve Officers' Training Corps (ROTC), he was called to duty as a ground officer in the US Army Air Corps on August 7, 1941. After graduating from flight training school, his request for duty in Europe or the Pacific was turned down. He first became a flight instructor at an air navigator's school, and later spent most of the war photomapping in South America, often flying over the Andes and the headwaters of the Amazon. He later said, "Flying long navigation training missions caused me to reflect about the Earth below in a generalized way" (Dietz, 1994: 21).

Dietz planned to start his own private photomapping company after the war, but he accepted an offer to form and direct a seafloor studies group at the new US NEL in San Diego, with the added inducement that he would be the oceanographer on Admiral Byrd's Navy-sponsored last expedition, Operation Highjump, to Antarctica (1946–7). He was based at NEL as a civilian scientist from 1946 until 1963. His tenure there was interrupted by a one-year (1953) Fulbright Fellowship to Japan and four years (1954–8) with the Office of Naval Research (ONR) in London, England. Much of Dietz's work at NEL was done in cooperation with Scripps, where he was an adjunct professor from 1950 to 1963. Dietz co-directed the 1950 MidPac Expedition, which was carried out jointly by NEL and Scripps. During the expedition, E. L. Hamilton discovered Cretaceous aged fossils atop guyots, which showed that Hess's speculation that they were drowned during the pre-Cambrian was not possible. Dietz made sure that Hamilton, then just a graduate student at Stanford University, got to work up the discovery for his Ph.D. (Menard, 1986: 54). While in Japan, Dietz discovered a colored 1940s Japanese Hydrographic Office "physiographic chart" that startlingly portrayed "relief" of a large sector of the Northwest Pacific. It featured a dogleg series of seamounts and guyots trending northwest from Midway Island to the Kamchatka Trench. Dietz named this progression the Emperor Seamounts, employing dynasty names for individual seamounts. This chart, with western notations substituted, and Dietz's explanatory text, was distributed very widely. During his assignment as Deputy Director of the London ONR office, Dietz himself wrote research reports on various European developments in the Earth sciences, but did little research. One covered the 1956 conference in paleomagnetism at Imperial College (II, 3.14, §5.2). Dietz was impressed with the paleomagnetic support for mobilism, which, I shall argue, first made him think seriously about mobilism. While attending a 1955 meeting in London on deep-sea diving, he met Jacques Piccard, who was testing his underwater, balloon-like bathyscaph, *Trieste*. Dietz convinced the US Navy to sponsor a series of dives off Italy in 1957, and then purchase *Trieste* and transfer it to NEL in San Diego. After improving the vessel, the US Navy sponsored a dive to the bottom of the seven-mile-deep Mariana Trench. Piccard wanted Dietz to accompany him on the dive, but the US Navy required a uniformed Navy officer, and Don Walsh, not Dietz, made the descent with Piccard. The book *Seven Miles Down: Story of the Bathyscaph Trieste* (1961) by Piccard and Dietz documented the dive.

Once his work with Piccard began winding down, Dietz, as he later remarked,

vowed to try to work alone: 1. eschewing team projects, 2. avoiding classified work which could not be published in open scientific journals, 3. avoiding contributing to the grey or internal Navy reports, and 4. selecting research on the leading edge of science. Although it was not in my position description to engage in scientific generalities, I chose to write about the marine geologic evidence for continental drift. A flurry of papers resulted concerning the origin and nature of continental slopes, the continental rise prism as nascent eugeoclines, and a mechanism for continental drift which I called seafloor spreading – a name I considered awkward but it has stuck.

*(Dietz, 1994: 15)*

Thus began the most productive period of Dietz's research career, work in marine geology that I shall later examine in detail. During this period, however, his work was not restricted to marine geology; he also wrote on meteorite impact structures. His work on meteorite impacts eventually intersected with his views about the origin of ocean basins, so I shall describe his work on them before that on marine geology, work that was relevant either to his coming up independently with the idea of seafloor spreading, or immediately appreciating its merits once he heard about it from Hess.

Dietz left NEL in 1963, and joined the USCGS in Washington, DC. He was involved in several oceanographic expeditions, including ones to the Indian Ocean, South Pacific, Antarctic, and Atlantic. He stayed with the Survey, and its two successors, the Environmental Science Services Administration and the National Oceanographic and Atmospheric Administration, until 1976 when he had completed thirty years of government service. During this time he held temporary positions at the University of Illinois (1974), Washington State University (1975–6), Washington University (1976–7), and Arizona State University. He accepted a tenured faculty position at Arizona State University (Tempe) in 1977, and continued on as an emeritus professor after 1985. He took the position at Arizona State because of its strong planetary group and Center for Meteorite Studies.

Toward the end of his career, Dietz engaged in one final controversy, the extra-scientific controversy about evolution versus intelligent design. Dietz described himself as a materialist and secular humanist.

I will end this account of my life and times with my worldview – my philosophy of life. This, of course, has been molded by my lifetime of experience and my scientific (and hopefully objective) outlook. I define myself as a naturalistic materialist, a no-nonsense scientist. The natural to me is sufficiently awesome with its three "infinites": the infinitely large celestial universe, the infinitely small world of the atom and quantum mechanics, and the infinitely complex realm of life. I reject the supernatural realm of miracles and faith. I accept the material – the reality of matter, but I reject mind and spirit. The mind is what the brain does. I am certainly a skeptic but I don't qualify as an agnostic – this, in my book is a gutless atheist. Nor am I an atheist as this term carries a lot of negative connotations; I regard the terms nontheist or ethical culturalist, as better. Besides an overt admission of atheism would disqualify one from holding public office in many states. I am a secular humanist, a term of derision coined by the TV evangelists, but it describes me well. Humankind must accept the here-and-now and the natural scheme of things in the real world by resolutely solving our own problems through reason and evidence. And since I have recently written a book entitled Creation/Evolution Satiricon: Creationism Bashed (Did the Devil Make Darwin Do It?) with illustrator John Holden, perhaps this qualifies me also as a secular humorist.

*(Dietz, 1994: 26)*

Dietz had little patience with those who tried to reconcile science and religion.

In 1981, during the early days of the resurgent creation/evolution controversy, the Council of the National Academy of Science stated in a resolution, "Religion and science are separate and mutually exclusive realms of human thought whose representations in the same context leads to misunderstanding of both scientific theory and religious belief." This official statement, apparently a peace offering in response to creationism parading as science, is off the mark. Science and religions are not separate and mutually exclusive realms of thought. They are overlapping and irreconcilable modes of thought with science being uniquely the way of knowing. The tenets of religion can and should be subject to exacting scrutiny and testing. Miracles, the power of prayer, life after death, heaven versus hell, and the god concept should not be placed off-limits for investigation. Religion should not remain exempt from criticism and science should not avoid confrontation.

*(Dietz, 1994: 27)*

He also was not very sanguine about the future of humankind. Environmental problems, he claimed, headed the list of human concerns.

Spaceship Earth is hurtling through space and no one, or any collective conscience, is at the helm. Time is of the essence and it is already late. Planets with biospheres where conditions are just right (the Goldilocks's paradox) for higher life are exceedingly rare in the cosmos. Planet Earth may even be unique. Nature is amoral and does not care.

*(Dietz, 1994: 31)*

His solution, what he called ZEES, was "Zero population growth, accepting Evolution as the ruling paradigm of life, Eugenics, and Secularization" (Dietz, 1994: 31).

Dietz died of a heart attack on May 19, 1995, at his home in Tempe, Arizona. He had hoped to die by being struck by a meteorite and then fossilized.

This would be a privilege. I would be the first human to meet such a heroic demise although this apparently happened to an Ordovician cephalopod in Sweden. Hopefully the meteorite might even be a fragment of the Phoceaid family asteroid #4666, so kindly named Asteroid Dietz by its discoverers Caroline and Eugene Shoemaker. As a memorial it is certainly better than any headstone.

*(Dietz, 1994: 32)*

Dietz was not elected to the NAS. He was, however, a Fellow of the AGU, the Mineralogical Society of America, the GSA, and Honorary Fellow of the Geological Society of London, Geological Society of Brazil, and Canadian Society of Petroleum Geologists. He also won several prestigious medals and prizes including the Walter Bucher Medal of the AGU (1971), the Francis P. Shepard Medal for Marine Geology of the Society of Economic Paleontologists and Mineralogists for Marine Geology (1979), the Penrose Medal of the GSA (1988), and the Alexander von Humboldt Prize (1978) given by the Federal Republic of Germany.

### 4.3  Dietz argues for meteorite and asteroid impacts, 1946–1964

Joanne Bourgeois and Steven Koppes (1988) have provided an excellent account of Dietz's involvement in the debate over impacts. Drawing primarily on their work, I shall describe how he made original contributions and that he was not one to shy away from controversy. Although Dietz was not allowed to pursue his Ph.D. studies on lunar craters, he did so on his own, and wrote a paper on the topic while still in the Army (Dietz, 1946a).

I wrote a paper entitled "Meteoritic Origin of the Moon's surface Features." Delays were inevitable, especially because R. T. Chamberlin, editor of the *Journal of Geology*, had recently written a paper on the tectonic origin of the lunar geomorphology; but eventually in 1946 he did publish my paper. It was the first paper in a geological journal to suggest neither a volcanic nor a tectonic origin since that of G. K. Gilbert in 1895. Although volcanic and endogenic explanations remained the consensus view until about 1970, the cosmic bombardment interpretation for lunar craters is now unquestioned.

*(Dietz, 1994: 21; the actual title of the paper was "The meteoritic impact origin of the Moon's surface features," which is my Dietz (1946a))*

Although correct about the meteoritic origin of lunar craters, Dietz seemed unaware of Wegener's similar and much earlier views (see Greene, 1998, and I, §2.5).[2]

Dietz also began thinking that certain structures on Earth were eroded ancient impact craters now variably eroded. This idea was even more controversial than the idea that lunar craters were impact craters. The competing and prevailing view that various crater-like structures were volcanic was championed by Walter Bucher, who, as we have seen, was an adamant foe of mobilism (§1.10). These structures, originally called cryptovolcanic by W. Branca and F. Fraas (1905) who studied one in the Steinheim Basin, southern Germany, are roughly circular, vary from less than 2 km to as much as 120 km in diameter, and have a depression surrounding a central uplifted area covered by sheared, shattered, and pulverized rocks. Because they contain neither meteorites nor obviously volcanic rocks, their origin, beyond being some sort of explosion, was a matter of speculation. Branca and Fraas suggested that the Steinheim cryptovolcanic crater had been caused by sudden release of trapped volcanic gases. Bucher (1936) agreed, and identifed six cryptovolcanic structures in the United States (Wells Creek Basin, Tennessee; Serpent Mound, Ohio; Jeptha Knob, Kentucky; Kentland Dome, Indiana; Decaturville Dome, Missouri; and Upheaval Dome, Utah) which, he proposed, had formed in a similar manner.

Dietz thought such structures were eroded meteorite impact craters. The idea came to him while viewing one of these structures from above during one of his low-flying training missions.

Flying long navigation training missions caused me to reflect about the Earth below in a generalized way. It is a form of remote sensing not unlike contemplating the Moon. One day in 1943, it occurred to me that the disrupted nest of lower Paleozoic strata in the Kentland quarry in Indiana might not be cryptovolcanic as commonly supposed but an asteroidal impact scar. Could not the orientation of shatter cones exposed there resolve this uncertainty? To test this idea I later stopped over at the Air Force Base in Rantoul, Illinois and hitchhiked to Kentland. There, indeed, was a preferred orientation such that, when the vertically-dipping strata were rotated to an assumed original horizontal position ... the cones pointed upward suggesting by Hartman's Law that the fracturing impulse came from above and hence was cosmic rather than volcanic.

*(Dietz, 1994: 21)*

The core idea was that if the craters on the Moon are impact craters, and Earth and Moon have shared roughly the same history, there should be impact craters on Earth. However, most Earthly craters will be less well preserved than lunar craters because they, unlike lunar craters, have been eroded and perhaps deformed tectonically.

Dietz was not the first to think that the "cryptovolcanic" structures in the United States were eroded impact craters. J. D. Boon and C. C. Albritton, both at Southern Methodist University, favored meteorite impacts. They (1937, 1938) argued that all cryptovolcanic structures identified by Bucher were really meteorite impacts. They also argued, contrary to Bucher, that Meteor Crater (thought originally to be volcanic) of Arizona was an impact crater, and identified two other cryptovolcanic structures in the United States as impact craters: the Sierra Madera dome, Texas, and Flynn Creek disturbance, Tennessee. Turning to other parts of the world, they noted that F. E. Suess (1936) had recently identified the Kofels crater in the Tyrolian Alps as meteoritic in origin, that others had suggested that the Steinheim cryptovolcanic structure and the Ashanti Crater of the West African Gold Coast (now Ghana) were meteoritic, and cited recent impact craters in Siberia to further their argument. They also suggested that the Vredefort Dome, South Africa, was a meteorite impact crater (see I, §8.12 for discussion of Suess's support of mobilism). Dietz discussed Boon and

Albritton's work with them, and sometimes referred to the meteoritic hypothesis as the "Boon–Albritton hypothesis" (Bourgeois and Koppes, 1998: 141).[3]

In 1944, the year after he flew over Kentland Crater, Dietz visited the Flynn Creek disturbance in Tennessee that Boon and Albritton (1937) already had identified as an impact crater. Armed with a mine detector to help find meteoritic debris, an effort he later described as naive, he found only nails and horseshoes. Undeterred by the negative results, he still "was inclined to believe that the eight cryptovolcanic structures described by Walter Bucher in 1935 (sic, 1936) were most likely impact structures" (Dietz, 1994: 21).

Dietz's first paper on terrestrial meteorite craters appeared in 1946, and another a year later; in them he made two original contributions. He argued (1946b: 465) that the term "cryptovolcanic" begged the question, and suggested that such structures should be referred to as "crypto-explosive." His really important contribution, however, pertained to the orientation of shatter cones (Dietz, 1947). Shatter cones form when rocks are fractured by intense and high-velocity shock waves. They range in height from 1 centimeter to 2 meters – those at Kentland are 2 meters. The orientation of shatter cones, he argued, determines the source of the shock waves. If their apexes face upward, they were caused by shock waves resulting from an above-ground explosion; if the apexes face downward, the shock waves came from below ground. Because the apexes of the shatter cones at Kentland faced upward, Dietz argued (1947) that they had been caused by meteorite impact.

Branca and Fraas (1905) first discovered shatter cones while examining the Steinheim Basin. About twenty-five years later, Krantz (1924) saw shatter cones in the Steinheim Basin, named them "Strahlenkalk" and argued (1924) that they were caused by an explosion, which he attributed to the rapid escape of volcanic highly pressured gases. Bucher reported all this in 1936 (p. 1070), agreed with Krantz, and noted the presence of shatter cones at Kentland. So Dietz's argument based on orientation of shatter cones turned on its head Krantz's explanation of their presence in crypto-explosive structures; it was a real advance.

Dietz returned to the origin of cryptovolcanic structures in the late 1950s and early 1960s. By 1959, he had observed shatter cones at three more sites, Wells Creek Basin, Flynn Creek, and Sierra Madera.

With the launching of Sputnik in 1958 there was a sudden surge of interest in space. I was invited along by Gerard Kuiper, along with Eugene Shoemaker and two other geologists, for four nights of Moon observation with the McDonald Observatory telescope in west Texas. Once there, I suggested we spend one day examining the Sierra Madera, a deranged mountainous structure, as a possible terrestrial analog of lunar crater. Claude Albritton, although he had never visited the site, suggested it to be a candidate astrobleme [meteorite impact crater] based upon its damped wave structural style ... We found it to be nicely shatter-coned, which already for me was a definitive criterion for astroblemes. Years earlier I had studied them at the Steinheim Basin, Wells Creek Basin, Flynn Creek, and Kentland. This started me on a worldwide search for shatter-coned structures.

*(Dietz, 1994: 21–22; my bracketed addition)*

A year later, he (1960a) listed three more shatter-cone sites, adding Crooked Creek (Missouri), Serpent Mound (Ohio), and the Ashanti Crater (Ghana), and he coined the term "astrobleme" from the Greek for "star" and the Latin for "wound by a thrown object" to refer to crypto-explosion structures on Earth caused by meteorite or asteroid impacts.

Dietz studied aerial photographs of the Vredefort Ring in South Africa in 1960, and following Boon and Albritton, and Daly, he argued (1960b) that it was an astrobleme. He predicted that shatter craters would be found there, and asked several South African geologists to look.

While searching for the terrestrial equivalent of lunar crater Copernicus on Earth, I was drawn to a giant bull's-eye in South Africa, the Vredefort Ring, a central Archean granite body 40 km across surrounded by a thick upturned collar of Precambrian sediments. It seemed a prime suspect as an astrobleme created by recoil uplift following a hypervelocity asteroidal impact. I wrote letters to South African geologists seeking to know if this structure was shatter-coned. Answers were at first negative but eventually Robert Hargraves replied "Yes!" and sent definitive photographs to prove it. With that information in hand I committed a cardinal sin: writing a paper before visiting the site entitled "Vredefort Ring Structure: Meteorite Impact Scar" espousing an impact scenario.[4]

*(Dietz, 1994: 21–22)*

With his prediction confirmed, Dietz (1961b, 1961c, 1963a) championed his astrobleme interpretation; he also debated the issue of whether crypto-explosions were impact or volcanic craters with Walter Bucher. Bucher (1963) wrote a lengthy review in which he continued to argue in favor of a volcanic origin, and he also invited Dietz to respond (Bourgeois and Koppes, 1988: 148). Dietz (1963b) obliged.[5]

Dietz next made another original contribution to the astrobleme debate; he proposed that the Sudbury structure in northern Ontario, Canada, is an astrobleme. Bourgeois and Koppes (1988: 139) noted that this was the only asteroid impact that Dietz proposed that had not been suggested by others previously. Dietz (1962b) presented the idea in December 1962 at an AGU meeting. It was extremely controversial. The astrogeologist and historian of geology, Ursula Marvin, told Bourgeois and Koppes (1988: 154) that around the time that he presented his AGU talk "she suggested, based on this and other work, that Dietz be invited to Harvard to give a talk on Sudbury. The students threatened to boycott; Dietz was not invited." Dietz recounted his work on Sudbury: what led him to think of Sudbury in terms of an astrobleme, how the idea was received, his finding of shatter cones at Sudbury, and the general acceptance of the idea in the early 1990s.

Following my Vredefort paper I turned my interest to searching for a terrestrial analog of a lunar mare – a "wet" impact with a central melt sheet or triggered volcanism. The large (35 × 60 km) kidney-shaped Sudbury Igneous Complex (S.I.C.) in Canada seemed a potential tectonized or squashed example. Of course, as an astrobleme one needs to scrap the classical model of an intrusive lopolith and consider the S.I.C. as an open pool of chilled magma crusted over with a fall back suevite (impact microbreccia) rather than a volcanic tuff. In the spring of 1962, to investigate my hunch I took leave from my position at the Navy Electronics Laboratory to visit Sudbury. A brief field study convinced me of the reasonable reality of my model based largely on geological relationships. It was not until the end of my visit that I discovered definitive shatter coning, as this fracturing is degraded in Precambrian rocks, unlike its development in limestone terranes. [Dietz discovered the shatter cones in August 1962 (Dietz and Butler, 1964: 280).] The resulting paper entitled "Sudbury Structure as an Astrobleme" eventually appeared in 1964 [my Dietz, 1964a]. Publication had been delayed by a "pocket veto" by the reviewer – who finally returned the manuscript only long after being prodded to do so. Even then the only comment was "nonsense." Fortunately the editor of the *Journal of Geology* decided to override this negative appraisal. In 1972, I amplified my model to argue that the Sudbury nickel ores found in the sublayer was cosmogenic. The view has yet to be accepted although geologic relationships, especially the emplacement timing, support this view. I participated in a NASA symposium at Sudbury on large terrestrial impacts with Sudbury being the type example. It was good to hear one's ideas become mainstream.

*(Dietz, 1994: 23; my emphasis and bracketed additions)*

Dietz (1964a) supposed that a 4 km diameter nickel-iron meteorite collided about 1.7 billion years ago with Earth at Sudbury, which was then a young granite and sedimentary terrain. A gigantic impact crater formed. The meteorite and country rock were partially liquefied and formed the Sudbury ore – in 1964, 75% of the Western world's nickel was produced at Sudbury. Magma triggered by the impact flowed from below into the crater bowl, forming a huge saucer-shaped body of magma that Dietz referred to as an extrusive, irruptive, or explosive lopolith in contrast to the standard intrusive lopolith. Ensuing tectonic activity and erosion reduced the Sudbury crater to its present state.

In his acceptance speech of the Penrose Medal, Dietz remarked:

My career has spanned an era of remarkable change; however, if I compare my college textbook with one of today, I find that few really new chapters have been added. I take a modicum of pride in having been involved early-on in three of these: marine geology, planetology, and plate tectonics.

*(Dietz, 1989: 988)*

The importance and originality of Dietz's work in planetology make it, I believe, his most original contribution to the Earth sciences. He correctly argued that shatter cones ruled out volcanism as the cause of craters. He kept hammering away at the idea that they were astroblemes, and made the daring hypothesis that Sudbury was one.

When Dietz recognized Sudbury as an astrobleme, it was not the first time he had made comparisons with lunar maria, as possible terrestrial analogues. In 1958, three years before his paper on seafloor spreading, he had proposed that asteroid impacts caused ocean basins, linking them with continental drift. Before describing this, I turn to Dietz's recollections thirty years later of his views on mobilism at the time, which he claimed were shaped by his work in marine geology during the first half of the 1950s.

## 4.4 Dietz recalls his pre-1954 attitude toward mobilism: a 1987 interview

In a 1987 interview with me, Dietz recalled that mobilism "was talked about but always in a derogatory way" adding that he did not "know of any supporters." Although he remembered reading Wegener and du Toit, he did not say when he first read them nor what he thought of their work.

I first learned about it from, I guess, Wegener's book, and du Toit, his book and also he wrote a special paper on his travel to South America, I guess, something with the Smithsonian, the du Toit paper was in the early 1930's. [It was actually the Carnegie Institution of Washington. Du Toit went to South America in 1923, and the resulting work, *A Geological Comparison of South America with South Africa*, was published in 1927 by The Carnegie Institution of Washington, which also funded his research (I, §6.6).] I also knew about the 1926 AAPG symposium.

*(Dietz, 1987 interview with author; my bracketed addition)*

I do not know whether Dietz had read Wegener and du Toit, and had looked at the proceedings of the AAPG symposium before his work in marine geology had sparked his interest in mobilism. In the interview he thought none of the classical arguments were incontrovertible, and that there were a lot of soft data. He had not read Holmes on mobilism. He thought Daly was the best North American geologist, but I do not know if he read *Our Mobile Earth* before arguing for seafloor spreading. Among the classical arguments, he thought the arguments based on Late Paleozoic glaciation were the best. He was also impressed by Carey's (1955b) fit of South America and Africa, and his ideas about oroclines and megashears (II, §6.7, §6.11).

In the interview he claimed that he first began thinking in terms of mobilism not because of continental evidence but because of his and others' findings in the Pacific, in particular the paucity of sediments on the seafloor, the youthfulness of guyots, the apparent descent of several guyots and seamounts into trenches, and apparent underthrusting of oceanic crust beneath trenches, which implied large horizontal movement. The idea of creating seafloor at ridges did not at this time seem to have crossed his mind. Dictz, unlike Hess, did not begin thinking about horizontal movement of seafloor because of already longstanding work on ocean ridges.

Here are some excerpts of what he said.

I became a mobilist after the MIDPAC Expedition and after my time as a Fulbright Fellow during 1953 in Japan and working on the seafloor of the Northwest Pacific and also upon the geologic history of Japan, apparently where some rocks which had an Asian provenance which are in Japan and they are separated by the Sea of Japan. But I guess my concern, my interest, was with all those surprises on the seafloor. So I was concerned with the ocean floor being mobile, being subject to large motions, because I was focused on the ocean floor. I guess I was a minimal drifter then.

I didn't support the idea until the MIDPAC Expedition in 1951 or so – something like that, 1952 I believe. I began to take it seriously ... when I went to Japan. I became involved in studying the northeast Pacific, then I became a fairly serious student of continental drift and I read all the literature there was. That was quite possible because maybe there would be two or three papers a year.

There were lots of surprises in the ocean floor, and most notably ... that the sediments were thin. We already had an inkling of this from the expedition of the Swedish *Albatross* Expedition. They did bottom reflection work, and the amount of sediment was one-tenth of that which Kuenen had proposed. He thought the sediment should be 10 km thick, and here it happened to be 1 km or a bit less than that. So this was a big surprise. Then the big thing was the guyots, which Hess proposed were Pre-Cambrian topography, a museum of ancient topography like on the moon, and we attempted to test this on the MIDPAC expedition by dredging the seamounts ... Hamilton had the data and the samples. Mainly we dredged up rudists. They're a mollusk which produces reefs and they died out at the end of the Cretaceous ... They look like a coral, but they are actually mollusks. But we had a collection of rudists as well as a body of microfossils, etc. This was a big discovery. We recognized that because we inferred that the seamounts were also Cretaceous age ... So we then realized that guyots were young. This was a big surprise. You can only have so many surprises before you realize your whole model is just a house of cards. So then we got to be kind of mobilistic. I finally went to Japan in '53 as a Fulbright Fellow, and I made a study of the Northwest Pacific.

Well then the idea goes in Japan when I began to get involved in looking at trenches and also earlier in the Gulf of Alaska with Menard we found these trenches with seamounts [and guyots]. We knew about them; we contoured them up from Coast and Geodetic Survey data. We found these lines of guyots, and one of them appeared to be going down in the trench. It had a tilted top we thought – this has been shown not to be justified. But anyway it could be that here is something moving down into the trench. So you have a mobilistic seafloor. Then I found this again in the Northwest Pacific. I saw seamounts in the Emperor Chain and also what I called the Magellan Seamounts to the east of the Mariana Trench where you had seamounts being consumed in the trench. And, of course, the idea was that trenches might be underthrusts. So I became rather wedded to that idea that they were underthrusting. I recognized that there must be considerable underthrusting, not minor. But I didn't think it was necessary to have seafloor spreading because [looking at Earth as a whole] ... you could take up transcurrent motion but have large [motion] here and small [motion] there. You could absorb it in a non-rigid crust. So this was an annoyance not having a place to generate new crust but it didn't really defeat the idea. Once seafloor spreading came along, I knew the mid-ocean ridge being mountains, being a wholly different type of mountain – not terrestrial mountains – being a rift there, then it is very natural to say, "Well, here is the missing part." Most people came to the idea, I think; they understood the spreading before they understood the consumption of crust.

*(Dietz, 1987 interview with author; my bracketed additions)*

## 4.5  Dietz's marine geological work before going to London in 1954 and his later recollections compared

I now want to review Dietz's papers on marine geology written before he worked as liaison officer for ONR in London. Most were co-authored; several were with Menard, who later remembered:

We [Dietz and Menard] then began [late 1949] a collaboration that was to result in coauthored papers on the origin of the shelf break; the marine geology of the Gulf of Alaska; the Mendocino escarpment; and the Hawaiian deep, swell, and arch, as well as a three-author paper with Ed Hamilton. Our practice was to identify a subject for a paper in the course of our discussions. The whole marine geology group was jammed into a few tiny rooms with a leaky roof, so communication was constant. When we had a subject, Dietz and I sometimes flipped a coin to decide who would be first author, with the understanding that whoever it was had to write the first draft of the manuscript. And so we produced the five papers [Dietz and Menard, 1951; Menard and Dietz, 1951; Menard and Dietz, 1952; Dietz and Menard, 1953; Dietz, Menard, and Hamilton, 1954]. I wrote the Gulf of Alaska and Hawaiian papers. [Dietz, however, was the first author of the Hawaii paper.] On the Mendocino paper I certainly identified the target, did all the fieldwork on two expeditions, and wrote the first draft and drew the figures, but what determined who would do these things I do not recall. In any event, the ideas in all four papers were the consequence of a thorough amalgamation of our discussions.

*(Menard, 1986: 58; my bracketed additions)*

Menard and Dietz discussed the origin of the Aleutian Trench and arc in 1950 (Menard and Dietz, 1951). Menard estimated that they contributed equally to the paper (Menard, July 1979 letter to author). They explained the origin of the Aleutian Trench and associated island arc by appealing to the tectogene or downbuckling hypothesis of Kuenen, Vening Meinesz, Umbgrove, and Hess. Citing works by Umbgrove and Hess, they (1951: 1277) proposed that Kodiak Island and the Kenai Peninsula arose during the Cretaceous as an ancient sediment-filled trench, originally formed by an earlier crustal downbuckle that was later uplifted. A second trench, the current Aleutian Trench, developed during the Tertiary, was filled with sediment, and is still being uplifted (1951: 1279–1282).

Dietz appealed to tectogenes while discussing island arcs and trenches in the northwestern Pacific.

It is now widely believed that the fundamental structure of arcs is the tectogene or a downbuckle of the earth's crust into subcrustal material ... Geanticlinal ridges appear to form on the active or concave side of the downbuckle which develops directly beneath or slightly on the concave side of the trench. The trench will remain as long as the crust is under horizontal stress but rises isostatically into an Alpine mountain range if the stress is relieved.

*(Dietz, 1954: 1205)*

There is here no suggestion that seafloor is consumed at trenches, no mention of continental drift. His appeal to tectogenes is essentially what Kuenen, Vening Meinesz, Umbgrove, and Hess made before the advent of seafloor spreading. It is a fixist explanation.

Dietz's explanation of guyots also was not mobilistic. He agreed with Hess about the formation of guyots; they are sunken seamounts decapitated by wave action. He claimed that their truncation occurred during the Cretaceous because of Ed Hamilton's finding of Cretaceous rudists fossils.

For reasons too lengthy to consider here, the writers do not consider Hess's reasoning regarding the extremely great age of the guyots to be compelling; and, in fact, the recent recovery of mid-Cretaceous shallow water fossils suggests that these were truncated in the Cretaceous (Hamilton, 1952).

*(Carsola and Dietz, 1952: 493–494; Hamilton reference is the same as mine)*

He gave two explanations for their sinking. Guyots that were not within or beside trenches sank through depression of oceanic crust, they sank under their own weight.

These ancient islands are now deeply drowned below the surface of the sea. What caused them to fall so far below the sea level? Our echo sounder provides a clue. Several of the table mounts are surrounded by depressions, or moats, around their bases; they look like a coffee cup turned upside down in a saucer. It is tempting to conclude that they sank because the mountain mass bent down the earth's crust, much as small body standing on thick ice bends down the ice over a large saucer-shaped area.

*(Dietz, 1952a: 22)*

Guyots found within trenches, he argued, were created before the trench formed. They sank when the crust downbuckled and formed a trench. He explained how the slanted-top guyot mentioned in the interview sank.

But movement and buckling of the crust itself has also contributed to the drowning of some of the guyots. As we drive north into the great undersea valley known as the Aleutian Trench, we can see evidences of this. On the south flank of the Trench is a table mount whose top is tilted in toward the Trench; evidently it was first decapitated and then tilted over when the Trench was formed by downbuckling of the crust.

*(Dietz, 1952: 22)*

Dietz envisioned a buckling seafloor that created trenches and drowned already truncated seamounts and guyots, not a moving seafloor that plunged into the mantle taking along seamounts and guyots.

Two years later, Dietz again appealed to local or regional adjustment to explain submergence of the Emperor Seamounts.

Five of the seamounts of the Emperor Seamounts appear to have flat tops and thus are guyots (Hess, 1946, p. 772) or tablemounts ... There are, of course, many possible ways in which the submergence of guyots may have taken place ... On the other hand, the present writer is inclined to favor local isostatic subsidence of large seamounts because of the great load they place on the earth's crust. Depressed zones or "moats" have been found around the base of some guyots which tends to support the latter idea (Menard and Dietz, 1951; Dietz and Menard, 1953). Soundings suggest such a moat at the base of Jimmu [one of the Emperor guyots].

*(Dietz, 1954: 1211; my bracketed addition; references are the same as mine)*

He related the submergence of the Magellan Seamounts beside the Mariana Trench to the trench's formation.

The seamounts along the western limb [of the Magellan Seamounts] are deeply submerged and are connected by a continuous ridge. Here they lie tangent to the Mariana Trench and possibly have been depressed by the formation of that foredeep.

*(Dietz, 1954: 1212)*

The Mariana Trench formed after the seamounts; they sank as the crust beneath downbuckled and became the Mariana Trench. His solutions to these various problems, proposed before leaving for London in 1954 invoked vertical not horizontal motion. They were not, contrary to his recollections, mobilist.

In his 1987 interview Dietz also mentioned that his favorable attitude toward large transcurrent motions fostered his interest in mobilism.

I was impressed with mobilism when I heard several papers. I was always inclined toward large transcurrent motions. I heard a lecture by Brouwer, a Dutchman, about transcurrent faults. I thought it was very good ... I also knew the work of [H. W.] Wellman in New Zealand. He was a strike-slip man; he first identified the Alpine Fault in New Zealand [and in 1956 he suggested a displacement of 300 miles along it].

*(Dietz, 1987 interview with author; my bracketed additions)*

Although Dietz did not specify which Brouwer he heard or when he heard him, it was likely Hendrik Albertus Brouwer (1886–1973). He favored mobilism during the second decade of the last century because of his work in the Dutch East Indies (I, §8.13). Brouwer retired in 1957 from the Geologisch Instituut der Universiteit van Amsterdam where he was its director, but remained active through the early 1960s (Egeler, 1973). He spoke on Earth movements along huge horizontal faults at the 10th Pacific Science Congress, held in Honolulu, Hawaii, August–September 1961, where Dietz also spoke on seafloor spreading, and both papers were published together (Dietz, 1961d; Brouwer, 1962) (§4.9). If this was the first time that Dietz heard Brouwer speak, then he heard him only after he had proposed seafloor spreading. Dietz most surely would have known about Wellman's work, but Wellman did not argue in favor of horizontal displacement until 1956. Although Dietz did not mention W. Q. Kennedy's 1946 work on the Great Glen Fault in Scotland, he knew of it because he and Menard referred to it in their paper on the Mendocino Escarpment (Menard and Dietz, 1952). (See II, §2.7 for R. A. Fisher's appeal to the Great Glen fault in his defense of mobilism.) Dietz was also familiar with the San Andreas Fault.

Dietz and Menard discussed transcurrent faulting in their studies of the Mendocino Escarpment and the Hawaiian region of the Pacific Basin.[6] Dietz's verbal recollection again does not agree with what he and Menard wrote. I begin with their explanation of the origin of the Mendocino Escarpment, the first of the great fracture zones in the northeastern Pacific to be discovered (Menard and Dietz, 1952). The escarpment trended westward from Cape Mendocino, was at least 1900 km long, and three times longer than originally thought (Murray, 1939; Shepard and Emery, 1941). The escarpment had a relief of 3200 m, and was located on the southern side of a long asymmetrical ridge; the seafloor to the north was approximately 0.8 km higher than that to the south. They suggested that the northern higher seafloor was less dense than the southern and the escarpment was caused either by thrust or transcurrent faulting. Thrusting better explained the relative elevations; transcurrent faulting, the escarpment's straightness. They also tried to link the escarpment to the San Andreas Fault. They acknowledged that both solutions faced difficulties, and that more data were needed before anything more than a working hypothesis could be formulated. Their account shows that Dietz was familiar with transcurrent faults, but does not confirm that he "was always inclined toward large transcurrent motions."

Their account (Dietz and Menard, 1953) of the origin of the Hawaiian Islands, and the associated deep and swell, was submitted approximately eight months after their escarpment paper. They rejected previous hypotheses and proposed a new one. They first considered the proposal of Betz and Hess (1942) that they had formed through strike-slip faulting (§3.4). They argued against it, citing the rarity of transcurrent faults (very large transcurrent faults were in fact not widely recognized globally at the time), especially when accompanied by massive volcanism, and the unlikelihood of them producing such a massive structure. On the rarity of transcurrent faults, they claimed:

> Several objections, however, can be pointed out. Positively identified major strike-slip faults are rare, which leads one to suspect that the crust does not often fail in this manner. Also, volcanism is not commonly associated with transcurrent faults.
>
> *(Dietz and Menard, 1953: 108)*

This does not sound like someone who "was always inclined toward large transcurrent motions" as he recollected thirty years later. Instead, they proposed that the Hawaiian region formed as a result of lateral tension between the ascending limbs of diverging convection currents. Movement was vertical. They argued that this avoided the above objections and was consistent with the gravity data; but nevertheless, they had "little confidence that the favored hypothesis proposed will prove to be entirely correct" and "hoped that [their new data], like all data, will serve to restrict (but not to inhibit) speculation" (Dietz and Menard, 1953: 112–113; my bracketed addition). Interestingly, they appealed to mantle convection, but the cells they proposed were smaller than those Dietz later invoked to explain seafloor spreading; their convection never surfaced, and merely fractured a static crust, forming fissures through which upwelling magma escaped and formed the volcanic Hawaiian Islands.

Dietz also offered the same explanation for other oceanic rises. Noting that seamounts were often associated with rises, he repeated his and Menard's solution, which was not mobilist.

> Seamounts [other than those associated with trenches] are not randomly disposed but, rather, are arranged in broad belts. The linear arrangement is not merely so perfect as that which characterizes the volcanoes associated with the arcuate structures; thus a different type of volcanism is indicated. Seamounts in the Pacific Basin usually seem to be developed on broad linear rises or swells of the crust. For this reason Dietz and Menard (1953, p. 110–112)

suggested that the swells and the seamounts are genetically related. Possibly the swells develop over thermal convection currents in the earth's mantle. Tension cracks may then develop in the swell along which magma finds access to the surface.

*(Dietz, 1954: 1208–1209)*

Dietz and his co-authors went on to consider the paucity of sediment beneath deep oceans. They did not conclude that the Pacific Basin must be very young because no pre-Cretaceous sediments had been found. Here is what they wrote.

It is clear that there is little basis for the common assumption that the seafloor is a smooth featureless sedimentary plain upon which sediments everywhere accumulate at a uniform rate. Instead it appears that seafloor topography is determined largely by tectonic movements and volcanic extrusions. The principal role of sedimentation is to fill in the topographic lows. The localization of sedimentation in the basins and the apparent absence of a draped cover of sediments over all the seafloor topography makes it evident that horizontal transportation of sediments takes place along the bottom. Such movement along the bottom and the localization of deposition in basins must be taken into consideration by those who attempt to determine the rate of deposition in the deep sea as a whole by radioactive or other measurements on cores obtained from the sites of deposition.

*(Dietz, Menard, and Hamilton, 1954: 272)*

Examination of Dietz's publications (usually with co-authors) during the first half of the 1950s shows that his solutions to the origin of trenches, sinking of seamounts, and lack of pre-Cretaceous seafloor sediments were not mobilist. He also rejected transcurrent faulting as an explanation of the Hawaiian Islands and swell, and favored a solution to the origin of the Mendocino Escarpment that did not invoke transcurrent faulting.

Despite what he recalled in 1987, his published record shows that in the early 1950s, before he left for London in 1954, he did not propose mobilist solutions to the problems he considered in marine geology. What he later said about his former views is inconsistent with what he had earlier written. It is as if he were recounting what he would have said had he been a mobilist while working on the origin of trenches and seamounts in the early 1950s.

Correspondence between Dietz and Hess in 1952 over the origin of guyots also provides evidence that Dietz did not present mobilistic solutions to the origin of guyots and seamounts during the 1950s. Ed Hamilton sent Hess his Ph.D. thesis on guyots. Hess responded on April 18, 1952. After praising Hamilton's work, and agreeing that guyots were not Precambrian, Hess added as his comment #7, that if guyots were not formed by some tectonic process they might be explained by rapid rise in sea level because of the addition of juvenile water to the oceans (Hess, April 18, 1952 letter to Hamilton; Hamilton papers at Scripps). Dietz replied to Hess on April 24 – Hamilton, already at sea when Hess's letter arrived, had given Dietz permission to open any of his interesting mail. Dietz, taking the liberty, wrote:

In regard to comment #7, Revelle is inclined to believe in drowning because of more rapid increase in juvenile water since the Cretaceous. We (Hamilton, Menard, and myself) are more inclined to believe in local subsidence of seamounts (the crust being "thin ice") plus semi-region subsidence – that is, the Mid-Pac Mts as a whole. This subsidence must mainly be isostatic but could also be due in part to cessation of divergent sub-crust currents.

*(Dietz, April 24, 1951 letter to Hess)*

This does not sound like someone inclined toward mobilism – the idea that the crust has undergone vast horizontal motions – rather it sounds like someone about to propose just the sort of fixist solutions Dietz had offered from 1952 through 1954.

Notwithstanding, Dietz did become receptive to mobilism before presenting his version of seafloor spreading. I believe that he was attracted to it not because of his marine geological work but because of what he learned about its paleomagnetic support.

### 4.6 In London, Dietz (1956) learns about mobilism's paleomagnetic support

I was favorably impressed with the work of Runcorn and others. I knew Blackett and I knew about his history and work when I was in London from 1954 to 1958. He was at Imperial College. I didn't interact with him because he was a physicist and I didn't know magnetics very well. I did meet him, and I met Clegg too. I knew Blackett and Clegg before and after [the meeting at Imperial College in January 1956]. [But] it is more important that I knew the Cambridge people – Runcorn and Bullard and a few other people who were there at the time. [Runcorn and Bullard were not at Cambridge at the same time, and Runcorn had already moved to Newcastle by January 1956.] I guess I met Neil Opdyke. He was a student of [Westoll and Runcorn]. I knew Westoll; he is still a good buddy and friend. He was a continental drifter early on, and he was a world authority on Devonian fish. Blackett was up there and I was down here. He was very well known. Runcorn was a person I could talk to and had enough geological general interest. I've known Runcorn very well over the years. But we didn't interact on a geological plane but in a general way. He is a very interesting person.

*(Dietz, 1987 interview with author; my bracketed additions)*

Dietz attended a conference on rock magnetism held at Imperial College in November 1956 while stationed in London as Deputy Director of the ONR office. The conference was sponsored by the International Union of Pure and Applied Physics and chaired by Blackett. Several speakers discussed the paleomagnetic evidence for mobilism. Dietz had landed in the hotbed of paleomagnetic research during the early stages of the global paleomagnetic test of continental drift.

Blackett announced his strong support for mobilism (II, §3.14). Clegg *et al.* (1957) discussed their work in India and Spain in favor of mobilism (II, §3.13). Runcorn (1957) argued strongly in favor of continental drift and polar wandering. He spoke about his own work in North America, Creer and Irving's work in Great Britain, Irving's Indian results, support for the GAD hypothesis, and the reliability of paleomagnetic data. Creer was in South America collecting paleomagnetic samples and did not attend (II, §5.6). Irving was invited but remained in Australia; his research fellowship there did not provide such extensive travel funds. Nonetheless, he accepted an invitation to present a paper, the one that appeared in the proceedings, and asked

John Clegg to read it for him. Clegg obliged (Irving, June 23, 2005 conversation with author). Irving (1957) proposed paleomagnetic reconstructions of the continents, discussed some of the problems in their construction, presented some of his and his student Ron Green's Australian poles, his Indian and Tasmanian poles, and argued for continental drift and polar wandering (II, §5.3). Phil Du Bois (1957), then at the Geological Survey of Canada in Ottawa and former student of Runcorn's, discussed his work in North America and compared his results to those obtained from rocks in Great Britain; he also argued in favor of continental drift and polar wandering (II, Chapter 5, Note 1). Ken Graham and Anton Hales (1957), both from the Bernard Price Institute in South Africa, tentatively concluded (1957) that their work on the Karroo dolerites supported continental drift and polar wandering (II, §5.4). They also mentioned Irving's new pole from the Tasmanian dolerites, and noted that their Mesozoic-aged pole from the Karroo dolerites was very different, as it was from roughly contemporaneous poles from Europe, North America, and India. Although Nairn (1957) discussed some results from his recent work in South Africa, and had argued in favor of mobilism, he restricted his comments to the difficulties caused by severe weathering of outcrops in tropical regions – a real barrier to work there (II, §5.4). There was, however, some opposition to mobilism. John Graham raised his magnetostriction difficulty, and Balsley at the USGS and Buddington at Princeton University argued that their study of metamorphic rocks from the Adirondack Mountains in New York supported Graham's concern (RS2, II, §7.4).

Dietz discussed mobilism's paleomagnetic support in a January 1, 1957 technical report to ONR, devoting two paragraphs to polar wandering and continental drift.[7]

The paleomagnetic case for continental drift and polar wandering, although favored by many English workers and especially championed by Runcorn, would appear from the consensus of opinion at the conference to be on shaky ground. Conflicting results seemed to be as prevalent as those which were in agreement. According to Runcorn the pole wandered from the central Pacific beginning in pre-Cambrian time, through Japan and Siberia, and finally to the present position. The theoretically reasonable assumption is made that the magnetic pole must always be near the geographic pole. Polar wandering could result either from a slippage of the entire crust of the earth with respect to the core or by spilling of the axis of spin – a hypothesis which seemed most unlikely until T. Gold recently showed that the earth's axis may be inherently unstable.

Regarding continental drift, Runcorn has marshaled evidence to show that North America has drifted 2500 miles from England between mid-Mesozoic and mid-Tertiary; however, drift of a landmass is paleomagnetically indistinguishable from its rotation, and the data can be equally well satisfied by assuming a rotation of England through about 10 degrees. Blackett also has evidence from the study of the Deccan traps that India has drifted several thousands of miles to the north. Although both drift and wandering are far from being firmly established, there is little doubt but that rock magnetism will eventually supply a definite answer to these intriguing problems.

*(Dietz, 1957: 8)*

This report differs from the review I have just given of the published proceedings. Few authors give me the impression of being on "shaky ground." Perhaps Dietz was reflecting audience opinion or that of his fellow Americans, John Graham, Balsley, and Buddington. Nonetheless, Dietz was certainly becoming familiar very quickly with some of the support paleomagnetism was providing from Europe, North America, Australia, India, and Africa for continental drift and the GAD hypothesis. He singled out Runcorn and Blackett. He did not specifically discuss the contributions by Clegg, K. W. T. Graham and Hales, and Du Bois. If he heard their talks, and Clegg's presentation of Irving's paper, he also would have known about work on the Iberian Peninsula and in Africa and Australia. Dietz also did not single out John Graham, and Balsley and Buddington, who raised difficulties with mobilism's paleomagnetic support. He certainly was correct about Runcorn, who was quickly positioning himself as a champion of mobilism: remember, he had come to accept drift in July 1956 only months before the conference (II, §4.3). Runcorn likely gave the most forceful presentation in favor of mobilism. So there is a lot missing from Dietz's report. Dietz noted mobilism's paleomagnetic support, but he was evidently skeptical and unwilling to change because of what he had heard at the conference. Most importantly however, he thought there was "little doubt that paleomagnetism" would "eventually supply a definitive answer." I do not know whether Dietz began at that point to educate himself about mobilism's paleomagnetic support and to follow new work including rebuttals of Graham's magnetostriction difficulty (II, §7.4). I suspect that Runcorn kept him abreast of new findings because they had become friends; in any event, Dietz, as I shall presently show, soon began to write favorably of paleomagnetism's support for mobilism.

## 4.7  Dietz proposes ocean basin formation by asteroid impact

Dietz's work in marine geology and astroblemes intersected in a paper he read on May 10, 1958, in France at a colloquium sponsored by the Centre National de la Recherche Scientifique (CNRS) on the topography and geology of the ocean deeps. The proceedings were published the following year, and included contributions by Heezen, Menard, Emery, and Dietz (1959b) who also wrote a summary of the colloquium (1959c). Dietz proposed that ocean basins formed by asteroid impact, which even by his standards was highly speculative. He later referred (1963a) to this in a critical review of a somewhat similar hypothesis by R. L. Callant, remarking on their similarity. Bourgeois and Koppes (1998: 144) go so far as to characterize Dietz's retrospective remark as confessional. Here is what he said.

I have indulged in this speculation myself, but with the more conservative approach that large impacts might have fractured the Earth's primeval sialic crust in "pregeological" time (that is, pre-rock record but post-Earth, or between 3.0 and 4.5 thousand million years ago). I suggested that the shattered fragments then may have completely rearranged themselves in what we now observe as the continents and the ocean basins. Under this scheme, the early impacts would only account for the discontinuity of the sialic crust and not the present shape or morphology of ocean basins … This idea still seems to me to have as much merit as, for example, another better-known catastrophic hypothesis – the breaking up of an initial crust by the Fisher–Darwin hypothesis of pulling the Moon out of the Pacific. If the Moon was cosmically bombarded, certainly it is reasonable to suspect that the Earth was similarly bombarded at some early time. But now I am inclined to accept a cold accretion model for the Earth under which proto-Earth had no sialic crust at all as we know it. The water, the air and the crust probably are all secondary, squeezed out of the mantle by processes related to thermal convection circulation as evidenced by seafloor spreading. Accordingly, early Earth impacts would have encountered no crust to fragment.

*(Dietz, 1963a: 39)*

Of course Dietz had, by the time he wrote this, abandoned his impact theory and adopted seafloor spreading. He (1994) made no mention of this in his personal account of his life and scientific career, and he must not have set much store by it.

This intersection between Dietz's work in marine geology and astrogeology merits attention because, crazy as it might be, it provides a vivid snapshot of his view of seafloor evolution immediately before he began championing seafloor spreading. Dietz (1959b) considered the problem of forming ocean basins as inseparable from that of continents. Continents could be bits of sial that had surfaced early in Earth's history by gravitational differentiation and thermal convection. He preferred, however, that Earth initially acquired a continuous sialic crust and ocean basins were created by a catastrophic event. He proposed instead asteroid impacts: an orbiting planet between Mars and Jupiter broke apart forming a belt of asteroids some of which collided with Earth, fragmenting the continuous sialic crust and creating ocean basins.

Dietz described the success he and others had had in interpreting cryptoexplosion features as huge asteroid impact craters. If terrestrial and lunar structures were analogous, he argued that ocean basins on Earth could correspond to lunar maria. Both are large and roundish.

For evidence of such cosmic collisions, one must look to the Moon, where in the absence of erosion, volcanism, and tectonism, the evidence of ancient events are still clearly preserved. Lunar maria, although waterless, strongly resemble ocean basins. These maria, as best shown in Mare Imbrium, are most reasonably explained by asteroid impacts. In fact, lunar craters and maria seem to belong to the same family of features whose physiographic variability is a function of the size of the impact.

*(Dietz, 1959b: 265, English summary)*

The impacts fractured the sialic crust, volcanism ensued, and magma covered the floor of craters forming ocean basins.

Aware that ocean basins do not appear, superficially at least, to be impact structures, Dietz argued, as he had done when interpreting cryptoexplosion structures as impact craters, that, unlike lunar impact craters, subsequent erosion,

volcanism, and tectonism had masked evidence of their origin, noting that the older
an impact crater on Earth the less well preserved it was. Because ocean basins, as he
claimed, have existed since the beginning of the Precambrian, their origins had been
progressively obscured. Little evidence remained of their explosive origin except for
their generally circular shape (Dietz, 1959b: 267).

The obscuring tectonism Dietz had in mind was continental drift. He did not
regard his appeal to continental drift entirely *ad hoc*; continental drift was supported
by findings of the new science of paleomagnetism.

And, although geologists agree generally that oceanic basins have been permanent features of
Earth since the first ages, they do not agree at all on the fact that the positions of the continents
were fixed geographically. In fact, the results of the new science of paleomagnetism tended to
reinforce belief in continental drift. For reasons which we will explain later, it is very improbable
that an unspecified continental drift took place on the Moon. Throughout its history Earth has
been prone to volcanism, tectonics and strong agents of erosion which mask evidence of its
former characteristics. On the other hand, because of the almost total absence of forces of
erosion, the lunar landscape must be old from the geological point of view, and events which
occurred several million years ago can be preserved in their primitive form, like the submerged
mountains of the Pacific.[8]

(*Dietz, 1959b: 267*)

He thought that paleomagnetism provided mobilism with enough support to make
his case reasonable. Nevertheless, because he singled out paleomagnetism among all
other evidence favoring mobilism, and had not yet – despite his later recollections
(§4.4) – offered mobilist solutions to problems about the origin and history of ocean
basins, he likely – and this is the important point in the present context – thought that
paleomagnetism provided mobilism with its strongest support. Thus it seems to me
that Dietz thought mobilism as a general notion was worthy of consideration before
Hess conceived seafloor spreading, and he did so by appealing to the paleomagnetic
evidence.

## 4.8  Dietz (1959) invokes continental drift and motions of seafloor
## to explain absence of pre-Cretaceous seamounts

Dietz wrote a semi-popular paper, "Drowned ancient islands of the Pacific," for
*New Scientist*, published on the first day of 1959. It provides a glimpse of his views
on the Pacific Basin after he learned of mobilism's paleomagnetic support but before
seafloor spreading was developed. His ideas on trenches, ocean rises, and the origin
and sinking of seamounts remained unchanged: none required mobilism, only vertical motions. He also wondered why there were no pre-Cretaceous seamounts in the
Pacific Basin, a problem he had not directly addressed before. Unfortunately, he
mentioned it only in passing without details.

Approximately a thousand seamounts were known in the Pacific Basin, and he
predicted that 9000 more would be discovered once the whole of the Basin had been

surveyed. He drew a distinction between islands and seamounts, the latter being either peaked or flat-topped (guyots), and focused on guyots. He again agreed with Hess that they were drowned islands, that their tops had been flattened by wave action, and that "for some reason have grown no coral cap to mark their grave" (Dietz, 1959d: 14). He mentioned Ed Hamilton's discovery of Late Cretaceous fossils atop two guyots, which meant that they sank after the Cretaceous. Noting that they are typically found on oceanic rises, he argued as he had earlier that wave action leveled them as rises were elevated by ascending convection currents. Seamounts and guyots then drowned as their parent ridge sank.

Nearly all the bands of seamounts are situated on great linear rises of the seafloor – and there are cogent reasons to believe that these are comparatively transitory features of the crust rather than permanent deformations. It appears likely that such rises form above ascending thermal convection cells in the Earth's mantle. Thus, a seamount developing on one of these swells could have been given a temporary boost to the surface of the sea only to be later drowned again.

*(Dietz, 1959b: 16–17)*

He said again that seamounts sank isostatically under their own weight. He even repeated the analogy between a guyot and "a small boy standing on thin ice," and added, "my colleagues and I have found seamounts surrounded by moat-like depressions – good evidence that our so-called terra firma actually has failed" (Dietz, 1959b: 17). He also discussed again the guyot in the Aleutian Trench with its purported tilted top, and offered the same explanation of its submergence and the formation of the trench.

Curiously, not all flat-topped seamounts are level, for some have tilted tops. An especially remarkable example lies on the south flank of the Aleutian Trench where the tilt can be clearly traced to the downbuckling of this great furrow. In the middle of this trench, there is also a tablemount [i.e., a guyot] with the unusual depth of 1400 fathoms. The few greatly drowned tablemounts that we know about owe their extreme depth to some such unusual structural position. For those in normal situations, 1100 fathoms seems to be about the maximum depth of drowning.

*(Dietz, 1959b: 17; my bracketed addition)*

Here as before there is no hint of horizontal motion of seafloor or of its destruction at trenches.

Then Dietz turned to the problem of why there are no pre-Cretaceous seamounts. He proposed continental drift and movement of ocean crust toward continents.

The question arises: What has become of the chains of volcanic seamounts which must have been formed in the early Mesozoic, the Paleozoic (between 180 million and 500 million years ago) or before that? Unlike mountains on land, seamounts are subjected only to extremely slow erosion, like mountains on the Moon, so that once formed they should persist through an enormous span of geologic time. When studying the Pacific, one finds that reasonable geological history can be pieced together subsequent to the beginning of the Cretaceous, but that prior to that time the record is completely erased – a blank page. The answer could lie in continental drift, a great crustal buckling or volcanic revolution, or migration of the sub-oceanic crust toward the continents.

(Dietz, 1959b: 17)

He kept to himself whatever he had in mind by "migration of the sub-ocean crust toward the continents." He did not elaborate on continental drift; he did not explain how either drift or migration of the seafloor toward continents would remove pre-Cretaceous but not Cretaceous or post-Cretaceous seamounts. He did not claim, for instance, that seamounts are always created atop oceanic rises, and move horizontally toward trenches. His movements of seafloor at trenches and rises were still vertical. He invoked continental drift and seafloor migration, but without integrating them with his explanations of ridges, trenches, and seamounts. If he had been thinking in terms of seafloor spreading, he would have realized, as he did two years later when espousing seafloor spreading (§4.9), that he could dispose of guyots as they and old seafloor disappeared on the backs of descending convection currents. He made no mention of huge convection cells that carry old seafloor and seamounts down into the mantle. Nor did he declare that seafloor is created along ridge axes. He mentioned only migration of seafloor, not its creation at ridges or destruction at trenches. When he wrote his 1959d paper, he was not, I believe, proposing or even anticipating seafloor spreading. What he did is not well integrated. Nonetheless, Dietz was willing to discuss continental drift for the third time before proposing seafloor spreading. During the next two years Dietz turned his inchoate thoughts into a coherent story.

### 4.9  Presentation and defense of seafloor spreading by Dietz, 1961–1962

Dietz wrote at least six papers in 1961 and 1962 on seafloor spreading (1961a, 1961d, 1962a, 1962c, 1962d, 1962e). In this section I shall refer to them respectively as (1), (2), (3), (4), (5), and (6). Reading them reveals a vivid geological imagination at play. Paper (1) was published in *Nature* on June 3, 1961; it has been the most widely discussed. Paper (2), a popular, clearly written account, appeared in the October 24, 1961 issue of *The Saturday Evening Post*. It contains the first extended use of "seafloor spreading" where "spreading" is a gerund rather than a participle as in "seafloor-spreading theory" or "seafloor-spreading concept." He had used "seafloor spreading" as a noun once before in the *Nature* paper (1961a: 857). Paper (4), "Ocean basin evolution by sea floor spreading," was received on March 26, 1962 for the 20th Anniversary Volume of the *Journal of the Oceanographical Society of Japan*. It includes his first extended discussion of mountain building as it relates to seafloor spreading. He presented paper (5) at the 10th Pacific Science Congress, where he probably heard H. A. Brouwer speak (1962d). The Congress was held in Honolulu, Hawaii from August 21 to September 6, 1961, approximately two months before *The Saturday Evening Post* version appeared. Paper (5), the published version of his Honolulu talk appeared in 1962, and I believe that Dietz revised it in the interim.

Paper (5) also includes Dietz's first public acknowledgment that Hess deserved priority for proposing seafloor spreading.

Dietz (1962a: 289) described another version, paper (3), the one included in *Continental Drift* edited by Runcorn (§1.6), as "a modified version" of that in *Nature*. Runcorn was impressed with Dietz's paper in *Nature*, and wrote him on November 1, 1961, asking for a contribution on his "recent thoughts."[9]

> Dear Bob,
>
> I am very interested in your ideas about movements of the earth's crust and it may interest you to know that I am editing, for publication early next year, a book on horizontal displacements of the earth's crust, and I enclosed a list of people who have written articles for it.
>
> It would be very nice indeed to have even a brief article from you (most of them are about 10 000 words), but because the book is in the final stages I am afraid the absolute deadline would have to be the end of the year. I would certainly be exceedingly pleased if you could write something on your recent thoughts. I think it would certainly be of great value to the book, which is intended to try and get people to look afresh at the problem of continental drift.
>
> > With kind regards,
> > Yours sincerely,
> > S. K. Runcorn

Runcorn's enclosed list of contributors shows that Dietz was asked last. Dietz had apparently sent Runcorn a copy of the *Nature* paper before he received Runcorn's request. Runcorn thanked Dietz for the paper in a letter dated November 15, 1961, mentioned his own forthcoming paper in *Nature* on convection, and suggested that he simply contribute an expanded version of his own *Nature* paper for the planned volume on continental drift.

> Dear Bob,
>
> Thank you very much for the "Nature" article. I expect you will be letting me have the discussion. I am exceedingly interested in your ideas which fit in very well with the theory of Continental Drift by convection currents which I am shortly publishing in "Nature."
>
> You will have received my letter asking whether you would contribute a chapter to the book on Continental Drift, which is to be published by Academic Press next spring. We will be most grateful indeed to have an article from you for it, and I would like to suggest very tentatively that an expanded version of this "Nature" article with slightly more detail in places and perhaps a few diagrams would make an exceedingly valuable addition to the chapter we already have.
>
> > With kind regards,
> > Yours sincerely,
> > S. K. Runcorn

Dietz added a few passages to and removed several from his *Nature* paper. He added no new diagrams.

Dietz was named a distinguished lecturer (1962–3) by the AAPG, and spoke on "Continent and ocean basin evolution by seafloor spreading, 'Commotion in the ocean,'" in mid-October 1962 (Anonymous, 1962: 1972). The manuscript he used for his talk is paper (6). Part of it was subsequently republished (Dietz, 1964b). Dietz also contributed to a discussion of his original *Nature* paper (1), which was published on October 14, 1961. J. D. Bernal (1961a) and J. Tuzo Wilson (1961) discussed Dietz's original paper, and Dietz (1961e) responded to Bernal, who (1961b) then replied to Dietz. (This discussion is presumably the one mentioned by Runcorn in his November 15 letter to Dietz.) I shall consider Bernal and Wilson's initial assessments of seafloor spreading and Dietz's response to Bernal when examining the development of Wilson's ideas (IV, §1.8).

I now describe paper (1), the one that appeared in *Nature* (Dietz, 1961a), note significant alterations or additions made to it in his later presentations, and draw comparisons with Hess's version. Dietz (1961a: 854) introduced seafloor spreading in his four-page paper as "an attempt to interpret sea-floor bathymetry," claiming that it was "largely intuitive" and required the following assumptions. First, the presence in the mantle of radiogenically produced large-scale convection cells. Second, an oceanic crust consisting of three layers: 0.3 km of unconsolidated sediments, a 2 km thick second layer of consolidated sediments and mixed volcanics, and a 5 km thick third layer probably gabbro but possibly serpentine. Most importantly, Dietz emphasized that crustal Layer 3 and upper mantle are petrographically different but chemically identical. The Moho is merely a phase difference. He first (1961a: 855) opted for an upper mantle of eclogite, and a Layer 3 of gabbro and not a phase change of peridotite to serpentine, but thought that seafloor spreading would work with either. He also suggested it made more sense to think of Layer 3 as uppermost mantle, "a sort of 'exomantle'," instead of the lowest crustal layer. Third, Dietz assumed a mechanical distinction between the lithosphere and underlying asthenosphere, separated by a region of no strain. He (1961a: 855) proposed that the lithosphere extends 70 km below both ocean basins and continents. Convection involves movement of both the lithosphere and oceanic crustal Layer 3. Fourth, he assumed that sialic continents are 35 km thick and float in the denser sima (eclogite) below. Convection currents in the sima do not invade continents. There is no significant difference between the relative strength of the oceanic and continental lithosphere, and Dietz (1961a: 855) added, "the continental lithosphere is no stronger than the oceanic lithosphere, so it is mechanically impossible for the sial to 'sail through the sima' as Wegenerian continental drift proposes."

Dietz's first model of oceanic crust and mantle is substantially the same as Hess's. Most importantly, both regarded the boundary between the upper mantle and ocean crustal Layer 3 as a phase change. However, they differed about what changed. Dietz identified the upper mantle as eclogite and Layer 3 as gabbro; Hess thought the upper mantle was peridotite and Layer 3 was serpentine. Hess was more adamant than Dietz about the choice, he strongly preferred the peridotite–serpentine transformation, and he considered serpentinization, one of his lifelong interests, as central to seafloor spreading. Dietz did not.

This change of phase may be either from eclogite to gabbro, or from peridotite to serpentine; its exact nature is not vital to our concept, but we can tentatively accept the eclogite-gabbro transition as it has more adherents.

*(Dietz, 1961a: 865)*

Although in paper (2), *The Saturday Evening Post* version, Dietz (1961d: 35) stayed with eclogite and gabbro, he changed his mind by the time he submitted his paper (3) for inclusion in Runcorn's edited anthology *Continental Drift*. Without giving a reason, he (1962a: 291) claimed that "Here we accept the opinion of Hess and others that the oceanic layer is serpentine, i.e. peridotite largely altered by hydration." He later supplied a reason in the published form of his talk at the Tenth Pacific Science Congress (paper 5), which did not appear until after his contribution to Runcorn's anthology, and in which he first began to articulate more fully the relation between his account of mountain building and seafloor spreading.

Although the Moho beneath the continents may, or may not, be a phase change from gabbro to eclogite, the writer believes with Hess that the oceanic Moho is a transition from serpentine to peridotite and related ultramafics. My reason for this belief is related to my conviction that eugeosynclines are equitable with continental rise prisms. As such they are laid down on the oceanic crust at the base of the continental slope. Eugeosynclinal mountains, for example, the Coast Range of California, contain abundant pods of ultramafics and serpentines which would be related to the mantle and the oceanic layer respectively.

*(Dietz, 1962d: 11)*

After presenting his model of oceanic crust and mantle, which included mantle convection of ocean crust Layer 3 or the "exomantle" as he preferred to call it, he introduced seafloor spreading.

Since the seafloor is covered by only a thick veneer of sediments with some mixed-in effusives, it is essentially the outcropping mantle. So the seafloor marks the tops of the convection cells and slowly spreads from zones of divergence to those of convergence. These cells have dimensions of several thousands of kilometers; some cells are quite active now while others are dead or dormant. They have changed position with geological time causing new tectonic patterns.

*(Dietz, 1961a: 855; see also 1962a: 292)*

Turning to major seafloor structures and eschewing details, he proposed that seafloor spreading explained the elevation of ridges, their high heat flow, their associated seamounts, the origin of trenches, and the origin of fracture zones that Menard had discovered in the eastern Pacific off the western coast of North America (RS1).

The gross structures of the seafloor are direct expressions of this convection. The median rises mark the up-welling sites or divergences; the trenches are associated with the convergences or down-welling sites; and the fracture zones mark shears between regions of slow and fast creep. The high heat-flow under the rises is indicative of the ascending convection currents as also are the groups of volcanic seamounts which dot the backs of these rises.

*(Dietz, 1961a: 856)*

Dietz (1961a: 856; see also 1962a: 292–293) turned to creation of seafloor at ridge axes to explain how seamounts were created atop ridges, why the Mid-Atlantic Ridge is characterized by a rough topography, and why there are areas of abyssal hills in the Pacific (RS1).

Great expanses of topography skirt both sides of the Mid-Atlantic Rift; similarly there are extensive regions of abyssal hills in the Pacific. The roughness is suggestive of youth, so it has commonly been assumed to be simply volcanic topography because the larger seamounts are volcanic. But this interpretation is no longer convincing . . . Actually, the topography resembles neither volcanic flows nor incipient volcanoes. Can it not be that these expanses of abyssal hills are a "chaos topography" developed as strips of juvenile seafloor (by a process which can be visualized only as mixed intrusion and extrusion) and then placed under rupturing stresses as the seafloor moves outward?

*(Dietz, 1961a: 856; see also 1962a: 292–293)*

He then explained the median position of ocean ridges, and without using the expression "continental drift" introduced it nonetheless as a consequence of seafloor spreading (RS1).

The median position of the rises cannot be a matter of chance, so it might be supposed that the continents in some manner control the convection pattern. But the reverse is true: conditions deep within the mantle control the convective pattern without regard for continental positions. By viscous drag, the continents initially are moved along with the sima until they attain a position of dynamic balance overlying a convergence. There the continents come to rest, but the sima continues to shear under and descend beneath them; so the continents generally cover the down-welling sites. If new upwells do happen to rise under a continental mass, it tends to be rifted. Thus, the entire North and South Atlantic Ocean marks an *ancient* rift which separated North and South America from Europe and Africa.

*(Dietz, 1961a: 856; emphasis added)*

He proposed three other incipient rifts. None were ancient; Dietz supposed they were just now beginning to rift continents.

Another such rift has opened up the Mediterranean. The axis of the East Pacific Rise now seems to be invading the North American continent, underlying the Gulf of California and California. Similarly, the Indian Ocean Rise may extend into the African Rift Valleys, tending to fragment that continent.

*(Dietz, 1961a: 856)*

Dietz's suggestion that the East Pacific Rise extended underneath North America, as he noted, came from Menard (1960). Dietz may have got the idea about African rift valleys as an extension of an oceanic ridge from Heezen, who had presented it at the 1958 CNRS meeting in Nice (§6.9) at which Dietz proposed that ocean basins were originally asteroid impacts (§4.7). Dietz probably got the idea that the Mediterranean was opening from Carey, not Argand. Argand proposed that after collision of Europe and Africa during the Oligocene, the two landmasses began rifting apart forming the modern Mediterranean and causing counterclockwise rotation of Spain and Italy relative to the rest of Europe (Argand, 1924/1977: 142–147) (I, §8.7). Dietz, however, did not mention that he was familiar with Argand's work. But Dietz (1987 interview with author) was familiar with Carey's work, and Carey (1955a, 1958) presented a scenario for opening the Mediterranean similar to Argand's with the counterclockwise rotation of Spain and Italy relative to Europe (II, §6.7). Regardless, Dietz erased all reference to an opening of the Mediterranean in his ensuing presentations of seafloor spreading (Dietz, 1961d, 1962a, 1962d, and 1962e). Perhaps, he took heed of J. Tuzo Wilson's remark that Dietz's "statement that a rift is opening the Mediterranean" is "enigmatic" because it "appears rather to be closing" (Wilson, 1961: 126).

Dietz ended his introduction of seafloor spreading by speculating broadly about its effects on continents. Without mentioning continental drift by name, he (1961a: 856) proposed that convection currents pass beneath sialic continents to which they are only partially coupled; he hinted at an explanation of mountain building, and, harkening back to Wegener, offered an explanation for Atlantic and Pacific-type coastlines (RS1).

The sialic continents, floating on the sima, provide a density barrier to convection circulation unlike the Moho, which involves merely a change of phase. The convection circulation thus shears beneath the continents so that the sial is only partially coupled through drag forces. Since the continents are normally resting over convergences, so that convective spreading is moving toward them from opposite sides, the continents are placed consequently under compression. They tend to buckle, which accounts for alpine folding, thrust faulting, and similar compressional effects so characteristic of the continents. In contrast, the ocean basins are simultaneously domains of tension. If the continental block is drifted along with the sima, the margin is tectonically stable (Atlantic type). But if the sima is slipping under the sialic block, marginal mountains tend to form (Pacific type) owing to drag forces.

*(Dietz, 1961a: 856)*

Dietz explicitly introduced continental drift only when he began considering the implications of seafloor spreading. Once again, he appealed to drift's paleomagnetic support, but, unlike Hess, he referenced no sources. He also exaggerated, perhaps for rhetorical reasons, its contemporary impact on the Earth science community. "Former scepticism about continental drift is rapidly vanishing, especially due to the palaeomagnetic findings and new tectonic analyses" (Dietz, 1961a: 856; see also Dietz, 1962a, 295). He did not specify the "new tectonic analyses." He was more informative in his *The Saturday Evening Post* article, where, before mentioning the paleomagnetic support for mobilism, and instead of invoking "new tectonic analyses," he claimed, "The jigsaw fit [between opposing continental margins of the New and Old World] practically demands continental drift. When correctly made by matching the bases of the continental slopes, the fit is essentially perfect" (Dietz, 1961d: 96; my bracketed addition). I do not know if this is what he had in mind by "new tectonic analyses," but he surely was also familiar with Carey's proper fit of South America and Africa (1955b), but obtaining an intercontinental jigsaw fit of the continents is only the first step in tectonic analysis.

Dietz avoided Wegener's mechanism difficulty that continents cannot plow through seafloor (RS1).

A principal objection to Wegener's continental drift hypothesis was that it was physically impossible for a continent to "sail like a ship" through the sima; and nowhere is there any seafloor deformation ascribable to an on-coming continent. Seafloor spreading obviates this difficulty: continents never move through the sima – they either move along with it or stand still while the sima shears beneath them.

*(Dietz, 1961a: 859; see also 1962a: 295)*

Like Hess, he counted this an important step forward. Like Hess, Dietz failed to note that Holmes' mantle convection theory also avoided the difficulty; Hess at least mentioned Holmes even though he did not reference any of his works. Dietz did neither.

Dietz, like Hess, raised two difficulties with theories of Earth expansion (RS2), or at least those versions that required large amounts of expansion such as Carey's (II, §6.14) and Heezen's (§6.9) – he ignored Egyed's more modest theory (II, §6.2–§6.4). He first raised a vague theoretical difficulty; both Heezen's and Carey's theories were radical, requiring huge increases in Earth's volume.

Geologists have traditionally recognized that compression of the continents (and they assumed of the ocean floors as well) was the principal tectonic problem. It was supposed that the Earth was cooling and shrinking. But recently, geologists have been impressed by the tensional structures, especially on the ocean floor. To account for sea rifting, Heezen, for example, has advocated an expanding Earth, a doubling of the diameter. Carey's tectonic analysis has resulted in the need for a twenty-fold increase in volume of the Earth. Spreading of the seafloor offers the less radical answer that the Earth's volume has remained constant. By creep from median upwellings, the ocean basins are mostly under tension, while the continents, normally balanced against sima creepage from opposite sides, are under compression.

*(Dietz, 1961a: 856; see also 1962a: 294)*

Dietz then argued as Hess had done that it would be highly unlikely that any increase in the amount of water would match volumetric increases so as to keep changes in sea level minor; Dietz claimed that fast expansion would cause a far greater drop in sea level than the geological record indicated.

The geological record is replete with transgressions and regressions of the sea, but these have been shallow and not catastrophic; fluctuations in sea-level as severe as those of the Pleistocene are abnormal. The spreading concept does no violence to this order of things, unlike dilation or contraction of the Earth. The volumetric capacity of the oceans is fully conserved.

*(Dietz, 1961a: 856; see also 1962a: 294)*

Seafloor spreading, unlike extreme dilation or contraction, required no volumetric changes.

Stated in this way his argument is not very satisfactory, but Dietz had an answer. The high stand of continents relative to sea level requires replenishment of the sial of continents to replace the losses through erosion. Erosion reduces the height of continents, which would cause them to rise isostatically and erode. Long-continued, continental roots would rise, and continents would eventually reduce to sea level.

A satisfactory theory of crustal evolution must explain why the continents have stood high throughout geological time in spite of constant erosional de-leveling. Many geologists believe that new buoyancy is added to continents through the gravitative differentiation from the mantle. Spreading of the seafloor provides a mechanism whereby the continents are placed over the down-wells where new sial would tend to collect even though the convection is entirely a mantle process and the role of the continents is passive. It also follows that the clastic detritus swept into the deep sea from the continents is not permanently lost. Rather, it is carried slowly towards, and then beneath the continents, where it is granitized and added anew to the sialic blocks.

*(Dietz, 1961a: 856; see also 1962a: 295)*

Continents stand high because ocean sediments are swept toward them and they become incorporated into continents. This was not subduction, which had not then been recognized, but it was a recognition that some such process was indicated.

Returning to the seafloor, Dietz (1961a: 857), like Hess, declared that seafloor appears to be young because it is young. "Marine sediments, seamounts, and other structures slowly impinge against the sialic blocks and are destroyed by under-riding them" (Dietz, 1961a: 856; 1962a: 295). Seamounts, created atop ridges, migrate away from ridges, and are eventually destroyed as they pass beneath continents or into trenches. Unlike peaked seamounts, guyots are planed down by wave erosion during their migration off ridges. There was also the problem of why there are so few peaked seamounts and guyots, if the ocean basins were Pre-Cambrian features. With seafloor spreading ocean basins are relatively young, and older seamounts are the first to disappear.

The puzzle dissolves if seafloor spreading has operated. Modern examples of impinging groups of seamounts may be the western end of the Caroline Islands, the Wake-Marcus Seamounts, and the Magellan Seamounts. All may be moving into the Western Pacific Trenches. Seamount GA-1 South of Alaska may be moving into the Aleutian Trench.[10]

*(Dietz, 1961a: 857; see also 1962: 296)*

Now Dietz definitely knew how to get rid of seamounts already moving into the Aleutian Trench – they continue descending and impinging against the Alaskan sialic block. He no longer had to claim that seamounts eventually sunk vertically into the mantle because of crustal failure. Turning specifically to the lack of sediments on the seafloor, he discussed and dismissed E. L. Hamilton's (1960) explanation that much of oceanic crustal Layer 2 is lithified sediment, which had lessened the difficulty of supposing that existing ocean basins were very ancient given acceptable rates of sedimentation. Dietz countered by noting:

If all layer 2 is lithified sediments, Hamilton finds that the ocean basins may be Palaeozoic or late Pre-Cambrian in age – but not Archaean. But very likely layer 2 includes much effusive material and sedimentary products of seafloor weathering. In summing up, the evidence from the sediments, although still fragmentary, suggests that the seafloors may be not older than Palaeozoic or even early Mesozoic.

*(Dietz, 1961a: 857; see also 1962a: 296)*

Having abandoned his earlier asteroid impacts theory of ocean basins, Dietz defended his version of seafloor spreading by arguing that it explained the origin of oceanic ridges, trenches, seamounts (including guyots), the lack of Pre-Cretaceous sediment on the seafloor, why Atlantic and Pacific-type coastlines differ, and how continents remain well above sea level even though erosion is ceaseless (RS1). He had begun to articulate a relation between seafloor spreading and mountain building, and felt that seafloor spreading offered a solution to mobilism's mechanism problem (RS1). He also claimed that mobilism, unlike fixism, explained paleomagnetic data and the fit between South America and Africa (RS1, RS2, and RS3). He argued that seafloor spreading had a greater problem-solving effectiveness than Earth expansion (RS3) against which he raised difficulties (RS2).

### 4.10  The priority muddle over seafloor spreading

Hess is credited with first proposing seafloor spreading. This is not in dispute. What, however, is still a matter of dispute is whether Dietz came up with his version of seafloor spreading on his own or whether he got it from Hess. The common view before Menard, who was "the only person thanked by both writers for critical discussions of the manuscripts before publication," was that Dietz had essentially stolen the idea from Hess (Menard, 1986: 158). Menard, however, disagreed and came to believe:

Dietz never meant to imply [when he acknowledged that Hess deserved priority] that he did not think of the concept of sea-floor spreading independently. He just meant that Hess thought of it first. Priority, however, is normally based on publication in a scientific journal, and there Dietz was first. Taken all in all it appears to me, for the first time in almost two decades, that there were two independent discoveries of this idea that Hess considered to be geopoetry and not worth rapid publication and that Dietz wrote up as a pot boiler.

*(Menard, 1986: 159; my bracketed addition)*

Menard also checked his interpretation with Dietz. Dietz agreed with Menard, and also told me that Menard's account of what happened was correct (Menard, 1986: 159; Dietz, 1987 interview with author).

I do not think there is enough information to definitely tell whether Dietz came up with seafloor spreading on his own or got the idea from Hess. There are the two versions of what happened; they more or less agree except for the crucial issues of whether Hess personally told Dietz about his theory before Dietz wrote his paper. Dietz denies that any meeting took place; Hess claims that they met and he told Dietz in detail about seafloor spreading. This is my reconstruction of the chronology of events.

1. Hess finishes his preprint "Evolution of ocean basins" in November 1960.
2H. Hess explains his ideas about seafloor evolution to Dietz in great detail in November 1960.
2D. They do not meet, and Hess does not discuss his ideas with Dietz.
3. Dietz submits his seafloor spreading paper to *Nature* (exact date unknown).
4. Hess presents a twenty-five minute talk on seafloor spreading in the Great Hall of the National Academy of Sciences on April 26, 1961 (Allwardt, 1990: 192).
5. About the same time Hess gives the public lecture, he sends his preprint to Hill, editor of *The Sea*, for publication in a forthcoming volume (Allwardt, 1990: 192).
6. Menard receives a copy of Hess's preprint probably at the beginning of May 1961.
7. Dietz gives Menard a copy of his paper (date unknown). Menard sees the incredible similarities, and telephones Dietz.

> I remember what then happened 23 years ago with perfect clarity. (Dietz remembers it exactly the same way.) I phoned him with news. He expressed surprise because he had not received or seen Hess's manuscript. He came over to my office to read the manuscript for the first time.
>
> *(Menard, 1986: 154–155)*

Dietz later confirms that this is what happened (Menard, 1986: Note 12, 317).

8. Fisher sends his co-authored paper on trenches with Hess to Hill for publication in a forthcoming volume of *The Sea* on May 15, 1961.
9. Menard writes to Hess about his preprint "Evolution of ocean basins" on May 25, 1961, but does not mention Dietz's paper (Menard, 1986: 159; part of the letter is reproduced in IV, §1.11).
10. Dietz's paper is published in *Nature* on June 3, 1961.
11. Dietz gives his talk on seafloor spreading at the 10th Pacific Science Congress at the end of August or beginning of September 1961.
12. Dietz, Bernal, and Wilson discuss Dietz's paper in the October 14, 1961 issue of *Nature*. In his discussion Wilson refers to Hess's preprint.

13. Dietz's popular piece on seafloor spreading appears in the October 24, 1961 issue of *The Saturday Evening Post*.

14. Runcorn asks Dietz on November 1, 1961 for an "updated" version of his *Nature* paper for inclusion in his forthcoming edited book *Continental Drift*.

15. Hess pulls his paper from *The Sea* on December 4, 1961, and arranges to have it published in the Buddington volume (Allwardt, 1990: 193).

16. At some point in 1962 before October, Menard, E. Hamilton, and perhaps others convince Dietz to acknowledge publicly that Hess deserves priority. Dietz adds the following note to the proof of the paper he presented at the Tenth Pacific Congress (Menard, 1986: 158).

> Note added in proof – The writer's attention has been drawn to a preprint by H. Hess also suggesting a highly mobile seafloor for any merit which this suggestion has.
>
> *(Dietz, 1962d: 12)*

17. Dietz is nominated as a Distinguished Lecturer for the AAPG, prepares his lecture and sends Hess a copy together with the following letter dated October 16, 1962, in which he privately concedes priority to Hess.[11]

> Dear Harry:
>
> I enclose a copy of a lecture guide for your perusal and retention.
>
> Last May I obtained and read for the first time from Bill Menard a manuscript preprint copy of your forthcoming paper in *The Sea* on the origin of ocean basins which established your priority to the general idea which I call "seafloor spreading." This is pointed out in the abstract and acknowledgments of the enclosed lecture guide. This is only a rough guide as I have decided to especially emphasize the geosynclines part as this is of special interest to petroleum geologists.
>
> I would like to assure you that I will endeavor to accord you full credit as a preface to each lecture in any future writings. My position will be that of seconding the notion of motion in the ocean.
>
> With my best regards.
>
> Sincerely,
> Robert S. Dietz

18. King Hubbert, outgoing President of GSA, writes in early May to Hess, incoming President of GSA, about a disciplinary matter concerning Dietz. Dietz, as co-chair of a session of asteroid impacts at the November 1962 GSA meeting, presented a talk on the Sudbury Structure as an asteroid impact site even though his abstract had been rejected (Allwardt, 1990: 194). Allwardt found the correspondence in Hess's papers at Princeton, and reproduced Hess's May 10, 1963 response. This letter, as Allwardt explained, documents Hess's private feelings about the priority muddle. It provides support for

(2H) above, and it is worth noting that Menard did not know of it when he wrote his version of what happened.

> If you want to pursue the Dietz matter further you will have to do it rather than I. If I did it, it would certainly look like pure animosity on my part.

> In November 1960 I explained to Dietz in rather great detail the ideas in my paper that appeared in the Buddington Volume [Hess, 1962]. I had just completed the manuscript. He rushed off and had these ideas published in *Nature* early in 1961 [actually in June; see Dietz, 1961a]. Even the phrasology which I used in describing the ideas is the same. There was no hint of any acknowledgment to me. Being as generous as I can I might suppose he forgot the source of "his" ideas. But I believe this is a bit too generous.

> Quite a few people who read my manuscript before Dietz published the substance are aware of this situation, so I would rather rule myself out of any action which might be taken against him.

There is also corroborating evidence that Hess met with Dietz in late 1960. Alan Smith, who was then a graduate student in geology at Princeton and later worked with Everitt and Bullard on a computer fit of the continents surrounding the Atlantic (IV, §3.4), initally recalled, "I knew Bob Dietz had visited Harry one day [in the afternoon] because as graduate students we had a sort of bush telegraph, but I cannot honestly say I saw the two talking, though I do remember Harry making some comment about Bob Dietz visiting" (Smith, April 7, 2010 email to author; my bracketed addition). After thinking more about what happened, and that he could not confirm the bush telegraph, Smith, as well as others then at Princeton, definitely believe that Hess and Dietz met.

> With regard to the Dietz meeting, I think it safest to say it was a meeting between Hess and Dietz, but not to be specific about an afternoon meeting, particularly if the story was told to me by a now deceased graduate student (i.e. you cannot verify its truth). What I am sure of was that there was a view amongst some of us that Harry had spent time explaining his views to Dietz and that later Dietz had published them. The precise sources of this view are not entirely clear, but mine are definitely second hand.
>
> *(Smith, May 24, 2010 email to author)*

19. Dietz sends off the first half of his AAPG Distinguished Lecture for publication in its Bulletin. Hubbert, an associate editor, receives the paper, and writes to Hess on May 22, 1963, just a dozen days after he had received the above letter from him. Allwardt (1990: 195) found it along with Hess's response in Hess's papers, and reproduced both letters.

> I have just received the enclosed manuscript entitled "Commotion in the Ocean," by Robert S. Dietz ... In view of the circumstance recounted to me in your letter of May 10, I would be most grateful, if you would take a look at this manuscript and advise me with

respect of its disposition … If this is a continuation of Dietz's plagiarism I should appreciate having a frank statement to that effect with a citation of specific instances.

*(M. King Hubbert, May 22, 1963 letter to Hess)*

Hess replies on May 28, 1963.

In this version Dietz does, albeit rather grudgingly, acknowledge that some of the concepts "seem" to have been first stated by Hess. So far as I am concerned this will do. I wouldn't want to be blamed for his whole manuscript because he has not thought through the implications of his hypothesis in many cases.[12]

*(Hess, May 28, 1963 letter to Hubbert)*

Dietz's paper is eventually published as part of a collection of essays honoring the Japanese oceanographer, Koji Hidaka (Dietz, 1964b). Dietz does not state specifically that Hess deserved priority for seafloor spreading, but he begins his abstract by stating, "Following Hess (1962), the belief in seafloor stability is discarded in favor of a mobile regime – one great commotion" (Dietz, 1964b: 465; Hess (1962) is the same as mine).

20. Hess, Dietz, and A. A. Meyerhoff revisit the priority issue in 1968. Meyerhoff, a diehard fixist, did so by arguing that Arthur Holmes (1931) should receive credit as originator of seafloor spreading (Meyerhoff, 1968). Hess disagrees and explains how his hypothesis differs from Holmes' but admits that he should have referenced his work pointing out these differences. Hess then turns to Dietz: "The cogent term 'sea-floor spreading,' which so nicely summed up my concept, was coined by Dietz (1961a) after he and I had discussed the proposition at length in 1960" (Hess, 1968b: 6569). Dietz also responds to Meyerhoff. He (1968b: 6567) agrees with Hess that Holmes had not "'fathered' sea-floor spreading, although perhaps he in some ways anticipated it," and again states that Hess deserves priority.

As regards sea-floor spreading, Hess deserves full credit for the concept, as correctly noted by Meyerhoff, by reason of priority and for full and elegantly laying down the basic premises. I have done little more than introduce the term sea-floor spreading, which now seems to have acquired a wide range and …

*(Dietz, 1968: 6567)*

Even with this chronology, Dietz's October 16, 1962 letter to Hess, Allwardt's discovery of Hess's May 10, 1963 letter to Hubbert mentioning his November meeting with Dietz, and Smith's supporting recollection, I still do not know with certainty whether Dietz came up with the idea of seafloor spreading on his own or got it from Hess. The evidence summarized above shows that Dietz very well *could have* thought of seafloor spreading on his own. Once Dietz returned to NEL in San Diego he already was sympathetic toward mobilism because of its paleomagnetic support. Even though he misremembered when and even why he became inclined toward mobilism, he (1959d: 17) tossed out the idea of "migration of the sub-oceanic crust toward the continents" to explain the lack of pre-Cretaceous seamounts. This was in

January 1959, at least twenty months before Hess said he told Dietz in detail about seafloor spreading. But Dietz was still a long way from seafloor spreading. Maybe he thought hard about what he had proposed, saw that it would not work unless the older seamounts had been located at the periphery of the Pacific Basin, and came up with seafloor spreading, and even neglected to mention that he had already proposed the inchoate idea of a mobile seafloor in 1959 when discussing the lack of pre-Cretaceous oceanic sediments.

Dietz's work on asteroid impacts shows that he was creative and willing to defend controversial views. Surely someone of his ingenuity *could have* come up with sea-floor spreading on his own. Menard correctly claimed that there have been many multiple discoveries. However, none of this shows that Dietz came up with seafloor spreading on his own; it shows only that he could have done so. Moreover, Menard neglected to discuss Dietz's 1959 paper in which he tossed out the ideas of a mobile seafloor and continental drift to explain why there were no pre-Cretaceous sea-mounts. There are so many similarities between the papers; Menard (1986: 155) also admits, "At the time [when we first saw the two papers] this seemed astonishing." Hess certainly believed that Dietz did not come up with the idea on his own. Even if Hess exaggerated how much he told Dietz, it seems extremely unlikely that he imagined his meeting with Dietz, especially with Smith's corroborating recollection. Yet Dietz claimed that they had not met for some time before he and Hess thought of seafloor spreading. Menard suggests:

I can only speculate that relatively early in 1960 Dietz and Hess talked, as we all did, about the stipulated data and current idea about their meaning. If Dietz had thought that Hess was proposing a revolutionary hypothesis, he surely would have remembered it when I told him his manuscript was so similar to Hess's. On the other hand, when Hess learned of Dietz's manu-script, he might well have attached more significance to their conversation than was warranted.

*(Menard, 1986: 158–159)*

Menard was friends with both of them. His charitable explanation could be correct, but I do not think so. Morover, Menard did not know about Hess's letter to King Hubbard, and Smith's recollection. The meeting was not, as Menard suggested, "early in 1960." I also do not understand why Hess would have been so upset with Dietz, unless he was quite sure that he had told Dietz in detail about seafloor spreading. Suppose Hess and Dietz had not talked about seafloor spreading. Dietz's *Nature* paper appeared in June. At the very least they would have shared priority because Hess's preprint already had been circulated widely enough for joint credit to be recognized. Menard (1986: 159) recalled that he did not mention Dietz's forthcoming paper in his May 25, 1961 letter to Hess in which he raised objections to Hess's seafloor spreading. Realizing that it seems peculiar, at least in retrospect, that he did not mention Dietz's version of seafloor spreading to Hess, Menard speculated, "I can only assume that I knew that Hess was well aware of the pending publication. If so, he had some opportunity for a preemptive note to

and Dietz co-authored four papers, and both wrote another with E. L. Hamilton. As explained already (§4.5), they examined the origin of the Aleutian Trench and Islands and guyots, and Hawaiian Islands, swell and deep. They proposed crustal downbuckling to explain the origin of the Aleutian Trench and Islands (Menard and Dietz, 1951). Menard and Dietz (1951) also considered Hess's 1946 explanation of guyots. They agreed with Hess that guyots are decapitated seamounts, truncated by wave erosion. Hess maintained, however, that truncation occurred during the Precambrian; the discovery of Cretaceous and earlier-aged fossils atop guyots by E. L. Hamilton led them to argue that truncation occurred during the Cretaceous and Cenozoic. Noticing that guyots are often surrounded by moats, they also maintained that guyots had sunk because of crustal downbuckling. They argued that the Hawaiian region formed as upwelling basalt broke through oceanic crust on the backs of ascending convection currents (Dietz and Menard, 1953). The convection cells were local; they remained in the mantle, and only fractured the crust.

Although his studies of the Aleutian Trench and Islands, guyots, and Hawaiian Islands and surrounding swell and deep affected Menard's later work, the effect of his discovery of the full extent of the massive Mendocino Escarpment was greater (Menard and Dietz, 1952). Menard estimated that he was 95% responsible for the discovery of the Mendocino Escarpment (July 9, 1979 response to author's questions). Within a few years, he had discovered three more escarpments, had reclassified them as fracture zones, and continued working on their origin for years to come.

Menard and Dietz (1952: 266, 271) proposed that the Mendocino Escarpment, which trended westward from Cape Mendocino, was at least 1900 km, possibly more than 2300 km long, at least four times longer than previously thought (Murray, 1939; Shepard and Emery, 1941). The escarpment reached a relief of 3200 m, and formed the southern side of a long asymmetrical ridge; the seafloor to the north of the escarpment was approximately 800 m higher than that to the south. They were surprised at the length and straightness of the escarpment, and noted that no current theory of the whole Pacific Basin "gives a clue to the origin of the Mendocino Scarp" (Menard and Dietz, 1952: 273). Its height and extent was a real surprise. They maintained that any adequate solution should explain why the seafloor is approximately 800 m deeper south of the scarp than north of it, why the scarp forms the southern slope of what is in effect a long asymmetrical ridge, and why other asymmetrical ridges run parallel to it.

They suggested as an explanation that the "scarp marks the transition between 'continental' rocks on the north and much more basic rocks to the south and is analogous to a continental slope" (Menard and Dietz, 1952: 273). This is how they explained the higher elevation of the northerly "Alaskan" province compared to the more depressed province directly south of the escarpment, but left unexplained how the escarpment and asymmetrical ridge formed. They thought that some sort of faulting was involved, and proposed reverse (thrust) or strike-slip (transcurrent or wrench) faulting. Under the first, the escarpment would have been caused by north–south compression, and perhaps linked to the north–south trending strike-slip San Andreas Fault. They saw the strike-slip extension of the San Andreas Fault being replaced to the north by north–south compression. However, they acknowledged that their explanation could be no more than tentative working hypotheses. They recognized a serious difficulty, their solution had the Mendocino Escarpment facing north, whereas it faced south (RS2). The authors suggested, without evidence as they admitted, that subsequent geomorphologic processes such as sliding and slumping could have altered the topography (RS1). They also were uneasy about extending 1900 km westward the effects of the north–south trending strike-slip San Andreas Fault (RS2). The possibility that the Mendocino Escarpment was a strike-slip fault was appealing because of its straightness and great length, but it less successfully explained the San Andreas Fault (RS2). Although the northwestward trending San Andreas Fault seemed to meet at the escarpment at its eastern end, seismic evidence, as it was then known, indicated that the San Andreas Fault continued northwestward instead of turning westward along the scarp. They also were uneasy about a strike-slip causing the huge vertical displacement across the escarpment. Neither of the then best known terrestrial strike-slip faults (the Great Glen and the San Andreas) had significant vertical displacement (RS2), but subsequent subaerial erosion and deposition could have masked them along these terrestrial faults, whereas such processes are likely to have had little effect at the bottom of the ocean (RS1).

In their first papers on the great oceanic escarpments, Menard and Dietz were groping for explanations of a truly new phenomenon, and I want to emphasize three points. First, although they slightly favored thrust over transcurrent faulting, they thought of both as tentative working hypotheses. Transcurrent faulting better explained the escarpment's apparent straightness (RS3), but they also noted that the distance between soundings was too great to establish whether or not the escarpment was really straight (RS2).

… the most definite conclusions possible are nothing more than working hypotheses. A few hypotheses will be outlined and tested, but the testing will be brief, because few facts are available for the purpose. It might be taken for a fact, for example, that the Mendocino Scarp is very simple and straight; and conclusions as to origin might be based on that "fact." All that is known about the topography in this respect is that the scarp appears to be straight between widely spaced sounding lines. A similar spacing of sounding lines across the west face of the Wasatch Mountains would also show a long, straight, simple scarp instead of the complexly branching group of faults and scarps which has been found by more detailed mapping.

*(Menard and Dietz, 1952: 273)*

Second, they did not yet view the Mendocino Escarpment as, what they later called, a fracture zone. The term "fracture zone" was not mentioned. They were impressed with the huge difference in elevation of the seafloor on opposite sides of the lineation. A much more detailed topographical analysis of the escarpment and surrounding area, or some other type of data, would be needed before they would come to think in terms of "fracture zones." Third, although Menard and Dietz noted in this first paper that they suspected the existence of what is now called the Murray Fracture Zone and said as such, the whole array of fracture zones lay undiscovered.

### 5.5  From the Mendocino Escarpment to fracture zones: Menard's 1953 solution

I was 32 years old, only three years past my doctorate, and I had discovered a whole new class of major geological structures. They resembled nothing known on land. Indeed, they seemed to display a bizarre combination of the classical features of both horizontal-slip and vertical-slip faults. Moreover, they were individually enormously long, and collectively they spanned 5% of the area of the earth. They had to be caused by some global or very large-scale phenomenon. What could they be? None of the existing hypotheses of global tectonics had predicted the existence of fracture zones. The mere fact of their existence, however, was hardly enough cause to propose a new hypothesis, so I attempted to explain the fracture zones in terms of the old ones.

*(Menard, 1986: 64)*

Menard discovered three more escarpments in the northeastern Pacific Basin that ran parallel to the Mendocino Escarpment (See Figure 5.1). He submitted a paper in October 1953, which was not published until September 1955, reporting his new discoveries, and offering a tentative solution to their origin (Menard, 1955a).

He now described the escarpments as fracture zones, naming them the Murray, Clarion, and Clipperton fracture zones – the Mendocino Escarpment became the Mendocino Fracture Zone. Acting on the basis of previous suspicions (Menard and Dietz, 1952: 270), he planned a reconnaissance south of the Mendocino Escarpment for summer 1951 and fall 1952. His prediction was confirmed; the Murray Fracture Zone is 2900 km long (Menard, 1955a: 1151). Menard also suspected the existence of a fracture zone further south of the Murray Fracture Zone; his prediction was confirmed in 1952 with the discovery of the Clarion Fracture Zone, which was further mapped in 1953, and shown to extend for 4800 km (Menard, 1955a: 1167). The 5300 km long Clipperton was discovered in 1952 (Menard, 1955a: 1171). All four are characterized by their asymmetrical ridges, escarpments, seamounts, and deep narrow troughs, and Menard was particularly impressed that all are *very nearly* arcs of great circles – a great circle tracks the shortest distance between two points on a spherical surface along the great circle passing through both points. He illustrated this by plotting them on a great circle projection on which great circles plot as straight lines (Figure 5.2).

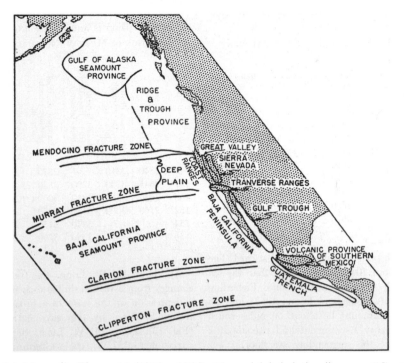

Figure 5.1 Menard's Figure 14 (1955a: 1166). Menard labeled the diagram, "Continental and Marine Geomorphic Provinces," and noted, "most provinces lie within or are bounded by fracture zones."

This discovery was accidental. Needing to find the shortest sailing route to the Marshall Islands from San Diego to plan how he could most economically use his allotted ship time on the second leg of the Capricorn Expedition of 1952–3, he used a great circle sailing chart to determine the shortest distance between San Diego and the Marshall Islands. He later recalled:

I was not on the first leg of Capricorn because little could be done. The ships had to meet a tight deadline in connection with the test of the first hydrogen bomb "device" in the Marshall Islands. The track from San Diego to Kwajalein [one of the Marshall Islands], however, was near the Murray fracture zone, and I was allotted surveying time of 12 hours or so on each ship in excess of that required on the most direct route. What was the most direct route? I obtained a great circle sailing chart, a special projection on which a straight line corresponds to a straight line on the special surface of the earth. Latitudes and longitudes look very odd. I drew a straight line between San Diego and Kwajalein and calculated the time the ships would need at cruising speed. Then I plotted the sparse data on the Murray fracture zone – they, too, fell on a straight line. The Clarion fracture zone was plotted; for 1700 miles [2700 km] it deviated from a great circle by no more than 15 miles [24 km]. The Clipperton zone followed a great circle for at least 3300 miles [5300 km]. There was nothing known on earth that was remotely

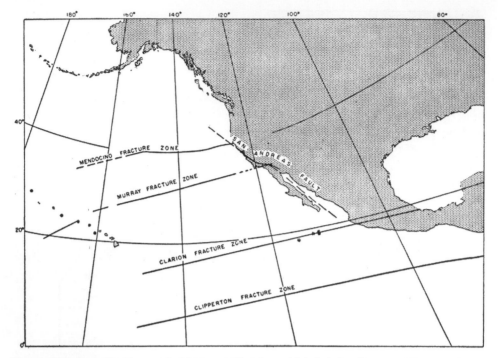

Figure 5.2 Menard's Figure 12 (1955a: 1163). Menard labeled the diagram, "Fracture Zones of Northeastern Pacific Plotted on Great Circle Projection," and added, "Straight lines on earth's surface are straight lines on this type of projection."

like these fracture zones. No two great circles can be parallel, but these were nearly so, and so was the western end of the Mendocino fracture zone once it deviated from the fortieth parallel. This totally unexpected discovery ultimately turned out to be grossly misleading with regard to the origin of the fracture zones.

*(Menard, 1986: 62–63; my bracketed additions)*

He continued to describe the fracture zones in terms of great circles. Fast-forward about fifteen years and there is Jason Morgan, while examining one of Menard's diagrams, noticing that the great fracture zones do not exactly fit great circles but trend accurately along small parallel circles (IV, §7.8). Morgan's modification of Menard's discovery played a pivotal role in his development of plate tectonics.

Menard identified four provinces (Figure 5.1) and another area within the north-eastern Pacific Basin, and claimed (1955a: 1152) that understanding their nature was "essential to an understanding of the great fractures." The northernmost Gulf of Alaska Seamount Province appeared to be topographically old because it possesses a thick apron of sediment, numerous guyots and no current seismic or volcanic activity. Next is the Ridge and Trough Province, extending southward to the

Mendocino Fracture Zone and is "block-faulted into long thin ridges which trend northeast or north" (Menard, 1955a: 1156); because of its rough topography and prevalent seismic activity, he argued that it is a young province. The seafloor just south of the Mendocino Fracture Zone is 800 m deeper than the Ridge and Trough Province, and he named it the Deep Plain Province; it extends south to the Murray Fracture Zone. The Baja California Seamount Province, which is between the Murray and Clarion fracture zones, is relatively mountainous near the shore and studded with what appear to be active volcanoes. Drawing on G. Arrhenius' earlier work on the 1947–8 Swedish Deep-sea Expedition, Menard divided the region between the Clipperton and Clarion fracture zones into the East Pacific Ridge and East Eupelagic areas, and claimed that the topography of the East Pacific Ridge "resembles the topography of the Baja California Seamount Province but is shallower" (Menard, 1955a: 1161).

Menard realized that the discovery of these fracture zones raised a new and important problem; they were a new kind of structure, and there were at least four of them. They were also huge, ranging from approximately 2300 km to 5300 km in length. He (1955a: 1173) wanted to find a unified solution that would relate them to the San Andreas Fault. Privy to Raitt's new findings based on his refraction studies of the Pacific seafloor, Menard (1955a: 1185) also realized that his and Dietz's previous explanation of the differing depths of the seafloor on either side of the Mendocino Escarpment was incorrect and could not be attributed to differences in crustal thickness and density (RS2). He would have to start afresh.

## 5.6 Menard's 1955 theory of fracture zones

He began by familiarizing himself with current ideas of global tectonics. He read the reports of a 1950 colloquium on plastic flow and deformation within the Earth, which had been jointly sponsored by the International Union of Geodesy and Geophysics and International Union of Theoretical and Applied Mechanics. The meeting brought together leading geophysicists and geologists involved in the mobilism–fixism controversy (I, §5.10). Beno Gutenberg, one of Menard's former teachers, chaired the special editorial committee and wrote a summary of the meeting. Griggs, Hess, Vening Meinesz, Birch, Umbgrove, Bullard, and several others discussed mantle convection. The general consensus was that mantle convection was still a speculative idea but that evidence in its favor had increased since the 1930s. Menard also took special notice of two other presentations. Arpad L. Nadai, an expert in plastic deformation from the research laboratories of Westinghouse Electric Corporation, discussed effects of temperature on creep. Menard learned of Nadai's idea that major features in Earth's crust are equivalent to Lüders' lines – lines that form when a metal sheet is compressed above its plastic limit. Menard also welcomed P. P. Bijlaard's ideas. Bijlaard (1951), then at Cornell University, discussed Lüders' lines in connection with plastic deformation, and developed an explanation for the formation of island arcs along the Pacific coast of eastern Asia in terms of plastic deformation caused by convection currents. Menard (1955a) viewed the fracture zones as equivalent to Lüders' lines brought about by plastic deformation through mantle convection.

Menard was especially influenced by Harvey Brooks' views on mantle convection beneath the Pacific Basin. Brooks (1915–2004), a Harvard-trained physicist, returned to Harvard in 1950, one year after Menard finished his Ph.D. Menard, I suspect, learned of Brooks' work from Griggs's favorable comment about his extension of Pekeris's study (I, §5.6) of mantle convection (Griggs, 1951: 527–528). Griggs discussed Brooks' idea of plastic deformation and his proposal that mantle convection beneath the Pacific Basin could have caused circum-Pacific mountains and island arcs. Describing his idea as "highly tentative and speculative," Brooks suggested:

If cooling begins, for example, over the center of the Pacific Basin, where the crust is both thin and low in radioactive content, then descending currents will be generated there, and a cell-system ... will rapidly be formed, the successive cells being arranged in the form of annular rings or zones surrounding the center ... from which the convective system originates.

*(Brooks, 1941: 550)*

Brooks calculated that the first ring, where currents ascend and diverge, will form 24° from the center; the second ring, where currents will descend and converge, will form 73° from center, and will lead to formation of a tectogene. He then added, "It is tempting to believe that this may be the origin of the great system of mountains which encircles the Pacific Basin" (Brooks, 1941: 550–551). Tectogenes where currents converge fail to form at the center of the Pacific because of the absence of continental crust. Not so, however, for the Atlantic, where Brooks applied his idea of annular convection, offering an explanation of the Mid-Atlantic Ridge.

No tectogene occurs in the center of an ocean-basin because of the absence of the light continental material on which the currents could act. In the Atlantic Basin there is a small amount of continental material, and this may be the origin of the well-known mid-Atlantic swell, which is then due to the segregation of continental material by the converging currents under the Atlantic Basin.

*(Brooks, 1941: 551)*

With the discovery that the crust of the Atlantic Basin lacks continental material, Brooks' solution was ignored. However, Menard found Brooks' speculative idea of annular convection when applied to the Pacific a worthy theory, and adopted it as his own.

Menard first reviewed and rejected three solutions, each invoking application of a general hypothesis that had already been proposed as a solution to another problem. The first depended on Vening Meinesz's hypothesis of polar wandering proposed in 1934 and revised 1947 (I, §5.6). Vening Meinesz (1947) argued that if the North Pole had migrated during the Precambrian from near Calcutta to its present position through slippage of the whole crust over the mantle, a worldwide shear net of compression and tensional strands would be produced, and that such a net correlated rather well with many of Earth's topographical features. Menard noted that three of the fracture zones were roughly parallel to Vening Meinesz's proposed shear net. Menard wrote him after the discovery of the Clarion Fracture Zone saying that given Vening Meinesz's hypothesis there could be at least ten fracture zones in the Pacific off the coasts of North and Central America. Vening Meinesz replied that he was "deeply stirred" and gave Menard his "hearty congratulations" (Menard, 1986: 62). Menard was initially attracted to Vening Meinesz's hypothesis because "The fracture zones of the northeastern Pacific were unknown in 1947 so that no element of bias enters into a comparison of the topography of these zones with the various shear nets proposed by Vening Meinesz" (Menard, 1955a: 1174). The Mendocino and Murray fracture zones fitted very well with Vening Meinesz's shear net.

The Mendocino fracture zone does not deviate from Vening Meinesz' preferred shear net by more than 12° anywhere along its 1400-mile length, and it follows the shear net exactly for about 700 miles. The correlation between the shear net and the Murray fracture zone is even more extraordinary. For 1700 miles the deviation between the two scarcely exceeds 5° although the bearing of the net itself changes by about twice that amount.

*(Menard, 1955a: 1175)*

However, the Clarion and Clipperton fracture zones did not (RS2).

Clarion and Clipperton fracture zones show a no less remarkable lack of correlation with the shear net. Neither trends within 12° of the net anywhere in a total distance of about 6000 miles.

*(Menard, 1955a: 1175)*

Reluctant to reject Vening Meinesz's notion of "shear nets" caused by polar wander, Menard suggested a modification that yielded a better fit with fracture zones and, as an added bonus, fitted the San Andreas Fault (RS1). Although he could only achieve the better fit by varying the direction and rate of the polar wandering, Menard was not yet ready to discard the idea. But, he raised another difficulty (RS2). Neither Vening Meinesz's original nor Menard's amended polar wandering hypothesis would produce enough stress to form fracture zones as huge as the four Menard had discovered.

Menard turned to another hypothesis that Vening Meinesz (1947) had considered and rejected, a worldwide stress pattern caused by flattening brought about by an increase in Earth's rate of rotation. Vening Meinesz rejected it because it would produce shear patterns with little resemblance to Earth's major topographical features, and so did Menard because it also failed to explain the orientations of the fracture zones and San Andreas Fault (RS2).

Menard examined J. Tuzo Wilson's 1951 contractionist view of mountain building. Wilson, a fixist at the time, later became a leading mobilist (IV, §1.3–§1.5). His view could not account for the orientation of the fracture zones and the San Andreas Fault.

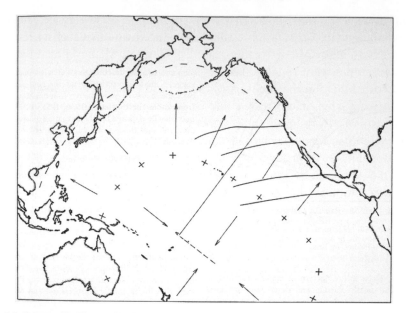

Figure 5.3 Menard's Figure 23 (1955a: 1181). Plus signs indicate upward diverging currents; minus signs mark downward currents, and arrows show horizontal movement of mantle convection. The four fracture zones are also shown.

Menard then introduced an explanation that combined features of Bijlaard's and especially Brooks' hypotheses of mantle convection beneath the Pacific, arguing that they explained the peripheral island arcs and surrounding mountain ranges – Bijlaard's crustal deformations as analogous to Lüders' lines, and Brooks' annular mantle convection with currents rising in the middle of the Pacific Basin and sinking along its periphery. Menard illustrated the dimension and orientation of Brooks' annular convection in his Figure 23 (Figure 5.3), which he labeled "Annular Convection Cells Produced by Cooling in Central Pacific Area." The orientation of the convection currents provided northeast–southwest compression, which he believed was required for the formation of the San Andreas Fault, placed the uprising convection currents beneath the Hawaiian Islands, and had the descending limbs located more or less beneath island arcs and trenches.

Brooks' theory, Menard maintained, offered solutions to problems that were unknown when Brooks first suggested it.

This theory [of annular convection] explains almost all the pertinent geological and geophysical features of the northern and southwestern Pacific basin, which is particularly impressive because many facts were not discovered until almost a decade after the theory was published.

(*Menard, 1955a: 1181; my bracketed addition*)

First and foremost was Menard's belief that this extension of Brooks' theory explained the origin of the fracture zones and San Andreas Fault (Menard, 1955a: 1181). He

(1955a: 1181) also claimed that this extension of Brooks' theory explained formation of island arcs in the western Pacific, the Guatemalan and Aleutian trenches, and the Hawaiian Islands, swell and deep. Menard (and Dietz) already had proposed descending convection currents to explain the origin of the Aleutian Trench and ascending currents to explain the Hawaiian swell, deep, and arch (Menard and Dietz, 1951; Dietz and Menard, 1953). He now thought he had a unified solution to both problems, which required mantle convection on a much larger scale than his (and Dietz's) previously envisaged solutions, and could account for other trenches and island arcs along Asia's Pacific margin.

Expanding the application of his ideas, Menard (1955a: 1181) claimed that he could also explain the "unexpectedly high heat flow through the floor of the Pacific Basin," and also (1955a: 1186) account for the topography of the fracture zones, and differences in elevation and volcanic activity of the oceanic provinces in the northeastern Pacific (RS1). Viewing fracture zones as welts caused by crustal shortening due to mantle convection, he claimed that fracture zones have ridges and roots, which, he argued, interrupt the main convective flow, creating small convective cells or eddies of different intensity between fracture zones. Further speculating, he claimed that eddies of greater intensity produce more heat, which expand the mantle, elevating the overlying province and producing volcanoes. Ridges, he (1955a: 1184) proposed, serve as dams to turbidity currents affecting sedimentation rates, thereby accentuating differences in elevation of provinces.

Very tenuously, he tried to justify the adoption of mantle convection by appealing to Elsasser's hypothesis that the geomagnetic field is generated by a self-exciting dynamo (II, §1.4); Menard speculated that rapid currents within the fluid outer core could produce slow counter-currents in the mantle (RS1). On firmer ground, he (1955a: 1179) cited Hess's (§3.8) early 1950s "compelling argument that convection can explain the persistence of tectogenes." Menard (1955a: 1179) also appealed to Hess's argument that only convection currents could account for the presence of negative gravity anomalies over trenches suppressing uplift (RS1). He admitted that Wilson's contraction theory could account for island arcs but not fracture zones, so he opted for convection (RS2 and RS3).

### 5.7 Menard's 1958 solution to the origin of mid-oceanic elevations

My publication of 1958 ... compared the position of the global "rift" of Ewing and Heezen with median line of the ocean basins. The median line of the mid-ocean ridge corresponded very closely to the median line of the ocean basins. It really was a *mid-ocean* ridge. The correspondence continued rather dramatically in previously unsuspected places such as around the southern end of Africa and between Australia and Antarctica. The only median line of an ocean basin lacking a mid-ocean ridge was in the western Pacific. This suggested that the guyots in the region were submerged because they had been on a mid-ocean ridge that had subsided, which in turn appeared to confirm Hess's idea that ridges are ephemeral. I conceived of an age sequence. The East Pacific Rise is seismically active, exceptionally broad, unrifted, and not centered in an ocean basin. Perhaps it was youthful; the Mid-Atlantic Ridge was mature; and the western Pacific was a dead ridge. The age sequence implied that ridges can be created anywhere, but in time they become centered in ocean basins between continents. How? Did the continents stay fixed and ridge move, or vice versa? In 1958 I thought that the marine evidence was not pertinent to the hypothesis of continental drift because the ridge circled southern Africa and ridges intersected continents in East Africa and western North America. Everyone, or at least Ewing, Heezen, Hess, and I, thought the ridge was fixed. If so, Africa could not drift away from the ridge, and yet the continents should have been torn asunder at the intersections with the ridge.

*(Menard, 1986: 140–141)*

Menard presented his first solution to the origin and development of oceanic rises and ridges in his 1958 paper "Development of median elevations in ocean basins." He plotted positions of oceanic rises and ridges and found that most bisected or nearly bisected oceanic basins. He classified these medially positioned oceanic rises into three morphological types, each of which he thought represents a sequential stage in their life-cycle.

He (1958a: 1181) claimed that young ridges are broad in extent, seismically active, have many seamounts but few guyots, and lack a central rift valley; he cited the southern East Pacific Rise as an example.[2] Middle-aged ridges are narrower and steeper, have a central rift valley, are seismically active, and are populated with volcanic islands and guyots; he selected the Mid-Atlantic Ridge as a prime example. Old ridges become steeper and narrower than middle-aged ones, are seismically inactive, have plentiful guyots and atolls but lack volcanic islands; he chose the Tuamotu Ridge and Mid-Pacific Mountains as examples of an old ridge. Grouping them together with the Christmas Islands Ridge, he suggested that they "probably would have been named the 'Mid-Pacific Ridge' if they had been discovered at the same time" (Menard, 1958a: 1181).[3] He also noted that heat flow over young and middle-aged ridges is twice that over old ridges and elsewhere in the seafloor.

He envisaged this evolution. Oceanic ridges start out as broad rises. They are seismically active, characterized by high heat flow and volcanic islands. They also seem to lack a central rift valley. As they age, they remain seismically active, a central rift valley forms, their flanks begin to subside, and some associated volcanic islands are truncated by wave action and become guyots. They eventually become narrower and narrower, and remaining volcanic islands become guyots and atolls. Menard also tentatively suggested that rises were impermanent:

The proposed sequence of development from rises to ridges implies that oceanic rises are not permanent features ... Observations do not rule out the possibility that broad rises may be temporary features which are elevated and then subside, leaving narrow ridges capped with guyots and atolls.[4]

*(Menard, 1958a: 1183–1184)*

Whether he also meant here that oceanic ridges are also not permanent is unclear from what he wrote in 1958. Much later he (1986: 140) retrospectively claimed that he had.

Considering the origin of oceanic ridges, he announced that any adequate theory of them must explain their sequential development and varied topography. It also must explain differences in petrology and structure of various ridges. Menard (1958a: 1183) learned from George Shor at Scripps that the crust of the East Pacific was not substantially different from ordinary ocean crust, but knew from the work of M. Ewing and others that the crust of the Mid-Atlantic Ridge was different. An adequate theory of ridge origin must explain their high heat flow and median position of oceanic rises: "It is at least not improbable that the agreement [over the median position of rises] signifies a causative relationship – that is, the rises are in the middle because it is the middle" (Menard, 1958a: 1180; my bracketed addition).

Following the same method of argument by exhaustion he used when considering the origin of the great fracture zones, he examined and dismissed previous solutions, and offered his own. He first considered and dismissed the explanation that John and Maurice Ewing had proposed at the 1957 Annual Meeting of the Seismological Society of America (RS2).

Ewing and Ewing (1957) suggest that the low maximum seismic velocities found under the Mid-Atlantic Ridge are caused by a physical mixing of mantle and crustal material. This process does not explain the relief of rises where the crustal section is of normal oceanic type.[5]

*(Menard, 1958a: 1183)*

He then turned to Hess's 1955a solution that ocean ridges were caused by serpentinization of mantle peridotite (§3.9).

Hess (1954, p. 346–347) proposes that serpentinization of the mantle is a common cause of oceanic epeirogeny. The reaction has the virtue of occurring wholly within the mantle and being reversible, and because it is exothermic it also offers an explanation for the high heat flow associated with rises.

*(Menard, 1958: 1183; the Hess reference is to my Hess, 1955a)*

Menard thought Hess's solution better than the Ewings'; it did not involve changes in oceanic *crust*, and was not subject to the above difficulty he had raised against their solution. Menard also acknowledged that Hess's solution (remember this was Hess before he proposed seafloor spreading) might explain the high heat flow of ridges, and their changes in elevation, but it failed to explain the median position of ocean rises and ridges. Although Menard did not explicitly raise this difficulty against Hess's solution, he immediately noted:

The process that centers rise in ocean basins has the following characteristics: (1) It is sensitive to the margins of the basins. (2) It is capable of acting at distances as great as half the width of the Pacific Basin – more than 5000 km or 50° arc. (3) It acts on both sides of a basin at one time. (4) It appears to be ephemeral or intermittent.

*(Menard, 1958a: 1183–1184)*

Hess's solution explained (4), but it did not explain (1) through (3), which Menard took as essential to explaining the centering of rises within ocean basins. However, Menard's classification of the northern extension of the East Pacific Rise as mid-oceanic was questionable. He did this because its northeastern extension from Easter Island intersects arcs of equal radii constructed from points on the western coasts of Mexico and Equator-Peru. But this hardly allowed him to classify as mid-oceanic the continued extension of the East Pacific Rise toward the western coast of Mexico and the Gulf of California. Regardless of whether or not Menard had exaggerated the significance and generality of their median position within ocean basins, he thought it needed explanation, and Hess's solution provided none.

Both Hess (1955a) and Menard (1958a) explained the sinking of guyots in similar ways; guyots, born as volcanic islands atop young ridges, were truncated by wave action as the underlying ridge sank. Thus, old ridges had numerous guyots while youthful ones had none, only numerous volcanic islands. Menard acknowledged that Hess had earlier offered the same solution (§3.9).

Menard argued that his form of mantle convection solved more problems than Hess's 1955 version (RS3). Both explained the high heat flow of oceanic ridges, their changes in elevation and ephemerality, and the sinking of guyots; only Menard's explained their high seismic activity and points (1) through (3) above. Menard's solution had another advantage.

One possible process is the convection current. If so, the high heat flow along seismologically active rises implies rising currents in the middle ... Continuity requires sinking currents near the margins of the basins with horizontal movement toward the margins under the crust. This is the direction of movement advocated by proponents of the convection origin of island arcs and Pacific fracture zones.

*(Menard, 1958a: 1184)*

He was in effect suggesting, without providing details, that his solutions to the origins of mid-oceanic ridges, great fracture zones, and peripheral island arcs could be combined. They required huge convection cells all with roughly the same orientation; the locus of the upwelling convection currents he proposed to explain fracture zones roughly coincided with the position of his newly named Mid-Pacific Ridge, the combined ridge formed by the Mid-Pacific Mountains, and the Christmas Island and Tuamotu ridges (see Figure 5.4).

Menard left unexplained why ascending convection currents center themselves within ocean basins. His form of mantle convection causes elevation and depression of ridges, and helped explain the formation of island arcs and trenches where oceanic crust is forced down by descending mantle convection currents, but said nothing about why such mantle convection causes little horizontal movement of overlying oceanic crust. There was only enough horizontal movement to rupture oceanic crust and create fracture zones and no more.

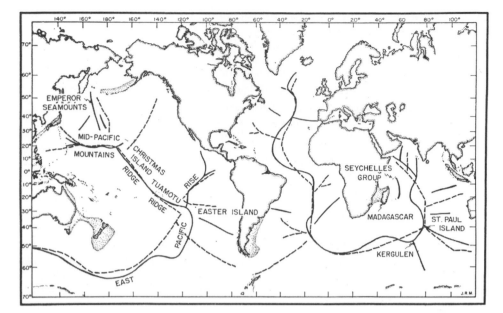

Figure 5.4 Menard's Figure 1 (1958a: 1180). Menard labeled the diagram, "Location of the Crests of Submarine Median Ridges and Rises (solid line) Compared with the Geometrical Median Line (dashed line) of the Ocean Basins."

## 5.8 Menard provides fixists with isthmian connections and rejects mobilism, 1958

Menard recalled that during the early and middle 1950s mobilism was not a topic of discussion at NEL in San Diego (§5.4). It was considered irrelevant to his work on the marine geology of the Pacific. Mobilism, or rather the lack of it, became relevant once he had presented his 1958a (p. 1184) explanation of the origin of mid-ocean ridges in which he derived "two of the most important … corollaries" of his version of mantle convection, corollaries that "can be used to test its validity and the validity of the other hypotheses." The first offered fixist biogeographers an updated version of Bailey Willis's isthmian connections (I, §3.6), the second an argument against mobilism. Here is the first.

*Faunal Migration.* Although not continental, if the median ridges of the Pacific were once elevated and were contemporary they would have produced many of the effects normally attributed to continents. Assuming approximate contemporaneity, the median line of the Pacific Basin was marked in Mesozoic time by an almost continuous deep bank capped by numerous closely spaced islands and shallow banks. The general picture closely resembles the isthmian links of Willis (1932) and is not more than a geographical extension of the dense island chain of the Mid-Pacific Mountains … Island stepping stones and almost continuous shallow water may have been available for faunal migration from Japan to Easter Island or even to South America. A branching structure reached almost to Kamchatka.

*(Menard, 1958a: 1184)*

In 1932 Bailey Willis suggested that isthmian connections served as routes for faunal migration, but mobilists objected that they did not appear to traverse deep oceans (I, §3.6). Menard disagreed, raising a difficulty against mobilism with his second corollary.

*Continental Drift*. Evidence in support of continental drift is derived from the similar shapes of the Mid-Atlantic Ridge and the adjacent shores of Africa and South America. This evidence can be dismissed as not pertinent to the hypothesis because both Africa and South America are almost circled by rises and ridges roughly parallel to the shore lines. If the Mid-Atlantic and Mid-Indian ridges are in part left by the drift of South America and Australia respectively, what are the rise and ridge structures of the Pacific? If the continents have only begun to drift in Early Tertiary time [referring to Wegener's timetable of events], how did the Cretaceous rise of the Mid-Pacific Mountains become centered relative to the present margins of the Pacific?

*(Menard, 1958a: 1184; my bracketed addition)*

He agreed that the Mid-Atlantic Ridge and Atlantic-facing margins of Africa and South America are similarly shaped but he explained their congruencies and the central position of the Mid-Atlantic Ridge by appealing to his own hypothesis, and thereby offered fixists their own solution (RS1). He also raised several related difficulties against the mobilist solution to the same problem (RS2). Mobilists maintained that ridges mark where continents became separated from each other. So Africa and South America separated from each other along the Mid-Atlantic Ridge. But both continents are almost entirely circled by ridges whose shapes are similar to facing continental margins. If ridges marked where continents separated, then what continents separated from the ridges in the Pacific? Mobilism required drift toward the Pacific, not away from it. If mobilists respond by delaying drift until the Early Tertiary, then why is the Cretaceous Mid-Pacific Ridge centrally located within the Pacific Basin?

Menard may have borrowed this difficulty against continental drift from Heezen. Heezen raised a version of it in a February 1, 1957 *New York Times* article, arguing that the position of the Mid-Atlantic Ridge supports the separation of Africa and South America, but the presence of ridges surrounding Africa implies that Africa drifted toward South America. Menard saw the *Times* article and even wrote Heezen about it (§6.7–§6.8).

Within two years, Menard had rejected his fixist explanation of the great fracture zones and mid-ocean ridges. He changed his mind because the discovery of magnetic anomalies in the northeastern Pacific implied large displacement of seafloor along the great fracture zones. I turn now to their discovery.

### 5.9 The discovery of magnetic lineations in the northeastern Pacific, 1952–1961

Ronald Mason, our Englishman who usually turns up in the lab [of *Spencer F. Baird* during the September 1952 – February 1953 Capricorn Expedition] about midnight to work on his magnetometer records. He is a lone wolf, working late at night when there is more space on which to spread out a twenty-three-foot-long record ... He pulled three hard-boiled eggs out of his pocket. These he had boiled in the galley after supper to fortify himself for the long night.

*(Raitt, 1956: 98; my bracketed additions)*

I started the project [of mapping the magnetic lineations] with no thought of its possible relevance to continental drift. The discovery of seafloor magnetic stripes was serendipitous: we were not looking for them, nor could we have been, because no one knew they existed!

*(Mason, 2001: 43; my bracketed addition)*

No more serendipitous relation can be imagined than the needs of the Navy and the discovery of magnetic stripes off the west coast of North America.

*(Menard, 1986: 72)*

Menard became head of the NEL's seafloor study group at San Diego, succeeding Dietz who left for Japan on a Fulbright Fellowship. After making ready to depart on the USCGS ship *Pioneer*, he was asked by King Couper and associates at the Bureau of Ships if there was anything else that could be done during the expedition that would not interfere with what already had been planned. Menard relates how he remembered that Ron Mason and Arthur Raff at Scripps had recently obtained interesting observations of remanent magnetization of red clays from marine cores they had collected on a previous expedition (Yo-yo Expedition, early 1954, on *Spencer F. Baird*), and he suggested that the Bureau of Ships

let us tow a magnetometer. Mason took on the job. The survey would be precisely located electronically by LORAN-C, an advanced system developed by the C and GS [USCGS], and the line spacing would be only 5 miles. No comparable survey existed anywhere else in the world, and the magnetics would be unclassified. In fact, no one in the Office of Naval Research or any of its advisors could see any point in towing a magnetometer on the *Pioneer* survey. The U.S. Geological Survey said it would be a waste of money, and, ONR refused our request to fund the magnetic survey. Thus it was that one of the most significant geophysical surveys ever made was wholly financed by the minuscule discretionary funds of the Director of Scripps Institution of Oceanography.

*(Menard, 1986: 72–73; my bracketed addition)*

Ron Mason, the man who assumed this huge, multi-year task, was English, a visitor at Scripps from Imperial College, London, where he was a lecturer in the Department of Geophysics, headed by Bruckshaw. Bruckshaw, recall, had discovered reversely magnetized Early Tertiary dykes extending southeastward from western Scotland to the northeast coast of England (II, §1.6). Mason (1916–2009) was born in Winsor, Hampshire. He attended Eastbourne Grammar School in East Sussex before studying physics at Imperial College, University of London.[6] After working with Army Signals Research during World War II, he returned to Imperial College, already having decided to work in geophysics because he did not want to spend his working life in a laboratory. In 1947 he was appointed lecturer in geophysics at Imperial College, under Bruckshaw. Mason juggled his commitment to Imperial College,

Scripps, and later the University of Hawaii throughout the 1950s. In 1964 he became reader in geophysics at Imperial College and, three years later, he was appointed to the chair of pure geophysics. In 1984, he retired from Imperial College but continued there as a senior research fellow. In 2002, he was honored with the Gold Certificate from the Society of Exploration Geophysicists.

Mason's interest in shipborne magnetic surveys seems to have been prompted by results from "Project Magnet," which, as a geomagnetist, he would have known about (Mason, 2001: 32). Up to that time, Earth's magnetic field had been observed on land at observatories scattered very unevenly over the Earth. One of the aims of Project Magnet was to fill in an observational void, by flying aircraft equipped with magnetometers over oceans to observe the geomagnetic field where there were no magnetic observatories (Alldredge and Keller, 1949). During these long flights, anomalies in the geomagnetic field associated with volcanoes and atolls had been observed; there were measurable magnetic features in the oceans. This interest, a conversation with Russ Raitt, and Roger Revelle's good hearing led to Mason's being asked to become "Scripps' magnetometer man."

In 1951 I took a year's sabbatical, which I spent at ... Caltech. While there, in the spring of 1952, I attended the annual meeting of the University of California Institute of Geophysics, held that year in La Jolla. The location of the meeting, right by the ocean, and the several presentations on marine seismology, a branch of geophysics new to me, set me thinking about the other geophysical techniques that had been or might be used for studying the oceanic crust. Apart from seismology, very little seemed to have been done ... Talking casually to Russ Raitt during the morning coffee break, I asked whether anyone had thought of investigating the magnetic anomalies associated with sea floor structures by towing a magnetometer behind a ship, an operation that could enable ships to obtain valuable data while engaging in other operations."What's that?" came a deep voice from behind me. Roger Revelle, Director of Scripps, had overheard the conversation. After the briefest of explanation, Roger, in his characteristically direct way, asked, "Well, do *you* want to do it?" To which I promptly said" Yes" and I became Scripps' magnetometer man.

*(Mason, 2001: 32)*

Airborne magnetic surveys were expensive. The beauty of Mason's idea was that ships could tow a magnetometer while engaged in other activities, and the cost of doing the magnetic survey would be minimal. Revelle recognized a good idea when he heard one.

The task of measurement and analysis that Mason and Raff, his technician-assistant, undertook was huge, and in character and background they were very well suited for it. Done before the advent of computers, they had to hand-plot the results. Because they used a fluxgate magnetometer, which, unlike the proton magnetometer, measures only relative values of the magnetic field, they had to adjust their readings to correspond to true values of the magnetic field. They did this "by overlaying" their "map on the map of the earth's magnetic field ... and subtracting the one from the other graphically" (Mason, 2001: 36). Their map is one of the icons of plate tectonics.

The next year, Mason got a trial run during Scripps' Capricorn Expedition, the same one during which he helped Fisher and Raitt to investigate the Tonga Trench (§3.10). Using a borrowed fluxgate magnetometer from Lamont, Mason met with success.

Lamont offered to loan us their magnetometer for Scripps' upcoming Capricorn expedition ... This presented a great opportunity for us to familiarize ourselves with the problems associated with operating a ship-towed magnetometer, and we gratefully accepted. After a scramble to get to the west coast in time, we towed it successfully over more than 8,000 miles (12,500 kilometers) of ship's tracks, during which we recorded magnetic anomalies associated with seamounts, atolls, scarps, and other features of the sea floor. Although the results had limited quantitative value, it was clear that there was a future in ship-towed magnetometry. We just had to acquire a magnetometer of our own.

*(Mason, 2001: 33)*

He based Scripps' instrument on Lamont's fluxgate magnetometer, itself based on Victor Vacquier's airborne fluxgate magnetometer which he invented in 1941 and was used to detect submarines.[7] Mason, with the help of Jim Snodgrass and Jeff Frautchy, improved the design, making it more suitable for towing behind a ship. Vacquier (1907–2009) himself, lured to Scripps by Revelle in 1957 after Mason agreed to head a Scripps sponsored IGY project in the equatorial Pacific, took over directing Scripps' magnetics program, and soon he and Robert E. Warren designed a proton precession magnetometer to use on future surveys (Shor and Sclater, 2010).

Mason and Raff's first survey after their trial run, the one promoted by Menard, began in August 1955 aboard *Pioneer*, cut across the Murray Fracture Zone, and covered the area between latitudes 32° and 36° N, longitudes 121° and 128° W (Mason, 1958). In all, it took twelve monthly cruises, the last completed in October 1956, followed by plotting of the results throughout the first half of 1957 (Mason, 2001: 36). They found positive (in which the strength of the magnetic field was enhanced), and negative (in which the field was diminished) anomalies, elongated north–south and perpendicular to the great fracture zones. Mason was

able to show that they [the anomalies] could be explained by shallow slablike structures, immediately underlying the positive stripes and more highly magnetized than the surrounding crust, but there was no geological model to support such structures.

*(Mason, 2001: 39)*

Although used as a basis for invoking displacements along the fracture zones, the origin of the lineations remained unexplained for four years. It was left to Vine and Matthews (IV, §2.13) and Morley (IV, §2.16) to provide that key component. Here I shall discuss only their arguments for displacement, and delay consideration of their attempt to explain the origin of the anomalies until IV, §2.3.

They found that the patterns of magnetic anomalies perpendicular to the great fracture zones were displaced either side of the fracture zones. The magnetic anomaly patterns could be matched across fracture zones only if it was assumed that there had been large relative horizontal displacement of oceanic crust along them, and the four researchers were unanimous about this. Mason (1958: 327) and Menard and Vacquier (1958: 5) argued that matching the pattern of magnetic anomalies north and south of the Murray Fracture Zone indicated a right lateral displacement of 84 nautical miles (154 km) along it. Mason (1958: 327) related the right lateral displacement to that of the San Andreas Fault, and suggested "a westerly movement of the North American continent relative to the floor of the Pacific." Menard and Vacquier were less daring; for them the displacement along the Murray Fracture Zone

raises interesting questions with regard to the relationship between the deformations of the oceanic crust and adjacent continental margins in response to mountain-building forces in the depths of the earth.

*(Menard and Vacquier, 1958: 5)*

The next survey undertaken by Scripps in December 1958, to determine if there is a displacement along the Pioneer Fracture Zone, revealed a left lateral slip of 138 nautical miles (256 km) (Vacquier, 1959). Vacquier was surprised and impressed by the limited amount of distortion exhibited by the magnetic profiles; the seafloor appeared to move without being distorted.

A left lateral fault with a displacement of 138 nautical miles is required to match the pattern. The goodness of fit shows that the amount of distortion accompanying this left lateral slip over a length of 90 miles was less than 10 per cent and implies a remarkable freedom of movement of one crustal block with respect to the other.

*(Vacquier, 1959: 453)*

The next two surveys in August 1959 and April 1960 extended westward and northward and revealed a left-lateral displacement of 640 nautical miles (1160 km) along the Mendocino Fracture Zone (Vacquier, Raff, and Warren, 1961: 1253). Vacquier and his co-workers were impressed by the extensive relevant movements of otherwise rigid blocks of seafloor, for which they cited physiographic and paleomagnetic evidence. They noted that these movements indicated mobilism, and they also raised the familiar difficulty with Wegenerian drift: if seafloor is rigid, then continents cannot plow through it (RS2).

The displacements along strike-slip faults in the ocean floor are thus of the same magnitude as the distances through which continents have been presumed to drift to accommodate physiographic and paleomagnetic data. Except for Carey (1958), who invokes a several-fold expansion of the whole Earth, continents are pictured in the literature as rigid plates drifting without distortion through the viscous ocean floor. Now we have evidence for the rigidity of the upper part of the oceanic crust that carries the magnetic pattern, which is just as good as the evidence for the rigidity of continents exemplified by the fit obtained by Carey (1958) between the coasts of Africa and South America along the 2000-m isobath. Since it is most unlikely that the magnetic or the physiographic correlations are accidental, the resolution of this paradox will be an important advance in geophysics.

*(Vacquier et al., 1961: 1257–1258)*

They submitted their paper in October 1960, two months before the publication of Menard's mobilist solution to the origin of oceanic ridges and the great fractures (§5.12) and the distribution by Hess of his preprint of seafloor spreading (§3.14). Hess's (§3.14) and, later, Menard's (§5.12) accounts resolved the apparent conflict of drifting continents and rigid sub-oceanic crust, as Holmes had done (I, §5.4, §5.8) without his work being appreciated in North America at the time.

The final and most northern survey, illustrated in Figure 5.5, extended from 40° N to 52° N latitude off the coasts of the northwest United States and southwest Canada. This last part of the overall survey attracted little immediate attention although it eventually played a key role in the acceptance of seafloor spreading; the magnetic anomalies observed there did not cross any of Menard's identified great fracture zones, and thus, at the time, appeared not to reflect any horizontal displacement of seafloor.

## 5.10 Menard's views in flux, 1959

From 1960 through early 1964, Menard developed, defended, and expanded a new hypothesis that he initially proposed to explain the great fracture zones and rise of the East Pacific, but later he applied it to other oceanic ridges, including the Mid-Atlantic Ridge, and the "Mid-Pacific Ridge," which he would later rename the Darwin Rise (Menard, 1964: 138–146). Menard's hypothesis, aptly called seafloor stretching, somewhat resembles Holmes' 1928/1931 version of mantle convection (I, §5.4). Menard, however, did not recognize the resemblance at the time he proposed his hypothesis, and only retrospectively noted that his was "unwittingly almost exactly the same as Holmes' idea" (Menard, caption Figure 20, 1986).[8] They both invoked mantle convection which caused horizontal movement of seafloor through crustal stretching. Neither proposed creation of new crust at ridge axes.

It was at this point that Menard became inclined to mobilism and moved away from his former fixism. He appealed to paleomagnetic studies supportive of mobilism, showed that his former objection to drift did not apply to his hypothesis of seafloor stretching, and argued that the formation of the Atlantic Basin naturally followed from it. He also applied his hypothesis to problems he had previously explained in other ways: the origin of guyots, island arcs, and trenches.

Menard's views had begun to change in 1959. That year he had a paper published that shows he was no longer confident of his 1958 explanations of oceanic rises and great fracture zones, recognizing that Mason and Raff's discovery of strong evidence for extensive relative horizontal movement of blocks of seafloor along the Murray Fracture Zone raised a severe difficulty. On his previous view, fracture zones were not faults, but ruptures in the crust caused by obliquely directed convection currents that tore apart the crust. Mason and Raff's discovery indicated that if convection were involved, opposing currents would have to run parallel to fracture zones. So in his 1959 paper he said nothing about his former solution and was content to report

Figure 5.5  Raff and Mason's Figure 1 (1961: 1268). Their caption was simply "Index anomaly map of the total magnetic field. The positive area of the anomalies is shown in black."

Mason's discovery and to repeat his and Vacquier's 1958 suggestion that the movement might somehow relate to the formation of coastal mountain ranges of California.

A seaborne magnetic survey of this region shows a north-trending pattern of anomalies with magnitudes of about $500\gamma$ [footnoted reference here to Mason (1958) as in press]. The anomaly pattern is disturbed along the east-trending line of the opposite sides of the disturbance and it appears that it has been offset 150 km which would make it one of the largest fault displacements known. The Transverse Ranges of California lie on the trend of the Murray fracture zone and, at least superficially, they seem to be an extension of the zone.

*(Menard, 1958b: 208; my bracketed addition)*

He still maintained that ridges develop sequentially, and that they are median-positioned within ocean basins. He even reproduced the figure he had presented in 1958 (Figure 5.4 above) to show their median position (Menard, 1958b: 207). Remaining silent about his former solution, he alluded vaguely to unspecified "very powerful deforming forces" under youthful oceanic rises (Menard, 1958b: 207). He again linked formation of seamounts and their eventual drowning to the subsidence of oceanic ridges through "some process in the mantle," but did not identify what process he had in mind; it is unclear to him whether or not they were caused by mantle convection. He still maintained that trenches and island arcs form above descending convection currents which "may be pulling the crust downward to form the trench" (Menard, 1958b: 208).

Menard also considered but, surprisingly, rejected the idea that the Pacific Basin is no older than Cretaceous on two grounds. The first was the *absence* of any dredged samples of pre-Cretaceous marine fossils and rocks.

The demonstrable history of the Pacific Basin based on samples is very brief. Cores penetrate only a small distance and could hardly be expected to reach very ancient sediments, even though Tertiary outcrops occur. Dredge hauls of seamounts might sample material of any age because most submarine elevations are swept clean of sediment. To date Cretaceous fossils are the oldest ones sampled but fossils of any kind are rare in dredge hauls and the basalt usually dredged is undateable. The potassium-argon method is capable of dating fresh basalts but the dredge can break off only weathered material unless the dredger has extraordinary luck. If present sampling is considered to be adequate, the Pacific Basin is no older than Cretaceous.

*(Menard, 1958b: 213; Menard referred to E. L. Hamilton, 1956)*

Notwithstanding this straightforward reading of the evidence, he then remarked, "However, further sampling may indicate greater age and it seems more logical to assume that the basin is as old as the continents" (Menard, 1958b: 213). He still rejected a Cretaceous age for much of the Pacific Basin. His second argument is equally strange. Estimating that there are 10 000 volcanic seamounts in the Pacific Basin and that they have an average life of one million years, he calculated:

Within the accuracy of the estimates it appears all the large volcanoes of the Pacific Basin could have been produced since Cretaceous times at the present rate of volcanism.

*(Menard, 1958b: 213)*

Yet again he did not accept the obvious inference. Interestingly, Hess (1962: 609–610) accepted both, arguing, reasonably, that ocean basins were young, and he referred to Menard when discussing them.

Menard revisited his idea that ridges and accompanying islands and seamounts might have served as migratory routes across the Pacific Basin for shallow-water and land organisms. He still thought it a good idea. This time, however, he did not couple it with a denial of mobilism. Instead, he suggested that the migratory routes would still be needed if mobilism had occurred, and even referenced mobilism's paleomagnetic support.

> Nevertheless, the biological evidence for former connections between what are now widely separated continents can be explained without violating geophysical evidence that continents have not subsided within ocean basins. The continents may have moved like ships through the ocean basins, the oceans may have been shallower, or the seafloor may have been elevated and then subsided. Paleomagnetic measurements suggest that some continental drift has occurred but not that the continents around the Pacific Basin have ever been much closer together. Consequently one of the other possibilities seems to be required. Numerous guyots and atolls indicated that island stepping stones may always have been present in the basin ... If the banks and islands were contemporaneous, they would have formed a bridge for shallow-water and land organisms that almost spanned the Pacific.
>
> *(Menard, 1958b: 213)*

Just how serious at this point was Menard about mobilism and its paleomagnetic support? Was he merely making a rhetorical point that even if continents had drifted, landbridges spanning the Pacific might still be needed to explain trans-Pacific biota? Had he come to view mobilism's paleomagnetic support favorably and was he, like Hess and Dietz, ahead of other North Americans, responding to the buildup of paleomagnetic support for drift in the mid and late 1950s? Did he no longer think that mobilism was impossible because Antarctica, Africa, and South America were almost entirely surrounded by ridges? I find it difficult to answer these questions. Perhaps being in a conflicted state he was rethinking his worldview in a piecemeal way, temporarily and selectively suspending judgment on certain issues.

### 5.11 Key factors behind Menard's shift in attitude about ridges and fracture zones

Within a year, Menard's indecision of 1958/9 had disappeared. To explain the differential movement of seafloor along fracture zones, he proposed that seafloor moves horizontally away from the East Pacific Rise. By extending his explanation of motions along the great fracture zones to the East Pacific Rise and by claiming, as explained in the following section, that the Americas separated from Europe and Africa as the Atlantic Basin was created, Menard became a mobilist.

To begin with there were three factors that led Menard to change his mind about the origin of oceanic ridges. Mason and Raff's discovery of the northeast Pacific anomalies was, I believe, the most important. It forced him to abandon his former solution to the origin of the fracture zones and to develop a new one that could explain large relative displacements of seafloor along fracture zones.

Second was his acceptance that the East Pacific Rise was not medial in the Pacific Basin.

> The position of the [East Pacific] rise along part of its length is at the edge rather than in the middle of the ocean basin, and the crest appears to pass under western America.
>
> *(Menard, 1960: 1745; my bracketed addition)*

Importantly for future developments, he claimed that the East Pacific Rise "reappears off Oregon and Washington as the Ridge and Trough province" (Menard, 1960: 1742).[9] As already described (§5.5), Menard had mapped and named the Ridge and Trough Province in his 1955a paper. (Retrospectively (1986: 249) he claimed that he had even realized at that time that in the Ridge and Trough Province, two extensions, now recognized as the Juan de Fuca and Gorda ridges, were connected by what is now recognized as the Blanco Fracture Zone.) However, in his 1958 paper on the median position of oceanic rises he had said nothing about the reappearance of the East Pacific Rise off Oregon and Washington; perhaps his belief that all oceanic ridges are medial had blinded him to the possibility that the East Pacific Rise might extend under North America. Without visualizing its extension under North America, he would not have thought of the ridge as *reappearing* in the northernmost northeastern Pacific. He realized that there were ridges and troughs in the northeast Pacific region, but he may not have thought them to be of the same type as those associated with the East Pacific Rise and Mid-Atlantic Ridge, which, unlike the short ridges of the Ridge and Trough Province, extended for thousands of miles. In addition, Menard's previous identification of ridges in the Ridge and Trough Province was based entirely on topography. Richard von Herzen, an oceanographer at Scripps, later observed high heat flows associated with ridges in the Ridge and Trough Province. Although some of von Herzen's work was still unpublished, Menard knew of it, and thought it important.

> The high heat flow characteristic of the East Pacific Rise south of Mexico has also been found by von Herzen in the Gulf of California, in the continental borderland off southern California, and in the Ridge and Trough Province.
>
> *(Menard, 1960: 1748)*

He also used it to construct a figure (his Figure 5) showing a strong correlation between high heat flows and topography of the East Pacific Rise (Menard, 1960: 1743).

The third factor that led to Menard's new hypothesis was his belief that the East Pacific Rise and great fracture zones are parts of the same system. "Although no single line of evidence is wholly conclusive, the rise and fractures probably are genetically related" (Menard, 1960: 1745). Actually, he had proposed in 1958 that there is a genetic relationship between oceanic ridges and fracture zones, albeit between the fracture zones and the Mid-Pacific Ridge (Darwin Rise). Because he thought that the East Pacific Rise extended no further north than the Gulf of California and appeared to cross only the Clipperton and Clarion fracture zones, I suspect that at that time he never even thought about a general genetic relationship between the great fracture zones and East Pacific Rise. Once he thought it possible that the East Pacific Rise might extend under the western United States and reappear in the Pacific just north of the Mendocino Fracture Zone, he was more likely to begin thinking of the East Pacific Rise and fracture zones as genetically related, especially because he now believed that the East Pacific Rise was in close proximity to the more northerly Murray, Pioneer, and Mendocino fracture zones, the very features along which seafloor had moved horizontally as shown by the magnetic surveys. Menard may also have been influenced by the discovery of more fracture zones (Galapagos, Marquesas, and Easter) that crossed the East Pacific Rise as it extended southward from the mouth of the Gulf of California through the Pacific Basin to New Zealand. He continued to maintain that mantle convection caused formation of ridges and the great fracture zones, but now he claimed that mantle convection forced seafloor to move horizontally away from ridges along fracture zones. Ascending mantle currents formed the East Pacific Rise; horizontally moving currents thinned oceanic crust on both its flanks as they forced seafloor away from the rise along fracture zones.

### 5.12  Menard's seafloor stretching hypothesis, 1960

Menard (1960) presented his new hypothesis in a paper entitled "The East Pacific Rise" that appeared in *Science* in December 1960, the same month Hess finished his preprint on seafloor spreading. He also wrote a less technical version for *Scientific American* (Menard, 1961); they have few differences. I shall deal mainly with the former, more technical paper.

After recounting the new discoveries about the East Pacific Rise and great fracture zones, and proposing that the East Pacific Rise reappeared in the northeastern Pacific after extending as he then thought under western North America, Menard argued the rise and fracture zones were genetically related. (The East Pacific Rise does not really extend under western North America; see Note 6.) He attributed the relative horizontal displacement of seafloor along the fracture zones to crustal stretching as the East Pacific Rise formed as a result of mantle convection (Menard, 1961: 57). After documenting these displacements based on the magnetic surveys of Mason, Raff, and Vacquier, he claimed that the fracture zones were wrench (transcurrent or strike-slip) faults. He discussed Raitt's and Shor's recent seismic refraction surveys in which they had determined an average crustal thickness of 3.8 km beneath the East Pacific Rise, approximately "1 kilometer less than that found at an average station elsewhere in the Pacific basin" (Menard, 1960: 1741). He attributed the thinning to crustal stretching brought about by mantle convection beneath the rise. He argued that the crust had been laterally stretched and, at its extremities, translated 600 km as the rise developed.

To turn to wrench faulting deduced from the structure of the rise, if the boundary between crust and mantle marks chemically different rocks, the thin crust of the rise probably has been stretched. Arching of the crust would produce some thinning, but the amount is trivial. The region which is now 3.8 kilometers thick has an average width of 2800 kilometers. If it once had a thickness of 4.9 kilometers, which is typical of the Pacific basin outside the rise, then the width was 2200 kilometers, and the crust on the flanks of the rise has been translated laterally for 600 kilometers.

*(Menard, 1960: 1744)*

In his second paper, the one published in *Scientific American*, he went so far as to describe his calculation as a "semiquantitative test."

The convection hypothesis seems to meet the only semiquantitative test proposed so far: a comparison of the horizontal displacements along the fracture zones with the amount of thinning of the crust on the crest. The thin region now averages 1750 miles wide. When it was as thick as the other parts of the Pacific basin, it would have been only 1400 miles wide. Assuming that the thinning is due to stretching, the crust has moved 350 miles. This agrees, at least as to order of magnitude, with the horizontal offsets of 95, 160 and 750 miles observed at the great wrench faults of the fracture zones.

*(Menard, 1961: 57)*

As expected on his hypothesis, the movement away from the rise on its continental (eastern) side is only 15 kilometers or so, much less than on the oceanic (western) side of the rise.

Menard (1961: 1745) listed the main features of the East Pacific Rise and fracture zones that needed to be explained: the minor ridges and troughs running parallel to the rise, the high seismic activity and high heat flow along the axis of the rise, the thin crust under the rise, the fracture zones which cut across and displace or change the trend of the rise, the relative displacement of crustal blocks along fracture zones, and the low heat flow found beyond the western flank of the rise. Claiming that his hypothesis (RS1) "offered a simple qualitative explanation of all the facts [which I have summarized] given above," he wrote:

A rising hot material, marked by high heat flow, produces an upward bulge of the mantle because of thermal expansion or physical chemical changes (Fig. 7). The mantle bulge arches the overlying crust and forms a system of tension cracks parallel to the rise. Arching stretches and thins the crust, but the observed thinning is so great that translation of the crust toward the flanks of the rise is also required. Accordingly, the horizontal limb of the convection cell moves the crust outward and thins it at the crest of the rise by normal faulting along the tension cracks. Blocks are displaced different distances by wrench faulting on fracture zones because of variations in intensity of convection along the rise. Displacements on fracture zones have opposite directions on the two sides of the crest because convecting material moves in all directions from a rising hot center. Farther out on the flanks of the rise the convection current

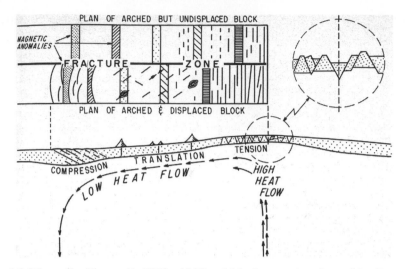

Figure 5.6 Menard's Figure 7 (1960: 1745) which he captioned as "A diagrammatic presentation of the convection-current hypothesis for the origin of various features associated with the East Pacific Rise." The blown-up part of the diagram shows the formation of small ridges and troughs surrounding the axis of the rise; relative displacement of adjacent blocks of seafloor and rise-segment separated by a fracture zone is illustrated in the upper level.

moves the crust between fracture zones as a unit (the displaced magnetic anomalies are not distorted as they are near the crest). On the outermost flanks the sinking convection current, marked by low heat flow, defines the outer limit of wrench faulting and, presumably, the crust is thickened by thrust faulting.

   *(Menard, 1960: 1745; Menard Figure 7 is reproduced as Figure 5.6; my bracketed addition)*

Menard offered no more details about his hypothesis as it applied to the East Pacific Rise. It was, he thought, a good working hypothesis.

   Turning to other oceanic ridges, he (1960: 1745) reiterated once more that they are "not all alike" and that their differences "mark different stages of development and times of origin." He again claimed that the East Pacific Rise was younger than the Mid-Atlantic Ridge, and after noting that the Pacific Basin lacked a medial ridge, reintroduced his notion of an ancient ridge, the Mid-Pacific Ridge, to explain the prevalence of guyots in the central Pacific.

All the ocean basins have median elevations except the Pacific, which, instead, has a line of atolls and guyots (different types of deeply submerged former islands) rising from a series of narrow, steep-sided ridges. If a broad oceanic rise, faulted at the crest and capped by volcanic islands, were to subside, the only evidence that it once existed might be such a line of drowned former islands and linear ridges (11). The guyots of the central Pacific were islands roughly 100 million years ago, and if they mark a former rise, the period of its subsidence must have been quite short compared with the age of the earth.

   *(Menard, 1960: 1745–1746; Menard's reference 11 is to Menard, 1958a)*

Although Menard still believed that many ridges are medial, young ridges such as the East Pacific Rise and the African Rift Valley, which he now categorized as a ridge, are not yet medially positioned within an ocean basin; the latter was not even within an ocean basin. In a regime of seafloor stretching, if ridges arise within a continent, new ocean basins form as mantle convection currents progressively drag overlying crust horizontally away from the ridge axis.

Turning to the Mid-Atlantic Ridge, Menard began to consider mobilism more, favorably appealing to its paleomagnetic support, which he had merely mentioned the previous year. He also found a way to remove the difficulty he had raised in 1958 against mobilism.

Continental drift, as suggested by the parallelism of the Atlantic Ridge, has been a very attractive concept for continental geologists, particularly since it was revitalized by the paleomagnetic evidence for polar shifts and possible drift. Marine geologists, on the other hand, have been reluctant to accept the concept of continental drift because they find no evidence for it in the geology of the seafloor. Indeed, the existence of rises centered in the Indian and Pacific oceans seemed to eliminate the possibility that Africa and South America had moved away from the Mid-Atlantic Ridge. However, if random distribution or relatively short-lived "oceanic" rises is accepted, the picture is entirely different. If all rises were in the center of ocean basins it would not be clear whether the convection current, or another agent, which produced the rise centered itself relative to the margins of the basin or created the basin. With rises bordering the Pacific and penetrating Africa, it appears more probable that most rises are centered because the margins of the basin have been adjusted by convection currents moving out from the center. If so, the African and East Pacific rises may mark relatively young or rejuvenated currents which have not yet had time to produce much continental displacement. Even so, east Africa is being torn by deep rifts and Baja California has almost been separated from North America along the crest of the East Pacific Rise.

*(Menard, 1960: 1746)*

By seafloor stretching Menard was able to turn the relation between rising mantle convection and the medial position of ridges on its head, and find a way to avoid the difficulty he had previously raised against mobilism. Oceanic margins do not control the positioning of rising convection currents; horizontally moving convection currents stretch seafloor and move oceanic margins away from newly created ridges. Ridges can form anywhere, and eventually they become medial. Arising within a continent, convection currents convey material away from the newly formed ridges at equal rates. Mantle convection that formed the Mid-Atlantic Ridge created the Atlantic by separating the Americas from Europe and the ridge always maintained its central position. Africa and South America, as he had formerly claimed, were not always surrounded simultaneously by active ridges at the time they separated because older ridges in the Pacific (Darwin Rise) and Indian oceans may have become inactive during the formation of the Atlantic Basin, and later the young East Pacific Rise and the even younger African rift valleys became active.

In 1958 Menard considered and rejected mobilism. A year later he noted mobilism's paleomagnetic support, but remained uncommitted. Introducing his version of seafloor stretching, thereby removing the difficulty he had raised against mobilism in 1958, he reverted, becoming less inclined toward mobilism. Menard's support of mobilism was not as steadfast as Hess's. Once committed, Hess never wavered. Hess became a mobilist in 1959 because of paleomagnetic evidence and he did so before he thought of seafloor spreading (§3.12, §3.14). Menard inclined to mobilism only after he developed seafloor spreading as the cause of ocean ridges. He continued to favor mobilism in his *Marine Geology of the Pacific*, which was written during the 1961–2 academic year while visiting the Department of Geodesy and Geophysics at Cambridge, but not published until 1964 (Menard, 1979 interview with author) (§5.12). Menard later rejected mobilism in March 1964 at the Royal Society of London's symposium (IV, §3.3) and proposed a fixist interpretation of ocean basin evolution a year later (IV, §3.6) even though mobilism's paleomagnetic support was rapidly becoming even stronger than it had been in 1960 with results flooding in from different continents and the gradual adoption of magnetic cleaning methods. Menard's work lacked a certain, forward motion. As already noted, Menard (1986) retrospectively suggested that he had been impressed with Munk and MacDonald's 1960 attack against paleomagnetism that they presented in *The Rotation of the Earth* (II, §7.11). Munk was a major figure and colleague at Scripps, and an eloquent foe of paleomagnetism. I suspect that Menard did not keep up with the influx of new paleomagnetic data supportive of mobilism.

### 5.13  Menard, the scientist

Menard, like Hess and Dietz, was not afraid to proffer explanations even if pertinent data were limited. He treated them as working hypotheses and was willing to change his mind if new seafloor data presented severe difficulties. With the discovery of several fracture zones, he rejected his and Dietz's 1952 explanation of what they then called the Mendocino Escarpment, and in 1953 proposed a new hypothesis, published two years later, to account for all newly discovered fracture zones (Menard, 1955a). Realizing in 1958 that Mason and Raff's first marine magnetic survey strongly indicated horizontal displacements along the Murray Fracture Zone, Menard questioned his 1955 hypothesis, which did not allow for them. Once evidence for extensive horizontal displacements along other fracture zones was discovered, he accepted them as key facts that must be explained, and in 1960 developed his crustal stretching model.

Menard typically used the three standard research strategies to defend his hypotheses, to attack opposing ones, and to argue that his solved more problems than others. Here are two examples. When he presented his 1955a explanation of the great fracture zones, which he had developed by adapting H. Brooks' hypothesis of mantle convection, he raised difficulties against Vening Meinesz's and J. Tuzo Wilson's ideas. He believed that the orientation of the Clarion and Clipperton fracture zones did not fit Vening Meinesz's model, and that Wilson's explained neither the orientation of the fracture zones nor the trend of the San Andreas Fault (RS2), whereas his own 1955a explanation of the great fracture zones could be extended to explain the origin of the Hawaiian Islands, the island arcs and trenches of the Pacific, and the trend of the San Andreas Fault (RS1). His was much broader in scope and explained more than theirs (RS3). Menard used the standard research strategies when defending his 1958 explanation of oceanic ridges. He argued that the solution proposed by the Ewing bothers failed to explain why the crust was normal (RS2). Menard, who then maintained, albeit erroneously, that all ridges are, eventually, median-positioned within ocean basins, argued that Hess's solution failed to explain this (RS2). He also expanded his explanation of ocean ridges by combining it with his 1955a explanation of the origin of the great fracture zones and circum-Pacific island arcs (RS1). Although Menard thought that Hess's solution solved more problems than the Ewing brothers' solution, he argued that his own solved more than Hess's did. Both explained the impermanence of ridges and the origin of guyots, but he believed that his, unlike Hess's, explained not only the median position of ridges but also the origin of the great fracture zones (RS3).

However, Menard did not employ the research strategies when presenting his 1960 crustal stretching hypothesis for the origin of the East Pacific Rise and the great fracture zones; at the time he said nothing specific about his own earlier solutions or those of others. Menard noted and explained that at that time, because of the flux of results and ideas, detailed comparison "does not appear warranted."

Inasmuch as the facts available about oceanic rises are few and new information is being accumulated very rapidly, hypotheses about the origin of the rises are modified frequently or abandoned for new ones. Consequently, an elaborate discussion of present working hypotheses does not appear warranted, but a brief outline may serve the useful purpose of relating otherwise disconnected facts.

*(Menard, 1960: 1745)*

However, Menard (1964: 148–149) did use the standard research strategies when, four years later, he revisited his 1960 hypothesis and argued that it offered a better explanation than some competing ones (IV, §3.3, §3.6).

Menard was particularly adept at offering geological explanations of geophysical data, and this also may be said of Hess and Heezen. This ability was precisely what was needed because most observations of the seafloor were geophysical and obtained by geophysicists less well placed to provide them. Others recognized this strength of Menard's, and thought it warranted his election to the NAS. Members from both the geology and geophysics sections supported his nomination. His proposers were James R. Arnold (geophysics), Birch (geophysics), Carl Eckart (geophysics), Elsasser (geophysics), Hess (geology), Hollis D. Hedberg (geology), Munk (geophysics), and

Hans Suess (geophysics).[10] Munk, Hess, Elsasser, and Revelle pushed the nomination. Munk, who was then at the Institute of Geophysics and Planetary Physics at UCLA, campaigned for Menard's nomination, wrote Revelle and Hess because he had heard that they earlier had attempted to get Menard elected. Munk (November 8, 1966 letter to Revelle and Hess) told them that he had "queried various people" at Scripps "and at the Geophysics Institute in Los Angeles as to whether they would support the nomination of Bill Menard to the Academy." Urey, Suess, Knopoff, Slichter, Eckart, and Bullard (a frequent visitor to Scripps) said that they would enthusiastically support his nomination; he also "received affirmative" endorsements from Henry G. Booker and William Rubey.[11] Revelle (November 14, 1966 letter to Munk with a copy to Hess) told Munk that he, Hess, and Elsasser "had talked several times about nominating Bill Menard jointly between geology and geophysics, but we never felt we had enough support to push it during that particular year." Apparently Menard's nomination was on hold when Elsasser and Hess later pushed the nomination for 1968. Elsasser wrote Carl Eckart, copying Hess and Revelle.

Harry Hess and I have agreed that it might be a very good idea to propose Bill Menard for the Academy next year on a joint Nomination between geology and geophysics. We both hope you and Roger [Revelle] would sign the original request for nomination. I personally feel that this should have been done maybe a few years ago, but I don't have to tell you anything about Bill.

I shall be very glad to do the work or eventual write-up unless somebody better acquainted with Bill and his work will do so, in which case I shall gladly yield. If you and Roger think I should prepare the details would you be good enough to send me all the necessary details, biographical and publications before the middle of June, as thereafter I will be away during most of the summer.

*(Elsasser, May 16, 1967 letter to Carl Eckart; my bracketed addition)*

I do not know who eventually wrote the nomination statement found in Hess's papers at Princeton, but I suspect it was Hess.[12] It made abundantly clear that Menard was skilled at developing hypotheses based on data from a variety of fields.

Outstanding explorer of the ocean floor, Menard has made an enormous contribution to the understanding of submarine topography, its history and processes in its development. He was the first to identify certain gross features of the oceans on a planetary scale, in particular the almost exactly median position of the great oceanic ridges. He pointed out that the gently sloping surfaces of vast extent around groups of small oceanic islands in the Pacific could not possibly be accounted for as debris eroded from the islands but rather that they must be huge submarine outpourings of basaltic lava. This in turn greatly increased estimates of the rate of generation and extrusion of volcanics on the Earth's surface.

By judicious extrapolation of data from many different fields he has evolved interesting hypotheses on the origin of the tectonic pattern found on the ocean floor. Of particular importance was his structural analysis of the great fracture zones of the Pacific.

The unusual success of his interpretations of ocean floor features stem largely from his ability to bring together coherently data from a diversity of disciplines geophysical, geological and biological. Thus he has used effectively the heat flow from the seafloor, magnetic anomalies, seismic investigations, and the flora and fauna of the bottom to make cogent deductions about bottom topography, the structure of the oceanic crust and processes by which it was developed.

*(Nomination letter for Menard to NAS, Hess papers)*

Menard was elected in 1968. Hess later congratulated Menard and added:

I am happy you are in. There is no question you deserve it. It is very difficult to make it from a splinter field like marine geology but your record is so outstanding, at least in my opinion, that you overcame this barrier.

*(Hess, May 22, 1968 letter to Menard)*

Of course it did not hurt Menard that he worked on features of the Pacific Basin that covers almost half Earth's surface. Yet, unlike Hess, Menard was not a globalist. Menard was an explorer of the Pacific Basin. He concentrated almost entirely on problems about the origin of its features. When he began working with Dietz at NEL, he worked on the origin of the Mendocino Escarpment (fracture zone), Hawaiian Islands and accompanying swell, and Aleutian Islands and Trench. With his discovery of other fracture zones, he became preoccupied with explaining their origin. He also offered solutions to the origin of guyots, essentially a Pacific Basin problem. Menard first concentrated on the oceanic ridges of the Pacific Basin before turning later to the Mid-Atlantic Ridge. He insisted that his solutions to the origin of oceanic ridges also explained fracture zones, and criticized Hess's 1955 solution because it said nothing about fracture zones. Menard himself thought it was his preoccupation with the Pacific that led him to believe that ocean ridges are ephemeral and that existing ridges were in different stages of development; the East Pacific Rise is a young ridge; the Mid-Pacific Rise is older, which he thought explained the origin of the many guyots in the central Pacific. Menard even suggested that his focus on the Pacific, Heezen's on the Atlantic, and Hess's concern for both, helps explain why they were led to develop the hypotheses they did and why they preferred them to others.

Our attraction to the various hypotheses was an almost inevitable consequence of the results of our individual research. The research, in fact, was narrow, mostly marine geomorphology, but the areas were hemispheric, and the conclusions correspondingly grand. Heezen worked in the Atlantic, an ocean created by continental drift according to Wegener and followers, but Heezen traced the median rift around Africa into the Indian Ocean. There was no intervening trench to absorb the crust being created at the rift. In the context of current thought, it was reasonable for Heezen to conclude that the earth was expanding and that the ridge was ancient and all of an age. I had accepted the conclusion that continental drift had opened the Atlantic. However, my own observations in the Pacific suggested that midocean ridges were ephemeral, so I was attracted to a sequential hypothesis. Hess had not been to sea for about 15 years but had explored both the Atlantic and the Pacific long before. In any event, he was really trying to explain mantle phenomena and trenches at the same time, and it was almost inevitable that if he produced an integrating hypothesis it would be seafloor spreading.

*(Menard, 1986: 133)*

Perhaps Menard is correct about himself. He was never adamantly opposed to mobilism. After all he had as a student heard lectures favorable to drift from Gutenberg and Daly. The classical arguments in support of mobilism had more to do with the Atlantic than the Pacific basin. I shall have more to say about this after examining why Menard later rejected mobilism at the Royal Society of London's 1964 symposium on continental drift and reverted to fixism (IV, §3.3, §3.6). I think, however, that he is not entirely correct about Hess. Hess became strongly inclined toward mobilism because of its paleomagnetic support. Once a mobilist, he formulated a mobilistic account of the evolution of ocean basins. Dietz was also impressed with mobilism's paleomagnetic support. Heezen also seemed more impressed than Menard with paleomagnetic support for mobilism (§6.11). No doubt, however, he was also very impressed by the median valley running along the axis of the Mid-Atlantic Ridge, in part, I suspect, because his co-worker Marie Tharp identified the median valley as she constructed profiles of the Mid-Atlantic Ridge based on Heezen's data (§4.7). Maurice Ewing, however, was also impressed by the median valley and invoked mantle convection to explain the origin of the Mid-Atlantic Ridge, but still rejected mobilism. Unlike Heezen, Ewing ignored mobilism's paleomagnetic support. It still seems to me that it was the very different attitudes that Hess, Dietz, Menard, Heezen, and Ewing had to the paleomagnetic evidence that significantly affected their views about the origin of oceanic ridges, and decided whether they proposed a fixist or mobilist explanation for the evolution of ocean basins.

## Notes

1 This brief account of Menard's life is drawn from Fisher and Goldberg's (1994) biographical memoir, from Menard's 1986 posthumously published *Ocean of Truth*, and from the author's July 12, 1979 interview with him.

2 Menard's denial of a central rift valley along the East Pacific Rise became a bone of contention between Menard and Heezen (§6.7).

3 Menard later renamed the "Mid-Pacific Ridge" the "Darwin Rise" (§5.6) as the name for the combined ridge area.

4 In an amusing aside, Menard (1986: 140) discussed an exchange he had with Hess about who first suggested that ridges are ephemeral. Menard wrote Hess a letter on August 22, 1966, in which he took issue with Hess over when they declared ridges ephemeral.

I am writing to point out an error in scholarship in your paper in the Colston papers. It doubtless arises from confusion in my previous papers, but in any event we might as well come out with the same version of the idea that mid-ocean ridges are ephemeral features. You state on page 327 of the Colston papers that "it was proposed by me in 1959 that mid-oceanic ridges were ephemeral features and Menard 1964 arranged them in an evolutionary sequence" and so on. This statement really is not correct. I had already proposed that mid-ocean ridges were ephemeral features in my publication in 1958, but you had proposed that they were ephemeral features in your publication in 1955. Thus there is not any question that you have priority on this idea but your references are wrong [Hess's Colston paper is Hess (1965); Menard's 1964 work is his *Marine Geology of the Pacific*].

Hess responded promptly on September 6, "I appreciate your corrections to my Colston paper. This was written during the first two weeks I was in England to meet a March 1 deadline and virtually without access to a useable library."

In analyzing Hess's relevant 1955 papers, I suggested that Hess implied that ridges are ephemeral, but did not explicitly state that they are until 1959 (§3.13). Menard, as noted above, did not explicitly claim that ridges are ephemeral. He did, however, claim that the process that centers oceanic ridges "appears to be ephemeral or intermittent" (Menard, 1958a: 1184). He said that rises are not permanent features, which, as he noted, was implied by his sequential analysis of ridges. Literally speaking, Hess was correct; he was the first to state explicitly that ridges are ephemeral, which he did in 1959.

5  The Ewing brothers presented their solution at the April 1957 Annual Meeting of the Seismological Society of America, and Menard referenced their presentation (Menard, 1958a: 1184). Their paper was published in 1959 (Ewing and Ewing, 1959), and is discussed in §6.12.

6  This brief biography of Mason is drawn primarily from his August 6, 2009 obituary that appeared in *The Times*.

7  Vacquier received AGU's Fleming Medal in 1973 for his work in marine magnetics. He won the Medal Award of the Society of Geophysicists in 1975 for his invention of the airborne stabilized fluxgate magnetometer.

8  Although Menard learned of Holmes' mobilism in a class he took at Harvard from Billings (§5.3), I see no reason not to take Menard at his word when he denied realizing the similarities between his seafloor thinning and Holmes' mantle convection. Perhaps he did not fully understand Holmes, seeing everything in such a different context. Menard's road to his seafloor thinning was very different from Holmes'.

9  The East Pacific Rise does not pass beneath North America; it is the San Andreas transform fault that does. But it would take several years before J. Tuzo Wilson developed the idea of transform faults, and he, Vine, and Hess recognized that it was a transform fault that went beneath North America connecting the Gorda Ridge with the East Pacific Rise. Moreover, it would also take the development of plate tectonics, the idea of triple junctions, and the work of especially Tanya Atwater in the 1970s to unravel the Cenozoic history of the Pacific Basin. This is just the first of several fast-forwards that I shall include about making sense out of the East Pacific Rise, San Andreas Fault, and the Cenozoic history of the Pacific Basin. Nonetheless, Menard was correct in thinking that there was a ridge within the Ridge and Trough Province off Oregon and Washington.

10  All documents pertaining to Menard's nomination as a member of the NAS cited here are from Hess's papers at Princeton.

11  Munk actually wrote that he had received an endorsement from someone named "Ruby." I believe that he simply made a typographical error and that the endorsement came from William W. Rubey. No one named Ruby has been a member of the NAS, and Rubey would have been an obvious person to ask. He and Munk were both at UCLA, Rubey was a member of the Academy, and he knew of Menard's work.

12  Comparison of the type-face of the nomination statement for Menard from Hess's file with letters that Hess typed and with Elsasser's letter to Eckart, suggests that the nomination was typed on Hess's typewriter and not on Elsasser's.

Holmes was harder on Bucher, because he "altogether ignores the hypothesis of sub-crustal convection currents; nor does he even mention any of the evidence which favors the 'drift' hypothesis" (Holmes, 1953: 670). Ewing also ignored Holmes' mantle convection hypothesis, perhaps not surprising as Bucher was his mentor, and continued to do so, even though he surely read Holmes' review (I, §5.11).

### 6.4  Ewing and Tolstoy's views about the origin of the Mid-Atlantic Ridge

As to my views concerning Wegener: as you know I never came out in print in favor of his theory. But I had read, while still a student in France, his monograph and had found it exciting; the congruence between the coastlines and especially the geology to both sides of the southern Atlantic was striking and undeniable. Even as a raw undergraduate I could see that. But while working at Lamont I also understood that if one were to take up the cudgels for Wegener one needed a physically sound model. This did not seem to be available at the time and I never defended him in print. (Was I being pusillanimous? Perhaps.) I tried to bring Wegener into the picture in one or two discussions, e.g., during my orals, but, while I do not remember the details, I do recall arousing Bucher's ire and being soundly slapped down!

*(Tolstoy, September 2006 email to author; my bracketed addition; Tolstoy took his orals in late 1949 or early 1950)*

The bathymetric studies of the Atlantic Basin and Ridge done by Ivan Tolstoy and M. Ewing were the first by Lamonters. Heezen was not involved until Tolstoy started working in seismology. Tolstoy was born in 1923 in Baden-Baden, Germany.[5] His family emigrated to France the next year. He received a Licenceès Sciences degree in 1945 from the University of Paris (Sorbonne). "I first studied geology at the Sorbonne where I obtained the corresponding *certificats* in geology and mineralogy (I also took a big dose of pure mathematics – the post-baccalaureat *Mathématiques Speciales* course)." Tolstoy recalled, "I became acquainted with mobilism on my own initiative from a French translation of Wegener's *The Origin of Continents and Oceans* (probably the 1929 edition)" but he did not discuss Wegener or any other geological issues with his professors at the Sorbonne.

Lectures at the Sorbonne were then an entirely more formal affair than in the USA. The professor of stratigraphy (Monsieur Jacob) always arrived in striped trousers and cutaway black jacket, gave us an elegant, beautifully crafted lecture and disappeared through a side door. You did NOT interrupt the flow by asking questions. Admittedly the lectures were well organized – remarkably high quality performances. In other words, we never DISCUSSED anything with our profs. They gave their formal, crystal-clear, well organized lectures and that was it. And no, no one ever mentioned Wegener (that I can remember).

*(Tolstoy, August 2006 email to author)*

Tolstoy spent part of 1944 in Switzerland where he sought asylum to avoid consignment to a labor camp in Germany.

In German-occupied Paris in late 1943 early 1944, the Germans began picking up young men of my age in street raids and, if they were not "usefully employed" shipping them off to work in Germany in labour camps. A cousin of mine got caught that way and I narrowly escaped a couple of "rafles" in the Metro. So I decided to cross the border into Switzerland and seek asylum. After an uncomfortably close call with a border patrol, I made it safely and stayed in Switzerland until Christmas of '44.

*(Tolstoy, August 2006 email to author)*

The next year he emigrated to the United States, becoming a graduate student in geology at Columbia University. He first studied under Bucher but switched to Ewing. None of his instructors mentioned Wegener.

After arriving in the USA in the fall of 1945 I enrolled as a grad student of Bucher's for the winter term of 1946. During my first year at Columbia I took Shand's course on petrology and Bucher's on structural geology and don't recall Wegener *ever* being mentioned. Bucher was good to me but I found being his assistant rather boring and, when I was approached by Worzel with a more interesting sounding offer, I switched to Ewing.

*(Tolstoy, August 2006 email to author)*

Like most of Ewing's recruits, Tolstoy was well versed in mathematics; unlike most, he also knew a lot of geology. Tolstoy joined Ewing's other students at WHOI in summer 1947, and Ewing put him in charge of the echo sounder on *Atlantis* cruise 150, Ewing's first expedition to the Mid-Atlantic Ridge.

During the summer of 1947 Ewing took the *Atlantis* on the first Mid-Atlantic Ridge cruise [Cruise 150]. Each of his grad students was given a particular area of responsibility. Mine was the echo sounder and bathymetry. That is how I got involved in bathymetry. That was when I spotted what is now known as the *Atlantis Fracture Zone* and made the first crude survey of it, using star fixes and dead reckoning (Tolstoy and Ewing, *Bull. Geol. Soc. Am.*, 1949). I have since seen modern accurate surveys and been rather surprised at how well our crude bathymetry and navigation had done.

*(Tolstoy, August 2006 email to author; my bracketed addition)*

Tolstoy and Ewing (1949) submitted their paper on September 7, 1948. They briefly reviewed previous hypotheses about the origin of the Mid-Atlantic Ridge, loosely identified different regions of the Atlantic Basin and Ridge, tentatively speculated about the origin of a section of the ridge adjacent to the Azores, and argued that more data were needed before anything beyond very speculative hypotheses could reasonably be proposed. They also seemed to presuppose fixism in their discussion of the Atlantic Basin's history.

They familiarized themselves with results from the *Meteor* Expedition, and superimposed the path taken by *Atlantis* on cruise 150 on a figure from that expedition (Tolstoy and Ewing, 1949, Plate 1, opposite p. 1527). They briefly mentioned former solutions to the origin of the Mid-Atlantic Ridge, including those proposed by the mobilists Taylor, Wegener, and Molengraaff, and by the fixists Umbgrove and Cloos. They raised no difficulties *per se* against previous hypotheses, but Tolstoy recalled that Ewing was not enchanted with Wegener's hypothesis.

The Tolstoy and Ewing Mid-Atlantic Ridge paper was written by us jointly, albeit most of the actual work was mine. I remember his scepticism concerning Wegener's idea, but that is all. I seem to remember too that our reading of the people you mention was somewhat perfunctory – or, at least, mine was.

*(Tolstoy, August 2006 email to author)*

They (1949: 1528–1529) claimed that the Mid-Atlantic Ridge is "the most conspicuous feature of the Atlantic Ocean basin," "is essentially a long narrow submarine mountain range," "parallels the American, European, and African coasts," and "rises gradually to Iceland." They discussed Gutenberg and Richter's 1949 association of shallow earthquakes and the ridge, and mentioned Wiseman and Seymour Sewell's 1937 (I, §6.12, §8.12) suggestion that the Mid-Atlantic Ridge might extend around Africa to the Carlsberg Ridge.

Turning to the area they surveyed of the central Atlantic on *Atlantis*, Tolstoy and Ewing (1949: 1537) divided the NE–SW trending ridge into a central mountainous region characterized by parallel ridges trending NE–SW, a flanking terraced region, and an outer mountainous region. They noted a seismically active area at roughly 30° N latitude with an E–W trending deep trough that lacked a surrounding terraced region. Shand (1949), a petrologist at Columbia University, identified rocks dredged near the area as metamorphosed ultrabasic (peridotites, olivine gabbro, and olivine basalt). They proposed that the deep trough "may be a graben" and that "it is a fairly recent feature, superimposed upon the typical Mid-Atlantic Ridge type of topography." They had actually found the Atlantis Fracture Zone, later recognized as a ridge-ridge transform fault that connects segments of the Mid-Atlantic Ridge.

They suggested an explanation for the Azores plateau, which they (1949: 1538) characterized as "an easterly bulge of the Mid-Atlantic Ridge." It trended E–W instead of N–S as the main part of the ridge. Appealing to Cloos (1939), they proposed (1949: 1530) that the plateau was caused by extrusion of magma through parallel E–W trending fractures caused by local doming of Earth's crust, and were sympathetic to Agostinho (1936), who suggested that the Azores plateau was part of an E–W succession of banks extending from the Great Banks off the coast of Newfoundland across the ridge through the Azores to the banks west of Gibraltar. They were unwilling, however, to extend their predecessors' ideas to the rest of the ridge.

Of all the hypotheses on the origin of sections of the Mid-Atlantic Ridge, perhaps the least speculative are those proposed by Hans Cloos and J. Agostinho for the origin of the type of topography on the Azores plateau. Though these hypotheses appear reasonable enough for the specific case of the Azores, it does not appear to us to apply necessarily to the rest of the Ridge, for, as we have already noted, the Azores plateau appears to be a feature distinct from the rest of the Ridge, characterized by its own particular type of submarine geology.

*(Tolstoy and Ewing, 1949: 1539)*

They concluded that too little was known about the Mid-Atlantic Ridge to warrant speculating on its origin.

Apart from a few highly interesting but scanty geological facts, the evidence about the nature of the Ridge obtained thus far by us is mostly topographic, and even this leaves much to be desired. We therefore cannot conclude much about the origin of the Mid-Atlantic Ridge.

*(Tolstoy and Ewing, 1949: 1539)*

More data were needed and Tolstoy and Ewing were quite prepared to collect it.

Obviously many more data in the form of hydrographic surveys and fathograms are needed. An accurate bathymetric chart of the Atlantic Ocean would contribute greatly to our knowledge of the structural history of the Earth. Such a chart would also be useful for navigational purposes, since the Atlantic sea floor is marked by numerous features which would provide excellent "landmarks." Such a project would be greatly facilitated by a more extensive use of continuously recording fathometers. Formation surveying – i.e., surveying by means of a flotilla of ships keeping radar formation and equipped with continuously recording fathometers – would enable this work to be carried out speedily and efficiently.

*(Tolstoy and Ewing, 1949: 1539)*

With the Cold War developing, new information about the world's oceans was becoming a premium and governmental agencies were willing to provide needed funding.

Tolstoy and Ewing emphasized the flatness and wide extent of the 2000-fathom plain (what would later be called the abyssal plains, on both sides of the ridge) and suggested that at least those parts of the basin had been tectonically inactive. They said nothing about how this might relate to mobilism or fixism.

The fathograms of *Atlantis* cruise 151 show a similar plain occupying the floor of the basin east of the Ridge, between the Ridge and northwestern coast of Africa (the northern Canary basin). It appears therefore that this 2000-fathom plain is of great regional significance. Its aspect on the fathograms suggests that vast portions of the ocean floor have remained tectonically undisturbed since their formation. The small knolls may be due to one of several causes: block faulting, lava escaping through fissures, or submarine volcanoes. Their true shapes and distribution pattern over the ocean floor remain to be ascertained

*(Tolstoy and Ewing, 1949: 1535)*

If continents had plowed their way through the seafloor, such apparently undisturbed regions were difficult to explain. They could have been construed as evidence for Holmes' theory of mantle convection, but Ewing already had ignored it, and neither he nor Tolstoy mentioned it. At the time Tolstoy did not know about Holmes' view. "In retrospect, I think our neglect of Holmes' writing was regrettable and due to ignorance (on my part, at least)" (Tolstoy, August 2006 email to author).

Tolstoy (1951) wrote another paper on the North Atlantic and Mid-Atlantic Ridge, submitted in December 1950. "It was the result of lengthy and tedious work by me on my own, which is why only my name appears on it" (Tolstoy, August 2006 email to author). Tolstoy showed a willingness to speculate. He (1951: 446) again characterized the Mid-Atlantic Ridge as the "most striking feature" in the Atlantic Basin, this time more emphatically, identifying and naming its three main topographic zones, the central Main Range, flanked on either side successively by the Terraced Zone and the outer Foothills. Again he mentioned the east–west trend of the ridge along 30° N, "which has some characteristics of a graben." He also noted that as a whole the ridge "roughly parallels the North American and African coasts," which, he added, was a reason why "Wegener (1924) considered that it favored his theory of continental drift" (Tolstoy, 1951: 436). He continued to suppose that there was a separate transverse trend within the Atlantic Basin. He offered a more complete description of a feature further north, and proposed that it might be a submarine mountain range connecting the Acadian orogenic belt of North America with the Variscides of Europe.

The transverse trend would either have been superimposed upon or superimposed by the Mid-Atlantic Ridge trends. This idea is further enhanced by the fact that the transverse trend of the Azores appears to be a local manifestation of a major transatlantic one, beginning at the Grand Banks, crossing the Ridge at the Azores, and going through the Gettysburg and Josephine banks and merging with either the Variscian mountains of western Spain, Portugal, and North West Africa, or the Cenozoic ranges of southern Spain and North West Africa. Since the structure in southern Newfoundland and in the Prince Edward and Cape Breton islands is mostly Acadian and Appalachian, the Grand Banks and indeed this whole transatlantic trend may be Variscan.

*(Tolstoy, 1951: 446)*

Thus he provided some evidence favoring a fixist explanation of a major trans-Atlantic tectonic disjunct.

Tolstoy (1951: 446) reconsidered the Ridge's origin, suggesting several possibilities without expressing a strong preference. Wondering whether the Ridge was caused by folding, faulting, or a combination of them, he said there was insufficient information to tell, but he did think that some parts resulted from faulting. There was, he claimed (1951: 449), "indisputable" evidence that an east–west trending trough at 30° N "is almost certainly a fault," and citing a conversation with Bucher, noted that the central Main Range could have been caused by block faulting like the mountain ranges of "East Africa," which, I believe, is the first comparison ever made of the Mid-Atlantic Ridge with the rift valleys of East Africa. He also suggested that the central part of Iceland was a graben, and again appealed to Cloos (1939) and Agostinho (1936) who proposed that the Azores Bulge was caused by the outpouring of magma through fissures caused by a doming crust. Noting that Iceland and the Azores both had experienced post-Miocene and recent volcanic activity, he suggested that the

origin of the Ridge might be associated with the Alpine orogenies. This idea is supported also by the relatively high degree of seismicity along the ridge, which is comparable to that of many sections of Gutenberg's Alpide Zone. Gutenberg and Richter (1949) have even tentatively suggested that the Mid-Atlantic ridge may be a belt of Tertiary folds, the high degree of seismicity of which would be due to block faulting following a redistribution of tectonic forces.

*(Tolstoy, 1951: 449)*

Combining these explanations for the ridge and its transverse features, he (1951: 450) speculated that the ridge "may have been superimposed upon" an older "possibly Paleozoic" transverse trend, causing "opening up of fracture through which huge masses of lava were extruded." He remarked that there were no known facts incompatible with the idea that the ridge had been caused by folding, and noted that structural histories of Iceland, the Azores Bulge, and the ridge might not coincide.

When Tolstoy submitted his paper on the North Atlantic and Mid-Atlantic Ridge, he had already begun working on seismics and ocean acoustics. He became ill and underwent a spinal operation. He could not go to sea and Ewing gave him a new project looking for Airy phases while he recuperated.

I was hospitalized after a major operation on my spine between December 1947 and March 1948 and kept in a cast until June. As a result I could not do any sea-going fieldwork. Ewing spent most of that spring and summer at sea and gave me a project to work on while he was away – visit the major seismological institutions, examine their collections of seismograms and see if I could find any of the Airy phases predicted by Press and Ewing's current work (AGU 1948).

I found a few Airy phases, which was my contribution to the Press, Ewing and Tolstoy paper (*Bull. Seism. Soc. Am.* 1950); I take no credit or responsibility for the maths here which, soon after, I found to be seriously flawed.[6]

*(Tolstoy, August 2006 emails to author)*

Airy phases are groups of waves generated by shallow earthquakes that travel through the ocean at a velocity of approximately 0.7 that of sound in water, and Tolstoy looked for them in earthquake records at the Dominion Observatory, Ottawa, Canada, the Jesuit seismological observatories at Weston College, Massachusetts, and at Fordham University, New York. Although he found evidence of Airy phases, he also found another acoustic phase, T-waves, the third group of waves to arrive after primary and secondary waves. Father Linehan of Weston Seismological Observatory first discovered and named them, but was unable to explain them (Linehan, 1940).

The T phase study was almost entirely my own work and the discovery was serendipitous. While searching for Airy phases at Weston I saw a multitude of late high amplitude signals generated by the great 1946 Dominican Republic quake and its numerous aftershocks. Nobody seemed to know what they were. The good Jesuit fathers at Weston had named them T for *tertius* but had no explanation. Simple arithmetic and geometry convinced me that these were water-borne sound waves transformed into seismic body waves when they impinged upon the shelf. I worked very hard on the data through that summer and when Ewing returned from his cruise I had the report ready for him – which, with minor alterations became the T phase paper. It became my Ph.D. thesis and Ewing, as my advisor, put his name on it too – as was, I suppose, customary at the time.

*(Tolstoy, August 2006 emails to author)*

Tolstoy argued that T-waves are compressional waves that travel from epicenters to seismic stations through the SOFAR (sound frequency and ranging) channel.

He also found a correlation between propagation of T-waves and tsunamis, and Ewing, once aware of the correlation, suggested that they may serve as a warning for future tsunamis (Tolstoy and Ewing, 1950).

Tolstoy obtained his Ph.D. based on his analysis of T-waves in 1950; its key parts were published the same year (Tolstoy and Ewing, 1950; Ewing, Tolstoy, and Press, 1950). Tolstoy also decided that he preferred seismics to marine geology.

As for why I eventually shifted from bathymetry to seismics, acoustics and wave theory I found the endless work of plotting ships courses, correcting it when possible, transferring depths from smudgy "fathograms" on to Mercator sheets very tedious indeed. Besides I had a good maths background and I had met and begun to discuss work on problems with M. A. Biot, one of the best applied mathematicians and engineering scientists of his day.[7]

*(Tolstoy, August 2006 emails to author)*

Biot was present at Tolstoy's Ph.D. thesis defense. Tolstoy had a distinguished career in seismics and ocean acoustics. After leaving Lamont in 1951 and working as a Senior Research Engineer for Stanolind Oil & Gas Co., in Tulsa, Oklahoma, he returned to Columbia University, where he engaged in classified research at Hudson Laboratories, becoming Assistant Director from 1965 until 1967. In 1967, he accepted a professorship of ocean engineering at Columbia, and took on the Associate Directorship of Hudson Laboratories. After one year, he accepted a position as professor of geology and geophysical fluid dynamics at Florida State University. This move may seem surprising, especially as he had been promised his own department, but there were difficult times at Lamont-Doherty in the late 1960s and Tolstoy did not want to become part of the attempt to force Ewing to step down as director.

The only move you might find curious is my leaving, after a year, a promising position as professor of Ocean Engineering at Columbia (where I had been promised my own department) for a professorship in geophysics/geology at Florida State. There were several reasons for this, one of which was that I felt uncomfortable at being set up by the Columbia powers to counter LGO's huge influence (I had had my doubts about Ewing's operations but I did not want to be one of the army of people who wanted to knife him; I owed him).

*(Tolstoy, August 2006 emails to author)*

He later taught at the University of Leeds, UK, and at the US Naval Postgraduate School, Monterey, California. He was awarded the 1990 Acoustical Society of Pioneers of Underwater Acoustics Medal by the Acoustical Society of America for his innovative studies in oceanic, atmospheric, and seismic wave propagation.

Before Tolstoy left marine geology entirely, he served as Chief Scientist aboard *Atlantis* on its cruise 160, which began in January 1950. During the cruise he undertook a more detailed and accurate survey of the E–W trending trough at 30° N latitude that he and Ewing had initially discovered and identified as a graben. After plotting the results, he was informed that he could not publish them because the US Navy had deemed them classified information.

On my last (1950) *Atlantis* cruise, on which I was PI (chief scientist on board), which lasted three months, I did a more detailed and accurate survey of a section of the ridge covering a few square degrees and containing the Atlantis Fracture Zone. I successfully used a new technique (first suggested either by Joe Worzel or Doc Ewing but I don't think ever used by them), consisting of anchoring a light buoy (beach balls in canvas sack) anchored to the bottom with the help of piano wire, to which we tied a hydrogen filled "kytoon" (a large balloon with fins) flying 40 or 50 meters above the sea surface on which we hung a radar corner reflector. Continuous radar readings gave us reliable navigation with respect to this fixed point out to distances of more than ten miles (the system usually survived the weather for two or three days). Thanks largely to the help of R. S. Edwards (later of Woods Hole), this Rube Goldbergish device worked surprisingly well and I obtained a far more accurate bathymetric survey of a portion of the central ridge than had been possible before. Imagine my disappointment when, after plotting up a beautiful new contour map of this region, the navy classified it and did not allow me to publish it (something about its being accurate enough to allow Soviet submarines to use it for navigation!).

*(Tolstoy, August 2006 email to author)*

Once the bathymetric results were declassified, Tolstoy wanted to publish them. Having left his only copy at Lamont, he asked Heezen for them.

A few years later a Navy guy called me to say the classification had been lifted and I was free to publish. However the only extant copy was at Lamont – guess where? Yes. In the Heezen & Tharp files. Bruce claimed himself unable to find it and this rather nice piece of work disappeared from view. Was it ever found? Did it appear somewhere else, unbeknownst to me? I don't know. I had other work in progress and lost interest. And I did NOT want to get into distasteful wrangles with Bruce (I did complain to Ewing who told me rather bitterly that he had little influence over the Heezen empire. I don't remember the year: late fifties or early sixties?).

*(Tolstoy, August 2006 email to author)*

Heezen and Tharp (1965) later identified and named Tolstoy and Ewing's graben the Atlantis Fracture Zone. Figure 6 (Bathymetric sketch of Atlantis Fracture Zone) in their paper, which they presented at the March 1964 Royal Society (London) meeting on continental drift, contains a contour map (p. 94) showing the same region surveyed by Tolstoy and Ewing during *Atlantis* Cruise 150 and by Tolstoy during *Atlantis* Cruise 160. Heezen and Tharp did not credit Tolstoy and Ewing (1949) or Tolstoy (1951) nor did they acknowledge Tolstoy's unpublished data. In his seminal work on transform faults (which the Atlantis Fracture Zone was later shown to be), Sykes (1967: 2141) noted that a fracture zone had been mapped in the area, and did acknowledge the work of Tolstoy and Ewing as well as Heezen and Tharp.

Once Tolstoy left marine geology, Heezen not only acquired his data but also his unofficial position as Ewing's choice as surveyor of the Atlantic Basin and the Mid-Atlantic Ridge. It is time to introduce Heezen properly.

## 6.5 Bruce Heezen, the making of a marine geologist[8]

Usually, in my experience ... chance meetings at the right moments in time, particularly when one is young, are more important than anything else. Of course, one has to have a certain number of attitudes towards the world and those who guide you in what directions you go. In my own case I was not good in high school, either academically or socially. I had no prospect of becoming either a businessman in the city I lived in, or being happy there socially. I had no chance to stay there. That was not a viable possibility. Therefore, all of the kind of pursuits that my childhood friends were able to go into were out for me. Those that involved staying home and not going to college or those that involved going to college and returning home to be a medic or lawyer or something else, these were all out because I did not have the mental capacity to get a degree in medicine or law or these other subjects. I knew I didn't, and therefore I could not fool myself with the idea that these were viable possibilities. I wanted to find something that was a little bit more fun and a little bit less buffeted by the everyday work and perhaps a little useless so it didn't attract much attention. So I puttered away with fossils when I went to college. Thought I might become a paleontologist, a field that was respectable, not too difficult if you were diligent, and so completely overlooked that you could become, if you wanted to be, the world expert on any particular type of fossil. Because at any one time in history there were never more than three or four people who were considered specialists in any particular field. It therefore, it had an attraction you could become THE specialist in late Tertiary nautiloids and know them all by heart. And when anybody found another one you could identify it and write a paper on it. Your scientific work was fairly well cut and dried ... after a few years [you] would become well known and therefore have a steady flow of rocks ... you didn't even have to go into the field, unless you wanted to, to find them. And you, and since no one else wanted to bother to become an expert in that field you would have virtually no challenge except perhaps another obscure man in your same field buried somewhere in Russia or France.

*(Lear, BCH 07: 3–4; my bracketed addition)*

Bruce Charles Heezen was born on April 11, 1924, in Vinton, Iowa, an only child. Later the family moved to a farm in Muscatine, Iowa. He graduated from high school in 1942 and immediately started at Iowa State College where he began an accelerated physics class in which a year's work was completed in two and a half months. He had trouble keeping up. Heezen was granted a deferment from the armed forces to help his father run the farm. Realizing that he definitely did not want to be a farmer or live in Muscatine, he returned to college in 1945, taking classes in nuclear physics, zoology, and geology. He liked geology, did well, and was asked by Walter Younguist, a graduate assistant in one of his geology classes, to be his field assistant during the summer. They collected over two thousand fossil cephalopods. Heezen decided to become a paleontologist.

Heezen then met Ewing. Ewing gave a Sigma Xi lecture in April 1947 at Iowa State College. He talked about seismic refraction at sea, and the unbounded opportunities to discover new things. After the lecture, Heezen saw Ewing standing by himself and began talking to him. Ewing asked if he wanted to participate in an expedition to the Mid-Atlantic Ridge during the upcoming summer. Heezen, momentarily speechless, was saved by Walter Trombridge, chair of the geology department, who said, "Of course, he will come."

Heezen spent summer 1947 at WHOI. He met Worzel, John Ewing, Officer, Wuenschel, and Northrop, a new graduate student at Columbia University. Ewing told him that he was to go on the *Atlantis* Expedition to the Mid-Atlantic Ridge, and sent him to Harvard's Museum of Comparative Zoology to look up papers on the ridge.

I looked up, in their rather good cross index catalogue, subjects such as the Mid-Atlantic Ridge, the Atlantic Ocean floor. The information on the North Atlantic ... was extremely skimpy. There were, however, papers written by George Wüst who later came to Lamont in his later years ... He wrote a very nice paper about the Azores Plateau ... And the principal thing I ran into was a full volume of the *Geologische Rundschau* ... and they devoted about a three or four hundred page issue to the Atlantic Ocean [see I, §8.4]. This was probably the most significant [collection of] papers I found. Also, I found very skimpy references to some work done by the METEOR in the late thirties in the North Atlantic ... One had also to wade through many hypothetical papers on continental drift which had mentioned the Mid-Atlantic Ridge in a hypothetical-theoretical fashion. We were not interested in theories and hypotheses. We were interested in what facts were available, in what we could do to add to a significant point to the Ridge. Working hypotheses might be useful, but as for the grander schemes, that had little importance since they were based on so little information that they could not serve as working hypotheses of our expedition. Unfortunately for us the most useful kind of working hypotheses were those related to drift and drift was totally unacceptable so we couldn't use those. And the other ones didn't work very well and weren't very clear cut whether it was a folded range or a subsided mass ... [and were based] on practically no information.

*(Lear, BCH 06B: 2; summarized by Tharp)*

Explanations involving continental drift were "totally unacceptable." Returning to WHOI, Heezen was told by Ewing that he was not to go on the expedition to the Mid-Atlantic Ridge – perhaps Tolstoy's decision to work with Ewing led to his supplanting Heezen aboard *Atlantis*. As a consolation prize, Ewing made Heezen chief scientist of an expedition aboard WHOI's R/V *Balanus* to survey the continental shelf along the eastern United States. Heezen designed a crude coring rig to which he attached a camera, and he, Northrop, and J. Ewing took over two hundred underwater photographs, often obtaining bottom samples, and, using a primitive echo sounder, the depth of the seafloor at each station (Northrop and Heezen, 1951).

By now Heezen knew he wanted to be a marine geologist. At Ewing's suggestion, he applied for and got a fellowship for graduate work in geology at Columbia. Returning to Iowa State College for his senior year, he obtained his A.B. (1948) degree with a major in geology, after which Ewing ordered him to go directly to WHOI. During that summer Heezen took part in the next expedition of *Atlantis* to the Mid-Atlantic Ridge (July–August, 1948). Ewing then appointed him Chief Scientist on *Atlantis* cruise 153 (November, 1948), the first to tow a magnetometer over the Mid-Atlantic Ridge (Heezen, Ewing, and Miller, 1953) (§6.4).

where he could get good advice. Not one of us could have done that work independently ... there was a true collaboration at that time: me, Doc and Ericson. I wasn't well enough trained in geology to identify the species and do the mineralogy. Eric was. He wasn't a man at that time broad enough; he was so conservative you know, that he wouldn't have seen, I think, the broad application of it, too conservative to expand, to expound it. Ewing and I weren't competent enough to be sure that some spook in paleontology would throw us off on mineralogy. Eric was so competent in those fields he could dispel those worries. It was a wonderful time and we worked very closely together.

*(Lear, BCH 06B)*

Heezen's next big success work was on the Mid-Atlantic Ridge. He went through the door Tolstoy had left ajar, and Marie Tharp with her drafting ability and attention to detail had begun working with him.

### 6.6 Marie Tharp, the making of an oceanographic cartographer

Marie Tharp was born in 1920 in Ypsilanti, Michigan. Her father was a soil surveyor with the US Department of Agriculture working across the country, and, as the family moved with him, she attended twenty-four different schools before graduating from high school. Tharp then went to Ohio University in Athens, Ohio, beginning as an art major, changed to English but her interests were not settled. Then she tried geology, which she liked, and

it was about a year after Pearl Harbor. There was a flyer that the geology professor showed me, that the University of Michigan was going to open its doors to women. A girl could go there and they'd guarantee women a job in petroleum geology. And the University of Michigan had a pretty good geology department.

*(Doel, 1996 Tharp interview)*

By this time she had enough credits to graduate with a major in English, but not wanting to become a school teacher, geology and the prospect of a job beckoned. Borrowing $400, she joined about a dozen other women at the University of Michigan in January 1944. Within five semesters, she had earned an MA in geology. She learned drafting while working part time for the USGS.

As promised in the flyer, Tharp got a job in the oil industry. She went to Oklahoma in 1945 to work for Stanolind Oil and Gas Company. Stationed in Tulsa at the company's headquarters, she became an administrative assistant to a senior geologist, mainly library work which she found boring. She started taking night classes in mathematics at the University of Tulsa. Over three years, she had amassed twenty-nine credits in mathematics, one fewer than needed for a degree. Her advisor told her that she could take a one-hour course in spherical trigonometry. Tharp (Frankel, 1986 Tharp interview) responded, "I guess I'll take spherical trig." to which her advisor replied, "Well most of us never have any occasion to do any great circle sailing!" She took the course and got her mathematics degree.

Still bored and not fully engaged, she left for New York City in search of another job. She enquired about petroleum companies with offices in New York City. Then she tried the American Geographical Society.

I went to the American Geographical Society and asked for a job in their office at 156th street. I read that they were writing a book on petroleum deposits in the world, and I thought, "Well maybe I can get a job, I can help with that. I know about oil" and this little biddy said, "Well, we don't need any file clerks here today."

*(Frankel, 1986 Tharp interview)*

How ironic that Heezen should receive the Cullum Medal of the American Geographical Society in 1973 for his and Tharp's mapping of the seafloor. She enquired about a research job in the geology department at Columbia University. Mentioning her degree in mathematics, the secretary suggested she should speak with M. Ewing when he returned in a few weeks. She did, it was November 1948.

I was ushered into Dr. Ewing's office. He asked me the standard questions about my background and he seemed to become more amazed as he heard about my assorted degrees and the order in which I had taken them. His courtly Texas manner could scarcely hide his bewilderment and finally, he blurted out, "Can you draft?" I remarked that when I was a student at the University of Michigan, I had a part time drafting job for George Cokee of the USGS.

*(Tharp, written addition by Tharp to Doel's interview)*

Ewing hired her. At first she did little drafting, spending most of her time punching numbers for Frank Press on a Monroe Calculator.

Tharp soon met Bruce Heezen

down in the basement of Schermerhorn [Hall, where the geology department was housed]. Came to work at eight o'clock, as I had to do in Tulsa, and he also came to work at eight, because that's the way he had been brought up, and he was sitting there. He had just gotten back from his National Geographic cruise in 1948 – the second one ... He and Doc were sitting there in the room of the office, working on the paper and the records, for quite a while.

*(Doel, 1996 Tharp interview, p. 42)*

Heezen recollected:

So I came back and just before Christmas in 1948 with this material [from *Atlantis* cruise 153] and this is the very same time that Marie showed up at Schermerhorn Hall. And, in fact, the very first time I met Marie was shortly before Christmas of '48. She was sitting in Ewing's anteroom there working ...

*(Lear, BCH 10A: 2)*

Tharp kept working almost exclusively for Frank Press until Ewing's group moved to Lamont, and she started drafting for others. Things soon got out of hand, however, and she quit.

When we all were at Schermerhorn, I worked for Frank Press. Then when Ms. Lamont gave us her estate in 1949 ... I was working for Frank, and then gradually I got to doing drafting for other people, and there were the geochemists, and there were quite a few people I was doing drafting for. They all pounced on me at once, and wanted all their work the same day. I got mad and quit.

*(Doel, Tharp interview, p. 50)*

Tharp returned to the family farm in Ohio. She recalled, "I received a nice letter from Dr. Ewing, handwritten from Brussels where he and Bruce were at a meeting [IUGG 1950 meeting]. Dr. Ewing urged me to come back and just consider my absence a vacation" (Tharp added note to Lear, BCH 09B; my bracketed addition). Once she returned, Ewing put Heezen in charge of her schedule. Heezen knew what to do, to have her work just for him.

I was the beneficiary because Ewing assigned me the responsibility of allocating her time. I ... decided it would be much better just to have her work for me full time ... I decided that it was ridiculous to have to pass out Marie's time to all these other people.

*(Lear, BCH 09B: 10)*

Heezen and Tharp would soon become a very successful team and remained together until he suffered a fatal heart attack on June 15, 1977, while investigating the Mid-Atlantic Ridge aboard submersible *NR-1*.


## 6.7 The discovery of the median rift valley

Tharp and Heezen, and Maurice Hill and John Swallow of the Department of Geodesy and Geophysics at the University of Cambridge discovered the median valley of the Mid-Atlantic Ridge. It was not immediately recognized as a rift. Wiseman and Sewell (1937: 220) discovered a "gully" associated with the Carlsberg Ridge and its extension into the Arabian Sea during the 1933–4 *John Murray* Expedition in the Indian Ocean. Tharp discovered the central valley on the Mid-Atlantic Ridge in late 1952 while plotting data collected during voyages of WHOI's *Atlantis I* from 1947 through 1952. Swallow and Hill discovered it during crossings of the ridge in 1953 aboard HMS *Challenger*. Tharp and Heezen did more with the discovery than the other two groups.

During the 1933–4 *John Murray* Expedition, R. B. Seymour Sewell, who headed the expedition, and John D. H. Wiseman, then at the British Museum of Natural History, armed with an "echo-sounding apparatus," mapped 22 000 miles of the floor of the Arabian Sea and Indian Ocean (1937: 219). They discovered a "gully" associated with the northern part of the Carlsberg Ridge and on an unnamed ridge in the Arabian Sea, which they considered an extension of the Carlsberg Ridge.

Commencing from the neighbourhood of Cape Monze, near Karachi, a double ridge, with an enclosed deep gully, runs towards the south-west across the entrance to the Gulf of Oman. On the north-west side of the gully lies the "Murray" Ridge ... while on the south-east side lies a second ridge ... As we trace these ridges and the enclosed gully to the south-west we find that the deep water of the gully seems to be continued for some distance ... Throughout the greater part of its [i.e., the Carlsberg Ridge's] length ... this ridge appears to be double. Along the two crests numerous soundings between 2291 and 3059 meters have been made and the least water detected over them is 836 meters near the northern end, and 1569 and 1752 meters on the two ridges respectively in about lat. 1°30′ N. where there is a depth of over 3383 meters in the enclosed gully.

*(Wiseman and Sewell, 1937: 220; my bracketed addition)*

They then pointed to the topographical similarity between this oceanic ridge with its "gully" and the rift valleys of east Africa, and proposed a connection between them.

We would here call attention to the apparent similarity which exists between the topography of the floor of the Arabian Sea and the region west of it that is characterized by the presence of the Great Rift Valley. The Rift Valley commences at its northern end in the deep tectonic rift of the Red Sea ... At its southern end the Red Sea opens into the Gulf of Aden which runs from west to east. At its eastern end the Gulf comes into close relations with the south-west extension of the Murray Gorge, thus forming, as it were, a connecting link between the Rift Valley of Africa and the deep gully that runs along the length of the northern part of the Carlsberg Ridge.

*(Wiseman and Sewell, 1937: 225–226)*

This seems to mark the beginning of the speculation that the "cracks" or "gullies" in ridges were in fact rifts, extensional features. They also drew attention to the recent work of Heck (1935), which had shown that the ridges in the Indian Ocean and the great African rift valleys as well as the Mid-Atlantic Ridge all were sites of earthquakes.

Of these secondary belts [of earthquakes] one is shown to extend southward along the Red Sea and then down the Rift Valley, and another can be traced across the Gulf of Oman and then down the middle of the Arabian Sea ... where we now know that the Carlsberg Ridge runs its course. A similar secondary belt can be traced down the length of the Mid-Atlantic Ridge.

*(Wiseman and Sewell, 1937: 227)*

Wiseman and Sewell did not at the time relate their discovery to continental drift. They (I, §6.12) said nothing about it, although Sewell later favored drift (I, §8.12).

Fifteen years later in late 1952, Tharp was the first to identify the central valley of the Mid-Atlantic Ridge. She and Heezen did most of the discovering of the world-wide central rift. Heezen, initially skeptical, did not become fully convinced until mid-1953. However, Heezen waited almost three years before publishing anything about it, first publishing it as an abstract with Ewing as first author (Ewing and Heezen, 1956a). In 1953, shortly after Tharp had identified it in her profiles of the Mid-Atlantic Ridge, Hill and Swallow found the rift valley while aboard HMS *Challenger*. However, Hill did not wait three years before announcing the discovery; presenting it at the September 1954 IUGG meeting in Rome. I first consider Hill and Swallow's discovery.

Maurice Neville Hill (1919–66), son of A. V. Hill, 1922 Nobel Prize recipient in Physiology or Medicine, and Margaret Keynes, sister of the economist Maynard Keynes, entered King's College, Cambridge, in 1938 as an Exhibitioner, reading mathematics, physics, geology, and physiology. Leaving Cambridge before graduating, he joined the Royal Navy. He returned in 1945 at the end of World War II and completed his undergraduate studies. In 1946 he became a research student in the Department of Geodesy and Geophysics, University of Cambridge, working originally under the supervision of Bullard and then after Bullard left Cambridge in 1948 under B. C. Browne. Working in marine geology, he developed an improved way of making seismic observations at sea using only one ship and sound-receiving buoys, and received his Ph.D. in 1951. Appointed an assistant in research in the Department of Geodesy and Geophysics in 1949, he remained at Cambridge for his career. He was elected a Fellow of the Royal Society (London) in 1962, and was awarded the Charles Cree Medal of the Physical Society in 1963. Sadly, Hill took his own life in 1966. "He believed that his brain was deteriorating and there is strong medical evidence that he was right" (Bullard, 1967: 202).

Hill was head of Cambridge's important marine geology group, supervising or working with students such as Swallow, Laughton, Matthews, Loncarevic, and Vine. According to Bullard, Hill was among the best in the world of marine geologists/geophysicists in designing instruments, getting them to work, and using them efficiently at sea to make observations. He was not, however, interested in "great general hypotheses such as continental drift, the spreading of the ocean floor, convection in the mantle or the origin of the earth's magnetic field. There is little speculation in his papers, the discussion keeps close to the facts" (Bullard, 1967: 200). Hill's abstract "The topography of the Mid-Atlantic Ridge" was short and to the point. He kept "close to the facts" and did not speculate about its significance.

During the survey work of the H.M.S. CHALLENGER in 1953, numerous east-to-west crossings of the mid-Atlantic ridge were made between 46°30′ N and 47°45′ N. The topography of the ridge in this region is very rough and only one clear feature can be observed which is common to all sections. There is a deep central trench running approximately N-S; the depth of the bottom of the trench is between 3300 and 7000 m. On either side the depth rises to about 2000 m. Apart from this trench there is no correlation between the various features in spite of the east to west lines being at most 15 miles apart.

*(Hill, 1954: 269)*

Hill next mentioned the central depression two years later. This time he (1956: 230) referred to it as "a deep steep-sided valley," adding that he would discuss it "separately in a future paper." The following year, he (1957) attributed the discovery of the valley to John Crossley Swallow. Swallow, one of his Ph.D. students at Cambridge, made the discovery while he and Hill made several cruises in the northeast Atlantic in 1953 aboard HMS *Challenger*. He also noted that Cambridge scientists had returned to the area in 1956.

> In the middle of the Ridge and roughly following its axis there is a deep valley approximately 1800–2200 fathoms in depth. The floor of the valley is about 5 miles across, and between the tops of the ridges forming its sides the width is about 10 miles. This valley was discovered in cruises of H.M.S. Challenger by Dr. Swallow in 1953 and since then some detailed surveys have been made of restricted parts of it. Its length is at least 85 miles and it seems to be blocked at its southern end in about latitude 47°20′ N. To the north its limit is not known. In the 1956 expedition the investigations were undertaken near its southern limit.

> *(Hill, 1957a: 11)*

Hill made the announcement about Swallow during a discussion at the Royal Society on May 2, 1957. At the time, as I shall show below, Ewing and Heezen were publicizing Tharp's discovery of the median valley, and made no mention of its independent discovery by Cambridge scientists. In their article in *The New York Times* of February 1, for example, it appeared as if Lamont scientists were solely responsible for the discovery. Perhaps Hill, who abhorred such publicity, was concerned that Ewing and Lamont scientists would receive all the credit, and wanted to make sure Cambridge scientists, and especially Swallow, his former student, received proper credit. Swallow had already left marine geology, becoming a leading figure in physical oceanography after designing an instrument to measure deep ocean currents. He was elected to the Royal Society (London) in 1968. He received WHOI's Bigelow Medal (1962), the Sverdup Gold Medal from the American Meteorological Society, and the Stommel Medal, WHOI (Charnock, 1997). Hill continued to survey the Mid-Atlantic Ridge, updating his earlier work.

Turning now to Tharp, Heezen, and Ewing at Lamont, Tharp recounted her 1952 discovery of the rift valley along the Mid-Atlantic Ridge.

> In 1952, after about six weeks of plotting, matching, and gluing, I completed six transatlantic profiles. I pieced together sections of data from eight cruises in order to make a more understandable set of west-east profiles from south to north ... Besides the general similarity in the shape of the ridge in each profile, I was struck by the fact that the only prominent matchup apparent when I compared the profiles was a V-shaped indentation located in the center of each. The individual mountains did not match up, but the cleft did, especially in the three top, or northernmost, profiles of the North Atlantic. Thus it seemed to me that V-shaped indentations represented cross sections of a valley that cut into the ridge at its crest and continued along its axis. Of course, I didn't know about the remainder of the ridge – the area beyond the most northerly and southerly profiles – and the V-shaped indentation was only obvious in the first three profiles.[9]

> *(Tharp and Frankel, 1986)*

The first (A-150) of the eight cruises of *Atlantis* from which she constructed her six profiles began in August 1947 with Ewing as chief scientist; the last (A-180) ended in October, 1952, with Heezen as chief scientist (Lear, BCH 21A: 1; WHOI's archival cruise data). Heezen claimed that the rift valley's discovery "can be unequivocally attributed to Marie Tharp," even though he thought that the rift valley did not reveal itself in "two or possibly three" of the profiles.

> She was examining the six trans-Atlantic profiles in the North Atlantic which was the total amount of information obtained by continuous echo sounders that were available in 1951. She recognized that at the crest of the Mid-Atlantic Ridge there was a valley although two or possibly three of her six profiles failed to confirm it.[10]
>
> *(Lear, BCH 02: 2)*

Tharp recalled that Heezen "groaned and said, 'It cannot be. It looks too much like continental drift'" (Tharp and Frankel, 1986: 53). Heezen remembered that he "discounted it as girl talk" (Wertenbaker, 1974: 144).

Heezen did not accept Tharp's supposition until mid-1953, approximately eight months after she had shown him the six profiles. Heezen, following in the footsteps of Sewell and Wiseman, found evidence that the earthquakes were located beneath the valley that Tharp proposed. Thanks to Gutenberg and Richter's 1949 *Seismicity of the Earth*, it was by then common knowledge among marine geologists that shallow earthquakes occurred along mid-ocean ridges in the Atlantic, Arctic, and Indian oceans.

> Much of the seismicity remaining to be discussed is concentrated along narrow belts which follow oceanic ridges. The best known of these belts passes centrally through the Atlantic Ocean, along the mid-Atlantic Ridge. It extends into the Arctic and across the polar area to the north coast of Siberia near the mouth of the Lena. A similar belt passes southward from Arabia through the central Indian Ocean to at least 30° S; hence there is a branching, and other ridges in the same region appear to be active. The principal branch extends southwest to connect with the Atlantic belt.
>
> *(Gutenberg and Richter, 1949: 74)*

Heezen then found that the earthquakes were not simply associated with ridges, "but were, in fact, associated with the rift valley" (Lear, BCH 10A: 11) at the center of ridges. While Tharp was working up the profiles, Howard Forster, a deaf graduate student from the Boston School of Fine Arts, had been plotting the epicenters of recorded earthquakes in the Atlantic and elsewhere. Heezen hired Forster with funds from Bell Laboratories, who planned to lay new cables, and contracted Heezen to help determine the best place to put them.

> But I was not skeptical for long. For at this point, it was about this time that I was approached by the Bell Laboratories to help in making the considerations for route layout for the new generation of telephone cables which was then being planned and were just beginning to be laid. These cables were to be much more costly; service on them was to be much more remunerative and they wished to lay them with less slack and perhaps without armor.
>
> *(Lear, BCH 02A: 2)*

Forster plotted the earthquake epicenters on a map of the same scale as that on which Tharp was constructing her profiles. Heezen superimposed Forster's plot onto

Tharp's, and found that the epicenters fell, within acceptable margins of error, within the rift zone. It was at this point that he became a believer in Tharp's median valley. Clarifying what happened, Heezen recalled:

Marie [Tharp] did not point out to me there was a close coincidence between the epicenters and the location of the rift. I pointed it out to her. She had found the rift valley from the profiles and convinced herself of the correlation and I was not convinced of the correlation until I took the map, which I had Foster plot, into the next room where Marie was working and plopped it on top of her rift valley plot, and we saw the correlation.

*(Lear, BCH 10A: 10)*

It is unclear if Heezen at this time had truly accepted that the median valley was in fact a *rift* valley. Although Heezen's recollection above suggests that he had, Tharp believed that he was not fully convinced until he saw the marked topographical and seismic similarities between the median valley of the ridge and the rift valleys of Africa. Here is what she told me in 1986.

Bruce keyed in on the terrestrial extension of the hypothesized oceanic-ridge system – the rift-valley system of the East African Plateau. He reasoned that since the system of rift valleys in east Africa appeared to be a landward extension of the oceanic-ridge system, he could learn about the oceanic part by studying the terrestrial. I made up profiles of some of the valleys in east Africa. Bruce noted the topographical similarities between the two sets of profiles: one across the ocean; the other across the land. He also saw that the belt of shallow earthquakes associated with the system of east African rift valleys stayed primarily within the confines of the valley walls. He decided to make the jump and endorse the existence of a central valley within the ridge itself that extended along the entire axis of the Mid-Oceanic Ridge system. As far as Bruce was concerned, the tightness of the analogy between the terrestrial and oceanic segments of the ridge system was the clincher. Before, even with the seismic data, he was not sure if the valley's presence on the original six profiles and other oceanic ones that we found was an accident or indicated a real feature of the ridge. Bruce stressed the importance of the analogy by calling the oceanic valley a rift valley, borrowing the term rift from the characterization of the African valleys as rift valleys. And with the "rift" designation came Bruce's suggestion that the central rift valley, whether in the oceanic or terrestrial part of the overall ridge system, was a huge tension crack in the earth's crust caused by a splitting apart of the earth's crust.

*(Tharp and Frankel, 1986: 54 and 60)*

Heezen had another hint that convinced him of the median valley's existence. He had profiles of the Mid-Atlantic Ridge for the South Atlantic from much earlier *Meteor* expeditions. Stocks (1932) and Stocks and Wüst (1933) showed profiles of the Mid-Atlantic Ridge from the *Meteor* expeditions of 1927 to 1929. Heezen knew about the profiles as an undergraduate. The profiles show a gully, although Stocks and Wüst did not single it out as anything special, as a continuous valley running along the axis of the ridge.[11]

I knew about them from an undergraduate time. The METEOR soundings were in no way obscure. They were well known information which any one who was interested in the ocean even though he may be only an undergraduate should have a strong interest in finding out about and know about. They were published widely, mostly in the German language but both in serious and in popular form. And so yes, we very much knew about them ... So I don't think that the METEOR soundings made little impression. They made a great impression. And a lot of people attempted to interpret them. But there were no clear clues; I mean no one happened to hit it right until Marie hit it right. The METEOR soundings were published ... They had lots of figures in them. Many of the profiles of the Ridge were shown. Despite the fact that many of them showed what Marie would identify as a rift valley, no one who ever looked at these, apparently ever got the idea, even though they were published and were available to look at from about 1927 or 1928 on until the day Marie noticed them on the North Atlantic profiles, on our profiles.

*(Lear, BCH 15A: 5–6)*

Ewing knew about them (§6.4). Menard (1986: 95) later judged that six of the fourteen *Meteor* profiles "definitely show a deep valley in the center of the Ridge" that is "far deeper than any place nearby on either side," and he agreed with Heezen that the profiles "were well-known within oceanographic circles." Menard (1986: 95) also noted that Stocks and Wüst "put no emphasis ... on the central valley or on the possibility that it might be continuous." Indeed, Wüst later remarked, after hearing Ewing and Heezen highlight the existence of the central rift valley, that he and other *Meteor* Expedition scientists had not "dared to connect ... this depression over the enormous distance of ... 400 miles."

As a physical oceanographer, I should like to say that we physical oceanographers are very much impressed by this important discovery, this surprising discovery of the continuous deep valley in the midst of the Atlantic ridge and the far-reaching continuations in the Indian Ocean and perhaps in the South Pacific Ocean. Naturally all oceanographers who have made cross-sections with continuous echo soundings through the Atlantic since 1925 when the "Meteor" expedition research seismic work began have sounded this deep depression in the center of the ridge but nobody, and also not these scientists of the "Meteor" has dared to connect ... this depression over the enormous distance of ... 400 miles.

*(Wüst, 1960: 36a–36b)*

Heezen again emphasized the importance of the *Meteor* profiles.

Well, it seems to me, now, I think then rather obvious when Marie and I discovered that the rift valley and epicenters coincided in the North Atlantic that the next thing to do was check to see where it went ... We did have, in fact, data for the South Atlantic from the METEOR expedition which we then did immediately check against the epicenters in the South Atlantic. And although there weren't very many epicenters there we did have a reasonable, what seemed to be a reasonable assumption, that the epicenters coincided with the median valley which we could see on a good number of the METEOR profiles. So, we then could see that this belt passed through the Indian Ocean and ran into the rift valleys of Africa. And that was a terribly important realization because that was how we got the name for the rift valley in the North Atlantic. We called it a rift valley not so much in terms of *rifting* as in terms of a genetic name relating it to the rift valleys of Africa ... So we drew up profiles which we published showing a profile of the Mid Atlantic Ridge with the rift valley, profiles of the rift valleys of Africa, and this was a very important development in our thinking because here we went to an extensional view.

*(Lear, BCH 10B: 2)*

Once Heezen became completely convinced, he told Ewing. Ewing took notice. Tharp (Tharp and Frankel, 1986: 61) recalled that he "would occasionally come around and ask me, 'How is the gully coming?'"

When Heezen told him about the "gully," Ewing had recently returned from the Royal Society of London's February 1953 discussion about the floor of the Atlantic Ocean at which Ewing had spoken even though he did not prepare a paper for publication (Bullard, 1954: 288). At this meeting Hess proposed convection and rising basalt as the cause of the Mid-Atlantic Ridge (§5.11), a view that Ewing would later adopt (§6.11), and Bullard and Hess applauded the efforts of those at Lamont, Scripps, and elsewhere in obtaining new data about the ocean floor and crust (§1.8). Ewing also heard J. P. Rothé, Director of the Bureau Central International de Séismologie at Strasbourg, discuss earthquake epicenter distributions in the Atlantic and Indian oceans, and suggest that the Mid-Atlantic and Indian Ocean ridges were connected.

The distribution of earthquake epicenters in the Atlantic and Indian Oceans is discussed ... the line of epicenters following the mid-Atlantic Ridge is continued round the Cape of Good Hope and joins the similar line marking the central ridge of the Indian Ocean. It seems, therefore, that these two ridges are related structures.

*(Rothé, 1954: 387)*

Heezen and Ewing now knew of additional evidence of the association between earthquake belts and oceanic ridges, and of a connection between the Mid-Atlantic and Indian Ocean ridges. Moreover, Gutenberg and Richter (1954) had just come out with an updated edition of *Seismicity of the Earth*.

Heezen, however, waited almost three years before publicly discussing the rift valley, and even longer before presenting details of Tharp's and his discovery of the valley. Ewing discussed the Mid-Atlantic Ridge at a symposium on Antarctica in Washington, DC, on April 26 and 27, 1956. Its aim was to help plan what should be studied in the Antarctic during the upcoming IGY. It was sponsored by AGU and held immediately before the annual meeting. Ewing gave the presentation at the symposium. Ewing (as senior author) and Heezen wrote the paper that appeared in the symposium's proceedings. Their paper (1956a) did not explicitly mention their own findings, but simply described the Mid-Atlantic Ridge as "being characterized by a median rift zone which follows the crest of the ridge very accurately" (1956a: 78). Drawing on Gutenberg and Richter's 1954 "magnificent studies on seismology" they proposed that the "belt of earthquakes" associated with the Mid-Atlantic Ridge should be used to locate its poorly

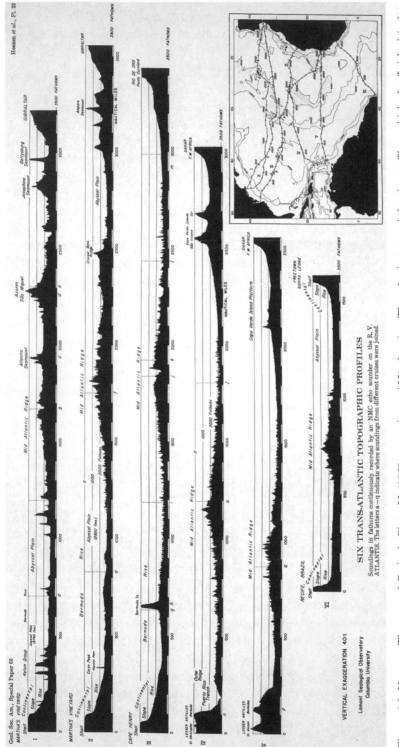

Figure 6.1 Heezen, Tharp, and Ewing's Plate 22 (1959: opposite p. 122) showing Tharp's six trans-Atlantic profiles, which she finished in late 1952 by piecing together sections of data from eight cruises. The rift valley is prominent in the more northerly three profiles but less so in the remaining three.

sounded segments as well as other seismically active ridges. They further suggested that seismically active ridges "may be a consequence of the rift" and might extend into a "continental area such as the East African rift zone, or into an area like the rift zone in the south island of New Zealand." They also dismissed Hess's 1953 (published 1954) explanation of formation of the Mid-Atlantic Ridge as caused by rising convection currents (§5.11) because it "apparently fails to account for the rift valleys and the close association of the earthquakes with them." Hess did know about the existence of the rift valley when he proposed his solution, and he could have proposed that they were the consequence of diverging convective currents. Ewing and Heezen cited Hill (1954), Wiseman and Sewell (1937), and Stocks' 1932 work from the *Meteor* Expedition. I do not know if Heezen knew of Wiseman and Sewell, or had wind of Hill and Swallow's findings before he accepted Tharp's suggestion.

Several days later at AGU's 1956 Annual Meeting in Washington, DC, Ewing and Heezen, again with Ewing as first author, introduced AGU members to Lamont's work on the "conspicuous depression." They again made no mention of Tharp's six profiles (Figure 6.1). Moreover, according to Heezen, Ewing listed himself as first author, and omitted Tharp's name (Lear, BCH 10B: 7).

*Mid-Atlantic Ridge Seismic Belt.* The Atlantic belt of earthquake epicenters follows the crest of the Mid-Atlantic ridge and its prolongations into the Arctic and Indian Oceans with a precision which becomes more apparent with the improvement of our knowledge of the topography and of epicenter locations. These are all shallow shocks. Their apparent departure from the narrow crest of the ridge seldom exceeds the probable error of location. The crest is 30 to 60 miles wide, very rough, and on a typical section shows several peaks at depths of about 800 to 1100 fathoms. There is usually also a conspicuous median depression reaching depths of about 2300 fathoms. This is interpreted as an active oceanic rift zone which continues through the African rift valleys.

*(Ewing and Heezen, 1956a)*

Because they did not then know if the Mid-Atlantic Ridge's "prolongations into the Arctic and Indian Oceans" possessed a "conspicuous median depression," their hypothesis was particularly bold. In contrast, Hill opted to stay close to the facts.

Prof. M. Ewing has suggested that this valley exists with occasional breaks in it throughout the length of the Ridge from 55° N. Lat. down to the latitude of Cape Town and that it there swings to the east and continues up the ridges in the western Indian Ocean. There is some support for this suggestion in the limited sounding information available, but its continuity is open to much doubt.

*(Hill, 1957a: 11)*

Four years later, Ewing and Heezen confirmed the rift valley's presence in the Indian Ocean, finding the ridge on five crossings of the Atlantic–Indian and Southwest Indian ridges, and one crossing of the Southeast Indian Ridge.

The existence of a continuous, rifted, mid-oceanic ridge in the southwestern Indian Ocean, previously predicted by us, has been confirmed by soundings taken by the research vessel *Vema* during the expedition now in progress.

*(Ewing and Heezen, 1960: 1678)*

Finally, Heezen and Ewing (1961: 629–630) reported that a rift valley had been found along the extension of the Mid-Atlantic Ridge in the Arctic Basin. When they first proposed the rift's worldwide continuity, they admitted that the evidence was inconclusive.

Heezen mentioned the median rift valley during a talk at the October 1956 Annual Meeting of the GSA in Minneapolis.[12] After characterizing the Mid-Atlantic Ridge as having a "rift valley" along its axis with an average depth of 1800 fathoms and width of twenty-five miles, he (1956: 1703) proposed that it and other ridges were "similar in topography and seismicity to African plateaus and rift valleys, suggesting a common origin and similar age." He did not speculate about what the common origin might be. That, Earth expansion, would come in April of the following year. Heezen also did not include Tharp as a co-author, and said nothing about her profiles. That would come two months after the GSA meeting in an announcement by Ewing at a press conference on January 31, 1957.

### 6.8  Lamont's "4500 mile undersea crack," is it continuous?

At the press conference Ewing overstated what Lamont scientists had discovered. The news release, which surely Ewing either wrote or carefully reviewed, issued from the News Office of Columbia University began as follows:

The existence of a continuous 45 000-mile undersea crack in the earth's crust, averaging twenty miles wide and one and a half miles deep, has been confirmed by Columbia University scientists. This crack coincides with a world-wide active earthquake zone along its entire length. The announcement was made yesterday (Thursday) by Dr. Maurice Ewing, director of the Lamont Geological Observatory, a division of the University.

As described above, the claim of having discovered "a continuous 45 000-mile undersea crack" went far beyond the published record in early 1957 when only valleys at crossings of the Mid-Atlantic Ridge had been reported, although at Lamont there might have been unpublished records of crossings with central valleys of other ridges. Ewing was arranging first publication through the news media rather than scientific channels, and was claiming that the topographic valley, the "undersea crack," was "continuous," which was not justified. Some critical segments of the ridge system did not have central valleys, and central valleys, far from being continuous, were frequently offset by what appeared to be large faults.

Notably at the news conference Ewing gave Tharp credit for the discovery saying, "We have been working on the hypothesis for about five years ... One day, Marie Tharp, a cartographer at the Observatory, was working on charts of the Atlantic Ocean bottom. She noticed that the deepest rifts in mid-Atlantic formed the locus of an oceanic earthquake belt." Heezen was also singled out in the press release.

When Miss Tharp noticed the mid-Atlantic rift-seismic zone, she pointed out her discovery to Dr. Bruce Heezen, a research associate at the Lamont Observatory. Drs. Heezen and Ewing and their associates examined other undersea topographical profile charts and unpublished data in the Lamont files.

Except for Gutenberg and Richter, however, who made available "a great deal of excellent and comprehensive information on the world's earthquake zones" in their *Seismicity of the Earth*, no other non-Lamont scientists were mentioned.

Ewing's press conference was a publicity coup. The next day (February 1), *The New York Times* displayed the "45 000-mile continuous trench," noted that "Columbia Teams Discover a 20-Mile-Wide, 2-Mile-Deep Fissure Circling Globe," and showed a photograph of Ewing. Heezen recalled the *Times* article:

The first announcement of the continuity of the rift around the world, I recall, was in *The New York Times* interview of 1957. *The New York Times* ran a little map. In fact, as I recall, the news reporter that showed up for the news conference was the merchant marine editor and wasn't even in the science department.

*(Lear, BCH 01A: 8)*

*The Times* correspondent, Ira Henry Freeman, made it seem that not a moment was being lost in Lamont's triumph over ignorance. "Columbia's three-masted schooner, *Vema* is exploring the 'rift' in the South Atlantic now. Dr. Ewing will fly down tomorrow to join her scientific crew in Buenos Aires." No scientists from other institutions were mentioned.

Ewing also dismissed continental drift "as impossible." He was quoted in the press release as saying:

The presence of a rift throughout the length of the North and South Atlantic Oceans, and the evidence that the two sides of the ocean basin are even now being pulled apart, may be taken as evidence to support the theory of continental drift. That is, North and South America and Europe and Africa were once together, and have gradually been separated by the full width of the Atlantic Ocean ... However ... the continuation of the rift into the Indian and Pacific Oceans makes this explanation impossible.

*The New York Times* report also credited Heezen with rejecting continental drift for the same reason.

Dr Heezen expressed the opinion that the tracing of the rift "tended to weaken" the "theory of continental drift." This theory holds that North America, South America, Europe and Africa were one immense land mass eons ago but cracked roughly in the middle and pulled apart. "But notice that the rift rounds the Cape of Good Hope and runs up along the East Coast of Africa too," Dr. Heezen commented. "If the Atlantic rift were caused by Africa moving eastward, how was the Arabian rift caused?"

Indeed, this difficulty was serious and Heezen (RS2) would continue to raise it against Hess and others who postulated huge sub-crustal or mantle convection currents that arise beneath fixed ridges and cause seafloor spreading. His inability to see through this difficulty was one of the principal reasons he later proposed Earth expansion as the cause of the break-up of Pangea.

Menard (1986: 102) later credited Heezen and Ewing for having "provided a new and stimulating way of thinking about mid-ocean ridges," and presenting "a target that unified global exploration during the International Geophysical Year." At the time, he also noted that non-Lamont scientists had heard of the alleged continuous, 40 000-mile-long, two-mile-deep world-girdling rift from in the press. Menard knew that the central valley as described by Lamont was not continuous and he wrote this to Heezen on March 20, 1957:

I just received a copy of your "Some problems of Antarctic submarine geology" and I am very gratified to find a discussion of the much publicized world-girdling rift. I have been increasingly distressed to read one account after another in the press and magazines of this fabulous rift 2 miles deep, 20 miles wide, and 40 000 miles long. I have had to say to more than one reporter that I believed that you were misquoted because I know no sounding existed to support such a continuous rift in the south Pacific and Indian Oceans. The constant repetition had begun to wear, however, and I was just reaching a point where I intended to write to you with the thought of issuing a contrary statement regarding the Pacific. After reading your paper this hardly seems necessary. I do not mean to imply that we have soundings to prove that the rift does not exist in the south Pacific; perhaps it does. On the other hand, north of N10° latitude in the eastern Pacific we can prove that it does not exist as you describe it to the press.

*(Menard, 1986: 103)*

Menard also reproduced Heezen's response of March 29, 1957.

The rift is not always 20 miles wide by 2 miles deep as the papers stated . . . the general depth below its rim is usually only 500–1000 fathoms . . . However, we have found no profiles across this belt which fail to show a topographic feature which could be interpreted as a rift. I would very much like to see your evidence that a rift does not exist north of 10° N in the Pacific. I believe that your published profiles can be interpreted to support its existence in this area. I realize, however, that you have a vast amount of unpublished data which I have not seen. If you can establish to our mutual satisfaction that you are right in this regard it might save me future embarrassment and the scientific literature further confusion.

*(Menard, 1986: 103)*

Menard (1986: 104) invited Heezen to join him on Scripps' Downwind Expedition of 1957–8. Heezen was unable to do so. Menard arrived home from the expedition on Christmas Day, and wrote Heezen on January 6, 1958.

I have the mournful honor of informing you that the median trough of recent fame does not exist along the crest of the East Pacific Rise between 48° S and 43° S. On our Downwind expedition, from which I have just returned, I managed to make seven crossings of the crust of the rise in that area with the express purpose of surveying the trough. I hesitate to dwell on the frustrations of staying up hour after hour on the original crossing waiting for a trough to appear so that I could survey it. In any event the crest is rather smooth except for the usual concentration of seamounts.

Heezen was not convinced by Menard's interpretation of his new bathymetric data from the East Pacific Rise. Menard showed him the echograms. Recalling what happened; Menard stated:

All I could see were a few low abyssal hills and a few small circular volcanoes. Bruce was convinced that one of the typically shallow valleys between elongate hills was "the rift valley."

*(Menard, 1986: 105)*

Respecting Heezen's point of view, Menard added the following footnote after stating, "A special survey during the DOWNWIND expedition did not find any apparent trough for 300 miles along the crest of the East Pacific Rise."

Doctor Heezen has inspected all the DOWNWIND echograms in this area with the writer in a preliminary attempt to standardize the topographic terminology in the Pacific and Atlantic basins. Heezen believes that the soundings may show a small median trough and certainly do not rule out the possibility of one.

*(Menard, 1958a: 1182)*

Menard turned out to be correct. Although there are exceptions, slower spreading ridges such as the Mid-Atlantic usually have rift valleys, but faster spreading ones such as the East Pacific Rise do not. Moreover, some earthquakes that occur near the crest of a ridge actually occur along what later came to be known as ridge-ridge transform faults connecting ridge segments. Heezen and Ewing thought oceanic ridges continuous; they did not view them as segmented, offset by faults.

Heezen first discussed Tharp's and his discovery of the rift valley during a talk at Princeton University on March 26, 1957. Hess had learned of their discovery, was impressed, and invited Heezen to give the talk. Tharp (Tharp and Frankel, 1986: 61) recalled, "Bruce brought along one of our globes with the rift valley outlined in bright red." After Heezen finished his presentation, Hess apparently said, "in essence: 'You have shaken the foundations of geology'" (Sullivan, 1974: 57). It would not be long before Heezen moved to Earth expansion.

## 6.9  Heezen's Earth expansion, 1957–1959

Expansionists: I first presented these views at the 1957 meeting of the American Geophysical Union (Heezen, 1957). I subsequently expanded these views in papers presented to a French National Research Council Colloquium in 1958 (Heezen, 1959[a]) and the recent International Oceanographic Congress in New York (Heezen, 1959[b]). I later discovered that Hildgenberg wrote and privately published a small in book in 1933 in which he expounded a similar concept. Halm, an astronomer, presented a similar theory in 1935 [See Note 1, Volume II,

Chapter 1]. Egyed, a geophysicist, published a similar theory a few years ago [II, §6.2–§6.4], and Professor Carey sent me a few months ago a paper in which he set forth his similar views [II, §6.13]. It is perhaps significant that each of these expansionists worked out the idea independently unaware of the others' work – each from a slightly different view point.

*(Heezen, 1959c: 5; my bracketed additions; Hildgenberg should be Hilgenberg, 1933)*

Heezen publicly presented Earth expansion as an explanation of the central rift valley of ocean ridges at the 38th Annual Meeting of the AGU, held in Washington, DC from April 29 through May 2, 1957. His abstract, "Deep-sea physiographic provinces and crustal structure," submitted no later than March 5, argued for a "considerable increase in the volume of the Earth's mantle throughout geologic time."

On the basis of detailed echo-sounding profiles, the North and South Atlantic Oceans have been divided into a series of longitudinal physiographic provinces. Each province exhibits similar topographic detail and sediments throughout its length, and available geophysical data indicate that each is the surface expression of a major crustal structure. On the basis of less detailed data these provinces have been traced through the Indian, Arctic, South Pacific, Norwegian, and Arctic Oceans. A mid-ocean rift valley of remarkable continuity and seismicity divides each of the aforementioned oceans into two equal parts. The province boundaries within each of the two major basins are strikingly parallel to the continental margins on each side and to the rift valley in mid-ocean. In explanation of this striking parallelism and symmetry an hypothesis of considerable increase in the volume of the Earth's mantle throughout geologic time is offered. Under this hypothesis the major structural features of the ocean are ascribed to tension and associated faulting.

*(Heezen, 1957)*

What, in his abstract, is most striking is what is left unsaid. He left unspecified both the rate and amount of expansion, said nothing about what is supposed to occur at trenches, did not compare the oceanic rift valleys with those of east Africa, did not mention continental drift or its paleomagnetic support, and did not speculate about the cause of expansion. He mentioned only the "striking parallelism" between continental margins and median oceanic rift valleys, and the topographic bisymmetry of the Atlantic, Indian, Arctic, South Pacific, Norwegian, and Arctic basins. I do not know if, in his actual talk, Heezen spoke about any of these "missing" topics. Perhaps, he thought he might get a better reception if he did not mention continental drift and its paleomagnetic support. Perhaps he simply had room in his abstract only for marine geology.

At the May 1958 colloquium in Nice on the topography and geology of the deep seafloor where Menard (§5.8) and Dietz (§4.7) presented papers, Heezen offered a robust defense of Earth expansion. Published in French, his paper appeared the following year. He now explicitly defended "displacements of continents" by appealing to its paleomagnetic support.

Studies of rock magnetism have brought new arguments to the hypothesis of continental displacement given by Wegener at the beginning of this century. Remanent magnetic measurements from a certain number of rock sample levels of different geological ages, in far away zones, suggested displacements of continents of several thousands of kilometers since the Cambrian. If submarine displacements do in fact exist, then the topographic shapes of the marine floor must in great part be the result of this displacement.[13]

*(Heezen, 1959a: 296)*

Heezen, however, provided no critique of mobilism's paleomagnetic support. He also introduced trenches, claiming that they are regions of extension, appealing to his 1955 paper with Ewing, and Ewing and Worzel's 1954 paper on the Puerto Rico Trench (§3.17).

Ewing and Worzel (1954) found that the explanation by a great "tectogene" of a downward bending of the crust of deep-sea trenches is not in accord with geophysical data. They show that gravity anomalies can better be explained as a result of the thinning of the crust. If all zones of category III [i.e., deep oceanic margins] have the same origin and the same structure, and if this structure is the result of stretching of the crust as stated by Ewing and Worzel (1954) and Ewing and Heezen (1955), we must conclude that the terrestrial crust has suffered major extensions.

*(Heezen, 1959a: 297–298; my bracketed additions; references are the same as mine)*

He said again that Earth expansion is indicated by the bisection of ocean basins into symmetrical topographic halves by the median valley. He went further than before: oceanic median valleys are tensional zones, just like the rift valleys of east Africa, and are tectonically young and currently active as indicated by their seismicity and the "high relief of the rift mountains" (1959a: 300). Comparing the topographic and seismic features of the East Africa rift valleys and oceanic ridge valleys, he presented profiles of Lake Tanganyika, Gulf of Aden, Norwegian Sea, and the Atlantic between England and Newfoundland, and between west Africa and Brazil, and argued (1959a: 302) that they form a "historical series which might represent the different stages in the development of an oceanic basin during successive continental displacements." Lake Tanganyika and other east African rift valleys widen to become narrow seas, which continue widening, and become ocean basins. He also proposed that rift valleys have a limit to their width, and once reached, a new rift valley forms in their center; this process preserves the bilateral symmetry of the widening ocean basins.

The symmetry of the physiographic provinces [i.e., continental margins, abyssal plains, and mid-ocean ridges] can be maintained if we envision the existence of a limit to the width of a rift valley. When the rift valley has reached a certain width, a new rift valley must be formed in the center of the old one. This critical width must be from 80.5 kilometers to 128.7 kilometers, if one judges the observed widths of the active rift valley.

*(Heezen, 1959a: 302; my bracketed addition)*

He compared Earth expansion with continental drift, and argued (RS3) for the superiority of the former because continental drift faced a difficulty that Earth expansion did not (RS2). Classical continental drift requires "compression along one of the edges" of a drifting continent and "expansion along the opposite edge," whereas there often is expansion on both sides as expected if Earth has been expanding.

These continental displacements can be effected in two very different manners. The manner considered generally is that of continental drift. In the drifting, the continental blocks float laterally across the upper part of the mantle. The continental displacements can also be affected by the expansion of the interior of the earth. In this last case, the original crust, solid and differentiated, would break and individual fragments would be displaced, each with respect to the other. In the continental case, one must expect compression along one of the edges with expansion along the opposite edge of a continent. In the case of displacement due to the expansion of the interior of the earth, expansion should be found in all oceanic zones.

*(Heezen, 1959a: 302)*

Heezen concluded that continental drift cannot explain in a general way "the evidences of submarine topography" while Earth expansion can (RS3).

It seems almost impossible to reconcile the evidences of submarine topography and continental displacement by lateral drifting. These same features may easily enough be explained by the continental displacement hypothesis due to internal expansion. If future paleomagnetic, paleoclimatological and other related studies require continental drifting as a solution, it would seem useful to continue studying the subject. Actually, we may express the opinion that the known features of submarine topography, if we are interpreting them correctly, must provide the most important key to the solution of the problem of continental drifting.

*(Heezen, 1959a: 302)*

He did not say what might have caused Earth expansion.

Heezen next spoke about Earth expansion at the August–September 1959 International Oceanographic Congress in New York, at which Hess was still offering a fixist explanation of the origin of ocean ridges (§3.13). His enthusiasm for the paleomagnetic support of mobilism had strengthened.

Reported measurements of remanent magnetism of rocks collected from a single continent generally show reasonably consistent pole positions for any one geologic period. Measurements of rocks from different periods indicated consistent variations with time of the positions of the poles relative to individual continents. However, individual continents give radically different paths for the wandering poles except for very recent times. This evidence has given new and impressive support to the controversial theory of continental drift originally postulated on purely morphological grounds.

*(Heezen, 1959b: 26, 28)*

He repeated his earlier argument, giving the same six ridge profiles, reinforcing his claim that they "may represent the 'typical' ocean basin in successive stages of development."

The six profiles in the figure may represent the "typical" ocean basin in successive stages of development. First a rift may form resembling Lake Tanganyika (1). Further extension of the crust might result in a form resembling the Gulf of Aden (2 and 3). Slightly more extension could produce a profile resembling that across the northern Norwegian Sea (4). Still further extension might then lead to a wider ocean basin (5 and 6), similar to the present Atlantic. In this sequence, tensional rifts would develop at successive later periods in the centers of the earlier rift as it became extended to widths exceeding 60 to 200 miles. A succession such as this might result if, as the earth's exterior expanded, the continental crust maintained a relatively constant area. The center rift valley shown on each of the six profiles, lies on the Mid-Oceanic Ridge seismic belt and is part of a world-wide belt of rift valleys.

*(Heezen, 1959b: 28)*

He repeated the same objection (RS2) to continental drift that he and Ewing had given *The New York Times* reporter (§6.8). He again acknowledged the radical nature of Earth expansion.

The submarine geologist has good reason to view these results carefully and critically, for, if continental displacements did occur, they must have been the basic cause of all major morphologic features of the ocean floor. To most proponents of continental drift all forms of sea-floor topography are negligible perturbations of the scum of the earth. However, if continental displacements did occur, the geology of the sea floor will inevitably specify the direction and the mode of displacement. The location of the Mid-Oceanic Ridge, oft cited as a remnant of the original continental rift, opposes continental drift since it seems to require that the continents drift in several directions at the same time.

A possible way out of this dilemma is to postulate an expanding earth; but, in view of the meager evidence now available, this may seem too drastic and may itself have other more serious objections of astronomical nature.

*(Heezen, 1959b: 28)*

He thought Earth expansion worth entertaining, and may have become inclined toward it, but was definitely not ready to embrace it.

### 6.10  Holmes and Heezen correspond about Earth expansion

Meanwhile, in Britain, Holmes had learned of Heezen's support of Earth expansion from an article in *The Sunday Times*. Already sympathetic toward Earth expansion, and in contact with Egyed (II, §6.3) and Carey (II, §6.13), he wanted to know more, and he wrote to Heezen on October 3, 1959, saying that he had written a letter to *The Sunday Times*, correcting the reporter's misunderstanding: Heezen was not the first to propose Earth expansion.

Dear Dr. Heezen,

I was delighted to learn from a short account of the International Oceanographic Congress published in the Sunday Times that you have joined the ranks of those few of us who think the hypothesis of an expanding earth worth following up. The reporter concerned described the hypothesis as if it were a new discovery, instead of giving any hints as to the evidence which led you to adopt it. I therefore wrote a short letter to the paper pointing out that so far as I know the hypothesis began with J. K. E. Halm in 1935. I also referred to the more recent work of Egyed and Carey, but they missed that part ... However, I have since then been inundated with letters asking for more information. All the writers are evidently fascinated by the idea and no one has produced a single item of adverse evidence.

*(Holmes, letter to Heezen, October 3, 1959)*

He then asked Heezen if he, like Egyed and Carey, planned to write a book on Earth expansion, that he planned to introduce Earth expansion in the next edition of his *Physical Geology*, and therefore would appreciate "copies of any papers you may be publishing." He also supposed that Heezen's view of trenches may be why he favored mobilism. Perhaps he had read Ewing and Heezen's 1955 paper in which they defended an extensional analysis of the Puerto Rico Trench.

As you probably know, both Egyed and Carey are thinking of writing books, and I am hoping to introduce the subject into the new edition of my "Physical Geology" – the revision of which, alas, goes very slowly. Possibly you too may think it worth while to write a book on the subject from your own special angle. Meanwhile, I am writing to ask if you would be good enough to let me have copies of any papers you may be publishing, as they become available, so that I may also be able to present your evidence, which I suppose to be the sub-structure of the oceanic trenches.

*(Holmes, letter to Heezen, October 3, 1959)*

Holmes also sent to Heezen a copy of his September 7, 1959 letter to *The Sunday Times*. Holmes strongly endorsed the new paleomagnetic support for large-scale lateral displacements of continents (§2.16).

It was gratifying to learn from Mr Peter Small's report of the International Oceanographic Congress that Dr. Bruce Heezen (Lamont Geological Observatory) has joined the small but growing band of scientists who realizes that there is now ample evidence for seriously considering the possibility that expansion of the interior of our planet may have played a predominant role in the evolution of its surface features. Most readers of *The Sunday Times* are likely to be familiar with the traditional idea that the earth has been slowly cooling from an originally higher temperature and therefore shrinking. But it is improbable that more than one in a thousand can have previously heard of the alternative possibility. The rest will inevitably have been left with the erroneous impression that the expansion hypothesis is a new one, announced only a few days ago. It is no discourtesy to Dr. Heezen to correct this impression. Dr. Heezen has made an invaluable contribution by showing that certain features of the ocean floor can be best accounted for by crustal stretching, that is by global expansion. But he is not the originator of the expansion hypothesis.

In 1935, Dr. J. K. E. Halm, a South African astronomer, challenged the current view by approaching the problem in the light of astro-physics and stellar evolution. He reached the conclusion that the earth was likely to be an expanding globe. So far as I know, he was the first to point out that terrestrial expansion would bring about the splitting and gradual separation of the continents as they moved radically outwards (leaving ocean basins between) without any necessity for the embarrassingly great lateral movements that were called for in Wegener's well known hypothesis of continental drift. Although it removed the only serious objection to continental drift by doing away with most of "drift," this revolutionary idea passed almost unnoticed and for all practical purposes remained unknown or was soon forgotten.

The amazing discoveries about the earth and her mysterious behavior that have rewarded post-war researchers have at last made the idea of continental drift respectable, as it was never before the war. From earthquake studies we now know, for example, that many parts of the lands encircling the Pacific are at present slowly moving round like the hands of a clock relative to the adjoining parts of the ocean floor. Palaeomagnetic studies leave little doubt of the reality of former world-wide movements on a far greater scale. Whereas in 1935 the climate of opinion was distinctly unfavourable to the development of Halm's unorthodox ideas, today it has become much more auspicious. But the ideas themselves have had to be thought out afresh against an entirely different background of experience.

In recent years there have been two leading proponents of the expansion hypothesis, both of whom originated it quite independently of each other and of earlier workers such as Halm: a Hungarian geophysicist, Professor L. Egyed of the Eötvös University of Budapest, where he is Director of the Geophysical Institute; and Australian geologist, Professor S. Warren Carey of the University of Tasmania. Egyed reached the idea from a consideration of the nature of the earth's core, where pressure ionization and consequent high density may formerly have been much greater than now. Carey reached the idea from a rigorous study of the earth's surface features ...

To me, as a geologist who has been actively interested in the measurement of geological time for nearly half a century, the expansion hypothesis is the more welcome because it goes far towards resolving a difficulty that has nagged at me for many years, and particularly since our recent realization that the crust of the earth has existed for some 4500 million years. Weathering and erosion for a fraction of this time would suffice to reduce the lands of the present day to sea level. How is it, then, that the land continues to keep its head above water, and does this so successfully that it stands today far higher than has usually been the case in former geological ages? The answer appears to be not so much that the land rises as that the sea level falls. In an expanding earth the area of the continents (including shallow seas) varies but little. It is essentially the ocean basins that increase in area, and consequently on balance the sea level falls relative to the land. Were the earth not expanding it is highly improbable that we should be here to discuss the matter.

Apparently Holmes did not yet know about the unexpected paucity of sediments on the seafloor, or, perhaps did not realize that accumulated seafloor sediments would be removed from ocean basins if his own theory of mantle convection were correct.

Heezen replied on October 6, 1959, sending a copy of the paper that he had given at the Nice colloquium.

Dear Dr. Holmes:

Thank you for your welcome letter of October 3, concerning the expanding earth. I inclose [*sic*] a reprint in which the concept is briefly outlined. I am, of course, aware of the work of Egyed, Halm, Hilgenberg, Dicke and the more recent writings of Carey. As each of these workers, I arrived at the conclusion that the earth is expanding without knowing of the work of the others. I, of course, now draw comfort from their views. It is indeed encouraging that you, too, view the concept favorably.

I was led to the concept through a study of the topography and the structure of the Mid-Oceanic Ridge System and its seismically active median rift which so closely resembles the East African rifts. As you will see by examining figure 5 in the accompanying reprint, I see an evolutionary series in a succession of profiles across this seismic belt. Previous studies on deep-sea trenches led us to the view that they were better explained by extension than by compression. Thus, over half the earth's surface seems to be covered with features suggestive of extension.

Paleomagnetic evidence favoring continental drift posed a difficult, if not insurmountable, problem in crustal structure. But with expansion, the continents are not required to float through the oceanic crust like icebergs in the sea. The oceanic crust has formed as the continents were displaced by expansion, thus avoiding the difficult mechanical problems of conventional drift. The paleomagnetic evidence of drift can be more easily interpreted as evidence of expansion. The concept says that the continental crust is old and the oceans are young, in contrast to the views of the accretion school. However, accretion may still occur to some degree.

I shall be glad to send a copy of my paper when it is completed.

Sincerely yours,
Bruce C. Heezen

Heezen either did not know of Holmes' theory of mantle convection or decided not to mention it. I say this because Holmes' theory did avoid the "difficult mechanical problems of conventional drift" because it, like Earth expansion, did not require the continents "to float through the oceanic crust like icebergs in the sea." Heezen also seems to have come to believe that the "paleomagnetic evidence of drift" not only can be explained by Earth expansion, but that it "can be more easily interpreted as evidence of expansion" than "conventional drift." However, he simply might have meant that the paleomagnetic evidence for mobilism can be better explained by expansion rather than Wegener's drift because the latter faced the notorious mechanism difficulty. Regardless, it is evident that Heezen was educating himself about the paleomagnetic support for the rifting and drifting apart of continents.

### 6.11 Heezen comes to rely on paleomagnetism as support for Earth expansion, 1959–1960

Heezen discussed Earth expansion at a symposium held at Columbia University, New York, on December 10, 1959. He defended his version of Earth expansion by arguing that it was required to explain the origin and development of mid-ocean ridges, abyssal plains, and oceanic trenches. He seemed certain that trenches were extensional: Lamonters had shown them to be so.

Refined geophysical investigations and interpretations of the marginal trenches have completely refuted the existence of a compressional down buckle (tectogene) (Ewing and Worzel, 1954). Marginal trenches are now interpreted as extensional features in no way attributable to crustal compression (Talwani, Sutton and Worzel, 1959). Thus the continental margins seem to be dominated by faulting and subsidence . . .

*(Heezen, 1959c: 1; references are the same as mine)*

As before, he seemed to equate refutation of the notion of the tectogene with refutation of downward convection. Fisher and Hess certainly disagreed with him; they supported convection but rejected the tectogene (§3.16–§3.20). So did Officer (§3.18), and Fisher, Mason, and Raitt (§3.19). Heezen did not mention their views.

He then went on to describe in a more restrained way what he believed to be the dominance of "tensional deformation" of the seafloor as indicated by "tectonic fabric." He clearly thought that new work on the seafloor had refuted contractionism, but he seemed ill-informed about the many difficulties that had already been raised against it long before, during the first half of the twentieth century; either that or he ignored them so as to amplify the importance of his and other Lamonters' work on ocean basins.

We have all been taught that the earth is shrinking and for a century few have offered any serious objections to this unfounded assumption. The above tentative and speculative conclusions that extensional deformation dominates the tectonic fabric of the sea floor seriously challenges the assumption of a shrinking earth. The sea floor after all covers over two-thirds of the surface of the earth so, even if every square mile of the continents conclusively indicated compression, firm conclusions based on sea floor geology should, by virtue of relative area, be given more weight in considering global tectonic evolution.

*(Heezen, 1959c: 3)*

He (1959c: 3) admitted that some of his colleagues at Lamont disagreed with him about the amount of extension needed to explain "the tectonic fabric of the sea floor." Refuting contractionism was one thing, establishing expansion quite another.

Although we have concluded from a study of deep sea topography that the oceans are dominated by extensional deformations, we cannot determine the amount of extension. Although many of my colleagues agree with the above conclusion they believe the amount of extension to be slight.

In the remainder of his Columbia 1959 talk, he expanded his appeal to paleomagnetism, repeated his defense of the claim that ocean basins evolve from rift valleys, and sidestepped what could cause Earth expansion.

I have of course considered the probable cause of global expansion but I do not wish to go into this aspect of the problem since I base my case solely on the morpho-tectonic and paleomagnetic data.

*(Heezen, 1959c: 5)*

His lengthened discussion of paleomagnetism is the most interesting part of his paper. Unfortunately what remains is incomplete.[14] Nonetheless, what is left shows that he had further familiarized himself with the paleomagnetic support for mobilism, thought it was essentially difficulty-free, and believed that it could be explained by continental drift or by Earth expansion. He (1959c: 3) strongly endorsed paleomagnetism's support of mobilism.

Continental drift has had a checkered career and was nearly laid peacefully away when students of paleomagnetism gave it a new lease on life. Studies of paleogeography and geology had produced so much evidence adverse to continental drift that the reappearance of this vexing hypothesis was not a happy sight for most earth scientists. Charges have been made that the paleomagnetic results are invalid due to effects of metamorphism, the former absence of a dipole field and the use of poor statistical methods by works [sic] in paleomagnetism. Although each effect may invalidate certain measurements, it is likely that the main patterns of the paleomagnetic results are valid.

Heezen knew the literature well enough. He realized that paleomagnetists avoided metamorphic rocks, that Ronald Fisher had provided the needed statistics, that there was strong support of the GAD hypothesis, and that by 1959 there were many reliable measurements clearly indicating that continents had changed their positions relative to each other (II, §8.7). Emphasizing the divergence among APW paths of different continents, he claimed:

At first paleomagnetic measurements on rocks of a similar age from England and from the United States seemed to agree. Continental drift, it was concluded, is not definitely disproved. However, as more measurements of paleomagnetism were made two separate polar wander curves were determined which showed increasing divergence backwards in time. It was then concluded that a displacement of a couple of thousand miles had occurred between England and North America since the early Paleozoic. When measurements were made in India and Australia still other polar paths were determined. It seemed that large scale continental drift must be accepted.

*(Heezen, 1959c: 3)*

He also showed a figure (my Figure 6.2), displaying very different APW paths for India, North America, and Europe. He labeled it "Ancient Pole Positions," and said it was "(after Deutsch, 1958, with modifications from Irving, 1958)" (1959c: Figure Captions page; Irving, 1958 is my Irving, 1958a). His references are ambiguous, but he very probably based it on Deutsch's (1958: 159) Figure 1. The Indian path was based on Clegg, Deutsch, and others of the London group (II, §5.2). Deutsch took the APW paths for Europe and North America from Irving (1956) (II, §3.12).

Heezen constructed his path of Australia on poles listed by Irving and Green.[15] It is peculiarly "M-shaped" unlike Irving and Green's (1958: 70) "C-shaped" path of Australia (reproduced as Figure 5.4, II); I believe that Heezen simply connected poles without regard to uncertainties as expressed by their ovals of confidence. Nonetheless, and more importantly, Heezen wanted to show the very large divergences of the APW paths of Australia and India from each other, and from European and North American paths.

Figure 6.2 Heezen's Figure 1 (1959c). Ancient pole positions. MIO, Miocene; EO, Eocene; CT, Cretaceous; J, Jurassic; TR, Triassic; P, Permian; CP, Carboniferous-Permian; D, Devonian. 1, Jurassic aged Indian pole; 2, Upper Cretaceous to Lower Eocene aged Indian pole; 3, younger Indian pole than 2; 4, probably Miocene aged Indian pole. Solid triangles, reversed magnetization; solid circles, normal magnetization.

His Figure 2, which he entitled "Latitude shift of Europe and North America (after Irving, 1958a)," is unfortunately lost. As the discussion below of his Figure 3 (my Figure 6.3) suggests, he was presumably interested in stressing the northward shift of Europe and North America, which, he thought, raised a difficulty for continental drift that was avoided by Earth expansion (RS2). He (1959c: Figure Captions page) entitled his Figure 3 "Jurassic to Recent Latitude Shift Plotted on a Continental Great-Circle"; his explanation of it is lost. However, four months later he said of a presumably similar figure at the meeting honoring Ewing for obtaining the first Vetlesen Prize:

If we were to draw a great circle on the globe which would pass through virtually the center of this great continent, it would lie along 90° E and 90° W. In the next slide (slide 3) we see a cross-section through the earth along this great circle. On this cross-section the latitude shifts as determined by paleomagnetism are shown for the interval Jurassic to Recent. In this slide we

Figure 6.3 Heezen's Figure 3 (1959c) showing northward change in latitude of North America, India, and Australia, and southward change in latitude of South America. Great Circle passes through parts of mega-continent that became North America, Asia, India, and Australia. Direction and length of arrows indicate direction and amount of latitudinal change in position of identified fragments.

are neglecting the changes in longitude and are considering only the latitude shifts indicated by paleomagnetic results. Paleomagnetic results indicate a 25° northern shift of Australia. Other parts of the great Eurasian-American continent have also shown northward changes in latitude in the same period. Measurements in Africa, which does not lie on this great circle and in Japan also show northward displacements of approximately the same value. The only continent which shows southern shifts in latitude for this period of the time is South America and this is based on but one measurement. If all these different sections of continent have been drifting northward in time, one would expect that they would collide somewhere near the North Pole and in this collision would produce folded mountains in a high northern latitude. However, a feature having these characteristics is not known.

*(Heezen, 1960a: 34)*

Heezen said nothing more about Figure 3 in his 1960 talk. The common northward drift of continental fragments would lead, he claimed, to a huge collision and "would produce folded mountains in high northern latitude"; because there are no such mountains, continental drift did not occur. This is a phantom difficulty. If these landmasses had moved northward from their past latitudes as determined paleomagnetically to their current latitudes, they would not have collided. Heezen never mentioned this phantom difficulty in any of his actual publications.

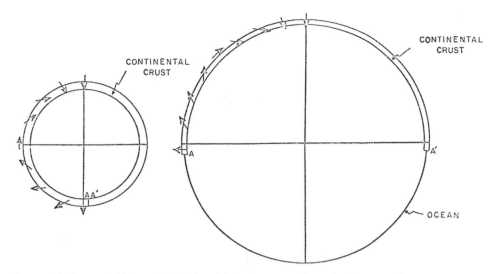

Figure 6.4 Heezen's Figure 5 (1959c) entitled "Interpretation of paleomagnetic measurements on an expanding Earth." Initially (left) Earth is completely covered by primordial continent; arrows represent direction of the geomagnetic field recorded in rocks at that time. Expansion causes formation of a rift valley at AA'. Continued Earth expansion (right) causes the rift valley to evolve into an ocean basin extending across the Southern Hemisphere. Arrows on the expanded sphere represent these initial paleomagnetic directions. It is assumed that the surface area of the original continent does not change during expansion, and the GAD holds. Paleomagnetic reconstruction would indicate that the landmass, which after expansion covers half Earth's surface, originally covered the entire surface. Thus Earth's radius has doubled.

He (1959c: Figure Captions page) entitled his next two figures (4 and 5) "Interpretation of Paleomagnetic Measurements as Evidence of Continental Drift" and "Interpretation of Paleomagnetic Measurements on an Expanding Earth." Figure 4 is lost, but his description of a similarly named slide from his March 24 talk indicates that it was taken from Deutsch (1958: 160), who gave an analysis of India's APW path in terms of continental drift. I believe Heezen wanted to pair it with his Figure 5 (my Figure 6.4), a schematic drawing of how he thought paleomagnetic results could be interpreted in terms of Earth expansion.

Delaying further discussion of Heezen's (1960a) presentation at the meeting honoring Ewing until §6.13, I want to insert here some relevant points of his *Scientific American*, October 1960, article. Heezen (1960b: 109) produced a diagram (Figure 6.5) displaying the formation and evolution of a rift valley into an ocean basin. He did not discuss the figure in the text, but from his lengthy caption (reproduced here) and Menard's (1986) discussion, he seems to have believed that the process was as follows: (1) continental region fractures because of Earth expansion; fault-blocks sink forming rift valley (top); (2) rift widens, material from upper mantle rises into the rift, and fault-blocks under extension rotate (second);

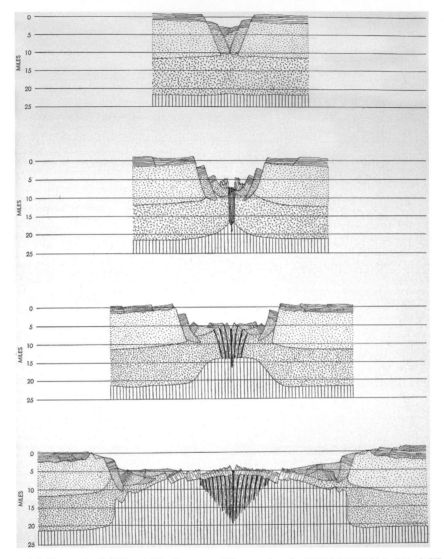

Figure 6.5 Heezen's (1960b: 109) diagram. His caption is "EVOLUTION OF OCEAN BOTTOM according to the expanding earth hypothesis is represented by these diagrams. Top layer of material is sedimentary rock of continents. Below it is continental crust. Beneath that is the type of material that makes up the crust of the oceans. Bottom layer (*vertical hatching*) is the earth's mantle. In top diagram continents are close together; rift between is just opening up. Next, material from mantle comes through rift, creating mid-ocean ridge seen in third diagram. Bottom diagram represents Atlantic Ocean bottom as it is today, with ridge and rift in center and continents at far right and left."

(3) ridge begins to form in center of expanding rift valley with dyke injection of mantle material mixed with oceanic crust (third); (4) rift valley evolves into ocean basin with formation of mid-ocean ridge, abyssal plains, and continental margins;

the mid-ocean ridge continues to grow with upwelling of more material and to elevate by isostatic adjustment; the rift valley forms in the ridge's crest above a wedge of mixed oceanic crust and mantle material, abyssal plains form as continental margins subsequently recede and collapse, and rotation of fault-blocks continues (fourth).

Heezen also lengthened his comparative assessment of continental drift and Earth expansion. He raised four difficulties against classical continental drift (RS2). They were not new, but he discussed them more fully. Here are the first three. He claimed that drifters did not have a reasonable explanation for the formation and nature of the Mid-Atlantic Ridge.

Ever since the discovery of the Mid-Atlantic Ridge, geologists have been debating theories of its origin. Probably the earliest explanation was "continental drift," which postulated that the continents were originally joined together in one, or two large masses that broke up and slowly floated apart. The familiar jig-saw-puzzle fit of the Atlantic profiles of Africa and South America was often cited in support of this idea. One school of drift theorists held that the Mid-Atlantic Ridge is a fragment of the original great continent, left behind by the edges of the new continents as they moved apart. Another group believed that the ridge was formed by a mass of sediment that filled the crack created by the breaking up of the original single continent. However, rock fragments from the slopes of the ridge, and the velocity of earthquake waves through the ridge, contradict both schools. If either of the drift explanations were correct, dredging should bring up the typical continental granitic, acidic-volcanic and sedimentary rocks. In actuality oceanographers have found serpentine, peridotite and gabbro – rocks generally thought to characterize the earth's mantle – as well as quantities of basaltic lava. As for earthquake waves, they travel through the Mid-Atlantic Ridge much too fast for it to be the abandoned fragment of a continent. The seismic-wave velocity would be still lower if the ridge were composed of sedimentary rock. Moreover, the large number of earthquakes associated with the ridge indicates that it is still an active feature of the earth's crust.

*(Heezen, 1960b: 9–10)*

Second, he (1960b: 12) repeated the by now hackneyed argument that if continents had plowed their way through the seafloor, they would have deformed it, but such "deformations are not found," nor, he added, is there evidence of "new crust forming somewhere."

If the granitic continents have been drifting through the basaltic floor of the ocean, they should be causing immense deformations in the floor. These deformations are not found … Moreover if the ocean floor were being subjected to such deformation, one would expect to find new crust forming somewhere on the floor. But virtually the only part of the ocean that is seismically active, aside from certain islands, is the mid-ocean ridge.

*(Heezen, 1960b: 108)*

Heezen had raised these two difficulties before at the Nice conference in May 1958. Both Hess's and Dietz's versions of seafloor spreading provided an answer to them, but neither had yet been published. However, there was a published fifteen-year-old version of continental drift that avoided them both: Holmes' theory of mantle convection that he presented in his *Principles of Physical Geology*. Holmes (1944) had argued that ascending convection currents bring basaltic magma up and into the gap created by the tearing apart of old crust (I, §5.8). Holmes also thought of continents as being carried on the backs of convection currents; and their movement would not deform the ocean floor. Heezen, like other Lamonters, seemed ignorant of Holmes' theory (§6.10) as did Ewing and Tolstoy when they wrote about the Mid-Atlantic Ridge (§6.4). Of course, Bucher knew about Holmes; but either he did not bother to tell Ewing and Heezen, or they did not heed what he said.

His last objection was a variation on the theme that every continent is surrounded by ridges, hence they cannot drift (Heezen, 1960b: 108). This difficulty, which Heezen and Ewing raised in 1957 with *The New York Times* reporter (§6.8), would remain a serious one for mantle convection theories of continental drift, be they Holmes', Hess's, or Dietz's. Rutland, for example, citing Heezen, raised this difficulty during discussion of Bullard's 1963 pro-drift address to the Geological Society of London (§2.15). Menard had raised it in 1958 (§5.8), and Heezen (1962) would continue to do so. Once Wilson adopted seafloor spreading, he (1961) sought to circumvent this difficulty by proposing that ridges migrate in order to circumvent the difficulty (IV, §1.9) and McKenzie, uneasy with Wilson's solution, would later develop a new solution within the framework of plate tectonics (IV, §7.6).

Heezen finally considered possible causes of expansion. Like Egyed (II, §6.4), he appealed to Dirac's (1937) idea, recently supported by Dicke (1959), that $G$ increases over time. He also mentioned, without naming him, Egyed's idea of density changes within Earth's interior. He (1960b: 110) proposed that together they "would produce a very large expansion." He estimated neither the amount nor rate of expansion. Meanwhile, also at Lamont, the Ewing brothers were arguing that the mid-ocean ridges were formed by convection in a fixist framework. Heezen correctly claimed that their solution "ignores the paleomagnetic data that indicate changes in the relative position of the continents" (1960b: 110). It is time now to tell what the Ewing brothers had to say about ridges.

### 6.12 Maurice and John Ewing's 1959 explanation of ocean ridges

On the Mid-Atlantic Ridge the sediments are underlain by two refracting layers with velocities averaging 5.6 and 7.4 km/sec respectively. The results indicated that the ridge has been built by the upwelling of great amounts of basalt magma along a tensional fracture zone. Presumably the extensional forces and the supply of basalt magma come from convection deep in the mantle. Measurements in the Norwegian and Greenland seas show results very similar to those on the Mid-Atlantic Ridge, and, from this and the extension of the belt of active seismicity, it appears that the ridge structure continues through the Norwegian and Greenland seas into the Arctic Ocean.

*(Ewing and Ewing, 1959: 291)*

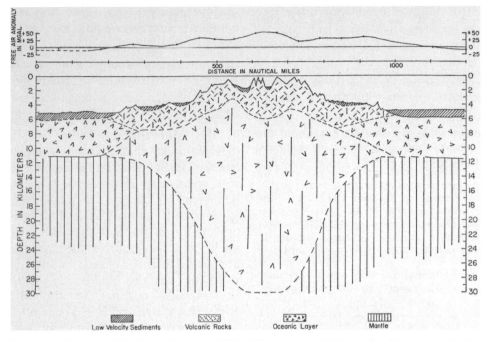

Figure 6.6 Ewing and Ewing's Figure 3 (1959: 308), captioned "Generalized Structure Section and Free-Air Anomaly Profile across the Mid-Atlantic Ridge South of the Azores."

Ewing and Heezen had championed the idea that oceanic ridges are characterized by a median rift valley, whose likely explanations were either Earth expansion or mantle convection. The Ewing brothers chose the second. Faced also with a choice between mobilism and fixism, they chose the latter.

Utilizing their own seismic data and earlier gravity studies by Vening Meinesz, they constructed a cross section of the Mid-Atlantic Ridge (Figure 6.6). Comparison with Figure 6.5 shows that Heezen and the Ewing brothers agreed about the general structure of a mature ocean ridge. The Ewings recognized an upper volcanic layer as olivine basalt.

They "tentatively identified" the root layer with the main oceanic crust, which yielded a density estimate of 3.07. Given Vening Meinesz's gravity results and the assumption of isostatic equilibrium, they estimated that the thickness of the deeper layer reached 30 km.

They dusted off Hess's 1954 solution to the origin of the Mid-Atlantic Ridge (§3.8) which they thought applied to the formation of all seismically active ocean ridges.

The formation of the ridge requires the addition of a great quantity of basalt magma and raises the question of its source. This is attributed to a rising convention current in the mantle, as described by Hess (1954, p. 345). We suggest that a convection-current system has contributed the basalt magma and has applied the extensional forces to the crust to produce the axial rift.

*(Ewing and Ewing, 1959: 307–308)*

Broadening the spatial and temporal scope, they adopted Vening Meinesz's view that the original convection system collected the continents into one hemisphere, and a secondary convection system carried continents to their present positions. This all happened, they added, "prior to the solidification of a competent oceanic crust." They were not adopting classical continental drift. They, like Vening Meinesz of 1957 and, for that matter, Jeffreys, thought that this secondary convective overturn that defined the present distribution of the continents occurred very early in Earth's history before "solidification of a competent oceanic crust." The current convective cycle rises and sinks in the same place as the one that moved continents initially, but with a solid oceanic crust, the continents remain fixed.

The rift and the earthquake activity indicate that the area is now under tension. This in turn indicates that the present convection system is such that material is rising along a belt, approximately following the axis of the north and south Atlantic oceans. The fact that the ridge follows the axis of the oceans has been discussed by numerous authors. The convection pattern envisioned here can best be related to this fact on the assumption that the positions of North America and South America relative to Europe and Africa have been determined by the current system rather than that the current system was determined by the location of the continents. To explain the relative location of the continents (with respect to the Mid-Atlantic Ridge) by this current system, we assume following Vening Meinesz (1957, p. 133) that all sialic crustal material was collected in one hemisphere by the initial current system. The second current system, whose pattern is assumed to persist to the present, contained a number of convection cells. These broke with the original continental mass into fragments which moved prior to the solidification of a competent oceanic crust, to areas of subsiding current in the present pattern.

*(Ewing and Ewing, 1959: 307–308)*

As Heezen (1960b: 110) noted a year later, they said nothing about the growing paleomagnetic evidence for continental drift.

### 6.13  Maurice Ewing's 1960 Vetlesen Prize talk: his fixist explanation of the Atlantic Basin

In 1959, Columbia University and the G. Unger Vetlesen Foundation established the biennial Vetlesen Prize "for scientific achievement resulting in a clearer understanding of the Earth, its history, or its relations to the universe." In 1960, Maurice Ewing was the first recipient. He was honored at a dinner on March 24, 1960. There were over 280 guests. Included among invited faculty and staff at Columbia were Bucher, Colbert, Dunn, Fairbridge, Heezen, Imbrie, Jardetzky, Kay, Luskin, Newell, Oliver, Tharp, Worzel, and Wüst. Inge Lehmann, who had worked with Ewing, and was at Lamont for part of 1960 was there also, as were Bullard, Hess, Press, Wilson, and Vening Meinesz.

   The festivities included an address by Ewing, given the next day at Columbia University's Men's Faculty Club. Worzel presided.[16] Ewing spoke on the "Mechanics of the mid-oceanic ridge and rift." I shall describe the responses, which, I believe, were prearranged, given by Vening Meinesz, Heezen, Wüst, Bullard, and Press in §6.14. Rhodes Fairbridge and Sidney Page also spoke; their remarks appear to have been extemporaneous, mostly directed at the respondents rather than Ewing. It is not clear if other dinner guests attended. If they did, they made no comments. Ewing's introductory remarks gave the impression that it was not a large gathering.

This talk was planned as a round table performance where I thought about a dozen of us would sit around and talk about difficult problems and I chose one that I thought was very problematic ... and now I am in the role of trying to give the answers to the problem, so you may find that the questions are raised rather sharply and answered rather poorly.

*(Ewing, 1960: 2)*

Ewing presented the evidence for the existence of a rift valley, his solution to the formation of mid-ocean ridges, and his defense of full mantle convection. Heezen could equally well have given the first part of the talk. Ewing documented the central rift valley along the ridges of the Atlantic and Indian oceans, discussed their association with earthquakes, and similarity with the East African rift valleys. He then turned to the Pacific, the questionable ocean, where Scripps geologists had shown that parts of its seismically active ridge system had no central rift valley (§6.7). Using as his guide the belt of shallow earthquakes associated with the valley, he looked through the literature for profiles that crossed it, and displayed them in an unnumbered figure entitled "Profiles across the rift belt: Pacific" (Figure 6.7). Ewing wanted to show that "mid-ocean" ridges of the Pacific have a central valley that followed the belt of shallow earthquakes and he found profiles of ridges with the telltale notch. Profiles I and II cross Menard's Ridge and Trough Province, which contains what became recognized as the Juan de Fuca and Gorda ridges.

Turning to oceanic trenches (and island arcs), Ewing did not claim that all trenches were extensional as Heezen had done. He (1960: 9) distinguished between the shallow earthquakes associated with mid-ocean ridges and "the circum-Pacific belt of shallow, deep, and intermediate shocks, which are usually associated with trench and island arc type of structure," and he (1960: 10) suggested that the circum-Pacific belt of earthquakes was compressional. He (1960: 10) also noted that compressional and tensional seismic belts "avoid each other properly except in that questionable part of western North America." He (Figure 6.7) thought the Pacific ridge system passed through Menard's Ridge and Trough Province and continued southward along the west coast of California and Baja California, but the presence of deep and intermediate earthquakes also indicated that the area was part of the circum-Pacific belt of earthquakes. At the time, the question of what happened to the East Pacific Rise in the western United States was unanswered. Menard (1958a) initially claimed that the East Pacific Rise ends abruptly off the western coast of Mexico (§5.7), but changed his mind two years later and argued that it passes under western California and reappears off Oregon and Washington in the Ridge and Trough Province (§5.11).

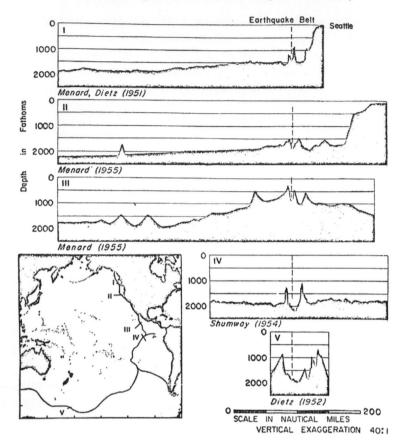

Figure 6.7  Ewing's non-numbered figure (1960: between pp. 9 and 10) entitled "Profiles across the rift belt; Pacific." Profile I (top) is from Menard and Dietz (1951: opposite p. 1628). Menard and Dietz's published profile stopped partway up the continental slope; Ewing extended and reversed it. Profile II is from Menard's (1955a: 1187) Figure 27, "A profile through the Ridge and Trough Province at 42° N Latitude, which cuts across the northern part of the Gorda Ridge." I am unable to identify the source of Profile III. It does not appear to be in either Menard (1955a) or (1955b), his two single-authored 1955 papers. Profile IV is after Shumway's (1954) Figure 7, "Profile across Cocos Ridge," which Ewing shortened and reversed. I cannot find the source of Profile 5; it is not from Dietz (1952a) or (1952b). Ewing indicated the position of the continuous system of Pacific oceanic ridges in his diagram of the Pacific Ocean.


Ewing now introduced mantle convection. He again aligned himself with Vening Meinesz, who spoke next, but dropped any reference to Hess.

The general pattern of alternating tensional and compressional belts is strong evidence in favor of a system of convection currents in the mantle, which has been discussed very ably by Professor Vening-Meinesz.

*(Ewing, 1960: 11)*

Following Vening Meinesz, he proposed that the first convective cycle, which occurred before Earth had a core, collected the lighter continental material into a super-continent. The next convective cycle, smaller than the first, began after the core formed and "temperature differences" in the cooling Earth "built up again."

Convection resumes at this time in smaller scale because of the interference of the core with rising currents in some regions and sinking currents in others, so that this primitive continent would be broken apart to form the several continents more or less as we see them now . . .

*(Ewing, 1960: 11–12)*

He also claimed that there is no "permanent record of" such "a convective pattern," presumably because oceanic crust had not then solidified, allowing the continents to move. He (1960: 12) argued (RS1) that his "suggestion" explained "the continental distribution," "the fact that the continents are generally judged by geologists to be in a state of . . . lateral compression," "the bringing of heat by the rising convection current to the mid-ocean areas," the "bringing [of] basaltic rocks to the same areas," "the continuity of the ridge system," and "the characteristic of the ridge of following more or less midway between the continental masses." He admitted, however, a difficulty with the last point, the median position of the ridges.

It is necessary, if you want to make that claim, to make an exception for the Central and North Pacific Ocean, and, as is customary, we simply say, "this is an odd place. We don't understand it. Too bad it is this way."

*(Ewing, 1960: 12)*

Ewing, as he and his brother had done two years before, showed a cross section of the Mid-Atlantic Ridge (Figure 6.6), and proposed that his version of mantle convection explained the wedge of rock beneath ridges whose low maximum seismic velocities suggested it was a combination of oceanic crust and mantle. He did not discuss Menard's 1958 objection that Ewings' model for origin of ocean ridges could not explain the formation of ridges like the East Pacific Rise that lacked the mass of crust–mantle rock found beneath the Mid-Atlantic Ridge (§5.7).

Ewing devoted the remainder of his talk to removing what he viewed as the chief difficulty to his mantle convection: discontinuities in the mantle (I, §5.10). Hess had also discussed the difficulty in 1950 when arguing for mantle convection (§3.8). Ewing did not mention paleomagnetism and its support for mobilism.

## 6.14 Responses to Ewing's Vetlesen Prize talk

I consider the prepared remarks by Vening Meinesz, Wüst, Bullard, and Press, and then those of Heezen and Fairbridge but not Page.

Vening Meinesz (1960) added nothing new to his earlier (§2.5) theory of mantle convection in which he incorporated Prey's study of surface elevations. He praised Ewing and his Lamont colleagues for their work on the seafloor, and, as might be expected, expressed general agreement with Ewing's solution to the origin and development of the Mid-Atlantic Ridge. Like Ewing, Vening Meinesz said nothing about paleomagnetism. It would be two years before he would adopt mobilism (§2.6).

Wüst, as already noted, acknowledged that he and other scientists on the *Meteor* Expedition had not recognized the central valley of the Mid-Atlantic Ridge (§6.7). He (1960: 37) graciously added:

We have not had the courage, the fantasy, the needed fantasy, and the physical argument, and we have not had a Maurice Ewing in the old Europe to find this convincing solution of one of the most important features of the ocean bottom.

*(Wüst, 1960: 37)*

Wüst wanted to know what counted as the median ridge in the Indian Ocean. He (1960: 37) also asked if the Mid-Atlantic Ridge is "crossed by deep trenches, as I have read in a paper of Tolstoy's some years ago?" The largest of these was an east–west trending trough at approximately 30° N (Tolstoy and Ewing, 1949; Tolstoy, 1951), later recognized as the Atlantis Fracture Zone (§6.4; IV, §5.7).

Press preferred Benioff's (1959, 1962) idea that the Pacific Basin has rotated in a counterclockwise direction relative to the surrounding continents to Ewing's mantle convection. Benioff, one of Press's colleagues at the Seismological Laboratory, Caltech, first proposed the idea in 1959 and returned to it in 1962 (§1.6). Press argued, as Benioff had done, in its favor because first motion studies of shallow circum-Pacific earthquakes overwhelmingly associated them with parallel, dextral strike-slip faults. Press (1960: 41) appealed to work by K. Aki (1960), who, working on the problem of "distinguishing earthquakes from nuclear explosions," and using surface waves generated by earthquakes to determine their mechanism "more or less confirmed the picture that Professor Benioff had of a rotating Pacific Ocean basin compared to the continental masses which surround it." Press (1960: 41–42) asked if there were a way to reconcile a rotating Pacific Basin with mantle convection.

Bullard, who had become sympathetic to mantle convection as a solution to the origin of mid-ocean ridges (§2.12), who at the time of Ewing's talk was no longer adverse to mobilism, and who, within three years, would openly support mobilism (§2.14), questioned Ewing about mantle convection. He raised four issues, which he probably had been asking himself for some time. He first queried Ewing on when the fragmentation of his primordial continent had occurred, and raised a difficulty with either an early or late breakup and suggested a way to decide.

In particular I notice that Professor Ewing carefully avoided any reference to the time scale of this operation and it does seem to me that it is extremely odd if this original world continent or whatever you call it waited there until the Lower Cretaceous perfectly happy and then suddenly at this later stage split itself up. On the other hand, there do seem to be many lines of evidence tending to suggest that the Atlantic Ocean isn't much older than that and this is obviously one of the really critical things to determine. Probably the most certain way of doing it is to drill right through the sediments and get a complete section and see what range of age there is there.

*(Bullard, 1960: 38)*

Bullard did not specify "the many lines of evidence tending to suggest that the Atlantic Ocean isn't much older than" Late Cretaceous. But, as Bullard quite reasonably emphasized, the idea that the "world continent or whatever you call it" did not break up until Early Late Cretaceous was "extremely odd." There was, however, a way out – propose that continental drift had begun early in Earth's history.

Bullard then turned to Ewing's argument that discontinuities in the mantle were not chemical but phase changes, which would not necessarily prohibit large-scale convection; convection through them still might not occur.

Then there is the question as to whether there is an inconsistency between the idea that Professor Ewing first pointed out, that his rather rapid increase in seismic velocity down to 900 kilometers indicates inhomogeneity and then saying that naturally it is all stirred up. This looks like a contradiction but, of course, it is not necessarily so because the inhomogeneity may not be chemical. It may be an inhomogeneity of phase change, but if it is a phase change, if we are to get a convection current, it will have to be very accurately reversible and I don't think we quite know how closely reversible it has got to be. One of the surprising things is, of course, the stability of these high-pressure phases, but I imagine they would probably reverse at these high temperatures.

*(Bullard, 1960: 38–39)*

Bullard questioned what should properly be inferred from his own heat-flow measurements on the East Pacific Rise and the Mid-Atlantic Ridge (§2.12). He cautioned putting too much stock in the measurements, and with a good dose of self-irony argued that they certainly did not prove convection.

Then I wouldn't like to hang too much on the high heat flows in the valley of the middle of the Mid-Atlantic Ridge. So far as I know we have the information collected by Scripps in the Pacific, which is on this rather atypical ridge. There is one measurement we have made in the Atlantic and I only hope you have some more. This one that we have got is bang in the middle of the valley and we have another one further out – still on the ridge but not in the valley – and this one, the one in the middle of the valley gives about seven microcalories per square centimeter per second and the one a little further out gives I can't remember what, I think about two, so that if this is the usual geophysical way of using one measurement and extending it to cover the whole Earth, it does suggest that heat flow, the anomaly in the heat flow, is rather narrowly confined to the crack. Well, this, I suppose means that actually where the crack takes place is where hot rock manages to get up but what we are observing really is just the volcanic heat from some intrusion or other and what we are really demonstrating is that in the Mid-Atlantic Ridge, in this crack, there is a fairly recent intrusion. You can't necessarily draw wider conclusions from this. We know this already, of course, because you can see it in Iceland.[17]

*(Bullard, 1960: 39)*

He concluded by remarking that we "must have a very careful look at" the Central Valley of Iceland.

Because it is alleged in the books that in the Central Valley of Iceland lavas come out under ice, whereas to each side the lavas are considerably older ... I think we may have in Iceland very direct evidence of what is happening in this Central Valley.

*(Bullard, 1960: 40)*

Good advice indeed. Lamont scientists eventually undertook an airborne magnetic survey of the Reykjanes Ridge just south of the Central Valley of Iceland, which was of critical importance for the conformation of seafloor spreading (IV, §5.4). Of course, this was not likely the kind of study Bullard had in mind.

Ewing, pleading, "I don't think my notes are good enough that I can recall all the questions," replied only to Press's question about reconciling mantle convection with a rotation of the Pacific Basin. He claimed (1960: 53) that Press had a similar problem reconciling the evidence for the circum-Pacific compression zone with a rotation of the Pacific Basin; rhetorically he asked Press, "What kind of action do you think will produce a rotation in the Pacific Basin?"

Heezen spoke next. Anticipating the unpopularity of what he was about to say, he attempted appeasement:

I suppose it is rather foolhardy to speak out in favor of a radically different hypothesis after hearing such eminent earth-scientists as the Vetlesen Prize winner, Dr. Ewing, and Professor Vening Meinesz so strongly support the convection current hypothesis for the origin of the mid-ocean ridge and rift. But my own studies, extending over a period of twelve years, have led me to radically different conclusions as to the origin and evolution of the mid-ocean ridge.

*(Heezen, 1960a: 30)*

After his talk, and unlike any other at the meeting, Heezen received no applause. His attempt to appease had failed; it was foolhardy and singularly inappropriate of him to attempt to promote Earth expansion at Ewing's celebratory meeting.

Heezen's remarks were much like those three months earlier (§6.11). He argued that continental rift valleys evolve into ocean basins. He outlined the paleomagnetic evidence, emphasizing the very different Late Paleozoic APW paths of Europe, North America, India, Australia, and Japan, claiming that they, and the paleoclimatic evidence, supported Earth expansion and continental drift. Indirectly attacking the fixist standpoint of Ewing and Vening Meinesz, he also noted correctly that paleomagnetism "bears very importantly on the evolution of the oceans," although it "hasn't been discussed this afternoon." Heezen said nothing about his and his colleague's extensional interpretations of both continental margins and ocean trenches.

Rhodes Fairbridge followed Press. Fairbridge, originally from Western Australia, received his undergraduate degree from Queens University, Canada, his master's from Oxford and (1944) a D.Sc. from the University of Western Australia, where he taught for many years. He moved to Columbia University in 1954, where he taught until retirement in 1982. He devoted most of his career to climate change, and the effect of solar variability on Earth's climate (Mackey, 2007). He (1956) publicly expressed his sympathy for continental drift, presenting a talk just in time for the IGY at the 37th Annual Meeting of the AGU in Washington, DC.

On this occasion and no doubt speaking in his usual flamboyant style, Fairbridge began by thanking Ewing for a letter he had sent after hiring him.

Dr. Worzel, Dr. Ewing: I should like to say straight away that some people wonder why I, as an Australian, like to live here in New York. About five years ago when I had a long distance phone call, would I like to come to Columbia, I said, "Nothing would give me greater pleasure," and five years later I do not take that back. The first letter I got after my acceptance of the appointment came from Maurice Ewing and I don't think any letter that I have ever received in my life gave me more pleasure.

*(Fairbridge, 1960: 43)*

But he disagreed with Ewing's fixism and his appeal to mantle convection. He, like Bullard, had misgivings about large-scale mantle convection.

I think that Sir Edward Bullard put his finger on one of the tricky things that both Dr. Ewing and Dr. Vening Meinesz mentioned today, the convection cells. They are like George Orwell's *Animal Farm* citizens. Some of the cells are a little more equal than others.

*(Fairbridge, 1960: 45)*

He agreed with Ewing and Heezen that the Mid-Atlantic Basin was symmetrical and bisected by the Mid-Atlantic Ridge, but argued that the symmetry of the Indian Basin was not nearly as clear as that of the Atlantic, and that the Pacific Basin was not only asymmetric but lacked an active mid-ocean ridge. He (1960: 45) also agreed with Bullard that the timing of the breakup of the mega-continent "is critical." But he then parted company with him; he agreed with Wegener that the continents had separated at different times during the Mesozoic, and closed his address (1960: 45) in his showy style, "I am prepared rather to agree with Bruce Heezen's ideas on this subject. I prefer a swelling." Unlike Heezen, Fairbridge was applauded.

Fairbridge (1960: 47–48), unable to contain himself and wanting perhaps to disassociate himself from Heezen, took issue with him. After Fairbridge had announced that he favored Earth expansion, Heezen was asked about its cause. He remarked, "I personally have no strong feelings concerning the specific mechanism of expansion," and, "I simply conclude from the morphological and paleomagnetic results that expansion has occurred." Heezen also appealed to Dirac's and Dicke's idea that "the needed expansion of the Earth would be accomplished partly by an overall effect of a decrease in gravity," (1960a: 47) and mentioned Egyed's idea "that material along the boundaries of the core and inner core are changing through a gradual decrease for the denser to the less dense state with a consequent increase in volume." But Fairbridge had the last word, making sure that the audience understood that he preferred Egyed over Heezen.

Just a brief word. I feel that Dr. Heezen hasn't done full justice to Egyed's theory. Egyed didn't approach this problem from the idea of the appearance of the external crust at all. He worked it all out from the interior and he deduced that this is a necessary condition. I felt that this added very greatly to the validity of his argument, that he wasn't thinking of the superficial features at all. It wasn't that he was looking for a cause for expansion. He merely looked to the thermodynamic history of the earth and said "And therefore it must expand."

*(Fairbridge, 1960: 47)*

## 6.15  Heezen's continued defense and later abandonment of Earth expansion, 1960–1966

During the early 1960s, Heezen defended Earth expansion twice more. He first summarized his views at the Helsinki, 1960 IUGG meeting (Heezen, 1960c: 15). His next defense of expansion, like his previous ones, still did not appear in a refereed journal but in Runcorn's edited 1962 *Continental Drift* (§1.6). In it, he again argued that ocean basins evolve from rift valleys, and he reintroduced his extensional analysis of continental margins that he had deleted from his presentation at the Vetlesen meeting honoring Ewing. He repeated his attack on conventional continental drift and extended it to seafloor spreading. There was only one substantial change in his defense; he excised his appeal to paleomagnetism, saying nothing about how he thought Earth expansion could explain the APW paths from different continents and accommodate other versions of mobilism. His attack on seafloor spreading I describe in IV, §1.13; I now want to explain why he expunged paleomagnetism from his defense.

In his 1962 article, Heezen waited until the penultimate paragraph before mentioning paleomagnetism, and here it becomes apparent why paleomagnetism no longer constituted a major part of his defense.

Cox and Doell [1961a], comparing paleomagnetic results from Europe and Siberia, concluded that the Permian radius of the earth was nearly identical with the present radius. But similar calculations based on data from other parts of Eurasia would clearly give radically different values.

*(Heezen, 1962: 285; my bracketed addition)*

Cox and Doell, using Egyed's paleomagnetic method for testing Earth expansion, had found that Carey's and Heezen's rapid Earth expansion of a 45% increase in Earth's radius since the Paleozoic was unlikely (II, §8.14); they also argued that Egyed's slow expansion was not ruled out by their test. Now Heezen had not specified a rate of expansion since the Paleozoic, but he had claimed that his version of expansion accounted for the divergent APW paths, which, as Deutsch (1963b: 12) later estimated, "cannot be much less than 50%." One might think that Heezen dropped his appeal to paleomagnetism because he had no convincing reply to Cox and Doell's critique.

This was not really true, but his reply was hardly convincing. Heezen did raise a difficulty against Cox and Doell, noting, "similar calculations based on data from other parts of Eurasia would clearly give radically different values" (RS2). So why did he omit what he had previously claimed to be paleomagnetic support? Perhaps he recognized the weakness of the difficulty he had raised. Cox and Doell had chosen Permian sites separated by regions thought by them to be tectonically inactive *since* the Permian, and their test depended on them being able to make such a felicitous choice. Heezen was correct in saying that results would be very different if they had been based on data from sites separated by a tectonically active region. But Heezen did not show that, and indeed they had not unwisely selected sites that were separated by a tectonically unstable region. Indeed, very promptly, Carey (1961) attempted to do just that, and Cox and Doell (1961b) successfully defended their case (II, §8.14). Their papers were published in *Nature* in May, giving Heezen ample time before submitting his paper, given that he did discuss both Dietz's paper on seafloor spreading, which appeared a month later in *Nature*, and an October discussion, also in *Nature*, amongst Bernal, Dietz, and Wilson on seafloor spreading (IV, §1.8, §1.9). I suspect that Heezen saw the Carey–Cox–Doell exchange, and chose to excise paleomagnetism from his defense of Earth expansion. So Heezen, like Menard (§5.12), enthusiastically appealed to paleomagnetism when he thought it supported his position, but disregarded it when it was shown not to do so.

Heezen now seemed less confident in his case for Earth expansion, although he still thought it worthy of serious attention.

It is just possible that the evidence for continental drift could instead be interpreted as evidence for continental displacement (without drift) due to the internal expansion of the earth accompanied by the growth of simatic oceans through emplacement of mantle material in the floors of the mid-oceanic rifts.

*(Heezen, 1962: 285)*

Looking ahead, in 1963, rapid Earth expansion suffered a serious setback through the global analysis of the Late Paleozoic and Mesozoic paleomagnetic data by Ward (IV, §3.10): Cox and Doell's analysis was of Permian data from northern Eurasia, the only data available to them at the time. Heezen abandoned rapid Earth expansion in 1966. With Fox as a co-author, he wrote an article entitled "Mid-Ocean Ridge" for an encyclopedia of oceanography in which there was no mention of his previous support of rapid Earth expansion (Heezen and Fox, 1966); they characterized mobilism's paleomagnetic support as "convincing." Readers of the article who were unaware of Heezen's previous fascination with rapid expansion would think that his earlier views were much like those of Dietz or Runcorn and opposed to those of the Ewing brothers.

As an outgrowth of the previously mentioned facts, most workers believe that the ridge is a result of tension and the material rising from the mantle beneath the crest of the mid-oceanic ridge is adding new rock to the floor of the rift valley. Some workers have interpreted this tension in terms of continental displacement (Heezen, 1959; Dietz, 1962; Runcorn, 1962) while others, although accepting the evidence of recent tectonic activity on the mid-oceanic ridge, prefer explanations involving permanence of the continents and ocean basins (Ewing, Ewing and Talwani, 1964). Largely due to the recent and convincing paleomagnetic evidence of large continental displacements, most workers prefer explanations featuring a type of continental drift, a gradual widening of the Atlantic, Indian, and South Pacific Oceans. Many workers thus believe that as the continents moved apart, intrusions beneath the crest of the gradually widening mid-oceanic ridge created most or all of the oceanic crust (Heezen, 1959; Dietz, 1962).

> *(Heezen and Fox, 1966: 515; their references are respectively my Heezen, 1959a;*
> *Dietz, 1962a; Runcorn, 1962; Ewing, Ewing, and Talwani, 1964)*

Heezen did not mention Hess's seafloor spreading (§3.14), and did not hold him in high regard. He also made sure that readers knew that the Ewing brothers continued to oppose mobilism in 1964. By the time Heezen had co-authored this encyclopedia entry, he and Maurice Ewing had, sadly, become estranged.

### 6.16  The Ewing–Heezen rift

For over a decade, Ewing and Heezen had a wonderful working relationship. They worked together on forty-four publications "in one of the best known and most widely cited collaborations in modern geology" (Menard, 1986: 100). They concluded their last joint publication (written in 1962) by noting "there are two schools of thought" concerning "upflow of mantle rock" in the center of oceanic rift valleys.

Most current workers believe that the mid-oceanic ridge is the result of tension, and that material rising from the mantle beneath the crest of the mid-oceanic ridge is adding new rock to the floor of the rift valley. However, there are two schools of thought concerning the reason for this upflow of mantle rock. The one favored by most workers is the convection-current hypothesis, which attributes the fundamental features of the ocean floor to the effect of convection currents which rise beneath the crest of the mid-oceanic ridge and then flow laterally toward each continent. It is supposed that the drag of these currents on the underside of the continents causes the continents to be compressed and the oceans to be stretched. The second school attributes the stretching of the ocean floors to a general expansion of the interior of the earth.

> *(Heezen and Ewing, 1963: 410)*

Both remained at Lamont until 1972, but never published together again. The rift was not caused by their scientific disagreements over the Earth expansion versus mantle convection or mobilism versus fixism. Menard (1986: 106) recalled, "Bruce complained that Ewing insisted on putting his name on reports to which he had made no contribution despite the fact that he did the same thing himself." Heezen and Tharp said that the rift began when they told Ewing that they were not going to put his name on their 1959 physiographic map of the North Atlantic. They included Ewing as an author in the accompanying text (Heezen, Tharp, and Ewing, 1959), but felt he had not contributed to the actual map. Heezen did not want to split with Ewing, and probably did not realize that omitting Ewing's name would end their working relationship.

I didn't really want to split from Ewing and not have any work with him. I thought that he might accept the idea that Marie and I would do this – these diagrams under our authorship and I could still work with him and ... co-author other papers with him. He was so upset by this that he didn't want to work with me on other papers.

*(Lear, BCH 08B: 10)*

Menard correctly added, "The split would lead to virtual banishment in 1966."

Ewing did not allow him on the *Vema* after Cruise 18 in 1962 (Tharp, May 29, 1986 letter to author). Heezen left the ship in Tahiti, and "disappeared for three weeks ... Ewing was furious" (Menard, 1986: 200). Heezen was not granted tenure and promotion to associate professor until 1964. Menard and J. Tuzo Wilson let Ewing know that Heezen merited both. Menard recalled:

In some important ways, 1964 was a good year for Bruce. He was awarded the Henry Bryant Bigelow Medal by Woods Hole Oceanographic Institution for his work on the physiography of the sea floor. Considering his international fame and this award, his status as an untenured Assistant Professor seemed highly anomalous. I so informed Maurice Ewing in a bit of unsolicited advice as we were strolling the streets in Washington. Tuzo Wilson gave the same opinion when queried by Ewing. In any event Bruce was promoted to an Associate Professor with tenure in 1964. He came home and told Manik Talwani that he would never again allow Ewing to put his name on one of his (Bruce's) manuscripts.

*(Menard, 1986: 200)*

Ewing found ways to make Heezen and Tharp's lives miserable. He cut Heezen off from Lamont's data. Even though he and Tharp found ways to see the data, Heezen could not include any in his publications (Menard, 1986: 201). Heezen was suspended in 1966, but reinstated in 1970. Ewing fired Tharp in winter 1967–8 (Tharp, "Outline of The Cradle of the Oceans"). After Ewing was forced to resign the directorship of what had become the Lamont-Doherty Geological Observatory and left, Talwani became director and gave Heezen ship time and access to the data. Even then, Heezen, refusing to acknowledge that protocols at Lamont-Doherty had changed since he had been chief scientist aboard Lamont-controlled ships, overstepped his bounds, courting further trouble.

### 6.17 Heezen's stress on collecting one's own data and resentment of Hess

Hess thought that collecting data was not an end in itself and thought that hypotheses should be used as a guide to decide what data to collect. He devoted much of his career to developing solutions to big questions about the seafloor (§3.2, §3.21). Except for his early work on trenches and guyots, his speculations were based on interpretations of data collected by others. Ewing emphasized collecting data over making grand hypotheses (§6.2). Heezen proposed grand hypotheses, but collected his own data, and believed that scientists should hypothesize only about data they had in part at least helped collect. He also said that Ewing felt the same way. Heezen had no respect for scientists who make a habit of explaining data collected by others, and that included Hess, Dietz, and Wilson.

Ewing and I are similar in one regard, perhaps many, but in this regard that we don't normally go out and talk in length about something we've never worked on. We don't discuss data at length which we've not collected or never seen. We like to work with information we obtain, we fully understand and that we have ... and also some data we could take some credit for having obtained. There is a big contrast between the feeling that Ewing and I had and those of people like, for instance, Hess, Dietz and Tuzo Wilson who like to sit in a library and read other people's papers, who rarely go to the field or ever look at any data, or obtain it. They like to synthesize other people's information. It is an important function but it is considered, it was considered by Ewing and considered by me something like of "not first rate" because, and in a way, it leads, very often to cheating. Because they tend to try to claim things that you did and you didn't understand and they only understood it, and people liked these pat solutions they came up with and give them credit for things they didn't even think of. They were probably basically too simple minded anyway. Anyway, this was a feeling he [Ewing] had and a feeling I got from him ... you should go out and do it, do the work, make a contribution, if you get the contribution then you have a right to pontificate about it a little bit. And if you ... haven't collected it, then you don't understand it, then ... your ... stuff can be classified as garbage and nobody has to pay attention to it.

(Lear, BCH 09A: 5–6)

This description of his (and Ewing's) attitude toward those who synthesize data of others probably is accurate. He resented the success that Hess, Dietz, and Wilson had achieved in explaining data collected by others, including his own.

Heezen's bitterness toward Hess appears to date from 1959 when he submitted a paper on seamounts that Hess rejected. According to Heezen, Hess was tired of Lamont papers that were heavy on data and light on interpretation.

[The paper] was sent to Harry Hess for review. And Harry wrote a one page review of it which hardly mentioned the paper at all but was a tirade against Ewing and his, the objectives of Lamont. He thought much more time should be spent sitting in the library, much less time spent out in the field making measurements so that the measurements could be better, more precisely located.

(Lear, BCH 09b: 7–8)

Heezen believed that the disagreement between Hess and Ewing was "not entirely divorced from people's green eyes ... of the Harry Hesses who didn't have any money being jealous of the Ewings who did" (Lear, BCH 09: 8). Heezen said that he had been surprised by Hess's review because they previously had enjoyed "an open and friendly relationship" (Lear, BCH 09: 8).

Heezen resented Hess deeply. In January 1966, Heezen and Hess were together aboard Scripps' ship R/V *Thomas Washington* investigating the delta area of

Colombia's Magdalena River. Francis Shepard was also aboard. Heezen wrote Tharp the note below dated January 26, 1966.

This has been an extraordinary but frustrating cruise. *The Great Oceanographer* HESS was aboard with about 8 students and one Asst. Prof. almost nothing done. He who criticized our methods made a 100% flop of his 6 days getting almost nothing done. He is revealed for the Bull Shit Artist he is. Fran [Shepard] is frantic and is buzzing about with dive ... and Navigator Plot but totally is ... of the results of Seismic Reflector. Dill is here and is a great boon to all.[18]

Not one to let things go, at least with regard to Hess, Heezen, in his review of Hallam's 1973 *A Revolution in the Earth Sciences*, called Hess a "landlubber" and tried to set the record straight about Hess and Lamont.

He [Hallam] takes some pains to point out that Holmes proposed a convection-current model of plate tectonics in 1929. Yet, curiously, he later attributes the convection-current model to Hess and gives 1962 as the date, noting that Dietz, who proposed his similar model in 1961, got the idea from a preprint Hess had distributed in late 1960. But from whom had the then-landlubber Princetonian geologist gotten it? Was it from his seagoing rivals at nearby Columbia University, who he often complained were so busy making discoveries that they took insufficient time to publish their findings or to contemplate the ramifications of their observations? An armchair is undoubtedly a great place to sort out thoughts; it is also without doubt that someone has to go fetch the data.

*(Heezen, 1974: 505; my bracketed addition)*

An armchair is also a great place to harbor disappointments and build up ill feelings toward others.

## 6.18  The effect of the Ewing–Heezen split on Heezen

In his interviews with John Lear in the 1970s, Heezen confessed that he had supported Earth expansion longer than was warranted by the evidence. He offered two reasons, both unflattering to himself. The first stems from his stubborn reluctance to admit that his interpretation of trenches as extensional was wrong. He claimed that he, Ewing, and Worzel believed that they would be more assured of obtaining further funding for seafloor exploration if they rejected trenches as compressional features, as advocated by Vening Meinesz and Hess. To accept trenches as compressional would mean that they were admitting that they were wrong and Hess was right. The second arises from his yearning for fame. Reflecting on how he had become known nationally and internationally by promoting the idea that turbidity currents supplied most of the sediment covering abyssal plains, he had reasoned that the best way to achieve fame is to promote a novel and initially unpopular view that later becomes established. Realizing the unpopularity of Earth expansion, he gambled that it would eventually be accepted. Heezen added that he may also have supported Earth expansion because Ewing rejected it.

Heezen suggested the first of the above in explaining why he did not propose subduction at trenches once he had adopted creation of seafloor at ridges. He began by describing Vening Meinesz and Kuenen's downbuckles, seeming to think them equivalent to subduction. He mentioned Hess's adoption of the downbuckle, naming it "tectogene."

OK. Now, why didn't we immediately come to the view that the earth was, in fact, subducting crust as fast as it was creating at the ridge? Well, there are several aspects to this. For instance, if one did come to this conclusion we had to abandon a position that we had taken on trenches for some time. In the 20's and 30's the tectogene idea had come about which was that the trenches consist of a large crustal down buckle in which the crust of the earth was shoved down some hundred kilometers or so. This would accommodate, of course, about a few hundred kilometers shortening of the crust in the vicinity and would give a light major material at depth where heavy should be which would account for the negative gravity anomaly which Vening Meinesz had discovered over the trenches. This theory was championed by Vening Meinesz and by Kuenen who made simple experiments of wax floating on warm water, on lukewarm water, which showed that, in fact, that in isostatic balance thin-sheets would, in fact, bend in the way that Vening Meinesz supposed. It was also supported, to some extent, by Harry Hess who adopted some of the aspects of the idea, gave the word "tectogene" wide distribution around, usage around the world and applied to it some ideas he had about the distribution of serpentines ...

*(Lear, BCH 06A: 1–2)*

Then there was the need for Lamont researchers to present a novel view in order to secure research funds.

Now ... the initial publications in the world in the field of gravity, interpretations of gravity and sea data, were made by Vening Meinesz and Kuenen. Kuenen was not a physicist; he made some rather geological deductions, and he made some models, some studies. Ewing then came along in the field to do some more work and after World War II he and his group became the predominant one in the world working in gravity at sea largely due to his great persistence and availability, for a while after the war, of large numbers of U.S. submarines ... So, now if you are in this position [of becoming the preeminent group doing gravity measurements at sea, and having learned from Vening Meinesz, who you respect enormously] what are you going to do? Well, are you going to come up and say, "Hey, what Vening Meinesz discovered in 1926 had not been surpassed, everything we've done since the war, and we have now taken four times as many measurements as the old man made during all of his primary work, and we have been unable to come up with any interpretation or anything else that was any more different that what he says, so what?" One doesn't do that? That's not designed to promote new research and new monies, new funding. One has to be in a position of upsetting something or revising it or coming to a new view. One has to, even if one isn't completely convinced that the other guy is wrong, if there, about all one has to do, has left, is to act as if he's wrong and give him a good test. So, part of the view that trenches were tensional rather than compression would have been ... a political position which Lamont would have to have in any case. Because otherwise one would have to say, "Oh, we're only just lacing up and polishing up what all these guys had said."

*(Lear, BCH 06A: 2–3)*

Hess's support of the tectogene and mantle convection was another reason for Heezen to reject them.

However, politically, Hess who was often called a marine geologist; he wasn't. Often called a geophysicist; he wasn't. He was a pretty good petrologist. By that I mean a man who knew his mineralogy and minerals and therefore, used them to try to interpret the origin of rocks by looking at rocks through a microscope ... Anyway ... the play of politics was complicated a little bit by the fact that Hess had been on one of the gravity expeditions more or less as a fifth wheel but he had written up some discussion of it and he had chosen to champion the tectogene idea. Therefore there was a little bit more than just a neutral feeling on it. Vening Meinesz was a greatly admired, loved man by Ewing and Worzel ... They were exceedingly kind and cautious in demolishing [the tectogene] so as not to discourage or in any way make the old man think that they were after him. Because they highly respected him and didn't want to discredit the idea in a way that would look like they were being mean.

*(Lear, BCH 06A: 3)*

But Heezen did not mean to imply that their rejection of Vening Meinesz's and Hess's views and adoption of their own was not without evidential support.

Well, anyway, what I'm trying to get around is that if, no matter who had been involved you may have been in a position at Lamont to see another view. So, in looking for another view on the trench, the only trench we had to look at in the beginning was the Puerto Rico Trench, it being near by ... It's a little bit different than some of the Pacific trenches because it is quite a bit wider, the bottom is almost 15 miles wide on the flat floor where most of the Pacific trenches are hardly more than 3 or 4 miles wide. It has a lot of sediment on the bottom, almost a mile of it. And it was quite obvious early on that a large part of the gravity anomaly in the trench could be accounted for by the large filling of mud. And that therefore turbidity currents, which were a subject we had championed, were brought into use as an explanation of the big gravity anomaly in the Puerto Rico Trench. And by implication perhaps that meant that the Pacific trenches would have an explanation, a similar explanation. Therefore, not having any necessity of having a downbuckle under the trench, simply crack the crust open, fill it up with mud and ... have a viscosity of the mantle sufficiently high that the mantle doesn't flow up in the bottom and therefore you have a gravity anomaly and you have a trench. So, this explanation seemed rather neat and Worzel and Ewing wrote papers on it. I joined in one paper in '54 which used this interpretation and we were fairly happy with it. It may not be the right answer. It may not have been the right answer. But it fit the facts certainly as well and probably better than the tectogene. Now the tectogene went out for several other reasons. They had determined the crust of the oceans was 20 kilometers thick. They went and measured, and it was 5. So, when you buckle down a 20 km crust you got a fair amount of low density stuff into the crust. When you buckle down a 5 km thick crust, it was a lot less. You had to buckle a lot more down to do it. And the models of the Earth's crust were so much in subject of revision at that time that the specific examples quoted by the tectogene people became so anachronous to what was known from the crustal structure after the War that the whole thing seemed do just sort of pass away ... So tectogene sort of went into limbo.

*(Lear, BCH 06A: 3–4)*

Heezen found it very difficult to admit at the time that he had been mistaken.

From his own account, it appears that Heezen incorrectly believed that he could not accept subduction without admitting he had been wrong about the tectogene. But he could have appealed to the work of Officer and company (1957); he could have maintained, given their new information about the depth of the Moho at the Puerto Rico Trench, that subduction seemed likely, and still, like Officer and company, have rejected the tectogene (§3.18). Heezen could not take this route because he mistakenly thought that the tectogene and the subduction were "virtually identical."

Now, as it was, as it turned out, of course, the subduction, as presently conceived, is over any short period of time of any geological period or any few millions of years, is practically indistinguishable from tectogene buckling over a short period of time. Because crust moves together, lighter material goes down into . . . the mantle of the earth and therefore subduction is virtually identical to tectogene. And tectogene was what we'd spent some years pretty soundly defeating – we thought. So, it was not the first thing one would grasp for. That is maybe we had written all these papers wrong. Maybe the down buckling under the trenches is what's happening and what we've said is all wrong. No, it is not something one would come to right away, probably never . . . So if we were to come to a conclusion that the expansion of the ocean was exactly compensated by subduction or destruction in the trench, we would have, at that moment, to accept defeat . . . Therefore it was easy to understand how Hess came to his position in '61 then saying, "Well, it's easy, we just changed the word tectogene to subduction and we have the same model do the same thing, and, see I was right all along."

*(Lear, BCH 06A: 4)*

Perhaps, it was his bitterness toward Hess that obscured his realization that the two concepts were different, and that caused him to muddle them in this way.

Regardless of how right Heezen was about himself, he most certainly was wrong about Ewing. Ewing adopted mantle convection and he did not change his mind, nor did he generalize from his tensional analysis of the Puerto Rico Trench to other trenches, seemingly regarding it as an exception. Perhaps he ended up agreeing with Officer *et al.* (1957) that the Puerto Rico Trench was compressional (§3.18). Perhaps the reason Ewing did not contribute to the paper on trenches that he, Fisher, and Hess were initially contracted to do for *The Sea* (§3.20) is that his views were in flux, perhaps he did not want to explicitly acknowledge that Fisher and Hess were probably correct.

The other retrospective explanation that Heezen gave for his insistent advocacy of Earth expansion was that he wanted to make a big splash; he yearned for fame. He wanted to show that Heezen and Tharp's theory was correct. According to Heezen, Hess and Dietz merely showed that Wegener's theory was correct, but that was not enough for Heezen. Heezen wanted more of a splash. If he had been right, he also would have bested Ewing, who never seemed to have even considered Earth expansion.

Anyway, you look at it from another point of view, as a gambler. If one, as important as it would be to lead America to believe in drift, it would be more important as a scientific discovery to prove an entirely different explanation of the facts that had been supposed to constitute drift. That's one of the reasons why today when I pointed out in the book that the definition of drift, plate tectonics, seafloor spreading, and everything, and the new global tectonics had become, by imprecise usage, had become virtually synonymous. Dietz objected and the reason why he objected is obvious. If seafloor spreading is not an absolutely new view and is simply a revision of continental drift then this supposed invention of it is not worth fighting for credit of ... because no matter who did it, it wasn't very important because it was just a revision of drift. If you looked at the point that it's a new concept then ... if expansion worked ... it doesn't become ... Wegener's theory of drift but it might become Heezen and Tharp's idea of, or Ewing, Heezen and Tharp and Ewing's expansion. But Ewing didn't like expansion, probably one of the reasons why I did. Even whether I believed it or not, it would give me a departure to publish on it without offending him because, since he didn't believe in it, he wouldn't want his name on it anyway. So, that was also a decided point. As it turns out, of course, these are not important to the final result, but at the time, these were considerations.[19]

*(Lear, BCH 06A: 5)*

Heezen had hoped that Earth expansion would have been received like his work on turbidity currents: ridicule followed by acceptance.

If one was known to have an explanation which finally became accepted and was controversial in the beginning, then one gets his name associated with it. As my name was associated very clearly and closely with the Grand Banks turbidity current story in ... the very dynamic and massive effects of turbidity currents in deep sea. The reason why that was so effective was that most everybody didn't believe the first paper and criticized it and lambasted it and then had to back down. That's the ideal sequence. On the other hand, if you make a paper and nobody believes it and they all ignore it, most of them ignore it, and a few of them copy it and republish it and nobody has lambasted it, you're ignored. So ... the sequence we did with turbidity currents was much more successful. We had so many attacks that the people who attacked us could not come today or come later and say, "Hey, we believed it all along," because they have their name written to a paper in which they severely criticized our view.

*(Lear, BCH 06A: 5–6)*

But, as Heezen admitted, Earth expansion was generally ignored because it was viewed as "ridiculous." Pushing Earth expansion was a bad gamble.

So, when the first criticism of the expansion came along I was not too upset. But not too many people criticized it because they took the view that it was so ridiculous that nobody would, nobody would even condescend to criticize it – not no one, but very few.

*(Lear, BCH 06A: 6)*

Heezen refused to admit that he was wrong about trenches, because of his bitterness toward Hess, and his yearning for fame seems to have led him to gamble on Earth expansion, and to promote it even after plate tectonics had been proposed. He would have been better served had he by 1959 acknowledged that work of Raitt *et al.* (1955) and Officer *et al.* (1957) showed that circum-Pacific trenches are probably compressional. Then, buttoning this on his belief that paleomagnetism's support of mobilism warrants acceptance of continental drift, he likely would have seen that seafloor is generated at ridges and destroyed at trenches. He then could have added that mantle convection in the forms proposed by Ewing or Hess cannot actively cause continental drift, the formation of mid-ocean ridges or the destruction of seafloor because Africa, mainly, and Antarctica entirely, are surrounded by active ridges. If he had, he would have left Ewing in the dust, and outshone Hess.

I really cannot be sure if my characterization of Heezen, based largely on his own testimony, is correct. Ewing's shameful treatment of Heezen (and Tharp) was psychologically as well as professionally devastating; surely it fed his bitterness and may have shaped his ambition. Heezen, reminiscing about Ewing after his death, likened Tharp and his relationship to Ewing as that of daughter and son.

Now, of course, it's a question why Marie and I should have stood there in such genuine honor of the deceased when he caused us so much grief, tried so hard to destroy us. But I suppose what many people there would not have realized was that we loved him as a father for seventeen years and the fact that for another five we had increasing qualms with his activities and the final five years bitter attack. It is easily forgotten, as is, I suppose, the family bickering between children and their fathers. But I had a most glorious relationship with him for fifteen to seventeen years, which were in many ways the happiest and most productive days of my life. Of course, in many ways he died for me years ago. But now he is dead and in going to the cemetery I was honoring not the being who lived last year but the immensely great man who I knew for over fifteen years.

*(Lear, BCH 05BM)*

Heezen still had good things to say about him. At the same time, he could not stop being bitter about the success of others, and thinking about what he could have done if he had not been "prevented access to the very records which were" instrumental in confirming the Vine–Matthews hypothesis and plate tectonics.

But he was a great leader, and a more perceptive one than perhaps his written record would suggest. And also people liked to point out how many of the things he purported to do he really didn't do … how he liked to take credit for things that were done by others. But also there are many cases, perhaps as many or more, where others took credit for what he did or gained much more benefit from his work than he did. A notable example, I think, is the magnetic interpretations which have been so important in recent years. They are very simple ones. There is nothing very sophisticated about magnetic anomalies or plate tectonics. Whether it is right or wrong it still is simple. Maybe that is the beauty of it. He liked simple things but perhaps not that simple. After all if it were that simple, anybody could figure it out. If it were a little more complicated it might require some serious physical thought of a real scientist … But in any case, it was not a sophisticated approach. It is an approach which I think I can understand and I would have done it. Perhaps I would have been the one to break the news if it were not for the political situation which again was part of his … which he helped to produce with prevented access to the very records which were …

*(Lear, BCH 05BM: 1; the ellipses are not mine, they are part of the transcription*
*of Heezen's remarks that he or Tharp recorded.)*

Menard (1986: 105) remarked that the Bruce Heezen of the 1950s was very different from Bruce Heezen of the early 1970s. Not only had he become extremely bitter, but his jealous assessment of the accomplishments of others shows a lack of understanding of the sophistication needed to develop plate tectonics and to construct interpretative models of marine magnetic anomalies in terms of remanent magnetization. Everybody did not figure it out, and everybody did not propose plate tectonics.

### 6.19 Ewing (1962) assesses the relative merits of mobilism and fixism in explaining sediments of ocean basins

In October 1962, Ewing gave a lecture intended for a lay audience at his alma mater, Rice University (Ewing, 1963). Ewing's lectures were usually technical and detailed, but, perhaps because it was to a lay audience, he wanted to bring out large general issues such as mobilism versus fixism. He particularly wanted to talk about ocean basin sediments, especially about Lamont's development of a method in early 1961 "for measuring the total thickness of the sediment layers on the floor of the deep sea and observing its stratification" (Ewing, 1963: 46). Lamont had begun choosing "the tracks of Columbia University research expeditions to favor the study of sediment distribution" (1963: 46). Proud that his brother John headed the Lamont investigation, he (1963: 46) let his fellow Texans know that all his studies after 1961 on ocean sediments "have been made in collaboration with him." (He and Heezen no longer collaborated.)

Explaining that "the unexpected paucity of the sediment cover is perhaps its most striking feature," which mobilists took to favor their view, he attempted to show that fixism could equally well explain the surprising thinness of the sedimentary layer. He spoke of classical continental drift theory in which continents plowed their way through the seafloor, and of seafloor spreading in which "mantle convection shifts the continents about, with outpouring of new crust occurring under the mid-ocean ridges" (1963: 44). He did not mention Earth expansion. When describing seafloor spreading, he did not mention Hess or Dietz by name, and continued oblivious to Holmes and his pioneering ideas on mantle convection. Producing a figure (his Figure 2), which resembled Hess's 1962 Figure 1 (Figure 1.8), he introduced seafloor spreading, and noted its failure to explain the lack of sedimentation in the South Pacific (RS2).

It has recently been suggested that convection currents in the mantle shift the continents about, with the outpouring of new crust occurring under the mid-ocean ridges. Now it has been further suggested that these currents extend outward across the floor of the ocean and sweep the sediment and crust under the continents (Fig. 2). This hypothesis for explaining the thinness of the Pacific sedimentary cover is, incidentally, much more difficult to support in the southern (or ocean) hemisphere than in the northern. In the southern hemisphere, continents are too small and too far apart to provide many subcontinental hiding places into which ocean-wide deposits of sediment could have been swept.

*(Ewing, 1963: 44)*

In the Atlantic Basin, sediments are "extremely sparse" on the Mid-Atlantic Ridge, and thickest near continental margins where they are derived from continents, terrigenous. He noted that (1963: 54) sediments generally extend from continental margins to the Mid-Atlantic Ridge, but sometimes "gradually become thinner and terminate before reaching" the ridge. This distribution, he claimed, eliminated the old mobilist idea that continents plow their way through the seafloor. He did not say that seafloor spreading, however, offered a straightforward explanation of the paucity of sediments in the Atlantic. What about fixism? Ewing maintained that the rate of sedimentation may have been much less in the distant than in the recent past, so fixism could explain the thinness of ocean floor sediments. Turning to the Pacific margins off Panama and South America, he noted the difference between them and the continental margins of the Atlantic, and explained why he favored fixism over classical continental drift and the later development of seafloor spreading. Mobilism apparently would not suffice.

But, on the assumption that these few samples [taken from the Pacific margins off Panama and South America] are typical, we may note an absence of indicators that the continent has been plowing through, or overriding, or is being undercut by the oceanic crust and any body of supposed sediments.

*(Ewing, 1963: 55; my bracketed addition)*

He explained the lack of sediments in the Pacific Basin by viewing the marginal trenches as sediment traps, which prevented the spreading of terrigenous sediments beyond their seaward limits.

These sections rather give the impression that a very small total amount of sediment has been delivered from the continent, that there is at the foot of the continental slope a trench whose floor has progressively tilted toward the continent, that the sediment was deposited by turbidity currents to provide a level floor, and that the sediments have a strong "continental influence" which provides several reflecting horizons and evidence of progressive tilting.

*(Ewing, 1963: 55)*

In this way he claimed that the distribution of sediments was consistent with fixism.

This entire situation is consistent with the idea that (a) the total sediment delivered to ocean basins from continents is very small, (b) throughout much of geologic time there has been a "sediment trap" along the Pacific coasts of the Americas which caught the sediments that now form the western mountain ranges, and (c) that the modern trench along South and Central America is trapping sediments in a way that probably represents the beginning of the next cycle of the mountain-building process.

*(Ewing, 1963: 55–56)*

He defended (a) by suggesting that, at bottom, at the time very little was really known about sedimentary rates throughout the geological past.

Even the present "rate of sedimentation" can be little more than a guess. The results of such guesses have been interesting, but necessarily meaningless as long as our real knowledge of the sedimentary process and rates is so fragmentary.

*(Ewing, 1963: 45)*

In (b) and (c), he essentially adopted the account of mountain building and accretion of continents given by Kay, his colleague from Columbia, but he may also have been influenced by Officer *et al.* (1957) who had adopted Kay's account of island arc and mountain formation in their compressional analysis of the Puerto Rico Trench.

In the following years, Ewing and his brother would continue discussing seafloor spreading in terms of its compatibility or incompatibility with the distribution and age of sediments they were observing on the ocean floor. Indeed, Maurice Ewing became obsessively concerned with deep ocean sediments, ensuring their collection, recording their characteristics, getting them catalogued and stored adequately, and may, I believe, have delayed his eventual acceptance of seafloor spreading (IV, §6.14). Very importantly, however, his obsession soon provided a rich dividend. The cores that he ordained should be collected as a routine on Lamont ships provided, under the active guidance of Neil Opdyke, a wonderful record of reversals of the geomagnetic field, a record that soon began making important contributions to the global reversal record, and hence to the confirmation of the Vine–Matthews hypothesis and to the development of plate tectonics (IV, §6.4). Ewing's greatness lay not in his geological theorizing but in keeping his ships at sea gathering and recording mountains of data from the world's oceans.

## Notes

1 This brief account of Ewing's life is drawn from Bullard's insightful biographical memoir (Bullard, 1980), Donn's discerning remembrance of Ewing (Donn, 1985), Worzel (1977), and some of the interviews conducted by Ronald Doel of Lamont scientists and spouses. Doel's interviewees include John and Betty Ewing, Officer, Tharp, Worzel, Charles Drake, Jack Oliver, Gordon Hamilton, and Neil Opdyke. Doel's interviews remain largely untapped. They provide an excellent resource for those interested in the development and evolution of Lamont. The interviews were conducted during the second half of the 1990s, and are part of the Lamont–Doherty Earth Observatory Oral History Project. There is also Wertenbaker (1974), but in it Ewing's and Lamont's contributions are somewhat exaggerated at the expense of other scientists and institutions.

2 Both Allyn Vine and George Woollard went on to have distinguished careers. Vine eventually took a position at WHOI in 1940 and remained there until his death. He helped design *Alvin*, the first US Navy deep sea research submersible, which was named after him. Woollard taught at Princeton from 1940 to 1947 and then moved to the University of Wisconsin where he retired in 1976. The George P. Woollard Award, established in 1983, is given by the Geophysics Division of the GSA in recognition of outstanding work in geology through the use of geophysical techniques and principles.

3 Nelson Steenland obtained his B.A. from Washington and Lee University in 1942 with a major in what was then called general science. He met Ewing at WHOI where he worked on the development of sonar. He received his Ph.D. under Ewing's supervision in 1949 and took a job the very day he passed his orals for his Ph.D. with Gravity Meter Exploration Company in Houston, Texas. Steenland's thesis was on magnetics, and he was involved in

the very early work on the subject by Ewing's group (Doel, Worzel interview, session 3: 391). Renee Brilliant obtained a B.A. from New York University majoring in physics and mathematics. She worked with Ewing on dispersion of Rayleigh waves (Brilliant and Ewing, 1954). Brilliant, however, left geophysics for medicine, frustrated by the lack of fieldwork in oceanography available to women at Lamont (Bell *et al.*, 2005: 27). She received her M.D. from New York University, and became a very successful pediatrician and pediatric hematologist. She married William Donn.

4  I can personally testify to the loyalty of one of Ewing's early students. I wrote a paper, "Ewing's reluctant switch to mobilism," in which I described, I still believe correctly, Ewing's unreasonable resistance to mobilism, and I did not discuss Ewing's great contributions to Earth science. I sent it to several former Lamonters. The one who had been an early student of Ewing's hoped that I would not try to publish the paper in his lifetime. I did not attempt to publish the paper; much of my account is found in IV, §6.14.

5  This brief account of Tolstoy's life and work on the Mid-Atlantic Ridge owes much to his willingness to answer the questions I asked him over several weeks during August 2006. Tolstoy himself not only has an interest in history of science but he (1981) wrote a delightful and insightful book on James Clerk Maxwell in which he discusses Maxwell's philosophical attitude toward scientific hypotheses and reluctance in reifying scientific theories.

6  The mathematical mistakes were pointed out by Leet, Linehan, and Berger (1951). Lamonters wrote a rejoinder (Ewing, Press, and Worzel, 1952).

7  Maurice A. Biot (1905–85), born and educated in Belgium, obtained his Ph.D. at Caltech (1932). He taught at Harvard (1934–5), Louvain (1935–7), Columbia (1937–45), and Brown universities. A Lieutenant Commander in the US Navy, he directed the structural dynamics section of the Bureau of Aeronautics (1943–5). He later served as a consultant for Royal Dutch Shell (1946–65), and held a similar position for Mobil Oil Co. beginning in 1969. An applied mathematician, he contributed to elastic theory, seismology, thermodynamics, soil mechanics, and geophysics. His papers are housed at Caltech.

8  This account of Heezen's life is derived from Tharp's efforts, beginning during the 1980s, to write a biography of Heezen, and Heezen and Tharp's extensive interview with the journalist John Lear during the middle 1970s. Lear planned to write a book on Heezen, but never did. At one point Heezen decided to write an autobiography, but died before he had done little more than make a few tapes. Tharp began to write the biography based on these tapes and her knowledge of him. I was fortunate to see her beginning efforts. Moreover, Tharp sent me transcripts of many of the interviews, and I have drawn on both in this brief account of Heezen's life, and development of his idea about the evolution of ocean basins. The tapes, I believe, are housed at the Smithsonian among Heezen's papers. I shall refer to transcriptions of the tapes using Tharp's system of BCH 01A, 01B, 02A, 02B . . . The information about Heezen's life before he met Ewing is drawn with her permission and encouragement from BCH 07A and BCH 20A, and Tharp's first few chapters of her planned Heezen biography.

9  Tharp and I co-authored this paper on her discovery of the rift valley along the axis of the Mid-Atlantic Ridge, on Heezen's initial skepticism and eventual acceptance of her suggestion, and on their decision to make physiographic diagrams of the ocean floors. Tharp recounted what happened, supplied me with notes, and we wrote the paper together in the first person. Part of this discussion of her discovery of the rift valley is based on our joint paper. In addition, I have greatly benefited from Menard's account of the discovery of the rift valley (Menard, 1986).

10  This recollection by Heezen conflicts with what he purportedly told Wertenbaker, who claims that one profile was made from observations made aboard *Kevin Moran* in 1952 (Wertenbaker, 1974: 137). Heezen suggested to Lear that Ewing changed what he, Heezen, had told Wertenbaker.

The statement that the sixth came from the seagoing tug, *Kevin Moran*, I think might have been inserted not by me but by Ewing. You probably know that the galley proofs of the

Wertenbaker book – that is the galley proofs of the first half of it at least were gone over very carefully by both Harriet and Doc [Ewing] and I know this from an independent source who has seen the very well marked up galleys of some little turns and twists in the book ... Why he might have wanted to mention *Kevin Moran* is that he was chief scientist on *Kevin Moran* while I was chief scientist on *Atlantis*. He was, of course, expedition leader. He controlled the movement of both ships but I was on *Atlantis* and he was on *Kevin Moran*. His success in getting good soundings was *not notable so we did not use – or even bother to work up the soundings from the Kevin Moran* for most of that voyage because for most of the voyage the two ships were either within sight of each other or no more than 20 miles apart because of the nature of the experiments we did ... So the *Kevin Moran* did not contribute anything significantly to the six profiles.

*(Lear, BCH 21A: 2)*

11 The profiles have been reproduced on the United States National Oceanic and Atmospheric Administration's (NOAA) website.

12 Menard in his otherwise excellent account of the discovery of the median rift valley mistakenly said that Heezen presented this talk at the 1955 Annual Meeting of the GSA. Heezen did give a talk at the 1955 meeting, but it was on "Turbidity currents from the Magdalena River, Columbia."

13 Although originally written in English (Menard, 1986: 313), this paper was published in French. I have an English version that was retranslated by Annette Trefzer of Lamont-Doherty. I have used her translation. I thank Tharp for giving it to me with permission to quote from it.

14 The written account of his talk (Heezen, 1959c) that Marie Tharp kindly gave me is incomplete as is the one she gave to the Smithsonian where Heezen's papers are housed. Page 4 and two figures are missing. Heezen began his discussion of paleomagnetism's support of mobilism on page 3, and, I believe, continued it on the missing page. A list of the "Figure Captions" indicates that Diagram 2, "Latitude shift of Europe and North America (after Irving, 1958a)," and Diagram 4 "Interpretation of paleomagnetic measurements as evidence of continental drift (after Deutsch, 1958)" are missing. Fortunately, Heezen, I believe, presented the same or very similar figures at a talk he also gave at Columbia University about on March 24, 1960, at a session honoring M. Ewing for being the first recipient of the Vetlesen Prize. Heezen's talk, as well as those by the other speakers (M. Ewing, Vening Meinesz, Worzel, Bullard, Wüst, and Fairbridge) were recorded and transcribed. I again thank Tharp for giving me a copy of the transcription. The transcription does not include copies of the figures he showed during his presentation, but it does contain discussion of them. I shall discuss key parts of his talk, and also use it to flesh out what he said during his December talk.

15 Examination of Deutsch's and Irving's publications of 1958 reveals that Heezen used Deutsch's Figure 1 from his paper "Recent Palaeomagnetic evidence for northward movement of India" (Deutsch, 1958: 159). Heezen drew his own APW path of Australia based on data from Irving and Green (1958: 70) displayed in their Figure 5, itself made up of two figures displaying on an equatorial projection APW paths since the Carboniferous of Australia, North America, and Europe (Irving and Green's Figure 5, is reproduced as Figure 5.4, II, §5.3). He also had read Irving's 1958a contribution to the Hobart symposium on continental drift. Given Heezen's reply to Holmes, he had read Carey's paper on Earth expansion (§6.10). The paper appeared in the proceedings of the Hobart symposium on continental drift, which also included Irving's paper. Moreover, Irving (1958a: 51, 53) also reported a single paleopole from South America, which Heezen mentioned in his March 1960 talk at the meeting honoring Ewing, and made use of in the December 1959 talk in question. Irving obtained this pole from rocks collected by Reinhardt Maack at Carey's request before Creer had obtained his much more extensive results from South America. Irving (1958b) reported Creer's results in his only other single-authored 1958 paper "Palaeogeographic Reconstruction from Palaeomagnetism."

16 The talks at the meeting were recorded. I thank Tharp for giving me a transcript of the recording. Barzun introduced Ewing, and Worzel served as moderator. The transcript also indicates when the audience applauded.

17 John Sclater (2001) told an amusing story about Bullard that couldn't be more appropriate.

When I was a graduate student at Madingley, Teddy Bullard jokingly complained that he had not accomplished very much in geophysics because his name had not been given to any hypothesis or law. To rectify this omission, the students and junior staff at Madingley, with support from Maurice Hill, created "Bullard's Law." This law asserted, "Never take one marine heat flow measurement within 50 kilometers of another measurement because it is likely that it will differ from the first by at least one order of magnitude."

18 I thank Tharp for sending me a copy of the letter. She also did her best in trying to read the letter. Each ellipsis is substituted for an illegible word.

19 In *The Face of the Deep*, Heezen and Hollister (1971: 541) claimed:

Even today the term "continental drift" evokes so many bad memories that modern advocates often support it under other names such as continental displacement, sea-floor expansion, sea-floor spreading, or plate tectonics. Although they originally had differences of meaning, these terms, through imprecise usage, have become synonymous.

Although some, including Heezen and Hollister *misused* such terms, they were and are not synonymous. Plate tectonics is a kinematic theory, continental displacement and seafloor spreading as then conceived included dynamical aspects. There were substantial differences between Wegener's, du Toit's, and Holmes' theories of continental displacement. Admittedly, an argument can be made that Holmes' revised 1944/1945 theory of mantle convection is very similar to seafloor spreading. As for the term, "sea-floor expansion," it was not widely used, and Heezen and Hollister's inclusion of it in their list suggests that they were attempting to show that Heezen's view about the formation of ocean basins was different from seafloor spreading. Heezen and Hollister were correct to emphasize that these terms were often incorrectly used.

# References

Agostinho, J. 1936. Volcanoes of the Azores. *Bull. Volcanol.*: 123–138.

Aki, K. 1960. Study of earthquake mechanism by a method of phase equalization applied to Rayleigh and Love waves. *J. Geophys. Res.*, **65**: 729–740.

Alldredge, L. R. and Keller, F. 1949. Preliminary report on magnetic anomalies between Adak, Alaska and Kwajalein, Marshall Islands. *Trans. Am. Geophys. Union*, **30**: 494–500.

Allwardt, A. O. 1987. On the "historical note" by Carl Bowin. *Geology*, **15**: 475.

Allwardt, A. O. 1990. *The Roles of Arthur Holmes and Harry Hess in the Development of Modern Global Tectonics*. University Microfilms, Ann Arbor, MI.

Andrews, H. N. 1961. *Studies in Paleobotany*. John Wiley & Sons, New York.

Anonymous. 1923. Wegener's hypothesis of continental drift: discussion. *Geogr. J.*, **61**: 188–194.

Anonymous. 1957. Minutes of the meeting. Transactions of the Rome Meeting. Laursen, V., ed., *IAGA Bulletin*, No. 15, 9–41.

Anonymous. 1962. At home and abroad. *Am. Assoc. Petrol. Geol. Bull.*, **46**: 1972.

Anonymous. 1963. Report of the Council to the One Hundred and Forty-third Annual General Meeting of the Society. *Q. J. Roy. Astron. Soc.*, **4**: 151–154.

Anonymous. 1964. Continental drift. *Q. J. Geol. Soc. London*, **120**: 27–33.

Anonymous. 2009. Obituary, Ronald Mason, *The Times*, August 6, 2009.

Argand, E. 1924. La tectonique de l'Asie. *Proceedings of the 13th International Geological Congress*, **1**, Part 5: 171–372. English translation by Carozzi, A. V. 1977. *The Tectonics of Asia*. Hafner Press, New York.

Arnold, C. A. 1947. *An Introduction to Palaeobotany*. McGraw-Hill, New York.

Atwater, T. 1970. Implications of plate tectonics for the Cenozoic tectonic evolution of western North America. *Geol. Soc. Am. Bull.*, **81**: 3513–3536.

Axelrod, D. I. 1952. A theory of angiosperm evolution. *Evolution*, **6**: 29–60.

Axelrod, D. I. 1963. Fossil floras suggest stable, not drifting, continents. *J. Geophys. Res.*, **68**: 3257–3263.

Axelrod, D. I. 1964. Reply. *J. Geophys. Res.*, **69**: 1669–1671.

Bailey, E. B. 1927. Across Canada with Princeton. *Nature*, **120**: 673–675.

Bailey, E. B. and Holtedahl, O. 1938. Northwestern Europe, caledonide. In *Regionale Geologie der Erde*, **2**, Part 2: 1–76.

Bailey, E. B. and Weir, J. 1939. *Introduction to Geology*. MacMillan and Co., London.

Bell, R., Laird, J., Pfirman, S., Mutter, J., Balstad, R., and Cane, M. 2005. An experiment in institutional transformation. *Oceanography*, **18**: 25–37.

Benioff, H. 1949. Seismic evidence for the fault origin of oceanic deeps. *Geol. Soc. Am. Bull.*, **60**: 1837–1856.

Benioff, H. 1955. Orogenesis and deep crustal structure: additional evidence from seismology. *Geol. Soc. Am. Bull.*, **65**: 385–400.

Benioff, H. 1959. Circum-Pacific tectonics. *Publ. Dominion Observatory*, **20**: 395–402.

Benioff, H. 1962. Movements on major transcurrent faults. In Runcorn, S. K., ed., *Continental Drift*. Academic Press, New York, 103–134.

Bernal, J. D. 1961a. Continental and oceanic differentiation. *Nature*, **192**: 123–124.

Bernal, J. D. 1961b. Response to Dietz. *Nature*, **192**: 125.

Bernard, E. A. 1964. The laws of physical palaeoclimatology and the logical significance of palaeoclimatic data. In Nairn, A. E. M., ed., *Problems in Palaeoclimatology*. Interscience, New York, 309–321.

Besse, J. and Courtillot, V. 1991. Revised and synthetic apparent polar wander paths of the African, Eurasian, North American and Indian plates, and true polar wander since 200 Ma. *J. Geophys. Res.*, **96**: 4029–4050.

Betz, F. and Hess, H. H. 1940. Floor of the North Pacific Ocean. *Trans. Am. Geophys. Union*, 21st Annual Meeting: 348–349.

Betz, F. and Hess, H. H. 1942. The floor of the North Pacific Ocean. *Geogr. Rev.*, **32**: 99–116.

Bigarella, J. J. and Salamuni, R. 1964. Paleowind patterns in the Botucatu sandstone (Triassic-Jurassic) of Brazil and Uruguay. In Nairn, A. E. M., ed., *Descriptive Palaeoclimatology*. Interscience, New York, 406–409.

Bijlaard, P. P. 1951. On the origin of geosynclines, mountain formation, and volcanism. *Trans. Am. Geophys. Union*, **32**: 518–519.

Billings, M. P. 1982. Ordovician cauldron subsidence of the Blue Hills Complex, eastern Massachusetts. *Geol. Soc. Am. Bull.*, **93**: 909–920.

Birch, F. 1952. Elasticity and constitution of Earth's interior. *J. Geophys. Res.*, **57**: 227–286.

Blackett, P. M. S. 1956. *Lectures on Rock Magnetism*. Weizmann Science Press of Israel, Jerusalem.

Blackett, P. M. S. 1961. Comparison of ancient climates with the ancient latitudes deduced from rock magnetic measurements. *Proc. Roy. Soc. London A*, **263**: 1–30.

Blackett, P. M. S., Clegg, J. A., and Stubbs, P. H. S. 1960. An analysis of rock magnetic data. *Proc. Roy. Soc. London A*, **256**: 291–322.

Blundell, D. J. 1962. Palaeomagnetic investigations in the Falkand Islands Dependencies. *British Antarctic Survey Reports*, **39**: 1–24.

Blundell, D. J. and Stephenson, P. J. 1959. Palaeomagnetism of some dolerite intrusions from the Theron Mountains and Whichaway Nunataks, Antarctica. *Nature*, **184**: 1860.

Boon, J. D. and Albritton, C. C., Jr. 1937. Meteorite scars in ancient rocks. *Field Lab.*, **V**: 53–64.

Boon, J. D. and Albritton, C. C., Jr. 1938. Established and supposed examples of meteoritic craters and structures. *Field Lab.*, **VI**: 44–56.

Bourgeois, J. and Koppes, S. 1998. Robert S. Dietz and the identification of impact structures on Earth. *Earth Sci. Hist.*, **17**: 139–156.

Bowen, N. L. 1927. The origin of ultra-basic and related rocks. *Am. J. Sci.*, **14**: 89–108.

Bowen, N. L. 1928. *The Evolution of the Igneous Rocks*. Princeton University Press, Princeton, NJ.

Bowen, N. L. and Tuttle, O. F. 1949. The system $MgO-SiO_2-H_2O$. *Geol. Soc. Am. Bull.*, **60**: 439–460.

Bowin, C. O. 1960. Geology of central Dominican Republic. Ph.D. thesis (unpublished), Princeton University.

Bowin, C. O. 1966. Geology of central Dominican Republic (a case history of part of an island arc). In Hess, H. H., ed., *Caribbean Geological Investigations. Geol. Soc. Am. Mem.*, **98**: 11–84.

Bowin, C. O. 1987. Historical note on "Evolution of Ocean Basins" preprint by H. H. Hess [1906–1969]. *Geology*, **15**: 475–476.

Branca, W. and Fraas, F. 1905. *Das kryptovulkanische Becken von Steinheim.* K. Preuss, Berlin, 1–64.

Briden, J. C. 1964. Palaeolatitudes and palaeomagnetic studies with special reference to pre-Carboniferous rocks in Australia. Ph.D. thesis, Australian National University.

Briden, J. C. 1965. Ancient secondary magnetizations in rocks. *J. Geophys. Res.*, **70**: 5205–5221.

Briden, J. C. 1966. Estimates of direction and intensity of the palaeomagnetic field from the Mugga Mugga Porphyry, Australia. *Geophys. J.*, **11**: 267–278.

Briden, J. C. 1967a. Recurrent continental drift of Gondwanaland. *Nature*, **215**: 1334–1339.

Briden, J. C. 1967b. A new palaeomagnetic result from the Lower Cretaceous of East-Central Africa. *Geophys. J.*, **12**: 75–380.

Briden, J. C. and Irving, E. 1964. Palaeolatitude spectra of sedimentary palaeoclimatic indicators. In Nairn, A. E. M., ed., *Problems in Palaeoclimatology.* Interscience, New York, 199–224.

Brilliant, R. M. and Ewing, M. 1954. Dispersion of Rayleigh waves across the U.S. *Bull. Seismol. Soc. Am.*, **44**: 149–158.

Brooks, C. E. P. 1949. *Climate Through the Ages* (2nd edition). Ernest Benn, London.

Brooks, H. 1941. Cyclic convection-currents. *Trans. Am. Geophys. Union*, **22**: 548–551.

Brouwer, H. A. 1962. Extrusive vulcanism with reference to earth movements. In MacDonald, G. A. and Kuno, H., eds., *The Crust of the Pacific Basin.* American Geophysical Union, Washington, DC, 87–91.

Brown, D. A. 1967. Some problems of distribution of Late Palaeozoic and Triassic terrestrial vertebrates. *Aust. J. Sci.*, **30**: 434–445.

Brunhes, B. 1906. Recherches sur le direction d'aimantation des roches volcaniques. *J. Phys.*, **5**: 705–724.

Brush, S. G. 1996. *A History of Modern Planetary Physics, Volume 1.* Cambridge University Press, Cambridge.

Brynjolfsson, A. 1957. Studies of remanent magnetism and viscous magnetism in the basalts of Iceland. *Adv. Phys.*, **6**: 247–254.

Bucher, W. H. 1936. Cryptovolcanic structures in the United States. *Int. Geol. Congr., XVI, Reports, II*: 1055–1084.

Bucher, W. H. 1952. Continental drift versus land bridges. In Mayr, E., ed., *The Problem of Land Connections Across the South Atlantic, with Special Reference to the Mesozoic. Bull. Am. Mus. Nat. Hist.*, **99**: 93–104.

Bucher, W. H. 1962. Descriptive palaeoclimatology. *Am. Sci.*, **50**: 296A–300A.

Bucher, W. H. 1963. Cryptoexplosion structures caused from without or from within the Earth? ("Astroblemes" or "Geoblemes?"). *Am. J. Sci.*, **261**: 597–649.

Bucher, W. H. 1964. The third confrontation. In Nairn, A. E. M., ed., *Problems in Palaeoclimatology*. Interscience, New York, 3–9.

Buddington, A. F. 1970. Harry Hammond Hess, 1906–1969. *Geol. Soc. Am. Proc. 1969*: 1–9.

Bull, C. and Irving, E. 1960a. Palaeomagnetism in Antarctica. *Nature*, **185**: 834–835.

Bull, C. and Irving, E. 1960b. The palaeomagnetism of some hypabyssal intrusive rocks from South Victoria Land, Antarctica. *Geophys. J.*, **3**: 211–224.

Bullard, E. C. 1951. Remarks on deformation of the Earth's crust. *Trans. Am. Geophys. Union*, **32**: 520.

Bullard, E. C. 1952. Heat flow through the floor of the Eastern Pacific Ocean. *Nature*, **170**: 199–200.

Bullard, E. C. 1954. The flow of heat through the floor of the Atlantic Ocean. *Proc. Roy. Soc. London A*, **222**: 408–429.

Bullard, E. C. 1960. Comments. Unpublished transcription of meeting held in honor of Maurice Ewing's winning of the first Vetlesen Award held on March 25, 1960 at the Men's Faculty Club, Columbia University, 38–40.

Bullard, E. C. 1961. The automatic reduction of geophysical data. *Geophys. J.*, **3**: 237–243.

Bullard, E. C. 1962. The deeper structure of the ocean floor. *Proc. Roy. Soc. London A*, **265**: 386–395.

Bullard, E. C. 1963. Review of *Continental Drift*, edited by S. K. Runcorn. *Geophys. J.*, **8**: 146–147.

Bullard, E. C. 1964. Continental drift. *Q. J. Geol. Soc. London*, **120**: 1–26.

Bullard, E. C. 1967. Maurice Neville Hill. *Biogr. Mem. Fell. Roy. Soc.*, **13**: 193–203.

Bullard, E. C. 1968. Conference on the history of the Earth's crust. In Phinney, R. J., ed., *The History of the Earth's Crust*. Princeton University Press, Princeton, NJ, 231–235.

Bullard, E. C. 1975a. The emergence of plate tectonics: a personal view. *Rev. Earth Planet. Sci.*, **3**: 1–30.

Bullard, E. C. 1975b. The effect of World War II on the development of knowledge in the physical sciences. *Proc. Roy. Soc. London A*, **343**: 519–536.

Bullard, E. C. 1980. William Maurice Ewing. *Biogr. Mem. Natl. Acad. Sci.*, **51**: 119–193.

Bullard, E. C. and Day, A. 1961. The flow of heat through the floor of the Atlantic Ocean. *Geophys. J.*, **4**: 282–292.

Bullard, E. C., Maxwell, A. E., and Revelle, R. 1956. Heat flow through the deep sea floor. *Adv. Phys.*, **3**: 153–181.

Byerly, P. 1960. Beno Gutenberg, geophysicist. *Science*, **131**: 956–957.

Cain, S. A. 1944. *Foundations of Plant Geography*. Harper and Bros., New York.

Carey, S. W. 1955a. The orocline concept in geotectonics. *Roy. Soc. Tasmania*, **89**: 255–288.

Carey, S. W. 1955b. Wegener's South America–Africa assembly, fit or misfit? *Geol. Mag.*, **XCII**: 196–200.

Carey, S. W. 1958. A tectonic approach to continental drift. In Carey, S. W., Convener, *Continental Drift: A Symposium*. University of Tasmania, Hobart, 177–355.

Carey, S. W. 1961. Paleomagnetic evidence relevant to a change in the Earth's radius. *Nature*, **190**: 35.

Carsola, A. J. and Dietz, R. S. 1952. Submarine geology of two flat-topped northeast Pacific seamounts. *Am. J. Sci.*, **250**: 481–497.

Chadwick, P. 1962. Mountain-building hypotheses. In Runcorn, S. K., ed., *Continental Drift*. Academic Press, New York, 195–234.

Chaloner, W. G. 1959. Continental drift. In Johnson, M. L., Abercrombie, M., and Fogg, G. E., eds., *New Biology*, **29**. Penguin Books, Baltimore, MD, 7–30.

Chamalaun, T. and Roberts, P. H. 1962. The theory of convection in spherical shells and its application to the problem of thermal convection in the Earth's mantle. In Runcorn, S. K., ed., *Continental Drift*. Academic Press, New York, 177–194.

Chandrasekhar, S. 1952. The thermal instability of a fluid sphere heated within. *Phil. Mag.*, **43**: 1317–1329.

Chandrasekhar, S. 1953. The onset of convection by thermal instability in spherical shells. *Phil. Mag.*, **44**: 233–241.

Charnock, H. 1997. John Crossley Swallow. *Biogr. Mem. Fell. Roy. Soc.*, **13**: 503–519.

Clegg, J. A. 1956. Rock magnetism. *Nature*, **178**: 1085–1087.

Clegg, J. A., Almond, M., and Stubbs, P. H. S. 1954a. The remanent magnetization of some sedimentary rocks in Britain. *Phil. Mag.*, **45**: 583–598.

Clegg, J. A., Almond, M., and Stubbs, P. H. S. 1954b. Some recent studies of the pre-history of the Earth's magnetic field. *J. Geomagn. Geoelectr.*, **6**: 194–199.

Clegg, J. A., Deutsch, E. R., Everitt, C. W. R., and Stubbs, P. H. S. 1957. Some recent palaeomagnetic measurements made at Imperial College, London. *Adv. Phys.*, **6**: 219–230.

Clegg, J. A., Radakrishnamurty, C., and Sahasrabudhe, P. W. 1958. Remanent magnetism of the Rajmahal Traps of North-Eastern India. *Nature*, **181**: 830–831.

Cloos, H. 1939. Hebung-Spaltung-Volcanismus. *Geologische Rundschau*, **30**: 506–510.

Colbert, E. H. 1964. Climatic zonation and terrestrial faunas. In Nairn, A. E. M., ed., *Problems in Palaeoclimatology*. Interscience, New York, 617–637.

Colbert, E. H., Cowles, R. B., and Bogert, C. M. 1946. Temperature tolerances in the American alligator and their bearing on the habits, evolution, and extinction of the dinosaurs. *Bull. Am. Mus. Nat. Hist.*, **86**: 331–373.

Coleman, A. P. 1907. A lower Huronian ice age. *Am. J. Sci.*, **23**: 187–192.

Collinson, D. W. and Runcorn, S. K. 1960. Polar wandering and continental drift: evidence from paleomagnetic observations in the United States. *Geol. Soc. Am. Bull.*, **71**: 915–958.

Collinson, D. W., Creer, K. M., Irving, E., and Runcorn, S. K. 1957. Palaeomagnetic investigations in Great Britain I: The measurement of the permanent magnetization of rocks. *Phil. Trans. Roy. Soc. London A*, **250**: 130–143.

Cook, A. H. 1990. Sir Harold Jeffreys. *Biogr. Mem. Fell. Roy. Soc.*, **36**: 303–333.

Coulomb, J. 1945. Séismes prefunds et grandes anomalies de la pesanteur. *Ann. Géophys.*, **1**: 244–255.

Cox, A. and Doell, R. R. 1960. Review of paleomagnetism. *Geol. Soc. Am. Bull.*, **71**: 645–768.

Cox, A. and Doell, R. R. 1961a. Palaeomagnetic evidence relevant to a change in the Earth's radius. *Nature*, **189**: 45–47.

Cox, A. and Doell, R. R. 1961b. Reply to Carey on palaeomagnetic evidence relevant to a change in the Earth's radius. *Nature*, **190**: 36–37.

Craig, G. Y. 1961. Palaeozoological evidence of climate. In Nairn, A. E. M., ed., *Descriptive Palaeoclimatology*. Interscience, New York, 207–226.

Cranwell, L. M. 1963. Nothofagus: living and fossil. In Gressitt, J. Linsley, ed., *Pacific Basin Biogeography*. Bishop Museum Press, Honolulu, HI.

Creer, K. M. 1955. A preliminary palaeomagnetic survey of certain rocks in England and Wales. Ph.D. dissertation, Queens' College, University of Cambridge, 203.

Creer, K. M. 1959. A.C. demagnetization of unstable Triassic Keuper marls from S.W. England. *Geophys. J. Roy. Astron. Soc.*, **2**: 261–275.

Creer, K. M. 1968. Arrangement of the continents during the Palaeozoic era. *Nature*, **219**: 41–44.

Creer, K. M., Irving, E., Nairn, A. E. M., and Runcorn, S. K. 1958. Palaeomagnetic results from different continents and their relation to the problem of continental drift. *Ann. Géophys.*, **14**: 492–501.

Creer, K. M., Irving, E., and Nairn, A. E. M. 1959. Palaeomagnetism of the Great Whin Sill. *Geophys. J.*, **2**: 306–323.

Creer, K. M., Irving, E. and Runcorn, S. K. 1954. The direction of the geomagnetic field in remote epochs in Great Britain. *J. Geomagn. Geoelectr.*, **6**: 163–168.

Creer, K. M., Irving, E. and Runcorn, S. K. 1957. Geophysical interpretation of palaeomagnetic directions from Great Britain. *Phil. Trans. Roy. Soc. London A*, **250**: 144–155.

Crowell, J. C. 1957. Origin of pebbly mudstone. *Geol. Soc. Am Bull.*, **68**: 993–1010.

Crowell, J. C. 1964. Climate significance of sedimentary deposits containing dispersed megaclasts. In Nairn, A. E. M., ed., *Problems in Palaeoclimatology*. Interscience, New York, 86–99.

Crowell, J. C. and Winterer, E. L. 1953. Pebbly mudstones and tillites. *Geol. Soc. Am. Bull.*, **64**: 1502.

Daly, R. A. 1936. Origin of submarine "canyons." *Am. J. Sci.*, Fifth Series, **31**: 401–420.

Daly, R. A. 1940. *Strength and Structure of the Earth*. Prentice Hall, New York.

Darlington, P. J. 1957. *Zoogeography*. John Wiley & Sons, New York.

Darlington, P. J. 1964. Drifting continents and Late Paleozoic geography. *Proc. Natl. Acad. Sci.*, **52**: 1084–1091.

Darwin, C. 1842. *Structure and Distribution of Coral Reefs*. Smith, Elder and Co., London.

David, T. W. E. 1907. Different geological epochs, with special reference to glacial epochs. *Compte Rendu: Xe session du congress geoloqique International, Mexico, 1906*, 437–482.

David, T. W. E. (edited and much supplemented by Browne, W. R.) 1950. *The Geology of the Commonwealth of Australia*. Edward Arnold & Co., London.

de Smitt, V. P. 1932. Earthquakes in the North Atlantic as related to submarine cables (abstract). *Trans. Am. Geophys. Union 15th Ann. Meeting, Part I*: 103–109.

Deutsch, E. R. 1958. Recent palaeomagnetic evidence for northward movement of India. *J. Alberta Soc. Petrol. Geol.*, **6**: 155–162.

Deutsch, E. R. 1963a. Discussion: polar wandering – a phantom event? *Am. J. Sci.*, **261**: 194–199.

Deutsch, E. R. 1963b. Polar wandering and continental drift: an evaluation of recent evidence. In Munyan, A. C., ed., *Polar Wandering and Continental Drift*, Society of Economic Paleontologists and Mineralogists, Special Publication No. 10, 4–46.

Dicke, R. H. 1959. Gravitation: an enigma. *Am. Sci.*, **47**: 25–40.

Dietz, R. S. 1946a. The meteoritic impact origin of the Moon's surface features. *J. Geol.*, **54**: 359–375.

Dietz, R. S. 1946b. Geological structures possibly related to lunar craters. *Popular Astron.*, **54**: 455–467.

Dietz, R. S. 1947. Meteorite impact suggested by the orientation of shatter-cones at the Kentland, Indiana, disturbance. *Science*, **105**: 42–43.

Dietz, R. S. 1952a. The Pacific floor. *Sci. Am.*, **186**: 19 23.

Dietz, R. S. 1952b. Geomorphic evolution on continental terrace (continental shelf and slope). *Bull. Am. Assoc. Petrol. Geol.*, **36**: 1802–1819.

Dietz, R. S. 1954. Marine geology of northwestern Pacific: description of Japanese bathymetric chart 6901. *Geol. Soc. Am. Bull.*, **54**: 1199–1224.

Dietz, R. S. 1957. Office of Naval Research London, *European Scientific Notes*, No. 11–1.

Dietz, R. S. 1959a. Shatter cones in cryptoexplosion structures (meteorite impact?). *J. Geol.*, **67**: 496–505.

Dietz, R. S. 1959b. Point d'impact des astéroides comme origine des bassins océaniques: une hypothèse. *Colloq. Int. CNRS, LXXXIII. La Topographie et la Geologie des Profondeurs Oceaniques*, 265–275.

Dietz, R. S. 1959c. Colloquium on the topography and geology of the deep sea floor. *Int. Geol. Rev.*, **1**: 113–122.

Dietz, R. S. 1959d. Drowned ancient islands of the Pacific. *New Sci.*, **5**: 14–17.

Dietz, R. S. 1960a. Meteorite impact suggested by shatter cones in rocks. *Science*, **131**: 1781–1784.

Dietz, R. S. 1960b. Vredefort ring structure; an astrobleme (meteorite impact structure). *Geol. Soc. Am. Bull.*, **71**: 2093.

Dietz, R. S. 1961a. Continent and ocean basin evolution by spreading of the sea floor. *Nature*, **190**: 854–857.

Dietz, R. S. 1961b. Astroblemes. *Sci. Am.*, **205**(2): 50–58.

Dietz, R. S. 1961c. Vredefort ring structure: meteorite impact scar? *J. Geol.*, **69**: 499–516.

Dietz, R. S. 1961d. The spreading ocean floor. *The Saturday Evening Post*, **234**(42): 34–35, 94, 96.

Dietz, R. S. 1961e. "Continental and Oceanic Differentiation," reply to discussion by J. D. Bernal. *Nature*, **192**, 124.

Dietz, R. S. 1962a. Ocean-basin evolution by sea-floor spreading. In Runcorn, S. K., ed., *Continental Drift*. Academic Press, New York, 289–298.

Dietz, R. S. 1962b. Sudbury structure as an astrobleme. *Trans. Am. Geophys. Union*, **43**: 445–446.

Dietz, R. S. 1962c. Ocean basin evolution by sea floor spreading. *J. Oceanogr. Soc. Japan, 20th Century Volume*: 4–14.

Dietz, R. S. 1962d. Ocean-basin evolution by sea-floor spreading. In MacDonald, G. A. and Kuno, H., eds., *The Crust of the Pacific Basin*. American Geophysical Union, Washington, DC, 11–12.

Dietz, R. S. 1962e. *Continent and Ocean Basin Evolution by Sea Floor Spreading, "Commotion in the Ocean."* American Association of Petroleum Geologists, Distinguished Lecture Series, 24 pages.

Dietz, R. S. 1963a. Reply to R. L. C. Gallant. *Nature*, **197**: 39–40.

Dietz, R. S. 1963b. Cryptoexplosion structures: a discussion. *Am. J. Sci.*, **261**: 650–664.

Dietz, R. S. 1964a. Sudbury structure as an astrobleme. *J. Geol.*, **72**: 412–434.

Dietz, R. S. 1964b. Commotion in the ocean: the growth of continents and ocean basins. In Yoshida, K., ed., *Studies in Oceanography: A Collection of Papers Dedicated to Koji Hidaka*, University of Tokyo Press, Tokyo, 465–478.

Dietz, R. S. 1968. Reply. *J. Geophys. Res.*, **73**: 6567.

Dietz, R. S. 1989. Response by Robert S. Dietz for awarding of Penrose Medal. *Geol. Soc. Am. Bull.*, **101**: 987–989.

Dietz, R. S. 1994. Earth, sea and sky: life and times of a journeyman geologist. Annu. Rev. *Earth Planet. Sci.*, **22**: 1–32.

Dietz, R. S. and Butler, L. W. 1964. Shatter-cone orientation at Sudbury, Canada. *Nature*, **204**: 280–281.

Dietz, R. S. and Menard, H. W. 1951. Origin of the abrupt change in slope at the continental shelf margin. *Am. Assoc. Petrol. Geol. Bull.*, **35**: 1994–2016.

Dietz, R. S. and Menard, H. W. 1953. Hawaiian swell, deep, and arch, and subsidence of the Hawaiian Islands. *J. Geol.*, **61**: 99–113.

Dietz, R. S., Menard, H. W., and Hamilton, E. L. 1954. Echograms of the Mid-Pacific expedition. *Deep-Sea Res.*, **1**: 258–272.

Dirac, P. A. M. 1937. The cosmological constraints. *Nature*, **139**: 323.

Dirac, P. A. M. 1938. A new basis for cosmology. *Proc. Roy. Soc. London A*, **165**: 199–208.

Doel, R. E., Levin, T. J., and Marker, M. K. 2006. Extending modern cartography to the ocean depths: military patronage, Cold War priorities, and the Heezen–Tharp mapping project, 1952–1959. *J. Hist. Geog.*, **32**: 605–626.

Donn, W. L. 1985. Memories of (William) Maurice Ewing: the little boy in the candy shop. *EOS*, **66**: 129–130.

Dorf, E. 1933. *Studies of the Pliocene Paleobotany of California*. Carnegie Institution of Washington, Publication 421. W.F. Roberts Company, Washington, DC.

Dorf, E. 1938. A late Tertiary flora from Southwestern Idaho. In *Miocene and Pliocene Floras of Western North America*. Carnegie Institution of Washington, Washington, DC, 73–124.

Dorf, E. 1942. *Upper Cretaceous Floras of the Rocky Mountain Region*. Carnegie Institution of Washington Publication 508, Lancaster Press, Lancaster, PA.

Dorf, E. 1959. Climatic changes of the past and present. *Contrib. Mus. Paleontol. Univ. Mich.*, **13**: 181–210. Reproduced in Charles A. Ross, ed., 1976, *Paleobiogeography, Benchmark Papers in Geology*, **31**: 384–411. Dowden, Hutchinson & Ross, New York.

Dorf, E. 1964. The petrified forests of Yellowstone Park. *Sci. Am.*, **210**, 107–114.

Dott, R. H., Jr. 1961. Squantum tillite, Massachusetts: evidence of glaciation or subaqueous mass movements? *Geol. Soc. Am. Bull.*, **72**: 1289–1305.

Du Bois, P. M. 1957. Comparison of palaeomagnetic results for selected rocks of Great Britain and North America. *Adv. Phys.*, **6**: 177–186.

Du Bois, P. M., Irving, E., Opdyke, N. D., Runcorn, S. K., and Banks, M. R. 1957. The geomagnetic field in Upper Triassic times in the United States. *Nature*, **180**: 1186–1187.

du Toit, A. L. 1924. The contribution of South Africa to the principles of geology. *S. Afr. J. Sci.*, **21**: 52–78.

du Toit, A. L. 1937. *Our Wandering Continents: An Hypothesis of Continental Drifting*. Oliver and Boyd, Edinburgh.

du Toit, A. L. 1954. *The Geology of South Africa* (3rd edition). Oliver and Boyd, Edinburgh.

Egeler, C. G. 1973. In memoriam, Prof. DR. IR. H. A. Brouwer. *Geol. Mijnbouw*, **52**: 253–256.

Einarsson, T. and Sigurgeirsson, T. 1955. Rock magnetism in Iceland. *Nature*, **175**: 892.

Elsasser, W. 1966. Thermal structure of the upper mantle. In Hurley, P. M., ed., *Advances in Earth Science: Contributions to the International Conference on the Earth Sciences, MIT, September, 1964*. MIT Press, Cambridge, MA.

Ericson, D. B., Ewing, M., and Heezen, B. C. 1951. Deep-sea sands and submarine canyons. *Geol. Soc. Am. Bull.*, **62**: 961–965.

Ericson, D. B., Ewing, M., Heezen, B. C., and Wollin, G. 1955. Sediment deposition in deep Atlantic. *Geol. Soc. Am. Special Paper*, **62**: 255–268.

Evans, D. A. D. 2006. Proterozoic low orbital obliquity and axial-dipolar geomagnetic field from evaporite palaeolatitudes. *Nature*, **444**: 51–55.

Ewing, J. and Ewing, M. 1967. Sediment distribution on the mid-ocean ridges with respect to spreading of the sea floor. *Science*, **156**: 1590–1592.

Ewing, M. 1931. Calculation of ray paths from seismic travel-time curves. Ph.D. thesis (unpublished), Rice Institute, Houston, TX.

Ewing, M. 1952. The Atlantic Ocean. In Mayr, E., ed., *The Problem of Land Connections Across the South Atlantic, with Special Reference to the Mesozoic. Bull. Am. Mus. Nat. Hist.*, **99**: 87–92.

Ewing, M. 1960. The mechanics of the mid-ocean ridge and rift. Unpublished transcription of meeting held in honor of Maurice Ewing's winning of the first Vetlesen Award held on March 25, 1960 at the Men's Faculty Club, Columbia University, 1–22 and 52–54.

Ewing, M. 1963. Sediments of ocean basins. In *Man, Science, Learning, and Education: The Semicentennial Lectures at Rice University*. William Marsh Rice University and Chicago University Press, Chicago, 41–59.

Ewing, M. 1964. Comments on the theory of glaciation. In Nairn, A. E. M., ed., *Problems in Palaeoclimatology*. Interscience, New York, 348–352.

Ewing, M. and Donn, W. L. 1956. A theory of ice ages. *Science*, **123**: 1061–1066.

Ewing, M. and Donn, W. L. 1958. A theory of ice ages II. *Science*, **127**: 1159–1162.

Ewing, M. and Donn, W. L. 1959. Reply to Livingstone. *Science*, **129**: 464–465.

Ewing, M. and Donn, W. L. 1961. Polar wandering and climate. In Munyan, A. C., ed., *Polar Wandering and Continental Drift*. Society of Economic Paleontologists and Mineralogists, Special Publication No. 10, 94–99.

Ewing, M. and Donn, W. L. 1966. A theory of ice ages III. *Science*, **152**: 1706–1711.

Ewing, M. and Ewing, J. 1957. Seismic-refraction profiles in the Atlantic ocean basins, in the Mediterranean Sea, on the Mid-Atlantic Ridge and in the Norwegian Sea. *Annual Meeting of the Seismological Society of America*.

Ewing, M. and Ewing, J. 1959. Seismic refraction profiles in the Atlantic Ocean basins, in the Mediterranean Sea, on the Mid-Atlantic Ridge and in the Norwegian Sea. *Geol. Soc. Am. Bull.*, **70**: 291–318.

Ewing, M. and Heezen, B. C. 1955. Puerto Rico trench topographic and geophysical data. *Geol. Soc. Am. Special Paper*, **62**: 255–268.

Ewing, M. and Heezen, B. C. 1956a. Some problems of Antarctic submarine geology. In *Antarctica in the I.G.Y.* AGU Geophysical Monograph, **1**: 75–81.

Ewing, M. and Heezen, B. C. 1956b. Mid-Atlantic Ridge seismic belt. *Trans. Am. Geophys. Union*, **37**: 343.

Ewing, M. and Heezen, B. C. 1960. Continuity of mid-oceanic ridge and rift valley in the southwestern Indian Ocean. *Science*, **131**: 1677–1679.

Ewing, M. and Landisman, M. 1961. Shape and structure of ocean basins. In Sears, M., ed., *Oceanography*. American Association for the Advancement of Science, Washington, DC, 3–38.

Ewing, M. and Worzel, J. L. 1954. Gravity anomalies and structure of the West Indies, part I. *Geol. Soc. Am. Bull.*, **65**: 165–174.

Ewing, M., Ewing J., and Talwani, M. 1964. Sediment distribution in the oceans, the Mid-Atlantic ridge. *Geol. Soc. Am. Bull.*, **75**: 17–36.

Ewing, M., Press, F., and Worzel, J. L. 1952. Further study of the T phase. *Bull. Seismol. Soc. Am.*, **42**: 37–51.

Ewing, M., Tolstoy, I., and Press, F. 1950. Proposed use of the T phase in tsunami warning systems. *Bull. Seismol. Soc. Am.*, **40**: 53–58.

Fairbridge, R. W. 1956. Geotectonic position of Antarctica. *Trans. Am. Geophys. Union*, **37**: 344.

Fairbridge, R. W. 1960. Comments. Unpublished transcription of meeting held in honor of Maurice Ewing's winning of the first Vetlesen Award held on March 25, 1960, at the Men's Faculty Club, Columbia University, 47–48.

Fairbridge, R. W. 1964. The importance of limestone and its Ca/Mg content to palaeoclimatology. In Nairn, A. E. M., ed., *Problems in Palaeoclimatology*. Interscience, New York, 431–476.

Fischer, A. G. 1963. Essay review of *Descriptive Palaeoclimatology*. *Am. J. Sci.*, **261**: 282–293.

Fischer, A. G. 1964. Growth patterns of Silurian tabulata as palaeoclimatologic and palaeogeographic tools. In Nairn, A. E. M., ed., *Problems in Palaeoclimatology*. Interscience, New York, 608–615.

Fisher, R. L. 1954. On the sounding of trenches. *Deep-Sea Res.*, **2**: 45–58.

Fisher, R. L. 1957. Geomorphic and seismic-refraction studies of the Middle America Trench, 1952–1956. Ph.D. thesis (unpublished), University of California, Los Angeles.

Fisher, R. L. 1961. Middle America Trench: topography and structure *Geol. Soc. Am. Bull.*, **72**: 703–720.

Fisher, R. L. and Goldberg, E. D. 1994. Henry William Menard. *Biogr. Mem. Natl. Acad. Sci.*, **64**: 267–276.

Fisher, R. L. and Hess, H. H. 1963. Trenches. In Hill, M. N., ed., *The Sea, Volume 3*. John Wiley & Sons, New York, 411–436.

Fisher, R. L. and Revelle, R. 1954. A deep sounding from the Southern Hemisphere. *Nature*, **171**: 469.

Fisher, R. L. and Revelle, R. 1955. The trenches of the Pacific. *Sci. Am.*, **193**: 36–41.

Flint, R. F. 1961. Geological evidence of cold climate. In Nairn, A. E. M., ed., *Descriptive Palaeoclimatology*. Interscience, New York, 61–88.

Frankel, H. R. 1980. Hess's development of his seafloor spreading hypothesis. In Nickles, T., ed., *Scientific Discovery: Case Histories*. D. Reidel, Dordrecht, 345–366.

Frankel, H. R. 1981. The paleobiogeographical debate over the problem of disjunctively distributed life forms. *Stud. Hist. Phil. Sci.*, **12**: 211–259.

Frankel, H. R. 1982. The development, reception, and acceptance of the Vine-Matthews-Morley hypothesis. *Stud. Hist. Phys. Sci.*, **13**: 1–39.

Frankel, H. R. 1984a. Biogeography, before and after the rise of sea floor spreading. *Stud. Hist. Phil. Sci.*, **15**: 141–168.

Frankel, H. R. 1984b. The Permo-Carboniferous ice cap and continental drift. *Compte rendu de Neuvième Congrès International de Stratigraphie et de Géologie du Carbonifère*, **1**: 113–120.

Friend, P. 2004. Walter Brian Harland, 1917–2003: obituary. *Geological Society, Annual Report for 2003*, **1**: 39.

George, W. 1962. *Animal Geography*. Heinemann, London.

Georgi, J. 1962. Memories of Alfred Wegener. In Runcorn, S. K., ed., *Continental Drift*. Academic Press, New York, 309–324.

Gilluly, J. 1963. The tectonic evolution of the western United States. *Q. J. Geol. Soc. London*, **119**: 133–174.

Girdler, R. W. 1958. The relationship of the Red Sea to the East African rift system. *Q. J. Geol. Soc. London*, **114**: 79–105.

Gold, T. 1955. Instability of the Earth's axis of rotation. *Nature*, **175**: 526–529.

Good, R. 1964. *The Geography of the Flowering Plants*. John Wiley & Sons, New York.

Gough, D. I. 1956. A study of the palaeomagnetism of the Pilansberg Dykes. *Mon. Not. Roy. Astron. Soc. Geophys. Suppl.*, **7**: 196–213.

Gough, D. I. 1989. Landmarks of a life in geophysics. *South African Geophysical Association Yearbook*, *1988*, 22–29.

Gough, D. I. and Brock, A. 1964. The paleomagnetism of the Shawa Ijolite. *J. Geophys. Res.*, **69**: 2489–2493.

Gough, D. I. and Opdyke, N. D. 1963. The paleomagnetism of the Lupata alkaline volcanics. *Geophys. J.*, **7**: 457–468.

Gough, D. I. and van Niekerk, C. D. 1959. On the paleomagnetism of the Bushveld gabbro. *Phil. Mag.*, **4**: 126–136.

Gough, D. I., Brock, A., Jones, D. L., and Opdyke, N. D. 1964. The paleomagnetism of the ring complexes at Marangudzi and the Mateke Hills. *J. Geophys. Res.*, **69**: 2499–2507.

Gough, D. I., Opdyke, N. D., and McElhinny, M. W. 1964. The significance of paleomagnetic results from Africa. *J. Geophys. Res.*, **69**: 2509–2519.

Grabau, A. 1940. *The Rhythm of the Ages, Earth History in the Light of the Pulsation and Polar Control Theories*. Henri Vetch, Peking.

Graham, J. W. 1956. Paleomagnetism and magnetostriction. *J. Geophys. Res.*, **61**: 735–739.

Graham, K. W. T. 1961. Palaeomagnetic studies on some South African rocks. Ph.D. dissertation, University of Capetown.

Graham, K. W. T. and Hales, A. L. 1957. Palaeomagnetic measurements on Karroo dolerites. *Adv. Phys.*, **6**: 149–161.

Green, Robert. 1961. Palaeoclimatic significance of evaporites. In Nairn, A. E. M., ed., *Descriptive Palaeoclimatology*. Interscience, New York, 61–88.

Green, Ronald. 1958. Polar wandering, a random walk problem. *Nature*, **182**: 382–383.

Green, R. and Irving, E. 1958. The palaeomagnetism of the Cainozoic basalts from Australia. *Proc. Roy. Soc. Victoria*, **70**: 1–17.

Greene, M. T. 1998. Alfred Wegener and the origin of lunar craters. *Earth Sci. Hist.*, **7**: 111–138.

Griggs, D. 1939. A theory of mountain-building. *Am. J. Sci.*, **237**: 611–650.

Griggs, D. 1951. Summary of the convection-current hypothesis of mountain building. *Trans. Am. Geophys. Union*, **32**: 527–528.

Griggs, D. 1954. Discussion, Verhoogen, 1954. *Trans. Am. Geophys. Union*, **35**: 93–96.

Gunn, R. 1936. On the origin of the continents and their motion. *J. Franklin Inst.*, **222**: 475–492.

Gunn, R. 1937. A quantitative study of mountain building on an unsymmetrical earth. *J. Franklin Inst.*, **224**: 19–53.

Gunn, R. 1947. Quantitative aspects of juxtaposed ocean deeps, mountain chains and volcanic ranges. *Geophys. J.*, **12**: 238–255.

Gunn, R. 1949. Isostasy extended. *J. Geol.*, **57**: 263–279.

Gutenberg, B. 1927. Die veranderungen der Erdkruste durch fliessbewegungen der kontinentalscholle. *Gerlands Beitr. Geophys.*, **16**: 239–247; **18**: 281–291.

Gutenberg, B. 1936. Structure of the Earth's crust and spreading of the continents. *Geol. Soc. Am. Bull.*, **47**: 1587–1610.

Gutenberg, B. (ed.). 1939. *Internal Constitution of the Earth*. McGraw-Hill, London.

Gutenberg, B. 1951a. Introduction. In Gutenberg, B., ed., *Internal Constitution of the Earth* (2nd edition). Dover Publications, New York, 1–7.

Gutenberg, B. 1951b. Hypotheses on the development of the Earth. In Gutenberg, B., ed., *Internal Constitution of the Earth* (2nd edition). Dover Publications, New York, 178–226.

Gutenberg, B. 1953. Response and acceptance of the Bowie Medal. *Trans. Am. Geophys. Union*, **34**: 354–355.

Gutenberg, B. 1959. *Physics of the Earth's Interior*. Academic Press, New York.

Gutenberg, B. and Richter, C. F. 1949. *Seismicity of the Earth*. Princeton University Press, Princeton, NJ.

Gutenberg, B. and Richter, C. F. 1954. *Seismicity of the Earth* (2nd edition). Princeton University Press, Princeton, NJ.

Hair, J. B. 1963. Cytogeographical relationships of the southern podocarps. In Gressitt, J. L., ed., *Pacific Basin Biogeography*. Bishop Museum Press, Honolulu, HI, 401–414.

Halm, J. K. E. 1935. An astronomical aspect of the evolution of the earth. *Astron. Soc. S. Afr.*, **4**: 1–28.

Hamblin, J. D. 2005. *Oceanographers and the Cold War: Disciples of Marine Science*. University of Washington Press, Seattle, WA.

Hamilton, E. L. 1952. Sunken islands of the mid-Pacific mountains. Ph.D. thesis, Stanford University Press.

Hamilton, E. L. 1956. Sunken islands of the Mid-Pacific mountains. *Geol. Soc. Am. Mem.*, **64**: 98.

Hamilton, E. L. 1960a. Ocean basin ages and amounts of original sediments. *J. Sediment. Petrol.*, **30**: 370–379.

Hamilton, W. 1960b. *New Interpretation of Antarctic Tectonics*. USGS Professional Paper 400-B: B379–380.

Hamilton, W. 1961. Origin of the Gulf of California. *Geol. Soc. Am. Bull.*, **72**: 1307–1318.

Hamilton, W. 1963. Polar wandering and continental drift: an evaluation of recent evidence. In Munyan, A. C., ed., *Polar Wandering and Continental Drift*. Society of Economic Paleontologists and Mineralogists, Special Publication No. 11, 74–93.

Hamilton, W. 1964a. Tectonic map of Antarctica – a progress report. In *Antarctic Geology, SCAR Proceedings 1963*. North-Holland Publishing, Amsterdam, 676–680.

Hamilton, W. 1964b. Discussion of paper by D. I. Axelrod, Fossil floras suggest stable, not drifting, continents. *J. Geophys. Res.*, **69**: 1666–1671.

Hamilton, W. 1965. *Diabase Sheets of the Taylor Glacier Region Victoria Land, Antarctica*. USGS Professional Paper 465-B: B1–B71.

Harland, W. B. 1958. The Caledonian sequence in Ny Friesland, Spitsbergen. *Q. J. Geol. Soc. London*, **114**: 307–342.

Harland, W. B. 1964a. Evidence of Late Precambrian glaciation and its significance. In Nairn, A. E. M., ed., *Problems in Palaeoclimatology*. Interscience, New York, 119–149.

Harland, W. B. 1964b. Critical evidence for a great infra-Cambrian glaciation. *Geologische Rundschau*, **54**: 61.

Harland, W. B. 2007. Origins and assessment of snowball Earth hypotheses. *Geol. Mag.*, **144**: 633–642.

Harland, W. B. and Bidgood, D. E. T. 1959. Palaeomagnetism in some Norwegian sparagmites and the Late Pre-Cambrian Ice Age. *Nature*, **184**: 1860–1862.

Harland, W. B. and Bidgood, D. E. T. 1961a. Palaeomagnetism in some East Greenland sedimentary rocks. *Nature*, **189**: 633–634.

Harland, W. B. and Bidgood, D. E. T. 1961b. Palaeomagnetic studies of some Greenland rocks. In Raasch, G. O., ed., *Geology of the Arctic: Proceedings of the First International Symposium on Arctic Geology*. University of Toronto Press, Toronto, 285–292.

Harland, W. B. and Rudwick, M. J. S. 1964. The great Infra-Cambrian Ice-Age. *Sci. Am.*, **212**: 28–36.

Harland, W. B. and Wilson, C. B. 1956. The Hecla Hoek succession in Ny Friesland, Spitsbergen. *Geol. Mag.*, **93**: 256–286.

Heck, N. H. 1935. A new map of earthquake distribution. *Geogr. Rev.*, **25**: 125–130.

Heezen, B. C. 1956. Outline of North Atlantic deep-sea geomorphology. *Geol. Soc. Am. Bull.*, **67**: 1703.

Heezen, B. C. 1957. Deep-sea physiographic provinces and crustal structure. *Trans. Am. Geophys. Union*, **38**: 394.

Heezen, B. C. 1959a. Géologie sous-marine et déplacements des continents. *Colloques Internationaux du Centre National de la Recherche Scientifique, LXXXIII. La Topographie et la Geologie des Profondeurs Oceaniques*, 295–304. Translation of paper and ensuing discussion by Annette Trefzer of the Lamont staff, 14 pages.

Heezen, B. C. 1959b. Paleomagnetism, continental displacements, and the origin of submarine topography. *International Ocean Congress preprints*. American Association for the Advancement of Science, Washington, DC, 26–28.

Heezen, B. C. 1959c. The tectonic evolution of the oceans. Unpublished paper presented on December 10, 1959, at Columbia University, 5 manuscript pages.

Heezen, B. C. 1960a. Comments. Unpublished transcription of meeting held in honor of Maurice Ewing's winning of the first Vetlesen Award held on March 25, 1960 at the Men's Faculty Club, Columbia University, 30–36a, 46.

Heezen, B. C. 1960b. The rift in the ocean floor. *Sci. Am.*, **203**: 98–110.

Heezen, B. C. 1960c. Tectonic evolution of the oceans. *Abstracts: International Union of Geodesy and Geophysics 12th General Assembly*, Helsinki, Finland, July 7– August 6, 1960, 15.

Heezen, B. C. 1962. The deep-sea floor. In Runcorn, S. K., ed., *Continental Drift*. Academic Press, New York, 235–288.

Heezen, B. C. 1974. Review of Hallam's *A Revolution in the Earth Sciences: From Continental Drift to Plate Tectonics*. *Science*, **183**: 504–505.

Heezen, B. C. and Ewing, M. 1952. Turbidity currents and submarine slumps and the 1929 Grand Banks earthquake. *Am. J. Sci.*, **250**: 849–873.

Heezen, B. C. and Ewing, M. 1961. The Mid-Oceanic Ridge and its extension through the Arctic Basin. In Raasch, G., ed., *Geology of the Arctic*. University of Toronto Press, Toronto, 622–642.

Heezen, B. C. and Ewing, M. 1963. The Mid-Ocean Ridge. In Hill, M. N., ed., *The Sea, Volume 3*. John Wiley & Sons, New York, 411–436.

Heezen, B. C. and Fox, P. J. 1966. Mid-Oceanic Ridge. In Fairbridge, R. W., ed., *The Encyclopedia of Oceanography, Encyclopedia of Earth Sciences, Series 1*. Reinhold, New York, 506–517.

Heezen, B. C. and Hollister, C. 1964. Turbidity currents and glaciation. In Nairn, A. E. M., ed., *Problems in Palaeoclimatology*. Interscience, New York, 99–107.

Heezen, B. C. and Hollister, C. D. 1971. *The Face of the Deep*. Oxford University Press, New York.

Heezen, B. C. and Tharp, M. 1965. Tectonic fabric of the Atlantic and India oceans and continental drift. In Blackett, P. M. S., Bullard, E., and Runcorn, S. K. eds., *A Symposium on Continental Drift, Phil. Trans. Roy. Soc. London A*, **258**: 90–106.

Heezen, B. C., Ericson, D. B., and Ewing, M. 1954a. Further evidence for a turbidity current following the 1929 Grand Banks earthquake. *Deep-Sea Res.*, **1**: 193–202.

Heezen, B. C., Ewing, M., and Ericson, D. B. 1954b. Reconnaissance survey of abyssal plain south of Newfoundland. *Deep-Sea Res.*, **2**: 122–133.

Heezen, B. C., Ewing, M., and Miller, E. T. 1953. Trans-Atlantic profile of total magnetic intensity and topography, Dakar to Barbados. *Deep-Sea Res.*, **1**: 25–33.

Heezen, B. C., Tharp, M., and Ewing, M. 1959. The floors of the oceans, I. the North Atlantic. *Geol. Soc. Am. Special Paper*, **65**.

Helsley, C. E. and Stehli, F. G. 1964. Comparison of Permina magnetic and zoogeographic poles. In A. E. M. Nairn, ed., *Problems in Palaeoclimatology*. Interscience, New York, 558–562.

Herries Davies, G. L. 2007. *Whatever is Under the Earth: The Geological Society of London 1807–2007*. The Geological Society Publishing House, Bath.

Hess, H. H. 1932. Interpretation of gravity anomalies and sounding-profiles obtained in the West Indies by the International Expedition to the West Indies in 1932. *Trans. Am. Geophys. Union*, 13th Annual Meeting, 26–32.

Hess, H. H. 1937. Geological interpretation of data collected on cruise of U.S.S. *Barracuda* in the West Indies – preliminary report. *Trans. Am. Geophys. Union*, 18th Annual Meeting, 69–77.

Hess, H. H. 1938a. Gravity anomalies and island arc structure with particular reference to the West Indies. *Proc. Am. Phil. Soc.*, **79**: 71–96.

Hess, H. H. 1938b. A primary peridotite magma. *Am. J. Sci.*, **35**: 321–344.

Hess, H. H. 1939. Island arcs, gravity anomalies, and serpentine intrusions: a contribution to the ophiolite problem. *17th International Geological Congress, Moscow Report, Volume 2*, 263–283. (Russian version is 279–399.)

Hess, H. H. 1940. Appalachian peridotite belt: its significance in sequence of events in mountain building. *Geol. Soc. Am. Bull.*, **51**: 1996.

Hess, H. H. 1946. Drowned ancient islands of the Pacific Basin. *Am. J. Sci.*, **244**: 772–791.

Hess, H. H. 1951. Comment on mountain building. In *1950 Colloquium on Plastic Flow and Deformation within the Earth*. *Trans. Am. Geophys. Union*, **32**: 528–531.

Hess, H. H. 1954. Geological hypotheses and the Earth's crust under the oceans. *Proc. Roy. Soc. London A*, **222**: 341–348.

Hess, H. H. 1955a. Serpentines, orogeny, and epeirogeny. *Geol. Soc. Am. Special Paper*, **62**: 391–406.

Hess, H. H. 1955b. The oceanic crust. *J. Marine Res.*, **14**, 423–439.

Hess, H. H. 1959a. Nature of the great oceanic ridges. *International Ocean Congress preprints*. American Association for the Advancement of Science, Washington, DC, 33–34.

Hess, H. H. 1959b. The AMSOC hole to the Earth's mantle. *Trans. Am. Geophys. Union*, **40**, 340–345.

Hess, H. H. 1960a. Scientific objectives of Mohole, and predicted section. *Am. Assoc. Petrol. Geol.*, **44**: 1250.

Hess, H. H. 1960b. Scientific objectives of Mohole, and predicted section. *Bull. Geol. Soc. Am.*, **71**: 2097. Reprint of (1960a).

Hess, H. H. 1960c. The AMSOC hole to the Earth's mantle. *Am. Sci.*, **47**: 254–263. Reprint of (1959b).

Hess, H. H. 1960d. Evolution of ocean basins. Preprint, 37 pages.

Hess, H. H. 1962. History of ocean basins. In *Petrologic Studies: A Volume to Honor A. F. Buddington*. Geological Society of America, New York, 599–620.

Hess, H. H. 1965. Mid-oceanic ridges and tectonics of the sea-floor. In Whittard, W. F. and Bradshaw, R., eds., *Submarine Geology and Geophysics*. Butterworths, London, 317–332.

Hess, H. H. 1968a. Response by Harry Hammond Hess, Penrose Medal. *Geol. Soc. Am. Proc. Vol. for 1966*: 85–86.

Hess, H. H. 1968b. Reply. *J. Geophys. Res.*, **73**: 65–69.

Hess, H. H. and Maxwell, J. C. 1953. Caribbean research project. *Geol. Soc. Am. Bull.*, **64**: 1–6.

Hibberd, F. H. 1962. An analysis of the positions of the Earth's magnetic pole in the geological past. *Geophys. J.*, **6**: 221–244.

Hilgenberg, O. C. 1933. *Vom wachsenden Erdball*. Giessmann and Bartsch, Berlin.

Hill, D. 1970. Clarke Memorial Lecture for 1970: the bearing of some upper Palaeozoic reefs on coral faunas on the hypothesis of continental drift. *J. Proc. Roy. Soc. NSW*, **103**: 93–102.

Hill, M. N. 1954. The topography of the Mid-Atlantic Ridge. *Procès verbaux, General Assembly at Rome, September 1954: 269*. Bergen Geofysisk Institutt, Bergen.

Hill, M. N. 1957a. Geophysical investigations on the floor of the Atlantic Ocean in *Discovery II*, 1956. *Nature*, **180**: 10–13.

Hill, M. N. 1957b. The sequence and distribution of Upper Palaeozoic coral faunas. *Aust. J. Sci.*, **19**: P42–P61.

Hill, M. N. 1959. Distribution and sequence of Silurian coral faunas. *J. Proc. Roy. Soc. NSW*, **92**: 151–173.

Hills, G. F. S. 1947. *The Formation of the Continents by Convection*. Arnold, London.

Hodgson, J. H. 1957. Nature of faulting in large earthquakes. *Geol. Soc. Am. Bull.*, **68**: 611–643.

Hodgson, J. H. 1962. Movements in the Earth's crust as indicated by Earthquakes. In Runcorn, S. K., ed., *Continental Drift*. Academic Press, New York, 67–102.

Hodgson, J. H. and Stevens, A. E. 1964. Seismicity and earthquake mechanism. *Res. Geophys.*, **2**: 27.

Hollingworth, S. E. 1962. The climatic factor in the geological record. *Q. J. Geol. Soc. London*, **118**: 1–21.

Holmes, A. 1931. Radioactivity and earth movements. *Trans. Geol. Soc. Glasgow (for 1928–9)*, **18**: 559–606.

Holmes, A. 1944. *Principles of Physical Geology* (1st edition). Thomas Nelson & Sons, Edinburgh.

Holmes, A. 1953. The South Atlantic: land bridges or continental drift? *Nature*, **171**: 669–671.

Holmes, A. 1957. Response by Arthur Holmes to presentation of Penrose Medal. *Proc. Vol. Geol. Soc. Am. 1966*: 74–75.

Holmes, A. 1959. A revised geological time-scale. *Trans. Edinburgh Geol. Soc.*, **17**: 183–216.

Holmes, A. 1965. *Principles of Physical Geology* (2nd edition). Thomas Nelson, London.

Hulley, J. C. L. 1964. Correlation between gravity anomalies, transcurrent faults and pole positions. In Nairn, A. E. M., ed., *Problems in Palaeoclimatology*. Interscience, New York, 224–226.

Irving, E. 1956. Palaeomagnetic and palaeoclimatological aspects of polar wandering. *Geofis. Pura Appl.*, **33**: 23–48.

Irving, E. 1957. Rock magnetism: a new approach to some palaeogeographic problems. *Adv. Phys.*, **6**: 194–218.

Irving, E. 1958a. Rock magnetism: a new approach to the problems of polar wandering and continental drift. In Carey, S. W., Convener, *Continental Drift: a Symposium*. University of Tasmania, Hobart, 24–57.

Irving, E. 1958b. Palaeogeographic reconstruction from palaeomagnetism. *Geophys. J. Roy. Astron. Soc.*, **1**: 224–237.

Irving, E. 1959. Palaeomagnetic pole positions. *Geophys. J.*, **2**: 51–79.

Irving, E. 1960. Palaeomagnetic directions and pole positions, part II. *Geophys. J.*, **3**: 444–449.

Irving, E. 1961. Paleomagnetic methods: a discussion of a recent paper by A. E. M. Nairn. *J. Geol.*, **69**: 226–231.

Irving, E. 1962a. Descriptive palaeoclimatology. *Geophys. J.*, **6**: 268.

Irving, E. 1962b. An analysis of the positions of the Earth's magnetic pole in the geological past. *Geophys. J.*, **7**: 279–283.

Irving, E. 1962c. Palaeomagnetic directions and pole positions, part IV. *Geophys. J.*, **6**: 263–267.

Irving, E. 1963. Paleomagnetism of the Narrabeen Chocolate shales and Tasmanian dolerite. *J. Geophys. Res.*, **68**: 2283–2287.

Irving, E. 1964. *Paleomagnetism and Its Application to Geological and Geophysical Problems*. John Wiley & Sons, New York.

Irving, E. 1966. Paleomagnetism of some Carboniferous rocks from New South Wales and its relation to geological events. *J. Geophys. Res.*, **71**: 6025–6051.

Irving, E. 1991. Citation for Kenneth M. Creer for the John Adam Fleming Medal. *Eos*, **72**: 54–55.

Irving, E. 2000. Continental drift, organic evolution, and moral courage. *Eos*, **81**: 546.

Irving, E. 2008. Jan Hospers's key contributions to geomagnetism. *Eos Trans. AGU*, **89**: 457–468.

Irving, E. and Briden, J. C. 1962. Palaeolatitude of evaporite deposits. *Nature*, **196**: 425–428.

Irving, E. and Brown, D. A. 1964. Abundance and diversity of the Labyrinthodonts as a function of paleolatitude. *Am. J. Sci.*, **262**: 689–708.

Irving, E. and Brown, D. A. 1966. Reply to Stehli's discussion of Labyrinthodont abundance and diversity. *Am. J. Sci.*, **264**: 488–496.

Irving, E. and Gaskell, T. F. 1962. The palaeogeographic latitude of oil fields. *Geophys. J.*, **7**: 54–63.

Irving, E. and Green, R. 1957a. Palaeomagnetic evidence from the Cretaceous and Cainozoic. *Nature*, **179**: 1064–1065.

Irving, E. and Green, R. 1957b. The palaeomagnetism of the Kainozoic basalts of Victoria. *Mon. Not. Roy. Astron. Soc. Geophys. Suppl.*, **7**: 347–359.

Irving, E. and Green, R. 1958. Polar movement relative to Australia. *Geophys. J. Roy. Astron. Soc.*, **1**: 164–172.

Irving, E., North, F., and Couillard, R. 1974. Oil, climate and tectonics. *Can. J. Earth Sci.*, **11**: 1–17.

Irving, E., Robertson, W. A., and Stott, P. M. 1963. The significance of the paleomagnetic results from Mesozoic rocks of eastern Australia. *J. Geophys. Res.*, **68**: 2313–2317.

Isacks, B., Oliver, J., and Sykes, L. 1968. Seismology and the new global tectonics. *J. Geophys. Res.*, **73**: 5855–5899.

Jeffreys, H. 1926. The rigidity of the Earth's central core. *Mon. Not. R. Astron. Soc. Geophys. Suppl.*, **1**: 371–383.

Jeffreys, H. 1929. *The Earth, Its Origin, History and Physical Constitution* (2nd edition). Cambridge University Press, Cambridge.

Jeffreys, H. 1939. *Theory of Probability*. The Clarendon Press, Oxford.

Jeffreys, H. 1948. *Theory of Probability* (2nd edition). The Clarendon Press, Oxford.

Jeffreys, H. 1952. *The Earth: Its Origin, History, and Physical Characteristics* (3rd edition). Cambridge University Press, Cambridge.

Jeffreys, H. 1959. *The Earth: Its Origin, History, and Physical Characteristics* (4th edition). Cambridge University Press, Cambridge.

Jeffreys, H. 1960. Beno Gutenberg. *Q. J. Roy. Astron. Soc.*, **1**: 239–242.

Jeffreys, H. 1961, 1967, 1983, 1998. *Theory of Probability* (3rd edition). Oxford Classic Texts in the Physical Sciences, Oxford University Press, Oxford.

Jeffreys, H. 1962. *The Earth: Its Origin, History, and Physical Characteristics* (4th edition. Reprinted with additions). Cambridge University Press, Cambridge.

Jeffreys, H. 1964. How soft is the Earth? *Q. J. Roy. Astron. Soc.*, **5**: 10–22.

Jones, D. L. 1965. Review of *Stratigraphy and Life History* by Kay and Colbert. *Science*, **148**: 488–489.

Jones, D. L., Duncan, R. A., Briden, J. C., Randall, D. E., and MacNiocaill, C. 2001. Age of the Batoka basalts, northern Zimbabwe, and the duration of Karroo large Igneous Province magmatism. *Geochem. Geophys. Geosyst.*, **2**, doi:10.1029/2000GC000110.

Kay, M. 1952a. Stratigraphic evidence bearing on the hypothesis of continental drift. In Mayr, E., ed., *The Problem of Land Connections Across the South Atlantic, with Special Reference to the Mesozoic. Bull. Am. Mus. Nat. Hist.*, **99**: 159–162.

Kay, M. 1952b. Modern and ancient island arcs. *Palaeobotanist*, **1**: 281–283.

Kay, M. and Colbert, E. M. 1965. *Stratigraphy and Life History*. Wiley Bibliography, New York.

Kent, D. V. and Irving, E. 2010. Influence of inclination error in sedimentary rocks on the Triassic and Jurassic apparent pole wander path for North America and implications for Cordilleran tectonics. *J. Geophys. Res.*, **57**: 227–286.

King, L. 1961. The palaeoclimatology of Gondwanaland during the Palaeozoic and Mesozoic eras. In Nairn, A. E. M., ed., *Descriptive Palaeoclimatology*. Interscience, New York, 307–331.

King, L. 1962. *The Morphology of the Earth*. Oliver and Boyd, Edinburgh.

Knopoff, L. 1999. Beno Gutenberg. *Natl. Acad. Sci. Biogr. Mem.*, **76**: 115–148.

Köppen, W. and Wegener, A. 1924. *Die Klimate der Geologischen Vorzeit*. Gebrüder Borntraeger, Berlin.

Krantz, W. 1924. *Begleitworte zur geognostischen Spezialkarte von Wurttemberg: Atlasblatt Heidenheim, Aufl.* **2**. Wurtt, Stuttgart, 52–105.

Kraüsel, R. 1961. Introduction to the palaeoclimatic significance of coal. In Nairn, A. E. M., ed., *Descriptive Palaeoclimatology*. Interscience, New York, 53–56.

Kuenen, Ph. H. 1936. The negative isostatic anomalies in the East Indies (with experiments). *Leidse Geol. Meded.*, **8**: 169–214.

Kuenen, Ph. H. 1937. Experiments in connection with Daly's hypothesis on the formation of submarine canyons. *Leidse Geol. Meded.*, **8**: 327–335.

Kuenen, Ph. H. 1952. Estimated size of the Grand Banks turbidity current. *Am. J. Sci.*, **250**: 874–887.

Kuenen, Ph. H. and Migliorini, C. I. 1950. Turbidity currents as a cause of graded beddings. *J. Geol.*, **58**: 91–127.

Lamb, H. H. 1964. The role of atmosphere and oceans in relation to climatic changes and the growth of ice-sheets on land. In Nairn, A. E. M., ed., *Descriptive Palaeoclimatology*. Interscience, New York, 332–348.

Laming, D. J. C. 1958. Fossil winds. *J. Alberta Soc. Petrol. Geol.*, **6**: 179–183.

Leet, D. L., Linehan, D., and Berger, P. R. 1951. Investigation of the T phase. *Bull. Seismol. Soc. Am.*, **41**: 123–141.

Lenk, C., Strother, P. K., Kaye, C. A., and Barghoorn, E. S. 1982. Precambrian age of the Boston Basin: new evidence from microfossils. *Science*, **216**: 619–620.

Linehan, D. 1940. Earthquakes in the West Indian region. *Trans. Am. Geophys. Union*, Part II, 229–232.

Longwell, C. R. 1945. The mechanics of orogeny. *Am. J. Sci.*, **243A**: 417–447.

Lotze, F. 1964. The distribution of evaporites in space and time. In Nairn, A. E. M., ed., *Problems in Palaeoclimatology*. Interscience, New York, 491–506.

Lovis, J. D. 1960. Discussion. *Proc. R. Soc. London B*, **152**: 669–670.

Lowenstam, H. A. 1964. Palaeotemperatures of the Permian and Cretaceous periods. In Nairn, A. E. M., ed., *Problems in Palaeoclimatology*. Interscience, New York, 227–246.

Maack, R. 1964. Characteristic features of the paleogeography and stratigraphy of the Devonian of Brazil and South Africa. In Nairn, A. E. M., ed., *Problems in Palaeoclimatology*. Interscience, New York, 285–293.

MacDonald, G. J. F. 1963a. Review of Runcorn, S. K., ed., *Continental Drift. Trans. Am. Geophys. Union*, **44**: 602–603.

MacDonald, G. J. F. 1963b. The deep structure of continents. *Rev. Geophys.*, **1**: 587–665.

MacDonald, G. J. F. 1964. The deep structure of continents. *Science*, **143**: 921–929.

MacDonald, G. J. F. 1966. The figure and long-term mechanical properties of the Earth. In Hurley, P. M., ed., *Advances in Earth Science*, MIT Press, Cambridge, MA, 199–245.

Mackey, R. 2007. Rhodes Fairbridge and the idea that the solar system regulates the Earth's climate. *J. Coastal Res.*, **50** (special issue): 955–968.

Marvin, U. B. 1973. *Continental Drift: The Evolution of a Concept*. Smithsonian Institution Press, Washington, DC.

Mason, R. G. 1958. A magnetic survey off the west coast of the United States between Latitudes 32° and 36° N, Longitudes 121° and 128° W. *Geophys. J.*, **1**: 320–329.

Mason, R. G. 2001. Stripes on the sea floor. In Oreskes, N., ed., with Le Grand, H., *Plate Tectonics*. Westview Press, Boulder, CO, 31–45.

Mason, R. G. and Raff, A. D. 1961. Magnetic survey off the west coast of North America, 32° N to 42° N. *Geol. Soc. Am. Bull.*, **72**: 1259–1266.

McElhinny, M. W. 1973. *Paleomagnetism and Plate Tectonics*. Cambridge University Press, Cambridge.

McElhinny, M. W. and Gough, D. I. 1963. The paleomagnetism of the Great Dyke of Southern Rhodesia. *Geophys. J.*, **7**: 287–303.

McElhinny, M. W. and Opdyke, N. D. 1964. The paleomagnetism of the Precambrian dolerites of Eastern Southern Rhodesia, an example of geologic correlation by rock magnetism. *J. Geophys. Res.*, **69**: 2465–2475.

McKee, E. D. 1964. Problems on the recognition of arid and of hot climates of the past. In Nairn, A. E. M., ed., *Problems in Palaeoclimatology*. Interscience, New York, 367–377.

McKenzie, D. P. 1987. Edward Crisp Bullard. *Biogr. Mem. Fell. Roy. Soc.*, **33**: 65–98.

McKenzie, D. and Parker, R. L. 1967. The North Pacific: an example of tectonics on a sphere. *Nature*, **216**: 1276–1280.

Menard, H. W. 1955a. Deformation of the northeastern Pacific Basin and the west coast of North America. *Geol. Soc. Am. Bull.*, **66**: 1149–1198.

Menard, H. W. 1955b. Deep-sea channels, topography, and sedimentation. *Bull. Am. Assoc. Petrol. Geol.*, **39**: 236–255.

Menard, H. W. 1958a. Development of median elevations in ocean basins. *Geol. Soc. Am. Bull.*, **69**: 1179–1186.

Menard, H. W. 1958b. Geology of the Pacific sea floor. *Experientia*, **15**: 205–213.

Menard, H. W. 1960. The East Pacific Rise. *Science*, **132**: 1737–1746.

Menard, H. W. 1961. The East Pacific Rise. *Sci. Am.*, **205**(6): 52–61.

Menard, H. W. 1964. *Marine Geology of the Pacific*. McGraw-Hill, New York.

Menard, H. W. 1971. *Science, Growth and Change*. Harvard University Press, Cambridge, MA.

Menard, H. W. 1979. Very like a spear. In *Two Hundred Years of Geology in America*. University Press of New England, Hanover, NH, 19–30.

Menard, H. W. 1986. *The Ocean of Truth: A Personal History of Global Tectonics*. Princeton University Press, Princeton, NJ.

Menard, H. W. and Dietz, R. S. 1951. Submarine geology of the Gulf of Alaska. *Geol. Soc. Am. Bull.*, **62**: 1263–1285.

Menard, H. W. and Dietz, R. S. 1952. Mendocino submarine escarpment. *J. Geol.*, **60**: 266–278.

Menard, H. W. and Vacquier, V. 1958. Magnetic survey of part of the deep sea floor off the coast of California. *United States Naval Research Office Research Reviews, June issue*: 1–6.

Meyerhoff, A. A. 1968. Arthur Holmes: originator of spreading ocean floor hypothesis. *J. Geophys. Res.*, **73**: 6563–6565.

Moores, E. M. 2003. A personal history of the ophiolite concept. In Dilek, Y. and Newcomb, S., eds., *Ophiolite Concept and the Evolution of Geological Thought*. *Geol. Soc. Am. Special Paper*, **373**: 17–29.

Mumme, W. G. 1962a. A note on the mixed polarity of magnetization in Cainozoic basalts in Victoria, Australia. *Geophys. J.*, **6**: 546–549.

Mumme, W. G. 1962b. Stability of magnetization in Cainozoic basalts of Victoria, Australia. *Phil. Mag.*, **7**: 1263–1278.

Mumme, W. G. 1963. Thermal and alternating magnetic field demagnetization experiments on Cainozoic basalts of Victoria, Australia. *Geophys. J.*, **7**: 314–327.

Munk, W. H. 1978. Dedication. In Jorna, S., ed., *Topics in Nonlinear Dynamics: A Tribute to Sir Edward Bullard*. American Institute of Physics, New York, v–vii.

Munk, W. H. and MacDonald, G. J. F. 1960a. *The Rotation of the Earth*. Cambridge University Press, Cambridge.

Munk, W. H. and MacDonald, G. J. F. 1960b. Continentality and the gravitational field of the Earth. *J. Geophys. Res.*, **65**: 2169–2172.

Murray, H. W. 1939. Submarine scarp off Mendocino, California. *Field Engineers Bull.*, **13**: 27–33.

Muttoni, G., Kent, D. V., Garzanti, E., Branck, P., Abrahamsen, N., and Gaetani, M. 2003. Early Pangea 'B' to Late Permian 'A'. *Earth Planet. Sci. Lett.*, **215**: 379–394.

Nagata, T. and Kobayashi, K. 1961. Palaeomagnetism and archaeomagnetism. In Nagata, T., ed., *Rock Magnetism (revised edition)*. Tokyo Maruzen Company Ltd.

Nagata, T., Akimoto, S., Shimizu, Y., Kobayashi, K., and Kuno, H. 1959. Paleomagnetic studies on Tertiary and Cretaceous rocks in Japan. *Proc. Japan Acad.*, **35**: 378–383.

Nairn, A. E. M. 1957. Palaeomagnetic collections from Britain and South Africa illustrating two problems of weathering. *Adv. Phys.*, **6**: 162–168.

Nairn, A. E. M. 1959. A palaeomagnetic survey of the Karroo system. *Overseas Geol. Min. Res.*, **7**: 398–410.

Nairn, A. E. M. 1961. The scope of palaeoclimatology. In Nairn, A. E. M., ed., *Descriptive Palaeoclimatology*. Interscience, New York, 45–59.

Nairn, A. E. M. 1963. Palaeomagnetic measurements on the Great Dyke, Southern Rhodesia. *Phil. Mag.*, **8**: 213–221.

Nairn, A. E. M. (Ed.). 1964a. *Problems in Palaeoclimatology*. Interscience, New York.

Nairn, A. E. M. 1964b. Introduction. In Nairn, A. E. M., ed., *Problems in Palaeoclimatology*. Interscience, New York, 11–13.

Nairn, A. E. M. 1964c. Palaeomagnetic measurements on Karroo and Post-Karroo rocks: a second progress report. *Overseas Geol. Min. Res.*, **9**: 302–320.

Nairn, A. E. M. and Thorley, N. 1961. Geophysical evidence of climates. In Nairn, A. E. M., ed., *Descriptive Palaeoclimatology*. Interscience, New York, 156–182.

Northrop, J. and Heezen, B. C. 1951. An outcrop of Eocene sediment of the continental slope. *J. Geol.*, **59**: 396–399.

Officer, C. B. and Drake, C. L. 1986. Cretaceous-Tertiary extinctions: alternative models. *Science*, **230**: 1294–1295.

Officer, C. B. and Drake, C. L. 1989. Iridium, shocked minerals and the Cretaceous/ Tertiary boundary. *Physics Today*, **42**: 545–546.

Officer, C. B., Ewing, J., Edwards, R. S., and Johnson, H. R. 1957. Geophysical investigations in the Eastern Caribbean: Trinidad Shelf, Tobago Trough, Barbados Ridge and Atlantic Ocean. *Geol. Soc. Am. Bull.*, **68**: 359–378.

Officer, C. B., Ewing, M., and Wuenschel, P. C. 1952. Seismic refraction measurements in the Atlantic Ocean, Part IV, Bermuda, Bermuda Rise and Nares Basin. *Geol. Soc. Am. Bull.*, **63**: 777–808.

Opdyke, N. D. 1959. The impact of paleomagnetism on paleoclimatic studies. *Int. J. Bioclimatol. Biometeorol.*, **3**, Part VI, Section A: 1–11.

Opdyke, N. D. 1961. The Palaeoclimatological significance of desert sandstone. In Nairn, A. E. M., ed., *Descriptive Palaeoclimatology*. Interscience, New York, 45–59.

Opdyke, N. D. 1962. Palaeoclimatology and continental drift. In Runcorn, S. K., ed., *Continental Drift*. Academic Press, New York, 41–65.

Opdyke, N. D. 1964a. The paleomagnetism of the Permian red beds of Southwest Tanganyika. *J. Geophys. Res.*, **69**: 2477–2487.

Opdyke, N. D. 1964b. The paleomagnetism of some Triassic red beds from Northern Rhodesia. *J. Geophys. Res.*, **69**: 2495–2497.

Opdyke, N. D. and Runcorn, S. K. 1959. Paleomagnetism and ancient wind directions. *Endeavour*, **18**: 26–34.

Opdyke, N. D. and Runcorn, S. K. 1960. Wind direction in the western United States in the late Palaeozoic. *Geol. Soc. Am. Bull.*, **71**: 959–972.

Opdyke, N. D., Roberts, J., Claoue-Long, J., Irving, E., and Jones, P. J. 2000. Base of the Kiaman: its definition and global stratigraphic significance. *Geol. Soc. Am. Bull.*, **112**: 1315–1341.

Oreskes, N. 1999. *The Rejection of Continental Drift*. Oxford University Press, New York.

Phillips, J. D. and Forsyth, D. 1972. Plate tectonics, paleomagnetism, and the opening of the Atlantic. *Geol. Soc. Am. Bull.*, **83**: 1579–1600.

Phinney, R. A. 1968. Introduction. In Phinney, R. A., ed., *The History of the Earth's Crust*. Princeton University Press, Princeton, NJ, 3–12.

Plumstead, E. 1961. Ancient plants and drifting continents. *S. Afr. J. Sci.*, **57**: 173–181.

Plumstead, E. 1962. Fossil floras of Antarctica. *Trans-Atlantic Expedition Reports 1955–1958*, **9**: 1–154.

Poole, F. G. 1957. Paleo-wind directions in late Paleozoic and early Mesozoic time on the Colorado plateau as determined by cross-strata. *Geol. Soc. Am. Bull.*, **68**: 1870.

Press, F. 1960. Comments. Unpublished transcription of meeting held in honor of Maurice Ewing's winning of the first Vetlesen Award held on March 25, 1960 at the Men's Faculty Club, Columbia University, 40–43.

Press, F., Ewing, M., and Tolstoy, I. 1950. The Airy phase of shallow-focus submarine earthquakes. *Bull. Seismol. Soc. Am.*, **40**: 111–148.

Radforth, E. D. 1966. The ancient flora and continental drift. In Garland, G. D., ed., *Continental Drift*. Toronto University Press, Toronto, 53–70.

Raff, A. D. and Mason, R. G. 1961. Magnetic survey off the west coast of North America, 40° N. Latitude to 52° N. Latitude. *Geol. Soc. Am. Bull.*, **72**: 1267–1270.

Rainger, R. 2000. Science at the crossroads: the navy, Bikini Atoll, and American oceanography in the 1940s. *Hist. Stud. Phys. Sci.*, **30**: 349–371.

Raitt, H. 1956. *Exploring the Deep Pacific*. W.W. Norton, New York.

Raitt, R. W. 1956. Seismic refraction studies of the Pacific Ocean Basin. *Geol. Soc. Am. Bull.*, **67**: 1623–1640.

Raitt, R. W., Fisher, R. L., and Mason, R. G. 1954. Tonga Trench. *Symposium on the Crust of the Earth*, Columbia University, October 13–16, 1954. Abstract.

Raitt, R. W., Fisher, R. L., and Mason, R. G. 1955. Tonga Trench. *Geol. Soc. Am. Special Paper*, **62**: 237–254.

Rastall, R. H. 1929. On continental drift and cognate subjects. *Geol. Mag.*, **66**: 447–456.

Read, H. H. 1949. *Geology: An Introduction to Earth History*. Oxford University Press, London.

Richter, C. 1962. Memorial to Beno Gutenberg (1889–1960). *Proc. Geol. Soc. Am.*, **1**: 93–104.

Rikitake, T. 1966. *Electromagnetism and the Earth's Interior*. Elsevier, Amsterdam.

Robertson, W. A. 1963. The paleomagnetism of some Mesozoic intrusive and tuffs from eastern Australia. *J. Geophys. Res.*, **68**: 2299–2312.

Robertson, W. A. and Hastie, L. 1962. A palaeomagnetic study of the Cygnet alkaline complex of Tasmania. *J. Geol. Soc. Aust.*, **8**: 259–268.

Roman, C. 2000. *Continental Drift: Colliding Continents, Converging Cultures*. Institute of Physics Publishing, Bristol.

Romer, A. S. 1961. Evidence of climate from vertebrates. In Nairn, A. E. M., ed., *Descriptive Palaeoclimatology*. Interscience, New York, 183–206.

Rothé, J. P. 1954. La zone seismique mediane Indo-Atlantique. *Proc. Roy. Soc. London*, **222**: 387–397.

Rudwick, M. J. S. 1964. The infra-Cambrian glaciation and the origin of the Cambrian fauna. In Nairn, A. E. M., ed., *Problems in Palaeoclimatology*. Interscience, New York, 150–155.

Runcorn, S. K. 1955a. Rock magnetism – geophysical aspects. *Adv. Phys.*, **4**: 244–291.

Runcorn, S. K. 1955b. The permanent magnetization of rocks. *Endeavour*, July: 152–159.

Runcorn, S. K. 1956a. Paleomagnetic survey in Arizona and Utah: preliminary results. *Geol. Soc. Am. Bull.*, **87**: 301–316.

Runcorn, S. K. 1956b. Palaeomagnetic comparisons between Europe and North America. *Proc. Geol. Assoc. Can.*, **8**: 77–85.

Runcorn, S. K. 1956c. Paleomagnetism, polar wandering and continental drift. *Geol. Mijnbouw*, **18**: 253–258.

Runcorn, S. K. 1957. The sampling of rocks for palaeomagnetic comparisons between the continents. *Adv. Phys.*, **6**: 169–176.

Runcorn, S. K. 1959a. On the hypothesis that the mean geomagnetic field for parts of geological time has been that of a geocentric axial multipole. *J. Atmos. Terr. Phys.*, **14**: 167–174.

Runcorn, S. K. 1959b. Discussion on the Permian climate zonation and paleomagnetism. *Am. J. Sci.*, **257**: 235–237.

Runcorn, S. K. 1959c. Rock magnetism. *Science*, **129**: 1002–1012.

Runcorn, S. K. 1960. Evidence for continental drift. *New Sci.*, **8**: 1689.

Runcorn, S. K. 1961. Climatic change through geological time in the light of the palaeomagnetic evidence for polar wandering and continental drift. *Q. J. Roy. Meteorol. Soc.*, **87**: 282–313.

Runcorn, S. K. 1962a. Towards a theory of continental drift. *Nature*, **193**: 311–314.

Runcorn, S. K. 1962b. Paleomagnetic evidence for continental drift and its geophysical cause. In Runcorn, S. K., ed., *Continental Drift*. Academic Press, New York, 1–39.

Runcorn, S. K. 1962c. Convection currents in the Earth's mantle. *Nature*, **193**: 1248–1249.

Runcorn, S. K. 1963. Satellite gravity measurements and convection in the mantle. *Nature*, **200**: 628–630.

Runcorn, S. K. 1964a. Paleomagnetic results from Precambrian sedimentary rocks in the western United States. *Geol. Soc. Am. Bull.*, **75**: 687–704.

Runcorn, S. K. 1964b. Paleowind directions and palaeomagnetic latitudes. In Nairn, A. E. M., ed., *Problems in Palaeoclimatology*. Interscience, New York, 409–419.

Runcorn, S. K. 1964c. Foreword. In Nairn, A. E. M., ed., *Problems in Palaeoclimatology*. Interscience, New York, v.

Runcorn, S. K. 1964d. Satellite gravity measurements and a laminar viscous flow model of the Earth's mantle. *J. Geophys. Res.*, **69**: 4389–4394.

Runcorn, S. K. 1965. Geophysics and palaeoclimatology. *Q. J. Roy. Meteorol. Soc.*, **91**: 257–267.

Runcorn, S. K. 1974. Geology ⊃ geophysics? *Nature*, **249**: 794.

Schlee, S. 1973. *The Edge of an Unfamiliar World: A History of Oceanography.* E. P. Dutton, New York.

Schmidt, P. W. 1976. The non-uniqueness of the Australian Mesozoic palaeomagnetic pole position. *Geophys. J. Roy. Astron. Soc.*, **47**: 285–300.

Schmidt, P. W. and Embleton B. J. J. 1981. Magnetic overprinting in southeastern Australia and the thermal history of its rifted margin. *J. Geophys. Res.*, **86**: 3998–4008.

Schove, D. J., Nairn, A. E. M., and Opdyke, N. D. 1958. The climatic geography of the Permian. *Geogr. Ann., Stockholm*, **40**: 216–231.

Schwarzbach, M. 1961a. *Das klima der Vorzeit* (2nd edition). Enke, Stuttgart.

Schwarzbach, M. 1961b. The climatic history of Europe and North America. In Nairn, A. E. M., ed., *Descriptive Palaeoclimatology*. Interscience, New York, 255–291.

Schwarzbach, M. 1963. *Climates of the Past*. English translation from the 2nd German edition by Muir, R. O. D. Van Nostrand, London.

Sclater, J. G. 2001. Heat flow under the oceans. In Oreskes, N., ed., with Le Grand, H., *Plate Tectonics*. Westview Press, Boulder, CO, 128–147.

Shand, S. J. 1949. Rocks of the Mid-Atlantic Ridge. *Geol. Soc. Am. Bull.*, **57**: 89–92.

Shepard, F. P. and Emery, K. O. 1941. *Submarine Topography Off the California Coast. Geol. Soc. Am. Special Paper*, **31**.

Shirley, J. 1964. The distribution of lower Devonian faunas. In Nairn, A. E. M., ed., *Problems in Palaeoclimatology*. Interscience, New York, 255–261.

Shor, E. N. and Sclater, J. G. 2010. Victor Vacquier (1907–2009), *Eos Trans. AGU*, **91**: 264.

Shor, G. G. and Fisher, R. L. 1961. Middle America trench: seismic-refraction studies. *Geol. Soc. Am. Bull.*, **72**: 721–730.

Shotton, F. W. 1937. The Lower Bunter sandstones of North Worcestershire and East Shropshire. *Geol. Mag.*, **74**: 534–553.

Shotton, F. W. 1956. Some aspects of the New Desert in Britain. *Liverpool Manchester Geol. J.*, **1**: 450–466.

Simpson, G. G. 1965. *The Geography of Evolution*. Chilton Books, Philadelphia.

Smuts, J. C. 1925. Presidential Address delivered July 6, 1925. *S. Afr. J. Sci.*, **22**: 1–19.

Socci, A. D. and Smith, G. 1987. Recent sedimentological interpretations in the Avalon Terrane of the Boston Basin, Massachusetts. *J. Atlantic Geosci. Soc.*, **23**: 13–39.

Steenis, C. G. G. J. van. 1962. The land-bridge theory in botany. *Blumea*, **11**: 235–372.

Stehli, F. G. 1957. Possible Permian climatic zonation and its implications. *Am. J. Sci.*, **255**: 607–618.

Stehli, F. G. 1959. Reply. *Am. J. Sci.*, **257**: 239–240.

Stehli, F. G. 1964. Permian zoogeography and its bearing on climate. In Nairn, A. E. M., ed., *Problems in Palaeoclimatology*. Interscience, New York, 537–549.

Stehli, F. G. 1966. Discussion: labyrinthodont abundance and diversity. *Am. J. Sci.*, **2664**: 481–487.

Stocks, T. 1932. Die Echolotprofile des Meteor, In *Wiss. Ergeb. Deutsche Atlantische Expedition, Schiff Meteor, 1925–1927, Volume 2*. Walter de Gruyter, Berlin.

Stocks, T. and Wüst, G. 1933. Die tiefenverhältnisse des offenen Atlantischen Ozeans. In Wiss. Ergeb. *Deutsche Atlantische Expedition, Schiff Meteor, 1925–1927, Volume 3*. Walter de Gruyter, Berlin, 1–33.

Stommel, H. M. 1994. Columbus O'Donnell Iselin. *Biogr. Mem. Natl. Acad. Sci.*, **65**: 165–186.

Stott, P. M. 1963. The magnetization of Red Hill Dike, Tasmania. *J. Geophys. Res.*, **68**: 2228–2297.

Stott, P. M. and Stacey, F. D. 1960. Magnetostriction and palaeomagnetism of igneous rocks. *J. Geophys. Res.*, **65**: 2419–2424.

Stott, P. M. and Stacey, F. D. 1961. Stress effect on thermo-remanent magnetisation. *Nature*, **191**: 585–586.

Strahan, A. 1897. On glacial phenomena of Palaeozoic age in the Varanger Fjord. *J. Geol. Soc. London*, **53**: 137–146.

Suess, F. E. 1936. Der Meteor-Krater von Köfels bei Umhausen im Ötztale, Tirol. *N. Jb. Min. Geol. Palaont.*, **72**: 98–155.

Sullivan, W. 1974. *Continents in Motion: The New Earth Debate*. McGraw-Hill, New York.

Sykes, L. R. 1967. Mechanism of earthquakes and nature of faulting on the Mid-Oceanic ridges. *J. Geophys. Res.*, **72**: 2131–2153.

Takeuchi, H. and Uyeda, S. 1964. *Chikyu no kagu (Science of the Earth)*. Nippon Hoso, Tokyo.

Takeuchi, H., Uyeda, S., and Kanamori, H. 1967. *Debate About the Earth*. Freeman, Cooper & Co., San Francisco.

Talwani, M., Sutton, G. H., and Worzel, J. L. 1959. A crustal section across the Puerto Rico Trench. *J. Geophys. Res.*, **64**: 1545–1555.

Taylor, F. B. 1910. Bearing of the Tertiary mountain belt on the origin of the Earth's plan. *Geol. Soc. Am. Bull.*, **21**: 179–226.

Tharp, M. and Frankel, H. 1986. Mappers of the deep. *Nat. Hist.*, **95**: 49–62.

Thompson, M. D. and Bowring, S. A. 2000. Age of the Squantum "tillite," Boston Basin, Massachusetts: U-Pb zircon constraints on terminal Neoproterozoic glaciation. *Am. J. Sci.*, **300**: 630–655.

Tolstoy, I. 1951. Submarine topography in the North Atlantic. *Geol. Soc. Am. Bull.*, **62**: 441–460.

Tolstoy, I. 1981. *James Clerk Maxwell, a Biography*. University of Chicago Press, Chicago.

Tolstoy, I. and Ewing, M. 1949. North Atlantic hydrography and the Mid-Atlantic Ridge. *Geol. Soc. Am. Bull.*, **60**: 1527–1540.

Tolstoy, I. and Ewing, M. 1950. The T phase of shallow-focus earthquakes, *Bull. Seismol. Soc. Am.*, **40**: 53–58.

Turnbull, G. 1959. Some palaeomagnetic measurements in Antarctica. *Arctic*, **12**: 151–157.

Umbgrove, J. H. F. 1947. *The Pulse of the Earth* (2nd edition). Nijhoff, The Hague.

Urey, H. C. 1952. *The Planets: Their Origin and Development*. Yale University Press, New Haven, CT.

Urey, H. C. 1953. On the origin of continents and mountains. *Proc. Natl. Acad. Sci.*, **39**: 933–946.

Vacquier, V. 1959. Measurement of horizontal displacement along faults in the ocean floor. *Nature*, **183**: 452–453.

Vacquier, V. 1962. Magnetic evidence for horizontal displacement in the floor of the Pacific Ocean. In Runcorn, S. K., ed., *Continental Drift*. Academic Press, New York, 135–144.

Vacquier, V., Raff, A. D., and Warren, R. E. 1961. Horizontal displacements in the floor of the northeastern Pacific Ocean. *Geol. Soc. Am. Bull.*, **72**: 1251–1258.

van der Gracht, W. A. J. M. 1928. Remarks regarding the papers offered by the other contributors to the symposium. In van Waterschoot van der Gracht, ed., *Theory of Continental Drift: A Symposium on the Origin and Movement of Land Masses both Inter-continental and Intra-continental, as Proposed by Alfred Wegener*. American Association of Petroleum Geologists, Tulsa, OK, 197–222.

van Hilten, D. 1963a. Reply to Dr. Deutsch's discussion. *Am. J. Sci.*, **261**: 200.

van Hilten, D. 1963b. Palaeomagnetic indications of an increase in the Earth's radius. *Nature*, **200**: 1277–1279.

Van Houten, F. B. 1961. Climatic significance of red beds. In Nairn, A. E. M., ed., *Descriptive Palaeoclimatology*. Interscience, New York, 89–139.

Vening Meinesz, F. A. 1941. *Gravity Expeditions at Sea, 1934–1939, Volume 3*. Netherlands Geodetic Commission, Delft.

Vening Meinesz, F. A. 1947. Shear patterns of the earth's crust. *Trans. Am. Geophys. Union*, **28**: 1–61.

Vening Meinesz, F. A. 1951. A remarkable feature of the Earth's topography, origin of continents and oceans. I. *Akad. Wet. Amsterdam, Series B Phys. Sci.*, **54**: 212–228.

Vening Meinesz, F. A. 1952a. The origin of continents and oceans. *Geol. Mijnbouw, Series 2*, **14**: 373–384.

Vening Meinesz, F. A. 1952b. Convection-currents in the Earth and the origin of the Continents. *Akad. Wet. Amsterdam, Series B Phys. Sci.*, **55**: 527–553.

Vening Meinesz, F. A. 1960. Continental and ocean-floor topography; mantle-convection currents. *Proc. Akad. Wet. Afd. Natuur. B*, **63**: 410–421.

Vening Meinesz, F. A. 1961a. Convection-currents in the mantle of the Earth. *Proc. Akad. Wet. Afd. Natuur. B*, **64**: 501–511.

Vening Meinesz, F. A. 1961b. Convection-currents in the mantle of the Earth. *Proc. Akad. Wet. Afd. Natuur. B*, **64**: 512–527.

Vening Meinesz, F. A. 1962. Thermal convection in the Earth's mantle. In Runcorn, S. K., ed., *Continental Drift*. Academic Press, New York, 145–176.

Vening Meinesz, F. A. 1963. Relative movements of continents. *Proc. Akad. Wet. Afd. Natuur. B*, **66**: 3–7.

Vening Meinesz, F. A. 1964. *Developments in Solid Geophysics, Volume 1, The Earth's Crust and Mantle*. Elsevier, Amsterdam.

Verhoogen, J. 1954. Petrological evidence on temperature distribution in the mantle of the Earth. *Trans. Am. Geophys. Union*, **35**: 50–59.

Vine, F. J. and Matthews, D. H. 1963. Magnetic anomalies over the oceanic ridges. *Nature*, **199**: 947–949.

von Herzen, R. P. 1959. Heat-flow values from the southeastern Pacific. *Nature*, **183**: 882–883.

Wade, A. 1934. The distribution of oilfields from the view-point of the theory of continental spreading. *Proc. World Petrol. Conf.*, **1**: 73–77.

Wade, A. 1935. New theory of continental spreading. *Bull. Am. Assoc. Petrol. Geol.*, **19**: 1806–1816.

Wegener, A. 1915. *Die Entstehung der Kontinente und Ozeane*. Friedrich Vieweg & Sohn, Brauhschweig. 1st edition, 1915; 2nd edition, 1920; 3rd edition, 1922; 4th edition, 1924; 4th revised edition, 1929; 5th edition, revised by Kurt Wegener, 1936.

Wellman, P., McElhinny, M. W., and McDougall, I. 1969. On the polar wander path of Australia during the Cenozoic. *Geophys. J. Roy. Astron. Soc.*, **18**: 371–395.

Wertenbaker, W. 1974. *The Floor of the Sea*. Little, Brown and Company, Boston.

Willis, B. 1932. Isthmian links. *Geol. Soc. Am. Bull.*, **43**: 917–952.

Willis, B., Blackwelder, E., and Sargent, R. H. 1907. *Research in China: Volume 1, Part 1*. Carnegie Institution, Washington, DC.

Wilson, J. T. 1949. The origin of continents and Precambrian history. *Trans. Roy. Soc. Can.*, **43**: 157–184.

Wilson, J. T. 1961. Discussion of R. S. Dietz: continent and ocean basin evolution by spreading of the sea floor. *Nature*, **192**: 123–128.

Wiseman, J. D. H. and Seymour Sewell, R. B. 1937. The floor of the Arabian Sea. *Geol. Mag.*, **74**: 219–230.

Wooldridge, S. W. and Morgan, R. S. 1937/1959. *An Outline of Geomorphology: The Physical Basis of Geography* (1st edition/2nd edition). Longmans, Green & Co., London.

Worzel, J. L. 1977. Memorial to Maurice Ewing, 1906–1974. *Mem. Geol. Soc. Am.*, **VI**: 5 pages.

Worzel, J. L. and Ewing, M. 1954. Gravity measurements and structure of the West Indies. *Bull. Geol. Soc. London*, **106**: 37–61.

Worzel, J. L. and Shubert, G. L. 1955. Gravity interpretations from standard oceanic and continental crustal sections. *Geol. Soc. Am. Special Paper*, **62**: 87–100.

Wüst, G. 1960. Comments at meeting honoring Ewing. Unpublished transcription of meeting held in honor of Maurice Ewing's winning of the first Vetlesen Award held on March 25, 1960 at the Men's Faculty Club, Columbia University, 36a–38.

Yang, J. Y. and Oldroyd, D. 2003. A Chinese palaeontologist, Ma Ting Ying (1899–1979): from coral growth-rings to global tectonics. *Episodes*, **26**: 19–25.

Zijl, J. S. V. van, Graham, K. W. T., and Hales, A. L. 1962a, b. The palaeomagnetism of the Stormberg lavas of South Africa (I and II). *Geophys. J.*, **7**: 23–39, 169–182.

# Index

*Index*